FOUNDATIONS OF NEUROSCIENCE

 📞 01603 773114
email: tis@ccn.ac.uk

21 DAY LOAN ITEM

Please return <u>on or before</u> the last date stamped above
A fine will be charged for overdue items

FOUNDATIONS OF NEUROSCIENCE

MARCUS JACOBSON

The University of Utah School of Medicine
Salt Lake City, Utah

PLENUM PRESS • NEW YORK AND LONDON

Library of Congress Cataloging in Publication Data

Jacobson, Marcus, 1930–
　　Foundations of neuroscience / Marcus Jacobson.
　　p.　cm.
　　Includes bibliographical references and index.
　　ISBN 0-306-44540-9 (Hardbound)　ISBN 0-306-45165-4 (Paperback)
　　1. Neurosciences. I. Title.
QP355.2.J33—1993　　　　　　　　　　　　　　　　　　　93-5742
612.8—dc20　　　　　　　　　　　　　　　　　　　　　　CIP

ISBN 0-306-44540-9 (Hardbound)
ISBN 0-306-45165-4 (Paperback)

© 1995, 1993 Plenum Press, New York
A Division of Plenum Publishing Corporation
233 Spring Street, New York, N. Y. 10013

10 9 8 7 6 5 4 3 2 1

All rights reserved

No part of this book may be reproduced, stored in a retrieval system, or transmitted in any form or by any means, electronic, mechanical, photocopying, microfilming, recording, or otherwise, without written permission from the Publisher

Printed in the United States of America

Acknowledgments

Now that I have completed this work, two years after its beginning, I reflect with gratitude on the advice and encouragement that I have received from so many friends. I am especially indebted to Dick Burgess, Peter Clarke, Pasquale Graziadei, Tom Parks, and Rob Williams, who have given me the benefit of their detailed critique of parts of this book. Transforming my handwritten manuscript into the final typescript I owe to the patience and skill of Julie Church and Ken Romney. I dedicate this book to my wife, Ruth Kurzbauer, whose affectionate encouragement has enabled me to overcome many of those frequent sticking points when my "sober intoxication" with the subject flagged, and when I had difficulty understanding it, and when I could not find anything new and interesting to add to the conventional view.

Introduction

Spring snow lingers on the Chung-nan mountains,
 lovely in the blue distance.
A thousand coaches, ten thousand horsemen
 crowd the dusty roads of Chang-an.
But not one turns and looks at the mountains.
                            Po Chü-i (772–846 A.D.)

I began this work as an attempt to think through neuroscience from the foundations, to arrive at a new understanding, leaving no assumptions unquestioned. The "known" facts kept striking me as less and less explained by the old reasons. New reasons are wanted, but I found that reasoning anew about facts that everybody "knows" is as difficult as discovering what no one yet knows. The known facts are necessary but insufficient. To know a fact we must understand how it is represented in a theory, and also know the probability of its existence as an object in itself. That requires the use of a critical apparatus which this work aims to provide.

Inevitably, necessarily, different people will have different theoretical concepts and will select different facts to support their theories. Our theories are only guesses at the truth. Even when theories are brought into consilience with facts, they remain incomplete representations of reality. They may be brought into successively closer approximations to reality by enlarging the base of facts on which they stand, and by a process of unremitting critique to detect new relationships and eliminate anomalies. We are good at inventing methods for making new observations, but we are less competent at exercising critique. Such a critical method is necessary to explain how our observations are related to theories and how theories lead to successful predictions of new observations. For those purposes, knowledge of the history and philosophy of neuroscience is indispensable.

I regard knowledge of the history and philosophy of science as the memory and consciousness of science. Just as consciousness and memory give us identities as human beings, our knowledge of history and philosophy of science give us identities as scientists. But histories of science are also incomplete and distorted representations of reality. This limits our ability to compare our versions of the past with the present. Historical accounts are highly selective; they present a past that is rearranged and tidied up and thus that seems to be quite unlike the messy and disorderly present. The truth is that the past two centuries, during which neuroscience has evolved rapidly, are quite similar to the present—disorganized, uncertain, and hard to explain.

The worst sort of history deals mainly with outstanding figures, with heroes, and sometimes with antiheroes. That sort of history makes us feel that we have come after the golden age, or at best, that we are located at the end of the truly creative age. That is nonsense, and it is also mischievous because it makes the present seem mediocre by comparison with the past. Many ask: Where are there heroes like Cajal today? We cannot find them because such heroes never really existed—they are fictions in the minds of the hero-worshipers.

These problems are not confined to the history of science but also infect histories of the arts, philosophy, ethics, religion, and politics. Much more than those histories, we must understand the history of science if we want to understand mankind. That understanding is preeminently necessary because only scientific knowledge is truly cumulative and progressive. I argue that a condition for the continued progress of science is that we understand the relations between theories and data, and how our values affect our perceptions of the best uses of scientific knowledge. My intention is to provide some materials for competent critique that can be used to advance our science to goals that are not only useful, but also just and humane.

Contents

Chapter 1. Making Brain Models

1.1. The Argument	1
1.2. Mental Models and the Search for Meaning	6
1.2.1. Definition of Mental Models	6
1.2.2. The Status of Descriptive Terms	16
1.2.3. Some Differences between Physical and Biological Theoretical Models	16
1.2.4. A Realist Theory of Meaning	17
1.2.5. "Principles of Neuroscience" Are Conceptual Models	23
1.2.6. Metaphorical and Analogical Components of Models of the Brain	27
1.3. Making and Breaking Neuroscience Theories	29
1.3.1. How Neuroscience Research Programs Progress	29
1.3.2. Conceptual Revolution and Evolution	32
1.3.3. Revolutionary Methods Promote Progress	35
1.3.4. Raising the Threshold for Errors Promotes Progress	37
1.3.5. Verification and Refutation	46
1.3.6. Relations of Observations to Theories	48
1.3.7. Illusion versus Reality	49
1.3.8. The Convenient Fiction	52
1.4. Maturation of Neuroscience Theories	54
1.4.1. Case Study: Dendritic Spines	58
1.4.2. Case Study: Maturation of the Neurotrophic Theory	66
1.4.3. Case Study: Myelination	74
1.4.4. Case Study: Neuroglia	81
1.5. Discovery and Rediscovery: Disputes about Priority	90
1.6. Conclusions	94

Chapter 2. Neuroreductionism

2.1. The Argument	97

2.2. Definition, Assumptions, and Goals of Neuroreductionism 101
2.3. Reduction by Analogy, Homology, and Comparative Extrapolation 105
2.4. The Appeal of Simplicity 107
2.5. Reduction Reveals Relations 109
2.6. Ontological Status of Neuroreductionism 112
2.7. Experimental and Intertheoretic Reduction 117
2.8. Are There Emergent Properties of Complex Neural Systems? 119
2.9. Mental-to-Neural Reduction 122
2.10. Can Identity of Mental and Neural Events Be Demonstrated? 128
2.11. Psychophysical Dualism: Its Place in Neuroscience 133
2.12. Opposition to Reductionism 140
2.13. On Causal Explanations in Neuroscience 143

Chapter 3. Struggle for Synthesis of the Neuron Theory

3.1. The Argument 151
3.2. Remarks on Historiography of Neuroscience 155
3.3. Facts Are Necessary, Explanations Essential 158
3.4. Construction of the Neuron Theory by Consilience from Lower-Level Theories 165
3.5. Neuron Theory in Relation to Microreduction of Nervous Systems 174
3.6. Neuronal Typology 184
3.7. The Dictatorship of Techniques 190
3.8. Theories of Nerve Fiber Development 197
3.9. Theories of Nerve Cell Microspecialization 201
3.10. Theories of the Nerve Cell Membrane 205
3.11. Theories of Nerve Connections: Continuity Theories 207
3.12. Theories of Nerve Connections: Surface Contacts and Synapses 213
3.13. Conclusions 223

Chapter 4. Hero Worship and the Heroic in Neuroscience

4.1. The Argument 229
4.2. The Cult of Heroic Scientists 232
4.3. Cajal's Case: "The Bonfire of the Vanities" 237
    4.3.1. Fast Track to the Neuron Theory 239
    4.3.2. The Making of a Neuroscientific Mind 245
    4.3.3. Science as Self-confession 247
    4.3.4. Morphology Cast in a Romantic Mold 251
    4.3.5. Metaphors and Models of the Microcosm 258
4.4. Sherrington's Case: "The Corrections and Restraints of Art" 262
    4.4.1. Ascent from Mechanism to Mind 266
    4.4.2. Reflexes Are Fractions of Behavior 269
    4.4.3. Behavior Is the Integration of Neural Events 273
    4.4.4. Reciprocal Innervation and the Concept of Active Inhibition ... 277

4.4.5. The Synapse: The Consummating Concept 279
4.4.6. Values of a Neuroscientist in a World of Facts 283

Chapter 5. Ethics in Science

5.1. The Argument 291
5.2. Neuroscientific Ethics—Ethical Neuroscience 295
    5.2.1. Integrating Science and Ethics 295
    5.2.2. Definition of Values Affecting Science 296
    5.2.3. Calculating the Values of Science 300
    5.2.4. Can Science Be Value-free? 301
    5.2.5. Can Scientists Claim Diminished Moral Responsibility? 304
5.3. "Moral Rules Need a Proof; Ergo Not Innate" 305
    5.3.1. Moral Statements Are Hypotheses 306
    5.3.2. Doubt and Certainty in Ethics 311
5.4. The Process of Iterative Reflective Judgment 312
5.5. Ethical Conduct for Reaching Consensus 315
5.6. Development of Ethical Consciousness 317

References 325

Index 375

CHAPTER 1
Making Brain Models

We make internal pictures or symbols of external objects, and the form which we give them is such that the necessary consequences in thought of the pictures are always the same as the necessary consequences of the external objects. In order that this requirement may be satisfied, there must be a certain conformity between nature and our thought. . . . When on the basis of our accumulated previous experience we have succeeded in constructing pictures with the desired properties, we can quickly derive by means of them, as by means of models, the effects which would only occur in the external world in the course of a long time or as a result of our own intervention.
        Heinrich Hertz (1857–1894). *Die Prinzipien der Mechanik*, 1894.

1.1. THE ARGUMENT

Neuroscience theories are mental models made to explain the meanings of our observations of nervous systems. Neuroscience is founded on the assumption that our models are more or less true representations of some objective reality. Theoretical models of nervous systems have their own internal structure and their own dynamics of restructuring. But theoretical models are not autonomous. They are part of larger social, cultural, and anthropological structures. At a deeper level, our theoretical models are embedded in the evolutionary process as it is expressed in our own neural and mental processes.

I emphatically reject the notion that scientific objectivity requires exclusion of the observer's mind from the world being observed. Rather than eliminating the mind in the manner of some modern neurophilosophers (e.g., P. S. Churchland, 1986; Dennett, 1991), I eliminate the boundary that they have raised between the subjective and objective worlds. Indeed, I argue that denial of the existence of the mind is certain to frustrate any attempt to understand how scientific theories can be made to explain the objective world. Objectivity is nothing more than can be negotiated by means of intersubjective dialogue between different observers whose intentions are to reach understanding by reconciling their nonequivalent mental models of the same phenomena.

Since any theory of neurobiology is no more than a mental representation of reality, we have to ask how closely such representations can ever approach reality, and how we can know that a theoretical model is true or false. The realist belief is that scientific explanation is possible only because human intelligence and language have evolved as part of an objective reality. This belief is closely related to our perceptions of the progress of scientific knowledge. The progress of science can be measured by the increasing coherence of observations in relation to theoretical models and by more accurate predictions that can be made from them. Progress of understanding would not continue to be made if scientific theories were only systematized delusions, and if our theoretical models of the nervous system were not approximate representations of the real thing. But the history of neuroscience is not a linear progression of ideas toward ever greater consistency with reality: There will always be some false ideas that have great appeal when their time is ripe.

Different mental models can be made to represent the same phenomenon. In principle, every possible model of any phenomenon can be constructed. Differences arise because people start with different guesses about the meanings of phenomena, because they use different methods of observation, and because their models include different observations of the same phenomena. Therefore, people arrive at a plurality of fictions as well as a plurality of approximations to the truth. Errors and anomalies always enter theoretical models, and one problem is to detect them and to eliminate them in order to bring theoretical models into closer approximation to objective reality.

One of our problems is to understand how different models, formed to represent and explain the same objects, can be unified to form a consensus model. A scientific communication is a rerepresentation of a mental model that is itself a representation of individual experience. Scientists who intend to build a consensus model actually communicate rerepresentations of their individual mental models, inevitably misrepresenting reality. That different people's mental models are nonequivalent is very important because it leads to construction of a consensus model that is different from the mental model of any individual. The choice between alternative theoretical models, made to explain the same phenomena, is seldom an all-or-none decision. One model is rarely replaced outright by its rival, but alternative models gradually converge or diverge as components are transferred from one model to another, as old components are replaced, and as new components are selectively included or excluded. This process of remodeling is accelerated by invention of powerful new techniques.

A powerful technique can provide new observations that can rapidly transform and even demolish long-cherished theoretical models. This has occurred when techniques were introduced such as electron microscopy, microelectrode recording from parts of single nerve cells, and molecular biological technology. No subtle dynamics of restructuring scientific paradigms occurred, as Kuhn (1970) has suggested, but radical transformations of theoretical models, and corresponding changes in their plausibility, have occurred following the use of revolutionary techniques.

What is a revolutionary technique? Obviously it is a method that enables a crucial experiment to be done to decide between two opposing theories. To qualify for that task the method must be neutral with respect to both theories. But in practice the promoters of rival theories have often used methods that were biased in favor of their own theory. Neuroscience techniques have rarely satisfied the criterion of methodological neutrality, and as expected, there have been few crucial experiments in neuroscience. That is why neuroscience theories are rarely decisively defeated in battle with their rivals. They slowly fade away.

The change is rarely an either-or-transformation of one theory into its opposite like the classical face-to-wineglass illusion and other figure-ground reversals. In such cases ideas flip between two states: The same data are perceived to have different meanings according to the viewer's point of view and state of mind. More often, it is a gradual transformation through almost imperceptible stages. Or it is a slow fade-out of the old theory, like the Cheshire cat in *Alice's Adventures in Wonderland* (Fig. 1.1), which "*vanished quite slowly beginning with the end of the tail, and ending with the grin, which remained some time after the rest of it had gone. 'Well! I've often seen a cat without a grin', thought Alice, 'but a grin without a cat! It's the most curious thing I ever saw in my life.'*" The phrase "*grin without a cat*" is not a bad description of a refuted theory. Such a theory tends to have a persistent historical grin, long after the loss of its scientific teeth.

A neuroscience research program can be regarded as a dialectical interaction between theoretical models and observations whose meanings are interpreted with respect to the alternative models. A theoretical model functions heuristically when it guides the direction of research and choice of methods, and when it explains observations and predicts new discoveries. This is not a closed process, restricted by the requirements of an ideal system of logical discourse entirely inside science. In reality the process of constructing theoretical models of nervous systems is widely open to sociocultural and historical influences, and therefore includes extrascien-

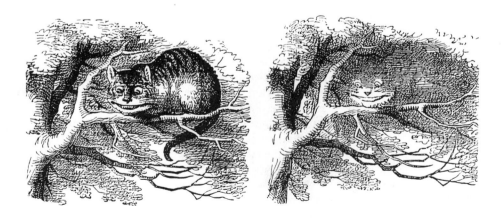

FIGURE 1.1. Neuroscience theories are rarely refuted outright. They often persist in the face of countervailing evidence, finally fading gradually away, like the Cheshire cat in *Alice's Adventures in Wonderland*, and remaining only as a historical grin without scientific teeth.

tific values. Values include emotions and beliefs that affect scientists' motives, intentions, and performances. There is no *homo scientificus* who lives in a sealed scientific culture, but every scientist is also a *homo economicus* driven by personal and professional interests.

Because more than one theoretical model can always be made to represent and explain the same phenomena, different scientists will be inclined to take different theoretical models as the starting points of their explanations. Value judgments about choices of research programs and their goals, and personal styles of working and thinking, have profoundly influenced the structure of theoretical models of nervous systems. A neuroscientist prefers one of the alternative models of nervous organization when observations can be given plausible meanings in relation to that model, and when that model can be included, with the least contradiction and conflict, in that person's sociocultural traditions.

Recently there have been two leading notions about how scientific theories are formed and how choices are made between theories that give different explanations of the same phenomena. One may regard the construction of a scientific theory as a growth of objective knowledge, entirely determined by requirements of logical consistency of explanation of observations, and by a theory's success at predicting new observations. In that view the choice between rival theories is a rational competition conducted entirely inside a closed scientific system. Alternatively, one may regard science as a process that is open to social and psychological influences. In that view scientific theories are social constructs, and choice between them is a struggle for acceptance and survival in a given sociocultural environment. Neither of those gives an adequate portrayal of neuroscience theories, as many examples will show.

The first is a more adequate characterization of theories of physical science than of neuroscience. Theories of neuroscience differ significantly from those of mathematics and physics. Although neuroscience theories may be reducible to physical and mathematical theories in principle, that reduction is very far from practical realization. With rare exceptions, the initial conditions and causal relations of many important nervous functions are poorly defined. That is one reason why it is rarely possible to make strong predictions from neurobiological theories. Therefore, the principal criterion of a successful biological theory is explanatory coherence as much as successful prediction from theory. A strong prediction is a logical deduction from theory, but that is not always possible in neuroscience. Instead, predictions in neuroscience have often been inductive inferences. These may take the form of conjectural entities, unobservable by the available experimental methods, but useful for explaining some phenomena. For example, synapses were unobservable entities when first conceived by Sherrington in 1897, and so were neurotransmitter receptor molecules when first conceived by Langley in 1906. Many powerful neuroscience theories were based on unobservable entities and thus could not be refuted. A good theory in neuroscience is not necessarily one that has been constructed so as to allow refutation, as Popper (1959) and Lakatos (1978) have contended.

The second way of portraying neuroscience theories, as social constructs, takes an excessively narrow view of the limits of autonomy of scientific theories, and an excessively broad view of the effects of extrascientific forces on construction of scientific theories. It fails to explain why scientific explanations have diverged progressively from folk explanations of natural phenomena.

The concept of progress is fundamental to all scientific research programs. It is inherent in the concept that theoretical models progressively approach reality, advance to higher levels of generalization and explanation, and are able to make more reliable predictions. The history of neuroscience shows a general progression of ideas toward ever greater consistency with one another and apparently with objective reality. In this chapter we shall discuss various ways of measuring progress, how different concepts of the nature of progress have been formed, and how they have affected the construction of neuroscience research programs, choices between alternative theories, and definition and selection of their goals.

An essential part of the character of modern neuroscience has been the elimination of metaphors from its theoretical models, and the diminishing use of allegorization in our modes of thinking about the brain. Allegorization is the representation of one thing under the image of another. It is a form of representation that evades the responsibility to grasp reality within the limits of our access to reality. If we cannot understand something because of the limitations of language or because we lack information necessary for building models that are approximations to reality, resorting to allegorization will not give us the power of understanding.

In past centuries things that were poorly understood in the brain were often explained by analogy with things that were well understood in other domains. Hydraulic systems, clockwork and other mechanical devices, railway networks, telephone systems, and computers have served as analogies for things in the brain. But do analogies and metaphors offer a proper basis for objective understanding of living systems, especially nervous systems? They may impede understanding by drawing attention to similarities that are not fundamental. By renouncing metaphor and analogy we hasten the possibility of recognizing the fundamental characters of nervous systems, and recognize the impossibility of giving them allegorical representations. Since the 1890s analogical-correlative thinking about the brain has been progressively replaced by analytical-reductive thinking. The effect has been to take theories about the mind away from metaphysics and metaphor, and to bring them into chemistry and physics, back into relation with nature. Allegorization persists in some theories of consciousness, the mind, and the structure of the psyche, the most extreme form in psychoanalytic theory, which never obtained a firm grip on reality. This would not have occurred if neuroscience had been able to arrive at a consensus theory of the relation of mental events to physical events in the brain. This failure is an indication of the incompleteness of neuroscience. Another indication is the inability or unwillingness to construct explanatory brain models of moral behavior and moral consciousness, thus leaving the metaphysical visions of morality unchallenged. Are these temporary holdups or will neuroscience always remain fundamentally incomplete?

Serious reflection on the process of scientific discovery, on its potentials and limitations, is too often overwhelmed by the chorus of voices that proclaim that neuroscience is on the threshold of a period of unprecedented and unlimited progress. But the history of our times is as much the story of the thwarting of such ambitions as of their fulfillment. Therefore, while we celebrate the limited victories, we should try to understand, however grudgingly, why we have failed to solve the central problems of neuroscience, such as the nature of thinking, of voluntary behavior, of consciousness, and of the relation of mind to matter.

I ask whether it is really necessary for us to continue doing so badly. If we measure the progress of neuroscience as the accumulation of paper confetti, on each of which is written a fact, we have indeed succeeded in arranging them to form ideas and even complete theories. But if the progress of neuroscience hinges on discovering the nature of perception, thought, volition, and consciousness, we have opened the door a fraction, but not wide enough to see clearly what lies on the other side. The desired view is obscured by blind faith in the progress of science. Optimism and pessimism equally obstruct the process of critique that is necessary for us to gain clear views of our goals. My main intention in writing this is to combat those obstructions to clearsightedness by providing materials for competent critique. Persistent critique is the most effective way to raise awareness of the uncertainties and limitations of the past and of the possibilities of the future of neuroscience.

1.2. MENTAL MODELS AND THE SEARCH FOR MEANING

1.2.1. Definition of Mental Models

The notion of a mental "picture" was given by Heinrich Hertz (1857–1894) as a means of understanding causal relations between parts of a complex mechanism. Hertz took a realist view of the relation of our mental models to the things that they represent when he wrote the epigraph to this chapter. He conceived of construction of a mental model as a process analogous to selecting materials and fitting them to a building. Also in his posthumously published *Principles of Mechanics* he concluded that *"the relation of a dynamical model to the system to which it is regarded as the model, is precisely the same as the relation of the image which our mind forms of things to the things themselves. . . . The agreement between mind and nature may therefore be likened to the agreement between two systems which are models of one another, and we can even account for this agreement by assuming that the mind is capable of making actual dynamical models of things, and of working with them"* (Hertz, 1894, Paragraph 428). This is similar to James Clerk Maxwell's view of the use of analogy and morphisms in thinking about complex systems. The notion goes back to the beginning of the 19th century in Georg Christoph Lichtenberg's conception of *paradigmata* (Greek: to show by the side of; hence models) that facilitate thinking about complex processes.

Building mental models is an adaptive perceptual process that requires extraction of information about properties and relations from the objects of experiences; it is a process of giving meanings to experiences. I shall argue that progress of neuroscience has been closely related to the successful construction of models that are representations of real nervous systems, from which the meanings of relations between elements of the system can be construed (Rosenblueth and Wiener, 1945; Rosenblatt, 1962; Hesse, 1966).

Mental models often have a strong representational character that is pictorial. In addition they may have conceptual and computational characteristics. Thus, a mental model may have the character of a topographical map in which positions and vectors of various functional operations are specified. The term "cognitive map" was introduced by Tolman (1948) to mean a mental representation of the spatial relations between perceived features of an animal's environment. The map is used to compute the animal's position relative to known points and to predict the optimal way to reach another position (O'Keefe and Nadel, 1978; Knudsen et al., 1987; Gallistel, 1989; Wehner and Menzel, 1990). Sensory and motor information can be represented in topographically organized arrays of neurons. The operational characteristics of neighboring neurons can vary systematically in at least one and possibly two or three dimensions of the map's space. Examples are cerebral cortical topographical maps of line orientation and of direction of movement of objects in visual space (Hubel and Wiesel, 1962, 1965, 1968, 1974a,b; Hubel et al., 1978; Albright et al., 1984); maps of auditory space in the brain stem and auditory cortex (Knudsen et al., 1987); and maps of motor programs in the superior colliculus (Wurz and Albano, 1981) and cerebral cortex (Humphrey, 1979; Georgopoulos, 1986). The motor map on its own is insufficient to achieve coordinated motor behavior—afferent signals, central pattern generators, and feedback control are also necessary. We conceive of a cognitive model as the coordinated binding together of several representations and subsystems in a functional relationship.

I shall argue that scientific theories are constructed in the form of cognitive models. In those models the objects of our experiences are represented by signs and relational rules. Thus, the objects may be released from reference to physical quantity, and may be defined by the methodological operations necessary to observe them and by their contextual relations. Symbolization can go in both directions—abstraction of objects and materialization of concepts. In one direction we can represent observed objects and events in the form of mental symbols and models. In the other direction we can actualize mental symbols and mental models in the form of our behavior and as physical products. Symbols always lag behind theories they serve. This is shown by the history of neuroscientific nomenclature, and even more strikingly by mathematical symbols which evolved slowly and capriciously for about 2000 years until the 17th century, and have continued evolving rapidly to this day (Cajori, 1928–1929). Symbols and terms serve to represent things, events, and operations in mental models. They can evolve only in relation to the models they serve.

When we say that "a picture is worth a thousand words" we mean that models are enormously rich in symbolic content because they can incorporate the evidence of all the senses (not only the conventional two dimensional static picture, but the full potentials of multivariate representation and of abstract symbolization), and they are not limited by the syntax of language. Observations made with different methods can be combined and can be given generalized meanings. A model of one domain, say spinal cord, can be compared with models of other domains. A model of a nervous system in one species can be compared with another species as a form of reduction by analogy, homology, or comparative extrapolation (see Section 2.3). A computerized model of one domain can be used to generate a model of another, as a kind of analogical method of solving problems (Holland *et al.*, 1986, Chap. 10). This is of special interest to us because of the frequent use of analogical problem solving in neuroscience.

The model is always simpler than the subject domain it represents. All efforts to make a simplified model with the aim of understanding something more complicated involve making decisions about significance of details of the subject domain and of the representation (see Tank, 1989, for the case of models of neuronal circuits). The tendency to simplify the world may become disadvantageous or even dangerously misleading when models lose touch with the complexity of reality. Realism is a continual struggle against the simplification of the world, whereas reductionism is a struggle to unravel the complexity, as we shall see in Section 2.4. I shall argue that the aim of modeling reality is not so much to achieve simplification as to achieve explanatory coherence, so that the model explains the largest possible subject domain. The subject domain can be the real nervous system or another model of the nervous system. A model of the nervous system can never be a complete representation of the subject domain, which may be an entire nervous system or even part of one. But the model must have sufficient resemblance to the subject to be meaningful. That criterion of resemblance between subject and model can only be established pragmatically, in the future, so to say, in association with other criteria of explanatory coherence and convergence.

Physical models are simulations that have been made to represent individual neurons or multineuronal systems (Tuckwell, 1988; Eekman, 1982; Segev, 1992). Such models are not direct representations of reality but are rerepresentations of mental models. Physical models are thus second- or higher-order derivatives of the objects of experience. For example, the object itself may be a part of a nervous system processed by the neuroscientist who forms a mental model of the fixed and stained histological sections or of any other empirical observations such as instrumental readings (Fig. 1.2). That model is rerepresented as a diagram, a picture, or some other symbolic form, and that is the primary object of experience of other neuroscientists. They may be pulled into the picture to reenact some of the original observer's processes of modeling, but each observer constructs his or her own personal mental models which are not totally congruous with the models of other observers.

During the process of mental representation and rerepresentation, objects in the real world may undergo three kinds of morphological misrepresentations: morphological distortions (anamorphic misrepresentation), elimination or addition of components (metamorphic misrepresentation), and translocation of parts or the assumption of a function by one part that properly belongs to another (homeomorphic misrepresentation). A well known example of anamorphic misrepresentation is Cajal's exaggeration of the width of the synaptic cleft in his drawings (Fig. 1.3A). The elimination of dendrites from drawings made to show that integration of inputs occurs mainly or entirely on the nerve cell body is an example of metamorphosis (Fig. 1.4). Misrepresentation of the forms of connections between neurons, showing them as nerve nets and large intercellular bridges, was the rule before 1870 and persisted for another 20 years (Fig. 1.5). That is an important case of homeomorphic misrepresentation. When Golgi attributed nutritional functions to the dendrites and depicted the nerve cell's inputs via axon collaterals

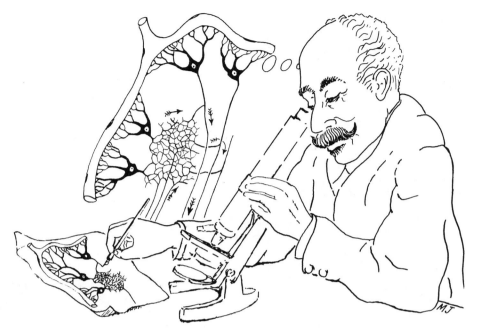

FIGURE 1.2. Theoretical models are formed as inductive inferences; often wrong guesses at the truth. Nerve net models, like Golgi's, were made on the basis of histological artifacts, often unavoidable at the time of the model's conception. They persisted because their supporters paid more attention to their models than to the realities they were supposed to represent.

> *And as imagination bodies forth*
> *The forms of things unknown, the poet's pen*
> *Turns them into shapes, and gives to airy nothing*
> *A local habitation and a name.*
> (Shakespeare, *A Midsummer Night's Dream*, V, i, 14)

he created a homeomorphic misrepresentation (Fig. 1.2). These three kinds of misrepresentation have been made repeatedly, and will no doubt continue to be made.

Good models are multivariate and show relations between variables.* Relations can express several things: relations of space, of time, of properties or attributes, of causes and effects, of functional states, and of purposes. Because relations are irreducible they can be represented well in diagrams and other models. Obviously a histological diagram cannot be a model of all those things. However, a histological drawing transforms a three-dimensional geometrical problem to a two-dimensional problem, and causal relations are lost in that transformation. A histological drawing, like any other two-

---

*Before the computer era, the graphical representation of theoretical models of nervous systems was severely limited by the available methods of graphical display of multivariate data. Some reference dealing with the history of graphical representation of multivariate data are Funkhouser (1937), A. H. Robinson (1967), Tilling (1975), Beniger and Robyn (1978), and Tufte (1983).

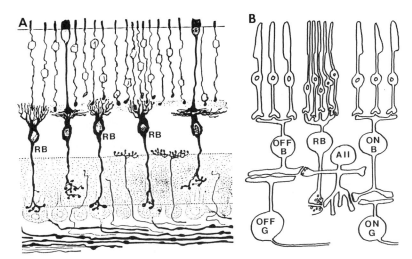

FIGURE 1.3. (**A**) Cajal's drawing of a Golgi preparation of the mammalian retina shows wide gaps between the distal processes of rods and dendrites of rod bipolar cells. The gaps were not observed but were part of a mental model constructed to explain how nerve cells connect with one another by surface contact, not by cytoplasmic continuity. The gap was an unobservable entity in Cajal's theoretical model. He also showed all rod bipolar cells connecting directly with retinal ganglion cells but he could only see the free nerve endings, not the contact of rod and bipolar cells on retinal ganglion cells. In reality, rod bipolar cells form complex synaptic connections in the inner plexiform layer. Panel **B** summarizes some electron microscopic observations showing that rod bipolar cells (RB) form excitatory synaptic connections with AII amacrine cells. This cell has gap junctions with ON-cone bipolar cells (ON B), which have excitatory synaptic connections with ON-ganglion cells (ON G). The AII amacrine cells also have inhibitory synaptic connections with OFF-bipolar cells (OFF B) and with OFF-ganglion cells (OFF G). Cajal never conceived of inhibitory synaptic connections. (A from Ramón y Cajal, 1893; B from Wässle *et al.*, 1991.)

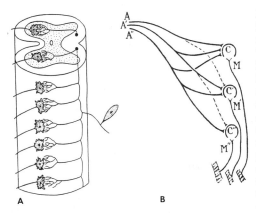

FIGURE 1.4. Elimination of dendrites in anatomical (**A**) and physiological (**B**) models of organization of the mammalian spinal cord. In **A** (from Waldeyer-Hartz, 1891) the spinal cord organization is represented in terms of the neuron doctrine. Dendrites were eliminated because they were considered to have nutritional functions, and because the most compelling evidence for the theory of connection by surface contacts came from observations of axosomatic contacts (Held, 1897; Auerbach, 1898; see Figs. 3.15 and 3.16). Panel **B** is a physiological model to explain how different synaptic densities serve to grade the synaptic excitatory effects on the spinal motoneurons and thus on the force of contraction of the muscles they innervate. Three afferents (A, A', A'') converge to form synapses exclusively on the cell bodies of three motoneurons (C, C', C''). Unbroken lines indicate greater synaptic density than broken lines. (From Sherrington, 1925.)

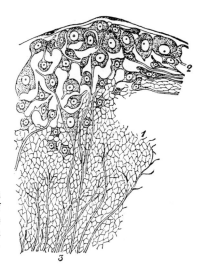

FIGURE 1.5. An early representation of connections between ganglion cells in the pleural ganglion of a snail. Network of nerves (1) connects the ganglion cells (2) with the afferent nerve fibers (3). Efferent nerve fibers are shown arising directly from some ganglion cells which are interconnected via large intercellular bridges. (Edinger, 1889; after Henle, 1879a.)

dimensional representation of a three-dimensional system, shows spatial relations well and temporal relations can be shown in a series of diagrams. Depiction of a series of events on the y-axis is not sufficient to show a causal relation. A histological diagram can model some properties such as form, color and texture, but not other properties like density, mass, and the like. It cannot show cause and effects, and plus or minus signs or arrows are inadequate analogies of functional states. Nevertheless, drawings were the principal means of making models of nervous organization until the invention of computer graphics.

We can trace the changes in a series of pictures of the same structure, say the cerebellar Purkinje cell, made over a period of time by different observers (Fig. 1.6). These changes are evidence of exchange between people of the rules of their models. People can communicate their individual mental models to one another, and one person can introduce a new rule or eliminate a rule from the model of another. This is how private mental models evolve as public explanatory models—by a process of reflection, communication, and critique, as we shall see in Section 5.4.

The well-known 19th century representations of the organization of nervous systems were just such models. Even the old Burdach, in the first volume of his *Vom Baue und Leben des Gehirns* (*On the Structure and Life of the Brain*) published in 1819, conceived of the task of neuroanatomy as a process of building a model. He wrote that *"gathering together material for the building is not all that is necessary. Every time a new supply is obtained, we should renew our attempts to fit it into the building . . . it is when we first obtain a view of the whole that we see the gaps in our knowledge and learn the direction which our investigations must take in the future."* The 19th century neuroscientists understood very well that their diagrams were working models. As Ludwig Edinger explained in the first of his *Twelve Lectures on the Structure of the Central Nervous System* (1st ed. 1885): *"A diagram is not always a picture of the*

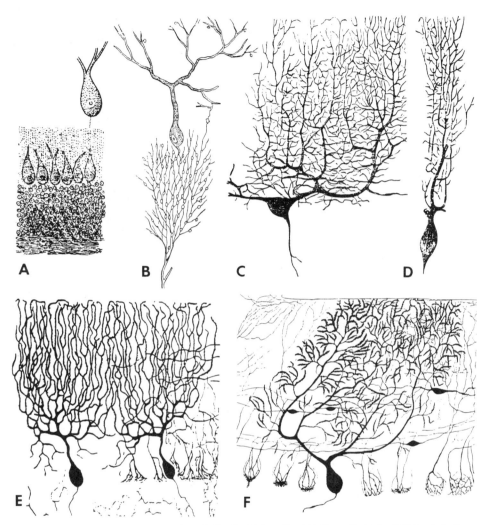

FIGURE 1.6. Evolution of models of cerebellar Purkinje cells. **(A)** From Purkinje (1837a), their first representation. The emphasis is on the cellular structure, especially the nucleus and nucleolus, in accord with the model of the cell proposed by Schwann (1839). Nerve fibers are shown unconnected to the cell bodies. Dendrites are not identified as such. **(B)** From Laycock (1860), after Gerlach (1858) showing the Purkinje cells connected to an axon which is shown clearly different from the dendrites. The representation is in accord with Gerlach's nerve net model of anastomoses between the axon and the dendrites. **(C and D)** Representation of Golgi preparations of Purkinje cells from Obersteiner (1888), emphasizing the morphology of the dendritic tree (C in a section transverse to the folium; D in a section in the long axis of the folium). The dendrites are shown devoid of spines. **(E)** From Ramón y Cajal (1888e). This was Cajal's earliest depiction of Purkinje cells in Golgi preparations. Dendritic spines were poorly impregnated. Cajal first showed dendritic spines unambiguously in 1891. **(F)** From Retzius (1892b). By this date Retzius had mastered the Golgi techniques and his depiction of Purkinje cells and other components of the mammalian cerebellar cortex was as good as it was possible to achieve with light microscopy.

demonstrated course of fibres; it is often enough only a graphic representation of the conclusions which have been drawn from great numbers of observations."

Part of the idea of forming models of nervous organization was to find structural and functional components which were each encapsulated within recognizable spatial boundaries and participated in a recognizable function. We would now term these "modules." They began to be recognized in the 19th century in the form of segments, ganglia, nuclei, cell clusters, columns, and laminae. The word "nucleus" (*"kern"*) in the modern sense of a strictly localized collection of cells in the central nervous system, was first used by Johan Christian Reil in 1809 with reference to the basal ganglia, but the general application of the word to all collections of nerve cells we owe to Benedikt Stilling (1846; see Schiller, 1969). Stilling's name is still attached eponymously to the red nucleus, which he first described as a collection of nerve cells, although others had earlier identified the gross structure (Schiller, 1969; Meyer, 1971).

Two concepts were related to this notion of modules: first, the concept of homology between modules in different organisms, for example between segments of invertebrates and vertebrates (Bateson, 1894; Gaskell, 1908; see Jacobson, 1991, pp. 6–18 for more on the concept of a basic modular pattern of the brain and the entire body); and second, the concept of structural connection and functional integration of separate modules. The concept that complex systems like the cerebellar cortex and cerebral cortex are formed by iteration of a basic functional module or by combinations of a small number of structural modules was clearly stated by Lorente de Nó in 1938c. He argued that *"the cortex is composed of an enormous number of elementary units, not simply juxtaposed but also overlapping."* This wonderful geometry is both satisfying and disturbing—the former because of its aesthetic qualities and the appeal of its apparent simplicity, the latter because of its intractability and its unwillingness to reveal its secrets. In more modern terms, modular structures are conceived of in terms of their capability of independent local processing as well as distributed parallel processing of information. Since the 1970s it has become quite conventional to represent the organization of the mammalian cerebral cortex in terms of neuronal modules, like printed circuits repetitively connected together in a computer (Szentagothai and Arbib, 1974; Szentagothai, 1976, 1978, 1979; Mountcastle, 1979). An obvious feature of cortical modules is their topographical periodicity, yet the periodicity is not uniform in the same species, and large interspecies variations exist.

A problem that is posed by all such organized arrays of nerve cells is the extent to which they are the cause or the effect of function. In Section 4.3.4 I shall consider the problem of the priority of structure or function and how it may be resolved. In brief, it is conventionally conceived as an either-or problem. Are modules a structural requirement for expression of certain functions such as processing of information in the cortex (Swindale, 1990)? Or are modules a secondary effect of the way the cortex functions (Purves *et al.*, 1992)?

People tend to get into a muddle about modules when they think about them in terms of the priority of either structure or function, and when they conflate the functional programs that operate in their development with the functions of the morphologically differentiated structures. For example, Purves *et al.* (1992) contend that cerebral cortical modules develop as *"incidental consequences of the rules of synapse formation,"* and "*not because the functional organization of the brain demands them.*" This is a case of the fallacy of misplaced functional efficacy. In this example it consists of locating functional efficacy entirely in the developmental process while denying it to the fully developed structure. In actuality there are no necessary similarities, correlations, or causal connections between the functions that are expressed during development and those later expressed by the fully developed structure. Even after a system starts functioning during the terminal stage of development, the functions expressed by a neuronal module remain distinct from the functions that were expressed during its development. In some cases the development of neuronal modules is contingent on positive feedback from relevant functions during a critical period. In those cases the two distinct levels of function enjoy the same region of time-space. However, the expression of functions of the morphological structure remains different from the expression of functions required for development of the morphology. We shall continue the discussion when considering the status of causal explanation in neuroscience (Section 2.13), and in historical context (Section 4.3.4).

Most neurohistologists recognized that their models of the cellular organization of the nervous system were simplified and distorted representations of reality and that they often deliberately represented purposeful causality (teleology). There is a fine line between representing a relation between structure and function and a relation showing the purpose of a system. Some neurohistologists deliberately simplified their representations and Cajal even claimed the right of selective exclusion. He wrote, "*A histological drawing is never an impersonal copy of everything present in the preparation. If that were true our figures would be far too complicated and almost incomprehensible. By virtue of an incontestable right, the scientific artist, for the purpose of clarity and simplicity omits many useless details*" (Ramón y Cajal, 1929a). All models are distortions of reality, Cajal's no less than others. Therefore, our task is to appraise the biases, account for them, and make the relevant adjustments to the model.

What kind of decisions are made when a detail is omitted or included in a model? Selection is a complex process, and different observers may not always agree on their criteria of inclusion or exclusion. But they can all agree that when they name an item that is included in a model their choice of names or terms is theory-laden. A scientist can selectively admit evidence to his or her pet theoretical model either because of total conviction or because of weak conviction: in the former case because of a hope that the countervailing evidence may prove to be faulty, in the latter case because of a fear that the counter evidence may demolish the model.

Selective inclusion and exclusion of evidence, and making predictions ahead of the evidence, are a normal part of the construction of theoretical models of the nervous systems. Cajal occasionally ignored evidence that conflicted with his pet theory of nerve cell contact. For example, in his studies of the structure of the vertebrate retina (1893), he noticed that the process of a bipolar cell sometimes appeared to fuse with the distal process of a rod, but he selectively discounted that evidence against the contact theory. We know now, from electron microscopy, that the bipolar cell process invaginates the rod process and a process of a horizontal cell to form a complex synapse (Dowling and Boycott, 1966; Boycott and Kolb, 1973; Sterling et al, 1986; Wässle et al, 1991). As seen with the light microscope, this would appear to be a confluence of the rod and bipolar cell, but Cajal selectively excluded such appearances in his drawings showing retinal cells always separated by a gap which was his manner of showing that they only come into contact and never fuse with one another (Fig. 1.3A). Cajal also showed all the rod bipolar cells of mammals terminating on the dendrites of ganglion cells. This was an error as we now know from electron microscopic evidence that the rod bipolar cells form synaptic connections mainly or exclusively with amacrine cells as shown in Fig. 1.3B.

Construction of a model sometimes involves a conflict between the attempt at a realistic representation and a deliberate distortion aimed at accentuating a part of the model which has special significance in the scientist's mind. A good example is Cajal's concept of the contact zone as a very tight junction and his depiction of it as a gap between the free nerve ending and the nerve cell that it contacts (Fig. 1.3B). Cajal conceived of the contact zone in the following terms, as he described it in 1906 in his Nobel lecture: *"A granular cement or particular conducting substance serves to connect the neuronal surfaces in contact very intimately."* He did not use the word "synapse" or refer to Sherrington. Cajal had little understanding of the functional significance of the synapse as we shall see in Section 4.3.4. He believed that the contact zones were structurally specialized for one-way excitatory transmission of nervous energy between neurons. However, the figures illustrating his Nobel lecture, like his other figures depicting nervous connections, show a substantial gap between the free nerve ending and the cell that it contacted. He did not intend such drawings to be taken literally to mean there was any cleft, however narrow, between neuron and neuron. But his dilemma was to avoid showing anything that might look like a confluence of two cells at the contact zone. His drawings were models designed to represent the difference between the contact theory and continuity theories of nerve cell connections, not realistic representations of synaptic clefts that were unobservable at that time.

Cajal was well aware of the potential danger of representing theoretical preconceptions. In his masterpiece *La rétine des vertébrés*, published in 1893, Cajal accused Max Schultze of falling victim to that danger: *"M. Schultze, [1866, 1873] in his works on the retina, shows the terminations [of the rod in the outer plexiform layer] as independent bulbs. Unfortunately, his prejudice in favor of the continuation of the visual cells with fibers stemming from the optic nerve and spreading out in the outer plexiform layer prevented him from admitting what he must have been able to see and thus draw. This is an eloquent example of the unfortunate influence of preconceptions, even on the most acute and calmest observers."*

Cajal would probably have admitted that the model could be misleading to the extent that it departed from a representation of the real thing. However, he would probably have agreed with me that inclusion of all possible observations is just as serious a departure from the truth as exclusion of all necessary observations. Good observation is always selective observation. The power of a model resides in the possibilities it provides for arriving at better understanding of the meaning of observations, and as a stimulus for making predictions. Inclusion of unobserved entities in the model is a form of prediction, as we saw in the case of Cajal's drawing of the retina. Inclusion of unobserved entities may be justified when it leads to reconsideration of the meaning of the model and to increase its explanatory and predictive powers.

1.2.2. The Status of Descriptive Terms

Descriptive terms are connected with empirical data by complex relationships and they become progressively more abstract as they are separated from the objects of direct observation by a large number of generalizations about those objects. For example, when we say that an object is a "neuron" we mean that it is a cell in the right place and has the right properties to qualify under the definition of a neuron in our dictionary of terms. The entry for "neuron" in the dictionary will include the past history and current status of the neuron theory, plus all the structural and functional characteristics of neurons that have been observed, plus any characteristics that can be inferred from theory but have not yet been observed. In addition a dictionary should give alternative definitions of the term "neuron" and state how they differ. The dictionary entry should give all the alternative theoretical models of "neuron" that have been constructed (and may, in addition, give some of the many more possible models that have not yet been constructed). Therefore, when we use a term like "neuron," we connect it to other terms (nerve cell, perikaryon, dendrite, axon, synapse, excitation, inhibition, facilitation, convergence, reflex action, afferent, efferent, and so forth), all forming a theoretical construct. The meanings of individual terms are not independent but are relative to theories in which they have been fitted. In principle I accept that the meaning of a term depends on the meaning of all the other terms and if a term changes so do all the connected terms. This limits the meaning of the individual terms but does not imply that scientists cannot communicate the meaning of the model as a whole, and use terms in relation to the meaning of the whole theoretical system.

All observational terms—glial cell, neuron, dendrite, inhibitory postsynaptic potential etc.—are related to other terms in a theoretical system. Regardless of how terms are defined by lexicographers or committees of experts, terms cannot be used outside a theory. That is true even when terms are contingent—"given so-and-so, that is a neuron"—or hedged with observational restrictions—"that is an object at this time and place which has observable properties a, b, . . . n." The principal reason for the frequent disputes over terminology is not so much about whether a new term muddles Greek with Latin. It is really about whether the term is biased toward their theory rather than ours.

1.2.3. Some Differences between Physical and Biological Theoretical Models

Theories in the biological sciences are said to differ from those in physics (Smart, 1963). In both, a theory is a conceptual model designed to explain observations, together with plausible inferences and the logical deductions drawn from it. But theories in physics are strongly predictive while those in biology are descriptive and weakly predictive, although they may permit retrodiction. For example, Newtonian mechanics allows the positions and movements of terrestrial bodies to be predicted from the initial conditions of mass, force, inertia, time, and space. By contrast, the Darwinian theory of evolution by natural selection may allow retrodiction but cannot predict the future evolution of species. Biophysical and biochemical theories of nervous transmission in axons, dendrites, and synapses approach the detail and rigor of theories in chemistry and physics and yet they do not allow prediction of nervous activities in large populations of neurons, or of any higher-level nervous functions.

The rigor of some biological theories approaches that of physical theory because those biological theories deal with relatively homogeneous systems in relatively steady states, and deal with systems with a high degree of structural order and a relatively low degree of complexity, such as the axon. Theories of nervous transmission in axons have a rigor approaching those of physics. But the rigor of biological theories departs more and more from theories of classical physics as they deal with functions of inhomogeneous systems like dendrites with many different synaptic inputs and with varying metabolic states, in which the accuracy of definition of initial conditions is often insufficient for prediction of future states.

Another difference is that the terms of physical theories must be (and are) defined completely, whereas those of neurobiological theories are often subject to alternative definitions or are not completely definable, for example, the terms "emotion," "thought," "consciousness," "the mind," "the unconscious," and so on. All who have attempted to define such terms have struggled to close the gulf between their concepts and their mangled verbalizations of them. This limits the rigor of discourse about theories in which those terms are used. One can defend the view that the differences between physical and biological theories is entirely the result of differences in their use of words (see Section 2.7). In that view, words used to communicate theories about mind are imprecise, but there is no reason to believe that the imprecision is inherent in the physical nature of mental events or that it would prevent us from reducing mental events to physical events. The idea of any fundamental difference between biological and physical systems is rejected in the reductionist tradition which we shall consider in the next chapter. That tradition is founded on the belief in the existence of objective reality that can be measured, and in the uniformity of nature and the unity of science.

1.2.4. A Realist Theory of Meaning

Realism of some sort is a necessary philosophical basis of empirical science and there seems to me to be no useful alternative to scientific realism (Bhaskar, 1975; Rosenberg and Hardin, 1982; Boyd, 1983; Smart, 1984; Rescher, 1987). Realism is considered again in Section 2.5 in relation to the method of reductionism. By realism we mean the belief that truth, meaning, and understanding of the world, including ourselves, are all attainable to the limits of observation of phenomena, and to the limits of our inferences from observations of things that cannot be perceived directly. Stated briefly, realism is the belief that scientific observations, corroborated by different methods, will yield an increasingly closer approximation to the real nature of the observed entities. Also, it is the belief that scientific observations can lead to increasingly true knowledge of theoretical entities that are not observable with available methods (like synapses which were theoretical entities from 1897 to the 1950s, when they were first observed with electron microscopes). Science would be no more than a systematized delusion; indeed it would be impossible, if scientific knowledge and theories were entirely or largely subjective, or were social constructs, as some recent philosophers of science have argued (e.g., Kuhn, 1962, 1970; Feyerabend, 1975; Goodman, 1979). Science has a privileged status as a process that provides true and useful knowledge. That is why science has been embraced by societies whose ideologies, like religious fundamentalism, are traditionally hostile to other aspects of realism and rationalism.

Duhem (1906) and Quine (1953, 1960, 1969) have raised the problem that there are an indefinitely large number of theories that can explain the facts more or less adequately. One of the consequences of this "underdetermination" of theory by empirical data is that the correspondence of empirical data with any one theory is not sufficient to make an absolute and final choice between theories, but in practice it is rare for the same data to be equally consistent with many theories. Many theories are potentially relevant to certain observations but that does not mean that they are all equally relevant. The question is whether it is possible to find criteria that can be used to decide between rival theories. Because we shall have to confront that question repeatedly, I should say that I have tried to find an eclectic answer, taken for the most part from realist theories of explanatory convergence and coherence (Rescher, 1973, 1987; Sellars, 1975; Rosenberg and Hardin, 1982; Boyd, 1983; Smart, 1984; Ziff, 1984). In this view, theories are preferred because, given the initial observations and conditions, they have the greatest predictive and explanatory powers. Data are preferred because they are corroborated by different observers; because they converge when they are obtained by different methods; and because they cohere within an explanatory theory. As we shall see from numerous examples, neuroscientists have to select their data and make choices between alternative theories. One of our problems is to understand how such choices are made. To what extent are they determined purely by the logic inherent in the scientific discipline and its methodology, and to what extent do extrascientific forces come into play? Estimates range from total dismissal of extrascientific factors in construction of scientific theories (e.g., Popper, 1959; Lakatos, 1978), to arguments that scientific theories are constructed and chosen largely by social conventions (e.g., Kuhn, 1962, 1970). These alternatives will be brought into our discussions repeatedly.

Meaning does consist to some extent in doing, and the truth is to some extent equated with that which works here and now, as pragmatists following William James and John Dewey have asserted. In addition to pragmatic and utilitarian criteria, the meaning of a scientific conception evolves as experimental evidence accumulates, and its effects are evaluated. *"The rational meaning of every proposition lies in the future"* was Peirce's way of saying that meaning depends on the results of an ongoing experiment (Peirce, 1905, in Peirce, 1958, *Selected Writings*, pp. 194–195).

Meanings are derived from knowledge of the relations between objects and from knowledge of the constancy or variability of objects and events in the environment (Quattrone and Jones, 1980; Mitchell, 1982; Thagard and Nisbett, 1982). It is not known how that is accomplished. I support the theory that meanings are derived from the construction of mental models that are neural representations of the objects and events of experiences. A mental model may be made by bringing together multiple representations of experience. This has been recognized since ancient times as shown by the fact that the words "cogitate" and "cognition" are based on the Latin *cogo*, meaning "I bring together." The ancients understood that there must be some place in the mind where separate experiences are brought together (e.g., St. Augustine, *Confessions*, Book 10). This concept is also found in the notion of the "*sensus communis*" which originated in the Middle Ages as the notion that all sensations were gathered together in the most frontal of three brain ventricles (Leyacker, 1927; Pagel, 1958) as shown in Fig. 1.7. The notion of the *sensorium commune* persisted into the first half of the 19th century (see Keele, 1957). In more modern theory, there must be processes to correlate, combine, and synchronize disparate representations. In recent times several models of functional binding have been proposed in which activity becomes synchronized in different populations of nerve cells that process different aspects of a complex stimulus (Milner, 1974; Freeman, 1975; Crick and Koch, 1990; Engel *et al.*, 1992).

Belief in the meaning of facts does not tell us what the true facts are, but beliefs are not independent of the facts. Facts and meanings are joined together in a theoretical model by flexible and elastic connections so that neither facts nor meanings are permanently fixed points of reference. Scientific terms and propositions have no meaning apart from their relationship to a scientific system consisting of theoretical models that, at the lowest level (i.e., less-inclusive models), are connected directly to

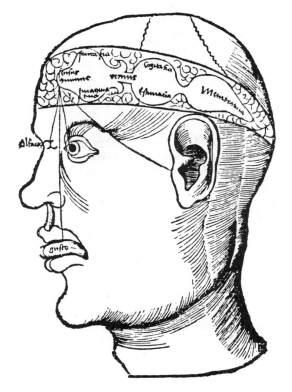

FIGURE 1.7. The idea that all sensations are united somewhere in the brain, the *sensus communis*, originated in the fourth and fifth centuries, especially with St. Augustine (354–430 A.D.). Its location was generally believed to be in the most frontal of three brain ventricles. In this figure the first ventricle (our lateral ventricles) is shown connected to the organs of special sensation and is labeled "*sensus communis, fantasia, imaginitiva.*" The second ventricle (our third ventricle) is labeled "*cogitativa, estimativa,*" and the third (our fourth ventricle) is labeled "*memorativa.*" This model of localization of different mental faculties to different ventricles persisted through the 16th century. This version is from the *Margarita philosophica* (1503) of Gregor Reisch. Such models also show the limits of introspection for gaining knowledge of mental processes.

empirical observations, and theoretical models at higher levels (i.e., more-inclusive models) are deduced from those at lower levels. Such a system stands as long as all its lowest-level theories have not been refuted. As we shall see in Chapter 3, a high-level theory like the neuron theory can survive refutation of some of the lower-level theories on which it stands. In such a hierarchical system the more-inclusive theory is connected to less-inclusive theories because it is determined by them, not absolutely but with some degree of probability, and with a certain degree of plausibility.

Since theoretical models are generated on observation only indirectly, we can never know in advance whether they are true or false. That kind of knowledge evolves in the future. That the genesis of a theoretical concept is independent of its truth seems obvious but needs to be explained because it is often misunderstood. To be useful, theoretical concepts should be the starting points for research programs (as defined in Section 1.3.1). This is what I mean when I say that in an effective research program, theories lead and empirical data follow. We recognize that theories in neuroscience are not fundamental: they can be reduced to more general theories and ultimately to universal theories of chemistry and physics.

The fallibility of scientific knowledge and the revolutionary overthrow of long-cherished scientific theories show that scientific knowledge is precariously related to the external world. Scientific knowledge implies only that we can have limited, qualified, contingent knowledge of an external world which exists independently of human minds. We can gain authentic knowledge because the human brain and mind have evolved, not as a passive reflection of the external world, but as an active part of it, subject to the same natural laws. However, the functions of the brain are only what they are because they have evolved by natural selection, and they do not guarantee that our brains give us *a priori* knowledge. The capacity to gain knowledge is genetically limited. Animals that can learn to find relations between objects in their environment can attach meanings to them, but the learning of higher animals involves a perception of relations that is beyond the capacity of the lower (Lashley, 1949). One of the aims of neuroscience is, or should be, to try to understand how our understanding of the external world is related to that reality, and how it is possible for us to decide which version of our understanding corresponds most adequately with real worlds, including the world of the human brain. That is one of the outstanding problems that we shall meet repeatedly in this work, especially in Chapter 2.

Mental models are useful when they provide the basis for behavior and generate the predictions that serve as the basis of inductive inference (Craik, 1943; Gentner and Stevens, 1983; Johnson-Laird, 1983; Holland *et al.*, 1986). The ability to make accurate predictions from a mental model is a requirement for conscious planning of future actions, and for intentions and motives. The capacity to form and modify mental models that enable the individual to solve the problems of everyday existence has adaptive and survival value. The natural ability to perceive systematic relations between items of experience and then to relate them to a mental model is vastly extended by the mastery of tools, especially language. The ability to perceive and understand the meaning of objects and events in terms of relationships of increasing generality develops in the child as part of the effort to use language (Vygotsky, 1962, p. 84). A model, like a map, can have words in it that perform a variety of different functions. Questions about the relations between thought and language remain unresolved, especially about how words can be both products and generators of thought. Similar difficulties arise when trying to define the relations between mental models and language. Whether one thinks as one talks, or one talks as one thinks, has long been disputed. Certainly, animals have neural representations even before they have the powers of thought, and thought precedes language in the apes and human infants.

The fact that the volume of scientific literature has been increasing exponentially for the past century shows the importance of verbal communication of scientific information, some of which is verbal definition and discussion of conceptual models. One problem, to be discussed in Section 2.9, is whether a mental model can be constructed by words alone without direct experience of the subject. How could a person who has been blind from birth form a mental model of the visual world from a complete description of visual experience and of the facts of neuroscience? How could a congenitally blind person's mental model of visual space differ from that of a person with normal vision?

Mental models are built up both from everyday experiences and from formal instruction. The child must reach a certain stage of development of mental models from everyday interactions with the environment before abstract ideas can be accommodated in the mental model. Meanings must be learned from ordinary experience before they are related to meanings learned from formal instruction (Vygotsky, 1962, p. 109). Mental models are gradually modified to bring the internal model into closer approximation with the outside world, but models of the world constructed only on the evidence of the senses are make-do compromises with reality. This limitation is shown by the tendency of children to misrepresent phenomena of daily experience (Fodor, 1985; Perner, 1991). Children do not have innate knowledge by means of which they can form substantive *a priori* judgments about space, time, and motion. Adults continue to have false beliefs about the nature of the physical world because those beliefs are not maladaptive and therefore they do not need to modify their misconceptions in order to survive (McCloskey, 1983; McCloskey and Kaiser, 1984). From this we conclude that scientific explanations of natural phenomena are not self-evident. That is quite different from the fanciful notion of *The Unnatural Nature of Science* (Wolpert, 1992). Science is no less "natural" than other skills that are culturally transmitted but could not exist without the evolution of the necessary neural equipment. One cannot understand a scientific theory as if there is nothing preceding it in the organization and function of the nervous system and nothing outside it in the sociocultural environment.

When one says that a model is "natural," one means that an untrained person can recognize the correspondences and the violation of correspondences between the model and the subject. The human brain has evolved so as to be able to recognize such correspondences easily, but feats of abstract logical analysis require hard work. That is probably why scientific theories have been originated most frequently by analogical procedures (Gentner and Stevens, 1983; Johnson-Laird, 1983; Holland *et al.*, 1986). The use of analogy is founded on the assumption that because two different things resemble one another in one respect they are more likely to resemble one another in further respects. As it relates to neuroscience, the analogical method is to explain something that is poorly understood in the nervous system by comparison and correlation with something that is well understood in another domain. Some of the things that have served as analogies for things in the nervous system are automata, railway networks, telephone systems, and computers, as we shall see in Section 1.2.5.

The logical-empiricist view of theory construction has had limited relevance to neuroscience. In the logical-empiricist view, as presented by Ernst Nagel (1961), for example, hypotheses are logical propositions that can be evaluated by a calculus of probability in the light of empirical data, followed by logical deductions drawn from the data, and so on, forming theoretical constructs of increasing generality. Critics of the hypothetico-deductive method, starting with N. R. Campbell (1920), pointed out the essentially conjectural nature of hypotheses, and the importance of induction and of analogical thinking in the formulation of explanatory theoretical models (e.g., Hesse, 1966). The hypothetico-deductive method in neuroscience has been limited mainly to constructing biochemical and biophysical models, and for making predictions from those models. The predominant form of conceptualization in neuroscience has been in the analogical-correlative mode and not the logical-analytical mode.

In the logical-empiricist view, unobservable entities must be excluded from theories, but in practice they have played an important part in construction of neuroscience theories. One can tinker with a model in ways that its subject would not permit under existing conditions. Models can be modified and extended, and components can be included in them that are not directly observable. For example, the synapse was conceived by Sherrington as a theoretical entity in 1897, before it was observed directly with the electron microscope 50 years later (see Section 4.4.5). During that time the concept of the synapse was forming increasing rich relations with anatomical and physiological observations in the context of the neuron theory. This is a good example of the importance of unobservable entities (i.e., entities that were not observable with techniques available at the time) in the construction of the central theory of neuroscience, as it is portrayed in Chapter 3. Sherrington did not hesitate to invoke unobservable entities when he found them convenient for the construction of a theory. A pointed example is his theory of central inhibitory and excitatory states discussed in Section 4.4.4. In his Ferrier Lecture of 1929, Sherrington tried to explain these states as changes in the stability of the surface membrane (see Creed *et al.*, 1932, p. 103). At that time the surface membrane of the neuron was an unobservable entity, inferred from the phenomena of unequal distribution of ions and molecules, and differences of electrical potentials across the cell surface.

Sherrington was not the first to infer the existence of unobservable entities for the purposes of a theory of nervous inhibition (Dodge, 1926). Several of the early theories of nervous inhibition were analogical and included unobservable entities, e.g., the theory of "drainage of nervous energy" proposed by William James (1890, p. 583) and William McDougall (1903, 1905). According to that theory, "nervous energy" existed in the form of a fixed amount of an imaginary entity called "neurin" that flowed into excited parts of the nervous system while its outflow from other regions caused inhibition.

McDougall (1903) illustrated this theory with the example of reciprocal innervation of antagonistic muscles (Fig. 1.8). While his model explained some aspects of the actions of antagonistic muscles it failed to account for several others such as the superactivity that often follows reflex inhibition, and the inhibitory action of nerves on the heart, blood vessels, and viscera. This lack of inclusivity, not the conjectural nature of "neurin," was

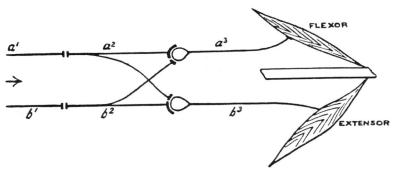

FIGURE 1.8. The drainage theory of reciprocal reflex action. It was supposed that on stimulation of any neuron an excitatory substance called neurin was generated and that transmission of neurin across synapses varied in direct proportion to the strength of stimulation. When stimulation of the afferent nerve to a flexor muscle was greater than to an extensor, neurin was supposed to drain from the extensor reflex pathway (where synaptic resistance was higher) to the flexor reflex pathway (where the synaptic resistance was lower), and vice versa. This theory was proposed by McDougall (1903).

recognized by Sherrington to be the weakness of the neurin theory. Sherrington (1906, p. 203) argued that it is *"unlikely that in their essential nature all forms of inhibition can be anything but one and the same process."* He did not question the legitimacy of admitting unobservable entities into a theory, only their usefulness for unifying all the observations in a single theory.

This example and many others that will be given later raise the question of how far inductive inferences and conjectural entities may legitimately go beyond observation. Logical empiricists have argued that nonobservable entities cannot be admitted into scientific theories. A theory that includes nonobservables cannot be refuted, and is thus an anathema to those whose criterion of a good theory is its refutability (see Section 1.3.5). But realists have argued the opposite namely, that theories are constructed on preliminary imaginative concepts that are inductive, and that knowledge of unobservables is possible (Maxwell, 1963; Boyd, 1983). Sharp distinctions cannot be made between what is observable and what is not. Many entities were inferred long before they were observed, e.g., atoms, molecules, blood capillaries, genes, viruses, and synapses. Receptor molecules were imaginary structures when they were first conceived by Langley in 1906, and membrane channels for different ions could not be seen when they were first conceived as hypothetical entities by Hodgkin, Huxley, and Katz. Nor could those who first started using those nonobservable entities as parts of their mental models have dreamed of the ways in which their physical reality would be demonstrated many years later by new techniques.

1.2.5 "Principles of Neuroscience" Are Conceptual Models

Without the belief in some principle of organization of the nervous system there can be no science of the nervous system. Neuroscience can be regarded historically as the search for evidence to support theories of the organizational principles of the nervous system. Holding such a theory is necessary but not sufficient: phrenology was a theory of organizational principle, and so was the reticular theory of nerve connections, but both were refuted. Nevertheless, both of those theories had heuristic value, stimulating research to corroborate or refute them.

Organizational principles are conceptual models that have for centuries been represented in the form of diagrams of the nervous system. In the process of making models of nervous systems, the modeler decides on the scale and level of the model and on the conventions used to represent things and events, structures and functions. Those conventions are almost entirely determined by the culture of the times; thus, we can now easily distinguish representations of the macroscopic anatomy of the brain made in the Middle Ages from those made in the 16th century, from those made in our time. There are significant differences between representations of neural circuits made before the roles of dendrites were understood and those made 30 or 40 years later. As late as the 1950s, diagrams of neural circuits tended to eliminate dendrites and to represent all connections as axosomatic (Figs. 1.4 and 1.8). The neuron was represented as a soma plus axon. Starting in the 1940s and increasingly since the 1960s the role of dendrites was finally taken into account more adequately (see Sections 1.4.1 and 3.9). The differences lie in the artistic conventions and in the cultural assumptions as much as in the representation of scientific knowledge. Models frequently contain unobservable elements, although model-makers are rarely explicit in their definition of the unobservable entities included in their models.

The oldest diagrammatic representation of the nerve fiber pathways in the central nervous system is found in Descartes's *De homine* which appeared in 1662. Descartes conceived of the nervous system as a mechanism controlled by a rational soul operating out of the pineal gland. The analogy of the human organism to a clockwork automaton was commonly used by 17th- and 18th-century writers—it is found in the introduction to Hobbes's *Leviathan* (1651), in Locke's *Essay Concerning Human Understanding* (1690), and it is the main idea of LaMettrie's *Man a Machine* (1747). One of the penalties paid by LaMettrie and his successors was to have eliminated mind from the mechanistic process of neural events. Instead they conceived of a sort of projection or representation of the external phenomenal world in the internal neural world.

The modern theories of neural representation and theories of topological mapping in the brain are well known, and we shall consider them again. It is not so well known that the idea began with LaMettrie's *L'homme machine* (1747): "*Thus judgment, reason and memory are not absolute parts of the soul, but merely modifications of this kind of medullary screen upon which images of objects painted in the eye are projected as by a magic lantern. . . .*" The problem that has been recognized for about a century is that this model requires a pontifical neuron, as William James called it, that looks at the screen and interprets the meaning of the image. Such a "theatre of the mind" metaphor is easily criticized. But alternatives are equally unsatisfactory. They merely distribute the mind from one central headquarters to many distributed substations: multiple parallel distributed processes (e.g., Dennett, 1991), or a "global workspace" (Baars, 1988), or a mechanism that synchronizes the activities of neurons to involve them in a global mental event (e.g., Crick and Koch, 1990). These are all models of the mind formed in the molds of conventional analogies.

From the mid-19th century on, the nervous system has been conceived as a kind of communications system (Fig. 1.9). Railway stations and train timetables were as emblematic of the culture of the late 19th century as airports and computer reservations are of our own culture. Karl Deiters (1865) introduced the concept of the nervous system as a railroad system with junctions at which traffic could be switched into different lines. Deiters conceived of nuclei of the brain as centers of nervous traffic and he called such a collection of ganglion cells a *Knotenpunkt*, namely a junction in the sense of the word used in railway engineering (Schiller, 1974). He used the term *Knotenpunkt* to describe the collection of large ganglion cells in what is now known as the lateral vestibular nucleus, or as Deiters's nucleus. In the 1870s and 1880s the analogy with the railroad network was preferred, but later the analogy with the telegraph exchange became more popular and persisted until the mid-20th century, when it was replaced by the computer analogy. Alexander Bain (1818–1903), in his book *The Senses and the Intellect* (1855), drew the analogy between the nervous system and a telegraph system: the nerves are wires transmitting currents; the brain is like a voltaic battery; the "mind" is the currents— *"No currents, no mind."* Thus, Bain is one of the founders of what has since become known as the identity theory of the mind-brain relationship, which is discussed in Chapter 2. Sherrington (1941, p. 282) used the telephone exchange analogy in his critique of the identity theory: *"Physiology has got so far therefor as examining the activity in the 'mental' part of the brain when activity there is in normal progress. The desideratum to carry observation into the telephone exchange itself with that exchange normally at work seems thus at last fulfilled. But has it brought us to the 'mind'? It has brought us to the brain as a telephone exchange. All the exchange consists of is switches.*

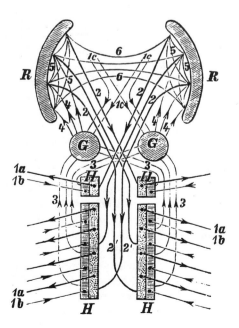

FIGURE 1.9. The nervous system represented as a communications network analogous with a railway system. Similar models of nervous function and dysfunction, in which nervous traffic flowed normally or the flow was altered or interrupted by disease or injury, persisted until quite recently. Arrows represent the direction of traffic. R = cerebral cortex; G = basal ganglia; H = spinal cord. (from L. Hermann, 1892.)

What we wanted really of the brain was, it would seem, the subscribers using the exchange."

Counter theories to the telephone exchange theory were soon constructed. One of the counter theories was based on the notion of redundancy—the nervous system has alternative structures for performing the same functions. Sherrington used that argument to counter the telephone exchange model of nervous organization. In his Nobel Prize Lecture, published in 1934, he emphasized that *"More than one way for doing the same thing is provided by the natural constitution of the nervous system,"* and he conceived of *"this luxury"* teleologically, as a means of functional compensation for injuries to the nervous system. Further discussion of this notion and of the significance of functional equivalence is deferred to the next chapter. Another rival to the telephone exchange model was Lashley's model of functional equipotentiality of all regions of the cerebral cortex, based on evidence that learning is not reduced by localized cortical ablation in rats (Lashley, 1924, 1929). The concept of "motor-equivalence," namely the ability of animals to perform a function in different ways even after localized destruction of the motor cortex (Lashley, 1924; Jacobsen, 1932; Sperry, 1947; Glees and Cole, 1950), was incorporated into models of cerebral cortical structural and functional plasticity. Large changes of cortical representation of peripheral structures and functions occurring in the intact cerebral cortex after peripheral nerve lesions, or after partial deafferentation, have inspired new models of cerebral cortical plasticity. Those models all required self-regulation and self-repair of the nervous system, things that telephone systems (and computers) are unable to do. These new models include plastic elements: activation of previously silent synapses, collateral sprouting of uninjured afferents, and formation of new synaptic connections in the adult mammalian brain (reviewed by Kaas, 1983; Devor, 1987; Wall, 1988; M. Jacobson, 1991, pp. 483–488).

The computer analogy is now most often given by functionalists who assert that mental processes are not identical with the material substance of the brain but with functional states of the brain (e.g. Putnam, 1975, 1980; Dennett, 1978, 1987, 1991; Shoemaker, 1984). Functionalism is the view that behavior and the mind are not explained by the physical composition of the brain but by its functions, that is, by its operational programs. Those imply causes, goals, and purposes (teleofunctionalism). The analogy is given of the relations between computer hardware and software. The computer and the nervous system are said to be functionally isomorphic if *"there is a correspondence between states of one and states of the other that preserves functional relations"* (Putnam, 1975). The important implication is that a mental event is not necessarily always the same as a particular neural event. The same mental event can be realized by a variety of different neural events. This concept does not require denial of the identity of neural and mental events, only denial of one-to-one physical isomorphism. Functionalists accept that there are "bridge rules" relating mental events to neural events and claim that mental events could be deduced from neural events if the rules were known. They like to represent nervous systems as black boxes connected by functional rules like "store," "retrieve," "compare," "execute," and so on. Functionalists belong to the long tradition of belief in the primacy of function over structure. Such people have also taken the view that there is no localization of function in the nervous system, or the less extreme view that there are only functional complexes related to structural complexes.

Functionalism assumes that brain modification is entirely instructional, and that it occurs by computation and modification in neural nets (e.g., Pellionisz and Llinas, 1986). By contrast, selectionist theories of brain plasticity make the assumption that variability in function (a criterion of fitness), selection, and heritability are the basis of modification of the brain as of any other organ. Selectionist theories have been proposed to explain development of neuronal projection maps (Jacobson, 1974b; Hirsch and Jacobson, 1975; Changeux and Danchin, 1976), and to account for learning (Changeux, 1986; Edelman, 1987).

The limitations of the intelligent machine analogy are obvious—such a machine is able to perform logical operations on the input and transform inputs into outputs without itself understanding anything about the inputs or outputs. This was understood by Descartes (*Discourse on the Method*, Part 5, pp. 56–57) as I point out in Section 2.11. Searle (1980) has illustrated this in his well-known Chinese room analogy, in which a person in a room has questions passed to him in the form of Chinese characters and produces meaningful answers by selecting from a dictionary of Chinese characters according to a set of rules, without any understanding of the meaning of the questions or answers. Further discussion of this problem is deferred to Chapter 2.

1.2.6. Metaphorical and Analogical Components of Models of the Brain

Analogy and metaphor have been very important in descriptive neuroscience in which formal-logical and mathematical criteria have played relatively little part in constructing theories and in making predictions from them. The analogical-correlative mode of explanation and model building (Holyoak, 1984) was much cultivated during the 19th century. We shall later discuss Cajal's use of metaphor (Section 4.3.5.). Note also that Darwin's theory was analogical-correlative. It was founded on two analogies: first, the analogy of improvement of domestic animals by selective breeding; and second, the analogy, supplied by Malthus, of human overpopulation followed by the struggle for survival of the fittest.

The most extreme examples of the uses of analogy and metaphor can be found in the writings of the German romantic natural philosophers including Jan Evangelista Purkinje (1787–1869) and Gustav Gabriel Valentin (1810–1883), and the French transcendental natural philosophers led by Étienne Geoffroy Saint Hilaire (1772–1844). The concepts that were common to both these schools of natural philosophy were "*the fundamental concept that there exists a unique plan or structure, the idea of the scale of beings, the parallelism between the development of the individual and the evolution of the race*" (E. S. Russell, 1916). These motifs in the world view of romantic *Naturphilosophie* reached far into the 19th century and even into the 20th. The leading motifs of German romantic natural philosophy were unity, uniformity, and analogy (Siegel, 1913; E. S. Russell, 1916; Lenoir, 1981, 1982). They saw unity in the diversity of living forms which they claimed to be able to trace to archetypal ancestral forms (*Urformen*) and to archetypal phenomena (*Urphänomen*) which recurred in many different situations. They recognized no fundamental differences between scientific discovery and artistic creation—both depended on imaginative perceptions of analogies, correspondences, and affinities between apparently different things. Those beliefs were held by many 19th century neuroscientists, notably by Gall, Burdach, Treviranus, Purkinje, and Valentin, and guided their research (Toellner, 1971; Lenoir, 1980, 1981, 1982; Clarke and Jacyna, 1987). In Section 3.5 we shall consider the effects of *Naturphilosophie* on the early development of the concept that the nervous system is formed of nerve cells.

An extreme case of the analogical-correlative mode of theory construction is Sigmund Freud's attempt at constructing a model of psychological processes in terms of the neuron theory. We shall again consider Freud's relations to the neuron theory in Section 3.9. What is important for our present purposes is that Freud conceived of the neuronal circuit depicted in Fig. 1.10 as a representation of the functions of the "ego," a metaphorical entity. He conceived of the "ego" as the routing of nervous energy or "cathexis." Cathexis is equivalent to facilitation of the quantity of nervous energy; repression is removal of cathexis; facilitation is storage of cathexis (Freud, 1895, pp. 380–384). Either the nervous energy flows in the main pathways or it is deflected and repressed in side pathways. Freud's theory is closely related to the drainage theory illustrated in Fig. 1.8, which was also based on unobservable entities. After 1895 Freud departed further from his neuronal model into more elaborate metaphors. Psychoanalytic theory lost its grip on reality. By 1900, when he published *The Interpretation of Dreams*, Freud had abandoned his earlier attempt at constructing a neuronal model of mental processes and had constructed his model entirely from metaphors (McCarley and Hobson, 1977; Hobson, 1988).

Analogical thinking can be helpful in arriving at solutions to scientific problems: the example is often given of the German chemist August Kekule (1829-1896) whose dream of the snake with its tail in its mouth led him to the solution of the ring structure of the benzene molecule. But metaphors can also deflect attention toward information that is misleading because of the total dissimilarity between the domains of the problem and the metaphor. The interdomain similarities in Freud's case are abstract whereas the differences are concrete. For the metaphor to have heuristic value, revealing meanings that are hidden in the problem domain, the two domains must share causally relevant conditions, as Mary Hesse (1966) first pointed out. One of the conditions necessary for success of a metaphor or analogy of nervous systems is that the metaphorical domain

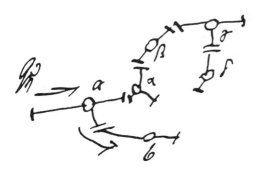

FIGURE 1.10. Freud's model of the neural circuit representing the "ego." Freud terms connections between nerve cells "contact barriers," and conceives of them conducting in both directions. He writes: "*Let us imagine the ego as a network of cathected neurones. . . . Then suppose a quantity ($Q'\eta$) enters an **a** neurone from the outside. If it were uninfluenced it would have proceeded to neurone **b**. But it is in fact so much influenced by the lateral cathexis in neurone α that it only passes on a quotient to **b**, or may not even reach **b** at all. Where then an ego exists, it is bound to inhibit psychical processes*" (Freud, 1895).

conserves relevant functional relations of the problem domain, although the structural components of the two domains may be entirely different.

In Freud's model there was no real relation between the functional or structural components of the neuronal and psychic domains. By contrast, Cajal's well-known metaphor of the growth cone, which he described as a *"living, flexible battering ram,"* was successful because the two domains shared enough causally relevant conditions. Cajal's metaphor was also misleading because of the conditions they did not share— different structures and different purposes, as we shall see in Section 4.3.5.

1.3. MAKING AND BREAKING NEUROSCIENCE THEORIES

A scientist who has never made a mistake has never made anything. Science advances when many theories are made to explain the same observations. Most theories are destined to be broken and discarded, although some of their fragments may be used to make new ones. The worst fate of a theory is to be ignored. The cruelest fate is for a useful theory to be discarded and later rediscovered, but its original maker forgotten.

A theory is rarely made by an individual working in isolation. A theory is generally made by more than one person working in a research program, as part of an intellectual tradition. In a progressing tradition the production of new theories is encouraged, theories are produced in profusion, and are vigorously discussed to decide whether they are worth the effort of experimental testing. In such a scientific tradition, there is merit in making testable theories even if they are finally broken.

A scientific research program or any intellectual tradition stagnates when praxis is more highly prized than theory, and when gathering error-free data is more highly valued than devising experiments to decide between alternative theories. The validity of that proposition will be examined in the following sections.

1.3.1. How Neuroscience Research Programs Progress

Scientific theories develop in the context of scientific research programs and research traditions. A scientific research program is a functional relationship between theories, techniques, observations and values. Imre Lakatos (1978), who developed the notion of the scientific research program, was primarily concerned with the history and philosophy of mathematics and physical sciences. His ideas have been adopted where they seem to me to apply to neuroscience, but their relevance to biological sciences is limited because there are significant differences between physical and biological sciences, some of which we have already discussed.

As Lakatos describes the structure of the scientific research program, it has a "hard core" of fundamental assumptions and central theories, which define the scientific research program, and which are stable, and a "belt of surrounding auxiliary hypotheses," which change continuously in response to new evidence (see Lakatos, 1978, pp. 48–52). Lakatos claims that assimilation of evidence into the scientific research program and changes in auxiliary hypotheses are determined by objective criteria with little or no subjective bias. In this chapter I give my reasons for disagreement with his opinion that *"most theories of the growth of knowledge are theories of the growth of disembodied knowledge: whether an experiment is crucial or not, whether a hypothesis is highly probable in the light of available evidence or not, whether a problem shift is progressive or not, is not dependent in the slightest on the scientists' beliefs, personality, or authority. Those subjective factors are of no interest for any internal history"* (Lakatos, 1978, p. 102). Later in this chapter I shall give more evidence showing that subjective judgments and irrational factors are inextricable components of any scientific research program, thus constituting valid matters for inclusion in an internal history.

I have departed from Lakatos in several important respects. I maintain that theories are mental models, therefore subjective, and that private mental models attain greater correspondence with objective reality by two processes: dialectical interaction between empirical observations and mental models, and intersubjective discourse between people holding different mental models of the same phenomenon, leading to construction of a consensus model. I include values as necessary components of any complete scientific research program. I show that it is not necessary or not even possible for the propounder of a theory to specify the conditions for refutation in advance. I propose that progress occurs when conflicts between opposing theories can be resolved, and that resolution of conflicts involves repeated cycles of reflection, communication, and critique, leading to appropriate actions (see Section 5.4). I also propose that conflicts between theories can be resolved most effectively when the rival theoretical models share causally relevant functional relations and share a significant set of explanatory terms and when they have the same practical and theoretical aims. When those conditions are fulfilled, the proponents of rival theories can engage in a dialogue leading to reduction, incorporation, or elimination of one theory by another. In Chapter 3 we shall examine an important example of that process—the construction of rival theories explaining the cellular organization of the nervous system, and the eventual triumph of the neuron theory.

In some respects competition between rival theories is analogous with competition between species in the struggle for survival—competition is keenest between those that are most similar and occupy the same territory. But to extend the analogy much further, to model a theory of theories of nervous systems on a theory of evolution is to put an unnecessary distance between us and the real nervous systems that we want to understand.

Popper (1959) and Lakatos (1978) follow in the logical empiricist tradition which argues that the primary function of a scientific theory is prediction. A theory is corroborated when it makes a successful prediction of an observable phenomenon and it is falsified when it fails to make predictions which other theories succeed in making. However, in addition to prediction, scientific theories have other functions: to draw further attention to phenomena and to explain them. A theory may survive refutation if it continues to explain phenomena even when it does not succeed in making accurate predictions. The logical empiricist view of scientific explanation is that it consists of logical statements which can be deduced from nomic generalizations (namely natural laws) and statements about initial conditions (Braithwaite, 1953; E. Nagel, 1961; Rescher, 1970). The adjective "nomic" was introduced by W. E. Johnson (1924) to express a regular, law-like connection between one observation and another (rather than a causal relation, implying necessary spatiotemporal connectedness). In that view we can explain the meaning of nerve conduction and predict its occurrence in a particular nerve by deductions from the Hodgkin-Huxley equations, and the initial conditions of ionic activities inside and outside the nerve, the membrane potential and some other well-defined starting conditions. But that does not give a complete explanation of the meaning of nerve conduction because it does not explain why the Hodgkin-Huxley equations are relevant or why the initial conditions came about. The explanation must include further conditions like the molecular structure of the membrane and more general relations such as the type of neuron, its synaptic connections, and any other relevant conditions. In the realist view, the "truth" of an explanation becomes increasingly more probable as a function of the number of logical, causal, and other relationships which can be perceived. We shall discuss this in greater detail in Chapter 2 when considering how things may be related by correlation, causal connection, or identity.

For Lakatos, a scientific research program is progressive if it continues to make novel predictions that are empirically confirmed and if it can assimilate the predictions of rival programs. When Lakatos says that a research program progresses he means progress of empirical knowledge only. He admits reluctantly that personal predilections may determine the choices of programs, the initial conjectures, and means of pursuing them. But he does not fully recognize the significance of subjective factors, even irrational ones, in originating the scientific research program, in sustaining its early stages, and in gaining consensus. He argues that the struggle between rival programs may be regarded as a rational competition in which the growth of objective knowledge inevitably occurs (Lakatos, 1978, *History of Science and Its Rational Reconstruction*, p. 102). He argues that nonscientific factors are not primarily effective in determining progress, but they may be effective in maintaining what he calls "tenacity," meaning willingness to continue the scientific research program in the absence of empirically confirmed predictions or even in the face of counterevidence.

Feyerabend (1970a) says of tenacity: "*Scientists must develop methods which permit them to retain their theories in the face of plain and unambiguous refuting facts.*" Others disagree, as shown by the statement by Crick (1988) to the effect that only amateurs stick to one theory, professionals try one theory after another until they find one that works. I agree with that, although Crick fails to explain how the choice between theories is ultimately made. According to Kuhn (1970) the choice is made by comparing a "paradigm" with alternative theories. Kuhn postulates that tenacity is characteristic of science during the periods which he calls "*normal science*" when scientists dig into entrenched positions and each holds a single paradigm (scientists behave like Cajal with respect to the neuron theory). But this is succeeded by a period of scientific revolution in which ideas proliferate (scientists behave in the way described by Crick) and new paradigms emerge. Kuhn's postulated periodicity has been criticized by Lakatos (1978) who proposes that proliferation and tenacity always work together, not successively. The examples which are given in this book also show the absence of successive periods of the Kuhnian kind.

Lakatos takes a larger conceptual scheme than the individual idea as the criterion of scientific progress. A scientific research program is either progressing or degenerating depending on its competitiveness with rival programs in explaining phenomena and in making testable predictions. I think that the individual idea can also be judged to be progressive if it has heuristic effects, even if it is anomalous and is eventually falsified. To state an idea in public that everyone is already thinking in private can have little heuristic effect. To have heuristic effects an idea must be stated in public and must come into conflict with existing ideas, but it need not be true. The idea that nerve cells form continuous networks was such an idea, as we shall see in Chapter 3.

Those who claim that science advances exclusively by logical discourse ignore the very important persuasive elements of scientific discourse. In Section 5.4 I argue that communication has a function of attracting attention to a problem as well as transmitting information about it. Critique has a function of persuasion or dissuasion, with the intention of gaining assent or fostering dissent. Our belief in progress depends on what we want. Progress does not mean only an increase in size or even of complexity but implies transformation. It implies that the new is either assimilated with or replaces the old—by evolution or revolution.

1.3.2. Conceptual Revolution and Evolution

For some 300 years great transformations in science have been hailed as revolutions—the Copernican, Vesalian, Harveian, Newtonian, Darwinian, Einsteinian, and so on, with the implication that the work of a single man produced a sudden break in the scientific tradition (reviewed by I. B. Cohen, 1985). I can confidently say that in those terms there have been very few revolutions in the history of neuroscience. To qualify as a "revolution" a change must be rapid and large. A succession of small incremental changes are evolution, not revolution. George Sarton in *The History of Science and the New Humanism* (1937, pp. 21–22) said that the view of science as "*gigantic stairs, each enormous step representing one of those essential discoveries which brought mankind almost suddenly up to a higher level*" is only a "*first impression of scientific progress.*" On more careful analysis we find "*the big steps . . . broken into smaller ones, and these into still smaller, until finally the steps seem to vanish altogether.*"

The stairway also serves as a model of the Marxist theory of transformation of quantity into quality if the new view that is gained at the top step is analogous to the qualitative change. In the Marxist theory the change emerges out of nothing but the succession of small quantitative steps. The following example shows how a Marxist proceeds to link the metaphysics of the quantity-to-quality transition to the problem of revolution versus evolution in science. In an attempt to explain the shift from morphology to experimental biology in America between 1890 and 1910, Garland Allen (1978, 1981) argues that "*all revolutionary change is dependent upon antecedent evolutionary change*" and that "*conversely all evolutionary change leads to revolutionary change*" (Allen, 1981, p. 173). He concludes that when such transitions from "*quantitative to qualitative*" occur slowly they are evolutionary, but when they occur rapidly they are revolutionary. For this appearance as a Marxist historian of science, Professor Allen wears a curious combination of modern American eagerness to embrace new ideas in front, and Marxist ideological stiffness behind. Allen makes the analogy between the progress of science and the punctuated equilibrium model of evolution of species (Gould and Eldredge, 1977) in which there are *"periods of rapid change, in which new species are formed or old ones die quickly, followed by stable periods of slow change, where adaptations are perfected."* His analogy is based on a misconception of the analogical argument, which is properly an argument from the function of one order of things to some other that functions in the same way. His analogy is false because the progress of science is not functionally like the evolution of biological species, by mutations and natural selection, which are both blind and unconscious, but science is promoted by people with different views of the truth who self-consciously assert their own views.

Constructivists like Thomas Kuhn (1962, 1970, 1977) reason that scientists work within theoretical traditions that determine which problems they tackle, which methods they use, and how they interpret their observations. So far I agree with those arguments. I part company with them at the point when they say that the reality which scientists study is entirely a social construct. On the contrary, I defend a realist position that the real world determines the content of theories but society influences their forms. Kuhn's thesis is that science advances by alternating short periods of "scientific revolution," during which criticism of theories flourishes, and long periods of "normal" science during which scientists become committed to one or another theory. The changes from normal science to scientific revolution are supposed to occur in relation to "paradigms" which embrace not only scientific theories but also all the extrascientific ideas and concerns of the time—the climate of opinion in which scientists work and with which their theories are continually interacting. Kuhn argues that the change from an old to a new theory involves such a radical break that the rival theories do not share any continuity of theoretical terms or methodology—they are "incommensurable."

I argue precisely the opposite: that a condition for rivalry leading to progress is that two theories share the same theoretical terms so that they can engage in a dialectical process, a constructive dialogue, leading to modification, assimilation, or replacement of one theory by another. Many examples will be given in this and the following chapter to show that prolonged coexistence of rival theories is the rule, and that revolutionary overthrow of one theory by another is the exception in the history of neuroscience. Moreover, there is a continuity of reference to theoretical terms used by earlier and later theories in neuroscience, at least since the early 19th century. Terms such as cell, reflex, synapse, excitation, inhibition, integration, and instinct have gradually changed their meanings within the same research program or research tradition, but without any revolutionary replacement of terms.

Construction of scientific theories is under the influence of sociocultural forces, but that alone is not sufficient to support the constructivist notion that experiments do not provide observations of reality but only provide social constructs. That notion leads to the conclusion, which I find absurd, that our most recent neuroscience theories are no more than social constructs of reality. That view would be more plausible if it were restricted to closed societies in which everyone shared the same sociocultural and ideological tradition and in which dissent from the accepted dogma were prohibited. In those societies the progress of science and even the progress of technology is retarded.

The constructivist notion is of little relevance to an open society in which a diversity of social, cultural, and ideological forces come into play, and in which free communication and criticism of different ideas is tolerated. Those are the societies in which science has flourished. I see no acceptable alternative to a realist view that science is able to arrive at knowledge that is close to the truth of things as they exist in the real world, provided that the cultural, social, and economic conditions are favorable (which include open communication and critique). Science progresses most rapidly under favorable socioeconomic and political conditions, in open societies as Popper defined them in *The Open Society and Its Enemies* (1962), and there is a correlation between the "openness" of a society and the progress of science in that society. Closed societies which do not foster scientific progress themselves may import scientific knowledge from open societies. In other words, closed societies may parasitize open societies. This raises the ethical issue of whether closed societies should be supported by transfer of scientific information from open societies, and whether transfer of scientific knowledge and scientific institutions may help to open a closed society. As we shall see in Chapter 5, those dilemmas are most effectively resolved when ethical values are well integrated with the other components of scientific research programs. Science alone can only decide whether knowledge is true, and we need ethics to enable us to decide what is good.

Kuhn's thesis has been effectively criticized from different theoretical positions by Watkins (1970), Feyerabend (1970a), Boyd (1983), and I. B. Cohen (1985). Lakatos (1978, Vol. 1, p. 9) remarks that: "*For Kuhn scientific change—from one 'paradigm' to another—is a mystical conversion which is not and cannot be governed by rules of reason and which falls totally in the realm of the (social) psychology of discovery.*" John C. Greene (1971, 1981) and Ernst Mayr (1976) have pointed out that Kuhn's theory of scientific revolutions does not fit the biological sciences as well as the physical sciences. Their critique relates mainly to theories of biological evolution. The merits of Kuhn's thesis when it was first published in 1962 were, first, that it pointed out the defects of the view of the history of science as the unwavering progress of knowledge, and provided an alternative view, and second, that it emphasized the importance of extrascientific factors in determining the construction of scientific theories.

From my position, Kuhn's paradigms, even in the form in which they have been modified to meet criticism (Kuhn, 1977), appear to be too vague and admit of too many interpretations to be usefully applied to the history of neuroscience. More significantly, the history of neuroscience does not show periodic revolutions as Kuhn's thesis requires. Preference for one theory over another is rarely sufficient to cause a revolution in neuroscience—most often preference is marginal at first, and the margin increases as evidence is obtained in favor of one theory but in conflict with the other, and as anomalies appear in one but not the other theory. In the history of neuroscience conflicting theories have often coexisted for decades, as numerous examples demonstrate. Important theories have been constructed by synthesis from different theories as often as by conflict between them. For example, the neuron theory was constructed by synthesis from many theories (see Chapter 3), and the modern neurotrophic theory was formed by confluence of theories regarding dependence of tissues on their nerve supply, with theories regarding flow of materials in nerves (see Section 1.4.2).

In the history of neuroscience large stepwise changes in the tradition have occurred when a new theoretical concept replaced its rivals, but the changes have always been gradual, often intermittent, and sometimes in the reverse direction. For example, the theory of mass action of the cerebral cortex (Flourens, 1824, 1842) cast doubt on the phrenological theory of localization of function (Gall and Spurzheim, 1810–1819). The mass action theory was in its turn overtaken by a neurophysiological-neuroanatomical theory of cerebral localization of function (Fritsch and Hitzig, 1870; Ferrier, 1876; Munk, 1881), which is discussed in Section 1.3.4. That "revolution" occurred as a result of refutation of the antilocalization theory by well-supported counterevidence. Nevertheless the two theories coexisted for almost a century. Near the end of that period, in the 1920s, theories of strict parcellation of functions in the cerebral cortex provoked a backlash from the antilocalizationists which resulted in construction of a revised version of the holistic theory of higher nervous functions (Lashley, 1929, 1938).

Another counterexample to Kuhn's theory is the replacement of the theory of electrical synaptic transmission (du Bois-Reymond, 1877, p. 700; Adrian, 1933; Eccles, 1937) by the theory of chemical synaptic transmission (Loewi, 1921, 1933; Feldberg and Gaddum, 1934; Dale, 1934, 1937-1938; Fatt, 1959; Katz, 1962). Even that "revolution" occurred over a period of more than 30 years during which both rival theories of synaptic transmission coexisted. The greatest "revolution" in neuroscience is conventionally taken to be the overthrow of the reticular theory of nerve cell connections by the neuron theory, but as I show in Chapter 3, elements of both theories coexisted for at least 70 years, which is almost half the duration of the history of modern neuroscience. In all these cases progress was made possible by the use of revolutionary techniques as much as by the advance of revolutionary ideas.

1.3.3. Revolutionary Methods Promote Progress

We must make a distinction between a revolution in thought and method. There are more cases of revolutionary methods than of revolutionary ideas. With good reason it can be claimed that the so-called Mendelian revolution was really a revolution of statistical methods applied to biology, and that the neuron theory resulted from a revolution of microscopic and electrophysiological techniques applied to the nervous system. The Golgi technique made it possible to see nerve fibers ending freely and revealed differences between dendrites and axons. Electron microscopy made it possible to resolve structures (e.g., the existence of intercellular spaces or synaptic clefts or myelin layers) that were not resolvable by earlier methods. Electrophysiological techniques showed that axonal conduction of nerve action potentials is different from electrotonus, and both are different from synaptic transmission mediated by chemical neurotransmitters. Powerful techniques can be called revolutionary when they can generate data which reveal serious anomalies in theories which cannot be saved by ad hoc modifications and therefore have to be abandoned. But even in cases such as the reticular theory of nerve connections or the theory of electrical synaptic transmission, abandonment of the refuted theory occurred gradually, over a period of decades.

New techniques may be regarded as revolutionary if their use enables a crucial experiment to be performed that demolishes one of the alternative explanatory models, but that has rarely happened. More often the technique only provides more reliable and more accurate data which can be added to a theoretical model to increase its explanatory and predictive powers. Such was the effect of introduction of the cathode ray oscillograph into neurophysiology. The 1922 paper by Gasser and Erlanger, "A study of the action currents of nerve with the cathode ray oscillograph" (*Am. J. Physiol.* 62:496–524), was, as far as a conceptual advance is concerned, only one link in a continuous chain, namely part of the evolution of understanding of conduction of the nerve impulse. That research program began in the 19th century, especially with the work of Carlo Matteucci (1811–1868), Emil du Bois-Reymond (1818–1896), and Hermann Helmholtz (1821–1894), and continued with the experiments of Ludimar Hermann (1838–1914), Julius Bernstein (1839–1917), Ernest Overton (1865–1933), and Keith Lucas (1879–1916). Those workers were severely limited by the mechanical and electrical artifacts of their measuring instruments—galvanometers of various types and the capillary electrometer as we shall see in Section 4.4.6. The capillary electrometer was combined with an electronic valve amplifier by Edgar Douglas Adrian (1926) and by Adrian and Zotterman (1926). Their technique made it possible to record trains of nerve impulses in single nerve fibers but not to record the shape of the action potential accurately (see Section 5.4). That advance was made possible by the use of the cathode ray oscillograph, a new technique that rapidly transformed electrophysiology and resulted in large advances, not only in understanding the mechanism of nerve conduction but also in knowledge of the electrical activities of nerve cells in general.

Microelectrode recording was a revolutionary technique invented by Ling and Gerard in 1949. This invention, and the availability of electronic amplifiers with a cathode-follower input stage and with high input impedance and low noise level, enabled Fatt and Katz (1950, 1951, 1952) to record end-plate potentials and miniature end-plate potentials (mepps) directly from the neuromuscular junction. The mepps occurred randomly (Fatt and Katz, 1952), and it was soon discovered that mepps were caused by release from the motor nerve ending of quantal packages of acetylcholine, each containing some thousands of acetylcholine molecules (Birks and MacIntosh, 1957; Fatt, 1959; Katz, 1962). In this research program, revolutionary techniques provided the evidence that decisively proved the theory of chemical transmission across the synapse, and consigned other theories, which had persisted for decades, to the trash can of history.

These examples of the progressive effects of revolutionary research techniques also show the interdependence of different fields of science, technology, economics, and politics. Thus, introduction of electronic amplifiers and of the cathode ray oscillograph into neurophysiology in the 1920s depended on prior advances in physics and electronics, and on commercial, military, and government support for technical implementation of those advances.

Another good example of such interdependence is the introduction into neuroscience of radioactive labeling and tracing techniques. A pointed example is tritiated thymidine autoradiographic tracing of the sites and times of origin and migration of neuron and glial cells in the central nervous system (Sidman *et al.*, 1959; Uzman, 1960; Angevine and Sidman, 1961; Miale and Sidman, 1961). This neurobiological application followed rapidly on earlier studies by cell biologists of the kinetics of tritiated thymidine incorporation into DNA in dividing cells (Quastler, 1959). Tritiated thymidine became available for scientific research because of the production of tritium in atomic reactors originally developed during the Second World War for production of isotopes for nuclear weapons. The so-called spin-off of the military and commercial production of radioactive isotopes was their availability at relatively low cost for biological research.

One of the old problems that was solved rapidly by the new technique was how neurons and glial cells originate and migrate in complex structures such as the cerebral cortex and cerebellar cortex. Thus, the model of assembly of the cerebral cortex as a result of migration to successive layers, starting with the outside layer, was initially proposed by His (1890), and accepted as the truth for 70 years. That model was overturned when the technique of labeling neurons with tritiated thymidine made it possible to show that the neurons assemble in the cortex from the deep to superficial layers in an inside-out temporospatial gradient (Sidman et al., 1959; Angevine and Sidman, 1961; Miale and Sidman, 1961). Here too, the old theory of His was not replaced by a new theory because of a paradigm shift in the Kuhnian scheme, but was eliminated by use of revolutionary techniques, tritiated thymidine labeling of DNA, and tissue autoradiography, which were imported into neuroscience from a different domain.

The new explanatory model has been progressively modified to include observations showing deviations from the simple inside-out sequence of assembly of neurons in the cerebral neocortex. Thus, in the dentate gyrus of the hippocampal formation, the granule cells assemble from outside to inside, and in the neocortex the cells of the most superficial layer and the deepest, subplate layer are formed first. The cortical plate, in which neurons assemble from inside to outside, develops between the superficial and subplate layers. Elsewhere I have reviewed the progress of explanatory models of cerebral cortical histogenesis and morphogenesis (Jacobson, 1991, pp. 41–93, 401–430). The purpose of this summary is to show that the most recent theoretical model has been constructed by inclusion of new evidence gained in the 1980s and 1990s without requiring a revolutionary demolition of the older model which was built in the 1960s and 1970s. What has actually happened is that several less-inclusive explanatory models dealing with neuronal birthdays, modes of neuronal migration, interactions between ingrowing afferent nerve fibers and migrating neurons, interactions between radial glial cells and migrating neurons, and so on, have been fitted in a more-inclusive explanatory model. In relation to the less-inclusive models, the more-inclusive model has been constructed at a higher level of explanatory generalization.

1.3.4. Raising the Threshold for Errors Promotes Progress

Error exists as a part of all observation. The notion that the "scientific method" can eliminate error completely was a misunderstanding that persisted until quite recently (e.g., in Beveridge's *The Art of Scientific Investigation*, 1951, which was a required textbook when I was an undergraduate, "truth" is "probable truth" arrived at by logical inferences from empirical observations, with a level of certainty decided by statistics). That positivist notion of "the scientific method" has been enlarged by recognition that wishful thinking and intellectual authority come into play in decisions about the room to leave for error in scientific knowledge (see footnote on p. 265). Ever since the narrow view of scientific truth was challenged by Michael Polanyi (1958) and John Ziman (1968), the 18th century view that all error is vice has gradually diminished, if not disappeared. Even in the 18th century, that view was satirized by Laurence Sterne in his novel *Tristram Shandy* (1760–1767). Walter Shandy, Tristram's father, affirmed that *"Knowledge like matter . . . was divisible in infinitum;—that the grains and scruples were as much part of it, as the gravitation of the whole world.—In a word, he would say, error was error,—no matter where it fell,—whether in a fraction,—or a pound,—'twas alike fatal to truth, and she was kept down at the bottom of her well as inevitably by a mistake in the dust of a butterfly's wing,—as in the disk of the sun, the moon, and all the stars of heaven put together."*

Now we know better than to deny that correct conclusions may follow from false data, and vice versa. Nevertheless, science aims to purge observations of the errors which exist in greatest abundance in the raw data. We assume that there are errors at that stage, but we also assume that anything studied by many people using different methods will be as free of error as the periodic table of the elements. We proceed on the realistic assumption that there is a steep gradient of error, maximum at the emerging front of scientific research and minimum in well-corroborated observations. It is obvious that errors must be accepted as a normal part of the raw data, for if the threshold for error was set so high as to exclude all error from observations, we could not make any. But how do we decide how far to lower the threshold for admission of errors? That decision may be difficult to make when the observations cannot be analyzed by statistics because they are qualitative and very variable, as they have generally been in neuroscience. Some examples will be given below to illustrate this point.

Progress of science can be correlated with the increasing accuracy with which observations can be made. This includes raising the threshold for admission of empirical evidence, thus eliminating artifacts and "noise." It also involves tightening of the logic of the experimental tests, and improved design of control experiments. As we shall see in Section 1.3.5, several criteria are used conventionally for admitting that data are either "true" or "false," and one of these criteria is the stringency of the conditions used for obtaining data. Increasing the stringency of criteria required for corroboration or refutation of a theory can also lead to collapse of consensus—some accepting and others rejecting the more stringent criteria. For instance, let us consider the problem of localization of function in the cerebral cortex.

Phrenology or cranioscopy was based on Gall's theory that different mental faculties are innately located in different regions of the cerebral hemisphere, and that excercise of those faculties results in growth of the corresponding cerebral region with consequent protuberances of the overlying regions of the cranium (Fig. 1.11). This theory of cerebral localization was as conjectural as Flourens's theory of the cerebral equipotentiality which replaced it. Although Flourens's experiments were meticulous and produced far-reaching results, they did not refute Gall's theory of cerebral localization as was generally believed. The counterevidence was actually not sufficient to refute the theory of cerebral localization, as we shall shortly see. Why then was Gall's theory regarded as subversive, leading to his expulsion from Vienna in 1807? One of the charges against him was that his theory was materialistic and undermined religious doctrines regarding the immaterial nature of the soul. Localization of the soul had been accepted

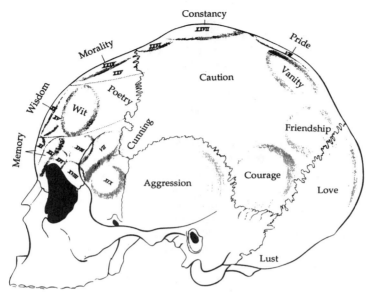

FIGURE 1.11. Cranioscopy was the correlation of bumps and depressions on the human skull ("cranial organs") with mental faculties—the first systematic attempt at reduction of the mind to matter. This was first expounded by Gall and Spurzheim in their four-volume work on the anatomy and physiology of the nervous system (1810–1819), with an Atlas in which this figure appears. I have labeled some of these "organs" as the original authors identified them.

dogma for centuries; it was localization of different mental faculties to different parts of the brain, division of the mind among different cerebral organs, unrelated to the abode of the soul, that was Gall's revolutionary idea. His idea was to reduce human mental facilities to physical entities—benevolence to bumps on the head. We see the weakness in that program, but is it any different from trying to reduce benevolence to genes?

Paul Broca was the first to publish evidence showing that speech and language functions are disturbed in patients who have lesions in and around the third frontal convolution of the left cerebral hemisphere (Broca, 1861a,b,c, 1864, 1888). Some questions have been raised about the validity of Broca's evidence and even about his priority as to discovery of the cerebral localization of language functions (see Penfield and Roberts, 1959; Schiller, 1979). However, Broca deserves the credit for starting the modern research program that rapidly refuted the old concept of cerebral cortical functional equipotentiality, and demonstrated the localization of different functions in separate regions of the cerebral cortex. Broca's work was the stimulus for the first experimental studies of the movements evoked by focal electrical stimulation of the cerebral convolutions by Fritsch and Hitzig (1870).

Before the report in 1870 by Fritsch and Hitzig giving evidence for localization of motor functions in the cortex, the cerebral hemispheres were believed to be inexcitable (e.g., Flourens, 1824; Magendie, 1839; Longet, 1842; Vulpian, 1866). Together with the notion that the functions of the cerebral cortex were exclusively sensory-psychic went the notion of cortex as the seat of the muscular sense, derived from Charles Bell and strongly advocated by H. C. Bastian (1888; see E. G. Jones, 1972). Those who regarded the functions of the cortex as purely sensory-psychic thought the motor centers were located subcortically especially in the basal ganglia (Schiff, 1875; Luciani and Tamburini, 1878; Bastian, 1888).

Fritsch and Hitzig (1870) showed localized excitability of the cerebral cortex of the dog, using bipolar platinum electrodes for electrical stimulation. Contraction of different voluntary muscles was seen on stimulation at five cortical positions (Fig. 1.12). They subdivided the cerebral hemisphere into an anterior motor and a posterior nonmotor region, and also conjectured that the motor region might be divided into centers causing movements of the muscles of the face, anterior and posterior limbs, and muscles of the trunk. Hitzig (1874) extended these to include centers for opening and closing eyes, movements of the tongue, the jaws, and the tail.

Fritsch and Hitzig stimulated the brain with a constant electric current delivered from a voltaic battery, which caused a muscular twitch on closing the circuit ("make" stimulus) and another on opening the circuit ("break" stimulus). When Ferrier (1873, 1876) repeated the experiments, he used an induction coil connected with an electrode to stimulate the surface of the cerebral cortex. The vibrating hammer of the induction coil produced repetitive stimulation resulting in prolonged muscle contraction which was much easier to see than a single twitch. This led Ferrier to conclude that the excitable motor cortex in the dog was considerably more extensive than had been observed by Fritsch and Hitzig (Fig. 1.12). That happened, we now know, because of the escape of his stimulus to distant regions of the cortex and to subcortical structures far from the point of application of his electrode. Ferrier also followed up by mapping the motor centers in the cerebral hemispheres of other animals—monkeys, jackals, cats, rabbits, guinea pigs, and rats.

Ferrier introduced more anomalies than he removed from the map of functional localization in the cerebral cortex. He (1874a,b) unwittingly cut the optic radiation and then concluded that the visual center was located in the angular gyrus, and he failed to find any loss of vision after removal of the occipital lobes because his incomplete removal of the occipital lobes spared substantial parts of the striate cortex representing peripheral vision and the extrastriate cortex subserving a variety of visual functions. In the summary of his work given in his book, *The Functions of the Brain* (1876), Ferrier completely mislocalized the visual, auditory, and somatosensory regions of the cerebral cortex (Figs. 1.12 and 1.13). He localized the visual center of the dog in the gyrus angularis, somatic sensation in the hippocampus, the auditory center in the gyrus temporo-sphenoidalis superior, the taste center in part of the temporal lobe, the center for smell in the uncus. In addition to those mistakes, he found the center for hunger and visceral sensation in the occipital lobes. Ferrier concluded that all those projections were completely contralateral, with the exception of the olfactory center which received inputs entirely from the same side.

While his English colleagues, including Horsley and Sherrington, politely overlooked Ferrier's errors, continental physiologists such as Hitzig (1874) and especially Hermann Munk were not so tactful. Munk gave a punishing critique of Ferrier's methods

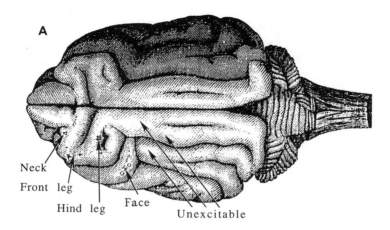

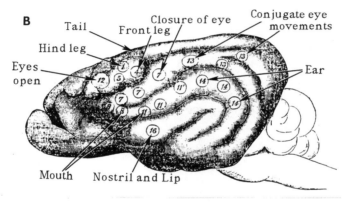

FIGURE 1.12. (A) The first demonstration of electrical excitability of the cerebral cortex and localization of regions from which brief constant-current electrical stimulation evoked movements of skeletal muscles on the opposite side of the body. A dorsal view of a dog's brain is shown (from Fritsch and Hitzig, 1870). (B) A left lateral view of a dog's brain showing positions at which alternating-current stimulation of the cerebral cortex evoked movements on the opposite side of the body (from Ferrier, 1876). Some of the movements were evoked from regions that later were found to have sensorimotor functions, like the frontal eye fields (12) and the prestriate area involved in visually guided movements (13).

and corrected his errors: "*Testing of the animals was without meaning: crudely operated on, crudely observed, crudely concluded. . . . When their value was appropriately considered in the ongoing controversy, Ferrier's researches gained no further attention from either friend or enemy of localization*" (Munk, 1877, in Munk 1881, p. 7).

Munk was correct on almost all points, but his work was largely ignored in England until some of his findings were "rediscovered" a decade later by Hughlings Jackson and Sherrington. For example, it was not the latter but Munk (1877, in 1881, p. 5) who first concluded that ablation of the motor cortex results in disturbances of voluntary move-

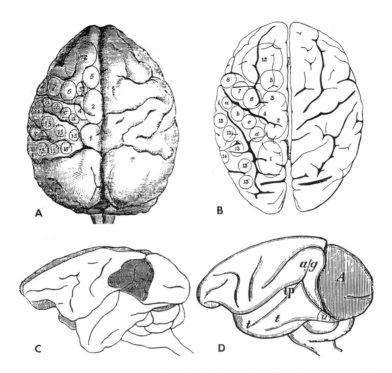

FIGURE 1.13. David Ferrier (1876) mislocalized the visual cortex based on observations of deficits in visual perception and visuomotor functions after cerebral lesions in apes. He concluded that visual functions were localized in the angular gyrus (labeled 13 in **A**, and shaded area in **C**). He extrapolated his findings on apes to the human brain (**B**). The first to map the visual projection to the occipital lobes of the cerebrum was Hermann Munk (1881), based on observations of deficits of visual function following localized cortical lesions (shaded area in **D**; also see Fig. 1.14). Functions indicated by numbers on the ape and human brain are the same as those shown by Ferrier on the dog's brain illustrated in Fig. 1.12**B**.

ments, not in paralysis of movements of individual muscles. Whether the cortex controls individual muscles or groups of muscles performing the same movement remained controversial and has not yet been decisively settled (reviewed by D. R. Humphrey, 1986). Munk's superior surgical skill and experimental techniques enabled him to obtain more accurate maps of functional localization in the cerebral cortex (Fig. 1.14). Nevertheless, his work was ignored in England because of his rough handling of Ferrier.

Technical limitations were the main causes of conflicting results that led to different theories of localization of functions in the cerebral cortex until well into the 20th century. The techniques of stimulating and making lesions in the brain, and observing their effects were inadequate, but the users did not take those inadequacies into account when assessing the meaning of their observations. The ablation techniques, pioneered by Flourens, led to much more extensive damage than people realized or were willing to

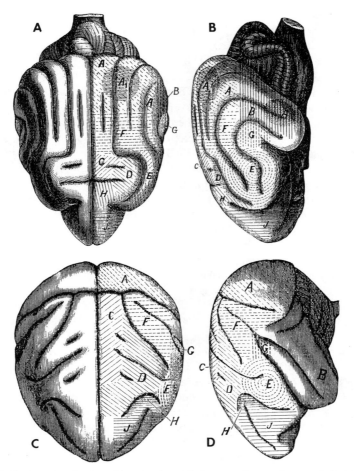

FIGURE 1.14. Regional localization of function in cerebral cortex (dog, A, B; ape, C, D: Munk, 1881). Functions of regions are as follows: A = vision; A1 = central vision; B = hearing; C–J = somatic sensation; C = hindlimb; D = forelimb; E = head; F = eye; G = ear; H = neck; J = trunk.

admit. Hemorrhage, infection, and interruption of nerve pathways, unknown to the operators, led to serious artifacts. A pointed example is the observation made by Flourens (1824) that removal of part of the cerebellum in both birds and mammals produced incoordinated movements on the *opposite* side of the body. We now know that the effects of lesions of the cerebellum are confined to the injured side. Flourens's error was probably the result of damage to the pons and medulla.

Localizationists disagreed among themselves, for example about the center for somatic sensation. It was localized in the cingular gyrus by Horsley and Schäfer (1888), but Hitzig (1874) and Munk (1881) localized somatic sensation in the precentral gyrus, coextensively with the somatic motor center. Lesions of the frontal lobes gave rise to complex symptoms which were interpreted differently by various observers. Ablation of the frontal lobes was reported to result in paralysis of the trunk (Munk, 1881), paralysis of head and eye movements (Ferrier, 1876), no symptoms (Schäfer, 1900), loss of perception, memory, and coordination of sensory and motor functions and in modifications of emotion and social behavior (Bianchi, 1922). The causes of the different conclusions included failure to define the precise site and size of lesions, and transneuronal degeneration effects in cases where the animals survived for longer than a few weeks. These effects were unknown before the work of Gudden (1870, 1889), Forel (1887), and Monakow (1885). As late as the 1950s some neuroanatomists failed entirely to take transneuronal degeneration into account when they estimated the extent of damage caused by "localized" brain lesions.

Another source of confusion of 19th century neuroscientists was their misunderstanding of the limits of extrapolation of their results from one species to another. They freely extrapolated their observations on brain functions of submammalian vertebrates to cerebral functions in mammals and even in man. The differences between the methods of reduction by analogy, homology, or comparative extrapolation were not understood correctly (see Section 2.3). The differences between analogous and homologous structures were defined by Richard Owen in 1848, and he defined those differences with respect to brain structures in 1868. He argued that certain cerebral convolutions and fissures are homologous because they can be recognized in all mammalian brains. This concept was eventually challenged (e.g., by Smith, 1902, 1910, and Brodmann, 1909), but during the mid-19th century Owen's arguments often led to overconfident extrapolation. Thus, Ferrier simply transferred the map of the localization of motor functions in the dog's cortex to the human cerebral cortex (Fig. 1.13). In doing so he failed to recognize the correct homologies of the sulci and gyri of the carnivore and human cerebral cortex. As is well known, the homologies of cerebral cortical convolutions are difficult to establish and have been the subject of much controversy among comparative neuroanatomists (for reviews see Ziehen, 1890; Turner, 1890; Smith, 1902, 1910; Brodmann, 1909; Ariëns Kappers *et al.*, 1936, Vol. 3, p. 1531 *et seq.*; Crosby and Schnitzlein, 1982).

Flourens (1824) fell victim to a similar fallacy when he arrived at his laws of the function of the mammalian nervous system from his observations of ablations of the cerebral hemispheres of birds (see Section 4.4.3). Was he consciously pushing his deductions as far as he dared, or was he simply ignorant of the limitations of such large leaps of extrapolation? Both, I think. The difference between homology and analogy was not understood even after it had been defined by Richard Owen in 1848: "*Analogue: A part or organ in one animal which has the same function as another part or organ in a different animal. . . . Homologue: The same organ in different animals under every variety of form and function.*" Flourens did not understand the difference between analogy and homology, and thus extrapolated his observations made on birds freely to mammals. That led to tremendous errors. For example, Flourens concluded that the cerebral lobes of birds are the centers for volition but not for movement and he extrapolated this to the mammalian cerebral cortex. His concept that the cerebral cortex has no motor functions, and all motor functions are subserved by the basal ganglia, cerebellum, and lower centers, persisted until it was demolished by Fritsch and Hitzig in 1870. In their hands, a powerful technique, electrical stimulation of the cerebral cortex, succeeded in providing data that could not be explained by the theory of nonlocalization of motor functions in the cerebral cortex, and that could be explained by the rival theory of cortical localization of motor functions. In other words, the motor effects of electrical stimulation of the cortex were anomalous in terms of one theoretical model but were explicable in terms of the rival model.

Anomalies in a theoretical model are observations and beliefs which are inexplicable in terms of the accepted theoretical model. Some historians of science (Kuhn, 1970; Laudan, 1977) have recognized that anomalies in theories, or in research programs, or in research traditions, may have heuristic value. However, both Popper (1959) and Lakatos (1978) have an aversion to the notion that anomalies may have heuristic value. As they see it, science progresses when scientists either ignore or exclude anomalous ideas and anomalous observations. They propose that anomalies can be recognized and excluded by strictly logical and rational procedures. But even the few examples so far given show us that unrecognized anomalies can enter theories, and it may well be impossible to construct a scientific theory entirely free of anomalies. However, that does not answer the question whether such anomalies may have heuristic value. Only those have heuristic value that are recognized and that demand explanations. Anomalies can have no effects on the construction of theoretical models if their existence is overlooked or if they are recognized only later by historians.

Anomalies that appear at first to be inconsistent with a theory may become explained later in corroboration of the theory. For example, the discovery of electrically coupled synaptic connections (Furshpan and Potter, 1957, 1959) appeared at first to be an anomaly that was inconsistent with the neuron theory and its auxiliary theory of chemical synaptic transmission. Later these electrically coupled synaptic connections were found to develop initially with a synaptic cleft between pre- and postsynaptic neurons, and the obliteration of the cleft was shown to occur later. The entire process can thus be seen as a corroboration of the neuron theory.

Another pointed example of a discovery that was first viewed as an anomaly, and later understood as a confirmation of a theory, is the discovery of retrograde transneuronal effects by Ruth Barnard in 1940. She found that spinal motoneurons lose the synapses which end on them following section of the ventral spinal root. Her finding was contested by Barr (1940) and Schadewald (1941) and was pronounced by Fulton (1943, p. 53) to be contrary to Cajal's theory of dynamic polarization of the neuron (Ramón y Cajal, 1891b, 1892). In time, the latter theory was overturned, but Barnard's findings were eventually corroborated (Blinzinger and Kreutzberg, 1968), and could then be viewed as fully consistent with the neuron theory.

The most useful sort of anomaly occurs when one theory cannot explain some data that can be explained by a rival theory. This is one of the criteria for choosing between rival theories. There are few examples of a theory in neurobiology that have been decisively refuted and abandoned only because of a single serious anomaly in the theory which could be explained by another theory. More often, accumulation of anomalies in one theory that could be explained by another led to the gradual abandonment of the first theory. This is explicitly stated in Chapter 3 which shows how the theory of cytoplasmic continuity of nerve connections accumulated anomalies that were explained by the rival theory of surface contact, leading to abandonment of the continuity theory.

1.3.5. Verification and Refutation

The question that arises in connection with the nature of scientific progress—either by evolution or by revolution—is how a theory or an entire research program evolves or undergoes a revolutionary change or total overthrow. Is it by a process of verification or by refutation or both? If it is impossible to prove any scientific theory, what level of verification or refutation is acceptable? Quine (1960) argues that scientific theories are *never* determined by the data in a logical and consistent manner. Braithwaite (1953, p. 19) declares that "*except for the straightforward generalizations of observable facts . . . . complete refutation is no more possible than is complete proof.*" Skeptics such as Paul Feyerabend (1957) claim that all theories are equal, and that claims to knowledge made by experts based on scientific data are no better than those made by laypersons based on common sense. To resist such extreme skepticism, Popper (1959) and Lakatos (1978) tried to establish rational criteria for preferring some theories above others. Popper's conception of scientific method is that only falsification of hypotheses is possible, verification is not (Popper, 1959). Popper's idea that evidence determines only one of two possible results is akin to the idea proposed by R. A. Fisher that "*the null hypothesis is never proved or established, but is possibly disproved, in the course of experimentation. Every experiment may be said to exist only in order to give the facts a chance of disproving the null hypothesis*" (Fisher, 1947, p. 16).

Popper totally rejected induction as a means of arriving at the truth. But to avoid skepticism which would put all theories on a par, Popper proposed that theories which admit of refutation are superior to others which do not. Theories which admit nonobservable entities, and thus cannot be refuted, are inferior theories. But we have shown that some of the most powerful theories were initially constructed by admitting nonobservable entities such as synapses, membrane receptors, and ion channels. According to Popper (1959), good theories have been correctly formulated to allow refutation. Without that provision, theories can be neither proved nor disproved. Therefore, he says that no theory can be disproved by experimental evidence unless the conditions for refutation are agreed upon in advance: any theory can be saved by ad hoc hypotheses or by reinterpretation of its terms (Popper, 1959; Lakatos, 1978).

It has been claimed by Popper that the criterion of a good theory is that it contains a statement about the conditions for its refutation. But many examples will be given to show that the methods and other conditions for refutation did not exist at the time the theories were constructed, and no one could have known in advance whether the methods would ever exist. In practice the invention of new methods is hardly ever predictable far in advance, and those that can be specified in advance often prove to be inadequate for the purpose of refutation. Many examples in support of this claim will be given later: valve amplifiers and oscilloscopes, intracellular microelectrodes, integrated circuits, computers, monoclonal antibody technology, molecular genetic technology, and many other techniques. Introduction of those revolutionary techniques in neuroscience research occurred so rapidly that their effects could not have been predicted sufficiently far in advance to satisfy the requirements of refutationists such as Popper and Lakatos.

The conditions for refutation must not be trivial. Lakatos (1978) makes the distinction between naive and sophisticated refutationists. For the naive refutationist a theory is refuted by well-corroborated evidence that is interpreted as conflicting with it. For a sophisticated refutationist a theory T1 is refuted if and only if another theory T2 has been proposed that predicts novel facts not predicted by or forbidden by T1; and if T2 explains the previous success of T1; and if T2 includes all the unrefuted content of T1; and if some of the excess content of T2 over T1 is corroborated. These stringent requirements for refutation allow time for proliferation of theories, and thus may help to sustain healthy competition between different research programs aimed at testing rival theories.

Theories in neuroscience have often been misunderstood and ignored. That is one type of premature theory which is discussed in Section 1.4. The worst fate of a theory is being misunderstood, not so much being refuted. Mendel's theory was ignored for that reason. The formulations of Popper and Lakatos leave no place for nonlogical factors leading to misunderstanding. Lakatos (1978, Vol. 1, p. 1) explicitly denies any role of subjective elements in deciding between rival theories: *"The cognitive value of a theory has nothing to do with its psychological influence on people's minds. Belief, commitment, understanding are states of the human mind. But the objective scientific value of a theory is independent of the human mind which creates it or understands it. Its scientific value depends only on what objective support these conjectures have in facts."* But my point is that scientific theories are no more than mental models which have no objective existence or value apart from the human mind. Even though theories may be more or less true representations of some objective reality, theories themselves are subjective. Lakatos evidently failed to see that two observers may agree about all or many of the available facts and may yet disagree on their theoretical interpretations of them. Rival theories can coexist because the observers may start with different assumptions and hold different values, and because there are unobserved facts as well as artifacts that are mistaken for facts. Illustrative examples of coexistence of opposing theories are: localization of brain function versus theories of distributed brain functions; electrical versus chemical theories of synaptic transmission; theories of myelin formation by axonal secretion versus myelin formation by glial cells; theories of the presence of a membrane enveloping the entire nerve cell versus theories of naked cells; theories of nerve cell connections by cytoplasmic continuity versus connections by surface contacts.

Conjecture and refutation are not symmetrical: it is always easier to make a conjecture than to refute it. Conjectures are quickly made, whereas refutation is not an all-or-nothing event but a gradual process. This view is supported by case analyses given in this and the following chapter showing that neurobiological theories have survived much counterevidence. This occurred in the case of the cytoplasmic continuity theory of nerve connections, which persisted in the face of counterevidence and after the rival theory of surface contact had made the successful prediction of the synapse which was forbidden by the continuity theory. It can be said that rival theories have to ignore one another, because if they took every bit of counterevidence seriously they would have to go out of business. As we shall see in Chapter 3, the cytoplasmic continuity theory was not rapidly refuted by the surface contact theory because new conditions for refutation were set up and because ad hoc modifications of the continuity theory were made. At first the entire cytoplasm of one nerve cell was claimed to be in continuity with the cytoplasm of other nerve cells, but later only certain cytoplasmic constituents such as neurofibrils were said to pass from one cell to another. The continuity theory became "degenerative" (Lakatos, 1978), meaning that the theory failed to account for new facts which could be explained by the rival neuron theory. For that reason the cytoplasmic continuity theory was finally abandoned by most neuroscientists long before it was finally refuted by the electron microscopic and electrophysiological evidence.

1.3.6. Relations of Observations to Theories

Logical relationships between data and theories take many possible forms. We shall not need to define those relations in terms of a calculus of relations or in the form of "truth tables." I reject the notion proposed by F. H. Bradley (1897) that everything which is less than the truth is not real or is false. Truth is not a single thing with a definite end that can be completed and perfected. It is not absolute but is incomplete, imperfect, and contingent. Truth is not quantifiable on any absolute scale of values but only relative to ignorance and falsehood. The truth content of the lie may be the best that we can estimate.

It is obvious that there are several relationships between data and theories in addition to the "normal" relationship between true data (generally, empirical data corroborated by independent observers using different methods) and true theories (generally, theories that have continued to explain new observations). Our case studies will show that true data can lead to false conclusions and also that theories in neuroscience have often been corroborated by artifactual data. It is impossible to exclude artifacts from the data generated by techniques working at the limits of accuracy. Small differences in technique can result in sufficiently large differences in data to support different theories. Conflicting theories in neuroscience have been supported by the same empirical data, and false theories have persisted for long periods in the face of counterevidence.

The historical occurrence of different theories that may be more or less consistent with the same data has several explanations: limitations of accuracy of the data may permit different interpretations; artifacts may be included in one interpretation but not the other; and the data may be open to different interpretations or may be used selectively by people holding different basic assumptions. This concept of the role of artifacts in construction of theories belongs to what James Frederick Ferrier (1808–1864), in his *Institutes of Metaphysics*, termed *agnoiology*, meaning theory of human ignorance, in contradistinction to *epistemology*, a term that he introduced to mean theory of knowledge.

The same data may appear to support conflicting theories, for example the various theories of myelination discussed in Section 1.4.3. The experimental findings are then open to different interpretations by each of the opposing sides in conformity with its own theory. Under those conditions refutation may fail because one theory does not predict facts that are forbidden by or cannot be predicted by the other theory. Even if counterevidence is produced, it may not be accepted as refuting evidence until corroborated by other observers and eventually accepted as "true data" by consensus: but consensus is not a guarantee of the truth—it may be arrived at by misinterpretation of data or by acceptance of false data.

Some case studies will be adduced in the following sections to show various relationships between theories ("true" or "false") and empirical data ("true" or "false"). These cases show that the relation between theory and data is often complex: "true" theories may be supported by "false" data and vice versa. In practice "false" theories are not always refuted in a direct and simple way by "true" data. A theory may be refuted even if it is true (depending on the stringency of the conditions required for refutation), and as a rule theories that remain unrefuted for long tend to be accepted as true theories. A well known example of the latter is the theory of "animal spirits" which was the accepted explanation for the flow of excitation in nerves from ancient times to the end of the 18th century, when it was refuted by Galvani's discovery of "animal electricity" (Brazier, 1959, 1984). Thereafter, "animal spirits" were replaced by "animal electricity" as the agent of nervous activity. That theory was in turn refuted after Helmholtz showed in 1850 that the speed of conduction of the nerve impulse is much too slow to be the same as the flow of current in an electrical conductor (Blasius, 1964). Another example of a theory that was accepted by consensus but was later refuted was the theory of cytoplasmic anastomoses between nerve cells. That theory, which is discussed more fully in Section 3.11, was believed by the majority from about 1840 to well after 1890, although the cytoplasmic anastomoses, as shown in Fig. 3.8A–D, were never observed directly. When improved histological methods did not show cytoplasmic bridges of the diameter proposed by the original theory, the theory was saved by ad hoc diminution of the conjectured cytoplasmic bridges to diameters below the resolving power of available microscopes.

1.3.7. Illusion versus Reality

Experimental artifacts face the scientist with the problem of discerning where reality ends and illusion begins. Science has developed methodologies for dealing with artifacts; principally the examination of the same phenomenon from different standpoints. This methodology would not satisfy the philosopher who is also concerned with appearance versus reality. The problem, as Nietzsche put it, is that "*it is possible to conceive of a reality that can be resolved into a plurality of fictions relative to multiple standpoints.*" This may be illustrated by the story of the six blind men of Indostan who each grasped at a different part of the elephant, and could report the nature of the beast as comprehended only from his own limited standpoint:

> *And so these men of Indostan disputed loud and long,*
> *Each in his own opinion exceeding stiff and strong,*
> *Though each was partly in the right, and all were in the wrong.*

Scientists should know this, but unable to comprehend the meaning of their disagreement, they blame the elephant.

There are two main criteria of the validity of observations: their consistency (or at least, lack of conflict) with other observations in relation to a theoretical model, and their repeated confirmation. Confirmation by means of one technique does not establish their plausibility as much as independent confirmation by different techniques. Artifacts that persistently result from one technique may not result from another, but the same may be said of facts, especially when techniques are pushed to their limits. That is one reason why experimental artifacts are probably the most pervasive cause of false conclusions in the history of neuroscience. Scientists have to identify and analyze the technical flaws in experimental design which give rise to artifacts and they have to recognize that artifacts are difficult to exclude from theoretical models. Histological artifacts have been the most common in the history of neuroscience. Whenever results are capricious and difficult to replicate one has also to think of the possibility of impure or contaminated reagents, badly designed or inaccurately calibrated instruments, and the operation of hidden variables.

The importance of bias of techniques cannot be overestimated in the construction of theoretical models. Numerous examples can be given to show that methodological neutrality is seldom attained. I mean that methods are not neutral with respect to alternative theories, but provide data that are biased in favor of one theory. Histological methods are notoriously biased. For example, the rapid Golgi technique, by staining a small percentage of neurons completely, showed them in apparent isolation, with their fibers ending freely, and was thus biased in favor of the theory of nerve cell autonomy and of connection by surface contact. This is discussed at length in Chapter 3. The Weigert method, which stained myelinated nerve fibers, and the Marchi method for tracing degeneration of myelinated fibers failed to show small unmyelinated fibers. Those methods were biased in favor of models in which the functions of large myelinated nerve fibers predominated. For example, there appeared to be no projection from the brainstem reticular formation to the cerebral cortex, and therefore no role for the reticular formation in cortical activities. Therefore, there was no theoretical model in which to include the finding by Moruzzi and Magoun (1949) that electrical stimulation of the reticular formation can change the EEG from slow, high-voltage alpha rhythm, to fast, low-voltage activity, similar to the change that normally occurs during arousal. The importance of reticular activation of the cortex could be understood only after it became possible to trace degeneration of small unmyelinated nerve fibers in the CNS by means of the Nauta silver impregnation technique (Nauta, 1957). Then the reticular formation was found to project to the cerebral cortex, mainly to layer 1, via intralaminar nuclei of the thalamus (Killackey and Ebner, 1972).

The importance of methodological neutrality in obtaining evidence with which to decide between opposing theories is shown by the history of the opposition between the continuity theory and the contact theory of nerve connections. The two theories are examined in detail in Chapter 3, and here we are only concerned with one of many examples showing that a theory can be supported by evidence obtained only by selected methods. Different histological techniques were favored by the supporters of the two theories because the techniques provided the preconceived results. The methods were not neutral with respect to the rival theories and therefore they failed to provide refutational evidence. Thus, what was considered to be refutational evidence by one side was regarded as artifactual by the other side. Ramón y Cajal (1933b) employs his most biting sarcasm when discussing these artifacts: *"Due to the existence of these and other artifacts . . . the reticularists, in their published figures dealing with Endfüssen, chalices, etc., have usually omitted the most common relationships whereas the most peculiar and spurious ones are shown with morbid delight."* This rhetoric conceals the truth that neither the proponents of the contact theory nor those of the continuity theory had the crucial evidence that was finally obtained with the electron microscope.

Both the opposing sides used the strategy of obtaining convergent evidence from independent experimental methods such as observations on outgrowing nerve fibers *in vivo*, as in the tail fin of the frog tadpole, and in histological preparations of embryonic nervous systems. But those results did not lead either to a consensus or to refutation of either theory. The methods were not neutral with respect to both the opposing theories, and so they yielded conflicting results. Held (1909) states: *"If I can summarize the various results of these new studies of the histogenesis of the nervous system, I must say that there is no observation which is not directly contradicted by another."* The fact that the same evidence could be construed both as corroboration and as refutation of the theory of cytoplasmic connections between neurons is consistent with the view that scientists do not abandon theories only because they accumulate anomalies, but they ignore anomalies or they adjust the theory to accommodate counterevidence.

It is clear to us now that the dispute about the way in which nerve cells form connections could not be resolved by making observations on intact organisms because there were too many unknown variables. The observations could be interpreted in conformity with a multiplicity of theories as Duhem (1906) and Quine (1969) have argued. The best that could be accomplished under those circumstances was to devise experiments that reduced the number of uncontrolled variables, for example, in tissue culture, as Harrison first showed in 1906.

Some theories such as the reticular theories of nerve connections, discussed in Chapter 3, and early theories of myelination, discussed in Section 1.4.3, were constructed almost entirely on foundations of artifacts. In those cases the theories were eventually toppled, not because of subtle dynamics of restructuring of scientific paradigms, in the manner proposed by Thomas Kuhn (1970), but simply by the use of better methods and invention of more reliable instruments. Techniques and instruments can move quite freely between different scientific disciplines without respecting boundaries between their different theoretical structures. This can be a potent cause of change in the structure of scientific theories. A new technique can very rapidly dispose of an old theory, and it can also provide the evidence to show that two conflicting theories are both partially true. An example of the former occurred when the electron microscope first revealed the rolled configuration and membrane layering of the myelin sheath (Geren, 1954; Robertson, 1955). This was a purely fortuitous discovery: it had not been predicted by any of the existing theories of myelin formation. All the old theories about myelin structure and development, which had been debated for a century, were so wrong that they were simply abandoned (see Section 1.4.3).

The use of a powerful new technique has provided evidence showing that two apparently conflicting theories were both contingently true, for example, theories of chemical versus electrical synaptic transmission. That occurred when invention of intracellular microelectrode recording and suitable electronic amplifiers made it possible to record synaptic potentials at the neuromuscular junction and from spinal cord neurons. The amplitude of the postsynaptic potential was found to be too big to be produced by electrical transmission across the synaptic junction, but the evidence of miniature end-plate potentials and the effects of pharmacological agents that selectively block synaptic transmission supported the hypothesis of chemical mediators of transmission across the synaptic cleft (reviewed by Eccles, 1959, 1964, 1990). The evidence refuted the old theories of electrical synaptic transmission, but no sooner had this happened than authentic cases of electrical synaptic transmission were discovered, by use of the same techniques (Furshpan and Potter, 1957, 1959).

At this point I wish to emphasize once more that claims to have refuted a theory may be made even if the theory is later confirmed. Several examples of such premature theories are discussed in Section 1.4. A familiar example, requiring little further comment, is the Golgi apparatus. For years after its discovery by Camillo Golgi (1898a,b, 1899a,b, 1900) in many different types of nerve cells there were claims and counterclaims concerning its authenticity (reviewed by Beams and Kessel, 1968). One can say that the Golgi apparatus was repeatedly killed by one technique only to be resurrected by another. Then, after its final resurrection, Ramón y Cajal (1901–1917) claimed that he had discovered the Golgi apparatus first. This bizarre priority claim is considered again in Section 4.3.3.

1.3.8. The Convenient Fiction

Reification or hypostatization occurs when conceptual entities are named as if they actually exist, or when ideas are treated as things. For example, "the life force" was treated by vitalists as a thing, a form of energy, until Helmholtz (1847) and du Bois-Reymond (1848) showed it to be a nonthing, merely a convenient fiction summarizing the diverse activities that constitute life. In some cases such as "the cell membrane" and "the synapse," the name was first given to unobservable theoretical entities which were much later identified as real objects. When "the mind" is conceived as a thing that thinks, feels, and wills, it seems to me to belong to the same category of reified concepts as "the life force." In other words they are non-things, and the "problems" in which they are treated as things are pseudoproblems. To avoid reification of the mind I prefer to designate states of feeling, willing, and thinking as mental states and to designate as mental processes those mental states that are prolonged, such as sadness, happiness, and fear, and I include subconscious as well as conscious mental states and processes. I deal with this in more detail in Section 2.10.

Another class of hypostatized entities were eventually shown to be purely imaginary, for example "the diffuse nerve net" of Golgi (see Section 3.11). There was at first no difference in the ontological status of the diffuse nerve net, the cell membrane, and the synapse—they were all hypostatized bodies invented for the purposes of certain theories. Proof of the existence of the cell membrane and the synapse was produced slowly, starting at the end of the 19th century and culminating in the electron microscopic evidence of the 1950s. Refutation of a theory supported by a hypostatized entity like the diffuse nerve net is very difficult because the entity can be re-invented in slightly different forms as required to serve the theory. The reticular theory of nerve connections was saved for more than 50 years by that strategy until it was finally refuted by the electron microscopic evidence.

Other hypostatized entities such as "the reflex" have persisted because of their convenience in facilitating communication and theory building, although they are only mental constructs. Such hypostatized entities may even enter the realm of objective knowledge called by Popper the "third world."* One way in which neuroscientists have dealt with such entities is to put them in the world of ideas in a separate category labeled "as though." Neuroscientists necessarily hypostatize when talking of "the glial cell," "the neuron," "the synapse," "the receptor," all of which normally exist as functional components only in relationship to a complex organization. These components are conveniently discussed as if they function independently. Their status will be discussed in Chapter 2 as part of the more general problem of reductionism.

Observations of isolated components of the nervous system have to be interpreted, either naively as if the conditions in the reduced state approximate those in the whole system, or recognizing more correctly that reduction should aim at revealing relations *between* as well as *within* different levels of organization. The problem has to do with how analytic and synthetic constructs are made in neuroscience, and also with how different levels of complexity of organization relate to one another. This is discussed more fully in Chapter 2 in relation to the reductionist methodology. However, it should be noted that the problem has not even been posed here in terms of dichotomy between holism and reductionism, because that is not the significant distinction to be made here. A more significant distinction can be made between analytic and synthetic methods.

The problem was already explicitly stated by Sherrington almost a century ago. In formulating his conception of *The Integrative Action of the Nervous System* (1906), Sherrington was compelled to isolate the reflex "*as though*" it exists independently, as "*a convenient if not a probable fiction.*"† He states that: "*A simple reflex is an abstract conception, because all parts of the nervous system are connected together and no part of*

---

*The first world is the material world, the second is the world of consciousness, the third is the world of what Popper (1972) calls objective knowledge.

†Hans Vaihinger (1911) made popular the concept that inferred entities are to be regarded "*as if*" they exist as "*starting points of calculation.*" The notion of convenient or conventional fictions was widely current at the end of the 19th century. Thus, Nietzsche wrote in 1886: "*One should not wrongly reify 'cause' and 'effect,' as the natural scientists do . . . one should use 'cause' and 'effect' only as pure concepts, that is to say, as conventional fictions for the purpose of designation and communication*—not *for explanation*" (*Beyond Good and Evil*, section 21). We shall take up this problem of causal explanation in Section 2.13. Modern philosophers of science disagree about the ontological status of unobservables. Logical positivists and other empiricists argue that knowledge can be based only on observations, so it cannot include unobservable entities (e.g. Ayer, 1959). Realists argue that there is no sharp distinction between observables and unobservables. There will always be entities, realists say, whose existence can be inferred but cannot be confirmed by sense experience and the best instruments. For more, see the excellent discussions by Maxwell (1963) and Boyd (1983).

it is probably ever capable of reaction without affecting and being affected by various other parts, and it is a system certainly never absolutely at rest. But the simple reflex is a convenient, if not a probable, fiction. Reflexes are of various degrees of complexity, and it is helpful in analyzing complex reflexes to separate from the reflex components which we may consider apart and therefore treat as though they were simple reflexes" (Sherrington, 1906, pp. 7–8). The artist has a similar problem of focusing on the parts of the composition while keeping the whole in view. In the words of Henry James written at the same time as those of Sherrington, the dilemma is that *"really, universally, relations stop nowhere and the exquisite problem of the artist is eternally but to draw, by a geometry of his own, the circle within which they shall happily appear to do so. He is in the perpetual predicament that the continuity of things is the whole matter. . . . That this continuity is never by the space of an instant or an inch, broken, and that to do anything at all, he has at once intensely to consult and intensely to ignore it"* (preface to *Roderick Hudson*, New York Edition, 1907).

What to consult and what to ignore in the continuity of the nervous system has increasingly become one of the main methodological problems of modern neuroscience. To reduce or not to reduce, that is the question, as we shall see in Chapter 2.

1.4. MATURATION OF NEUROSCIENCE THEORIES

It now remains to consider the conditions for the heuristic functioning of a theoretical model in a research program. I call this condition the "maturity" of a theory. The maturity of a theory is the concept that relates the functions of the theory to the progress of the research program. A theory is mature when it fits into a progressing research program and operates effectively in guiding techniques to obtain new facts. In addition, a mature theory must have explanatory and predictive powers and it should be possible to test its validity or refute it by means of the available techniques.

The theory that the nerve fiber is an outgrowth of the nerve cell is an example of a mature theory which is discussed in detail in Chapter 3. This theory was progressive, in advance of the facts, from the time of its introduction by Bidder and Kupffer in 1857 for almost a century (His, 1886; Ramón y Cajal, 1890a,b,c,d; Harrison, 1910). After the 1950s it became related to a number of theories that could be included in the axonal outgrowth theory (theories concerning axonal transport, the cytoskeleton, growth factors, extracellular matrix interactions with axons, and so on). These theories are less-inclusive when ranked in relation to the more-inclusive axonal outgrowth theory, and the latter was less-inclusive in relation to the more-inclusive neuron theory. This hierarchy of theories is discussed more fully in Section 3.4. The validity of the original, more-inclusive theory was not necessarily affected by refutation of some of the less-inclusive theories. For example, the axonal outgrowth theory was not affected by refutation of the theory of bulk flow of axoplasm (Weiss *et al.*, 1962; Weiss, 1963).

A premature theory is too far in advance of the available techniques and the accepted facts. By premature I do not mean only that all scientific theories are tentative and subject to change—persistently immature. Rather, by premature I mean that the scientific theory cannot be accepted into a research program. That may happen because the techniques to test the theory have not yet been invented, because the theory is opposed by well-accepted theories, or is in conflict with prevailing values.

Many scientific problems are neglected or abandoned for some time before being taken up again when relevant techniques are available. Investigations detour around some problems because they are insoluble or appear to be so. Or research on a problem is abandoned because other problems appear to be easier to solve or otherwise more attractive. Old problems are also abandoned because the invention of new techniques provides opportunities to work on new problems. Theories are abandoned after they are refuted, but false refutation may occur. Quantitative and qualitative deficiencies may exist in the data that result in false refutation of a theory. There are many examples showing that neurobiological theories have been corroborated or refuted by histological artifacts produced as a result of unreliable techniques. The descriptive and weakly predictive character of neurobiological theories makes them vulnerable to admission of artifactual data. When opponents of a theory expose artifacts and other anomalies they may claim refutation of the theory even when it is finally shown to be true.

Those intrinsic scientific anomalies may be augmented by extrascientific factors which result in neglect or abandonment of a theory. The most common extrascientific factor is that the premature theory originates outside the scientific establishment or from somebody of low rank in the scientific hierarchy. Those factors are especially potent when a premature theory conflicts with another theory held by scientists of greater authority.

Theories may be premature for a period and then become mature, that is, the techniques and data catch up with the theory. Ramón y Cajal (1928, p. 750) recognized a category of premature theories, and understood that they are based on conjectures supported by few facts: "... *we are still in the phase of collection of materials. As a consequence our hypotheses are premature. . . . They are conceptions to point the way, conjectures thrown out to excite and keep up investigation.*" I should add that if they "*keep up investigation*" they become mature theories and are then open to refutation. But as it happened the theories concerning regeneration in the CNS, to which Cajal's statement refers directly, were premature and they have only recently been repeated and accepted in mature research programs (e.g., Aguayo, 1985; Aguayo *et al.*, 1983). This example is discussed in more detail later in this chapter.

You may well ask what is the point of finding anticipations of modern scientific discoveries in older work. It shows that if one scientist does not make the discovery another is almost certain to do so. It also challenges us to explain the different conditions that denied priority to the one but permitted it to the other. We can explain those conditions in terms of differences in maturity of research programs. At the same time I agree with Merton (1961) that one must show that the two discoveries are indeed equivalent or almost so, and be careful not to make excessively liberal interpretations of what was discovered then and what was discovered later.

Mendel's research on inheritance is the most notable example of a premature scientific research program in which both the statistical method and theory were premature (Fisher, 1936). Mendel first described his experiments in 1865, and wrote about them to Carl Naegeli over a period from 1866 to 1873, but Naegeli failed to recognize their significance (Stubbe, 1963, 1972; Olby, 1966). Mendel's findings of the independent recombination of different characters were unintelligible in terms of the prevailing belief in great variability of hybrids. Mendel found a much more regular pattern of inheritance than Darwin's theory had led people to expect, and it was rejected mainly for that reason. The relationship between meiosis and heredity began to be discovered just as Mendel's findings were rediscovered after 34 years of neglect. By then the way had been prepared by Weismann who showed that the soma is a product of the hereditary material, and not vice versa, as had been accepted in previous theories (Weismann, 1892, 1904). Thus, the conjecture of Maupertuis that particles originating from all parts of the body direct development (Glass, 1959), or Charles Darwin's similar theory, made the hereditary determinants the products of the soma. Mendel, supported by Weismann, first conceived of the soma as the product of the hereditary determinants, a premature theoretical concept that could not be fitted into any research program in the late 19th century.

The inductive leap is a characteristic of many premature theories. Mendel's theory of inheritance was an induction from relatively few empirical observations. Nowadays, Mendel's theory is freely generalized to all genetically inherited characteristics in plants and animals, yet relatively few have been subjected to genetic analysis. Darwin's theory of natural selection began as a premature theory based on a conjecture which was followed by more than 25 years of accumulation of facts before the theory was published in 1859. As Darwin recognized, the theory starts as a conjecture which can be generalized, by induction: *"In scientific investigations it is permitted to invent any hypothesis and if it explains various large and independent classes of facts it rises to the rank of a well-grounded theory"* (*The Variations of Animals and Plants Under Domestication*, Vol. 1, p. 8, 1868).

Inductive leaps occur not only at the beginning but also in the middle of some research programs that are largely constructed on the empirical plan. For example, Harvey's experiments demonstrating circulation of the blood, published in 1628, are held up as excellent examples of the empirical method, as they indeed are. Nevertheless, Harvey made a bold inductive leap when he predicted the existence of invisible pores or passages connecting the arterial with the venous system. These pores were nonobservable entities necessary for Harvey's model of circulation of the blood through a closed system. More than 30 years elapsed before the capillaries in the frog's lung were observed by Malpighi in 1661 and communicated by him in the form of two letters addressed to Borelli. The existence of capillaries was confirmed by van Leeuwenhoek in 1668. It is worth pointing out that Harvey's discovery of the circulation of the blood indirectly gave a new impetus to neuroscience—it left no other place for the soul but the brain, and no other place for the flow of animal spirits but in the nerves, as Descartes and Willis were quick to see.

Another example of a premature discovery that could not be accommodated in the structure of a neuroscience research program is the report by Ruth Barnard (1940) that synaptic terminals become anatomically separated from spinal motoneurons undergoing chromatolysis after axotomy. She correctly deduced the existence of a retrograde transneuronal stimulus. This was promptly denied by the authorities (Barr, 1940; Schadewald, 1941). In his influential book *Physiology of the Nervous System* (1943), Fulton disagreed with Barnard's conclusions. But we know that when the meeting of a book and a head produce a disagreeable hollow sound the head is not always to blame. Fulton (1943, p. 53) dogmatically asserted that *"Miss Barnard's deductions are wholly contrary to the neuron doctrine."* According to that doctrine, synaptic action can occur in the anterograde direction only, and therefore retrograde transsynaptic effects must be impossible. The unfortunate premature discoverer was consigned to oblivion by a majority vote. Her work remained uncited when the phenomenon of loss of synaptic contacts by axotomized neurons was rediscovered (Blinzinger and Kreutzberg, 1968; Hamberger *et al.*, 1970; Chen *et al.*, 1977). That correct findings are dismissed because they are in conflict with an erroneous theory goes to show that theories lead and facts follow. Premature discoveries will be taken seriously only by the minority who do not have fixed ideas, and the majority of scientists will ignore findings that do not fit their preconceived theories.

Scientific research programs are postmature when their empirical data cannot be explained by their theories, and their theories fail to predict new discoveries. Eventually the accumulation of data far exceeds the heuristic limits of theories, and data are collected merely because techniques make it easy to do so. Prolixity is the present predicament of many research programs in neurobiology. Under the prevailing conditions, research techniques generate vast amounts of data that are not relevant to any large theory. Prolixity is most serious in molecular biology, biochemistry, and cell biology. In those fields the most prolific authors average several research publications per week (C. Anderson, 1992). We are showered with paper confetti, on each of which is written a small piece of an idea, and we are expected to catch these and arrange them in some order to express complete concepts! Scientists seem to be unwilling to restrain themselves from publishing confetti-like papers and have become increasingly unable to limit themselves to a few well-considered reports each year. In addition, when scientists publish in collaboration, they often fail to disclose their respective contributions and responsibilities for the results and conclusions. These practices have increased the number of publications, diminished their significance, and reduced the responsibilities of the authors. Instead of journals being filled largely with reportage of the latest techniques and data, there should be greater scope for discussion of values and for elaboration of theoretical concepts. If the number of papers were to be reduced, some scientists would reconsider their definition of an original observation or a conceptual innovation, which are the best justifications for publication.

Scientists tend to ignore theories whenever they can easily apply techniques to obtain data copiously. In those circumstances the data lead and theories may follow. This may be unavoidable at certain stages in any research program. Under those conditions, theories may be modified to fit the data, or auxiliary theories may be constructed to conform with some of the data. Very rarely a new theory is produced that includes all the data and has additional predictive and explanatory powers that the previous theories lacked. The research program matures when that occurs. Theories guide the acquisition of data in a mature scientific research program. It cannot be sufficiently emphasized that scientific theories may lead a research program to practical goals, but to complete it, moral values must lead the research program toward just and humane goals. I defend that thesis explicitly in Chapter 5.

Postmature theories are based on well-corroborated data. They are empirical and deductive and avoid inductive leaps in advance of the evidence. They are made when facts are produced very rapidly and theories must be made to fit the facts. Therefore, they are likely to be proposed simultaneously by several people. They meet with universal approval at their first appearances. Postmature theories often have little heuristic value because they merely state what the majority are already convinced is true or at least what they fully expected. But they help to promote the careers of some scientists who know how to say what their audience likes to hear. That may be good politics, but is not good science. Postmature theories are formulated when powerful techniques produce many facts that are at first unrelated to any theory, but later become fitted into a theoretical construct—the theory comes after the facts. The theory may then start operating as a mature theory, leading to new facts, or it may persist as a postmature theory, merely explaining old facts, or it may serve as a modification of an old theory to protect it from refutation. Examples of these modes of operation of theories in neuroscience research programs will be given in the following sections.

1.4.1. Case Study: Dendritic Spines

A more detailed case analysis of the discovery of the dendritic spines shows how empirical data are related to theories. This case illustrates some of the factors that are involved in construction of different theoretical models explaining the same phenomena: different world views and extrascientific factors lead scientists to different initial conjectures; relatively small differences in technique can give rise to sufficiently large differences of data to support alternative theories; refutation is delayed for various reasons which will be considered below.

Dendritic spines were discovered and correctly described by Philip Owsjannikov in 1864: *"The processes of the Purkinje cells have a dense covering of short, fine hairs, which are somewhat longer towards the periphery."* We now know that there are about 200,000 dendritic spines on one Purkinje cell. Spines are extremely small and variable in size and shape (length 0.8–1.6 μm; diameter of neck 0.4–0.5 μm). Spines were close to the limits of resolution of the achromatic objectives available to Owsjannikov. It is not surprising that his premature discovery of the spines in cerebellar preparations stained with carmine and hematoxylin was followed by a long intermission before the rediscovery of the dendritic spines in the cerebral cortex in 1891 by Cajal, using the Golgi and methylene blue techniques and apochromatic objectives. In his very first paper in 1888 on the cerebellar cortex, Cajal depicted some irregularity of the surface of dendrites of Purkinje cells (Fig. 1.6E), and mentioned them without further comment. His drawings in the first paper were not very convincing demonstrations of spines, as if he lacked confidence in their reality. Anyhow, it was known that such irregularities often appear on dendrites stained with the Golgi techniques, but they were regarded as artifacts by Golgi and others who, therefore, depicted dendrites with smooth surfaces (Fig. 1.6E).

One should not underestimate the importance of the Golgi techniques only because they were liable to produce artifacts (see Section 3.7). Golgi's first publications of his potassium dichromate-silver technique appeared in 1873, his mercuric chloride method in 1879 and his rapid method in 1886, but were largely ignored until 1887 when Koelliker made their merit more widely known. Cajal's double impregnation modification of Golgi's potassium dichromate-osmic acid-silver nitrate technique appeared in the 1891 paper in which he first showed the dendritic spines. The various "Golgi methods" were first compared, one might even say demystified, in the useful paper by Hill (1896). Nevertheless, confidence in the Golgi techniques was reduced by their capriciousness and their liability to produce artifacts. Most significantly there were no means of telling whether the metallic precipitates, the so-called *reaziona nera*, were intracellular or on the cell surface. Not much could be added to the debate about what was stained by Golgi techniques until much later studies with the electron microscope showed that the metallic deposits are intracellular and often completely fill the cell (Blackstad, 1965; Stell, 1965; Spacek, 1989). In the 19th century, the authenticity of the picture of nerve cells revealed by the Golgi techniques was questioned by those using the gold chloride methods (Apáthy, 1897) and reduced silver methods (Bergh, 1900). These showed nerve cells much smoother and more delicate than they appeared when impregnated with the Golgi technique, which was, therefore, said to produce artifacts consisting of precipitates on the surface of neurons. Golgi himself refused to accept the authenticity of dendritic spines for that reason. The dendritic spines were at first explained away as artifacts by supporters of reticular theories, but later, when their authenticity was no longer in doubt, they explained spines as the sites of cytoplasmic continuity between dendrites—as a form of dendro-dendritic connection (Bethe, 1903; Held, 1929).

Dendritic spines were shown to be authentic structures, not artifacts, by Ramón y Cajal (1891a,b) in pyramidal cells of the cerebral cortex of newborn mammals, stained by his double impregnation version of the Golgi technique or vitally stained with methylene blue as shown in Fig. 1.15 (Ramón y Cajal, 1896a). Retzius (1891c) was the first to show the dendritic spines in cerebral cortical cells of the human brain and to confirm their existence in other mammals. Camillo Golgi showed perfectly smooth dendrites in all his publications before 1903 and later regarded spines as artifacts, an opinion which was also held at first by Koelliker (1896, p. 647), although he later accepted their existence as a fact. Our question is why Cajal accepted their authenticity while Golgi and Koelliker did not. The answer is complex: Cajal's theoretical model of connections between nerve cells could include dendritic spines whereas Golgi's model and Koelliker's model could not include them.

The criteria given by Koelliker for the artifactual character of dendritic spines were as follows: (1) Dendrites lack spines and are quite smooth in neurons dissected out of macerated CNS tissue, including Purkinje cells, spinal motor neurons, and cerebral cortical pyramidal cells. This was also shown by Deiters (1865), Ranvier (1878), and Max Schultze (1870, Fig. 28) and by Vignal (*Développement du système nerveux*, 1889). (2) There is a large variation of the presence, number, and size of dendritic spines seen in Golgi preparations. (3) Structures that are like dendritic spines are present on some glial cells and some axons and axon collaterals in Golgi preparations. In short, the variability of occurrence of dendritic spines and the lack of any known functions for them, make it likely that they are artifacts.

Cajal's criteria of the authenticity of dendritic spines were as follows (Ramón y Cajal, 1896a, 1897b, Vol. 1, p. 54): (1) Spines stain with the Golgi method as well as the Golgi-Cox method; (2) they are always found on the same regions of the dendrites and

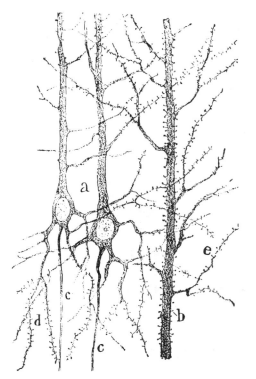

FIGURE 1.15. Pyramidal cells of the cerebral cortex of an adult rabbit stained with methylene blue, showing dendritic spines (1.40 N. A. Zeiss apochromatic objective). a = two medium size pyramidal cells; b = dendritic spines on the apical dendritic shaft of a giant pyramidal cell; c = axons; d = spines on basilar dendrites; e = spines on collateral branches of the apical dendritic shaft. (From Ramón y Cajal, 1891a.)

are always absent from certain places such as axons, cell bodies, and the thick stems of dendrites; (3) viewed with high-power objectives, spines do not resemble crystals or ill-defined aggregates, but they always resemble fine unbranched hairs that extend without a break from the substance of the dendrite; finally, dendritic spines are as well stained with methylene blue as with the Golgi methods. In short, the invariability of their occurrence in certain positions and not others and the convergence of evidence obtained by different techniques, make it likely that they are authentic structures.

Regarding the functions of dendritic spines Cajal conjectured that they increased the surface of the dendrites and thus increased electrical transmission between nerve cells. Cajal (1906) conceived of intracellular transmission within each nerve cell, with the neurofibrils acting as the conducting elements, as shown in Fig. 1.16. He depicted a single neurofibril in each of the dendritic branches on which the majority of dendritic spines were found. Such an arrangement might conceivably subserve summation of excitatory nervous activity, but Cajal did not conceive of integration of excitatory and inhibitory effects. Until the 1960s both neuroanatomists and neurophysiologists failed to understand the possible importance of dendritic spines in nervous integration. Thus, Sherrington simply eliminated the dendrites from his diagrams of integration of pre-synaptic inputs on the spinal motoneurons (Figs. 1.4 and 4.5). Diagrams of neural

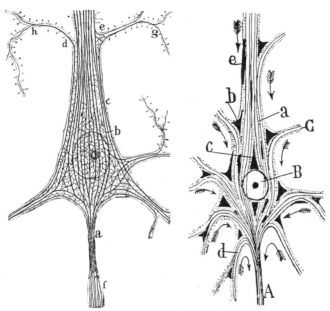

FIGURE 1.16. Anatomists conceived of neurofibrils as conductors of nerve impulses. This concept was held despite physiological evidence that generation and propagation of nerve impulses occur at the surface membrane. The notion that intracellular neurofibrils are the conducting elements was held by neuronists and reticularists. An important difference was that neuronists like Cajal believed that intracellular neurofibrils were confined to individual neurons, whereas reticularists like Held and Nissl believed that neurofibrils extended between different neurons and formed the conducting pathways through the nervous system. These figures were drawn by Ramón y Cajal (1906) to illustrate his concept of the neurofibrillary network restricted to the individual neuron (A); and the polarity of neurofibrils determining the direction of nerve impulse conduction—Cajal's concept of "dynamic polarization of the neuron."

circuits often omitted dendrites and showed all afferents ending directly on the neuron soma. Physiologists were especially prone to collapse the entire dendritic tree into the cell body for the purposes of showing the integrative action of the afferents that converge on a single neuron, as shown in Figs. 1.4 and 3.17. Neuroanatomists allowed the dendritic tree to perform integrative functions even before the functions of dendritic spines were properly understood, starting in the 1960s (Fig. 1.17).

We now know that spines are the main postsynaptic targets of excitatory synaptic inputs (Gray, 1959; Spacek, 1985; Coss, 1985). Their peculiar shape and evidence that they can change shape (Fifková, 1985) suggest that they may function to modulate the passage of current from the synapses to the dendritic shaft (Rall and Segev, 1988). It has also been suggested that they may act as metabolic compartments modulating calcium levels (Gutherie *et al.*, 1991; W. Müller and O'Connor, 1991), and may control the level of protein synthesis on ribosomes at the base of each spine (Steward, 1983; Koch *et al.*, 1992). While Cajal admitted that their functions were unknown, he also conjectured that

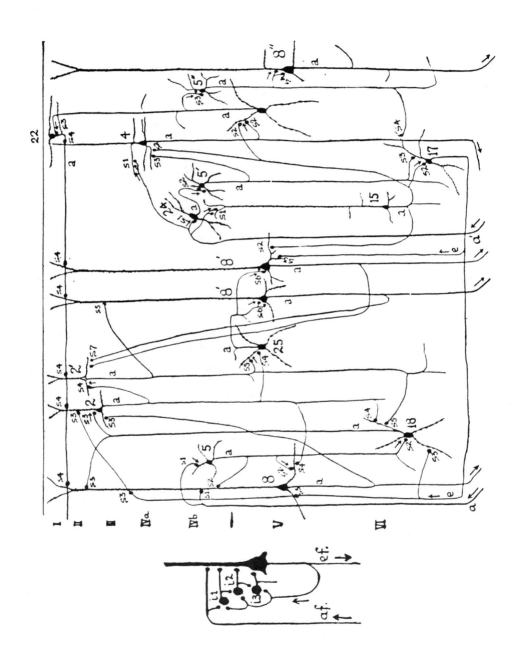

dendritic spines might be involved in transport of nutrients to the dendrites. Not least, he regarded the free endings of dendrites as good evidence of the contact theory of nervous connections by surface contact and evidence against the cytoplasmic continuity theory of nervous connections. In his final work ¿*Neuronismo o reticularismo?* published in 1933, Cajal outlines the history of his discovery of dendritic spines and refers to them as the sites of synaptic connections. He argued that the spines receive synaptic connections (by then he was belatedly using the term "synapse") mainly from the axon collaterals of cortical pyramidal cells.

In the late 19th century several different theoretical models were constructed to explain how connections formed between nerve cells. As we shall see in Chapter 3, some components of these models were included in all or most models while other components were uniquely held by one or another model. One of the latter was the form of connections. They were conceived either as cytoplasmic bridges between the insides of nerve cells or as contacts between membranes on the outsides of nerve cells. We can call these continuous or discontinuous models. These two forms of connections were not necessarily mutually exclusive, and for a time some neuroscientists were able to conceive of models that included both forms of connections. However, as we argue in Chapter 3, use of different techniques and selection of data led to the construction of models that exclusively incorporated either the continuous or discontinuous forms of connections between nerve cells. In Golgi's theoretical model, nerve cells were connected by means of a network formed by axon collateral branches. His theory ascribed purely nutritive functions to dendrites in relation to blood vessels and glial cells. He proposed that the nervous activity leaves the nerve cell via the main axonal trunk but enters through axon collaterals, thus excluding the dendrites and cell body from the conducting pathway. Golgi showed the dendrites with smooth surfaces in the spinal cord (Golgi, 1880a, 1883), cerebral cortex (1886), and cerebellar cortex (1886). There were no obvious functions for dendritic spines in Golgi's model, and he therefore thought they must be artifacts.

Golgi's model was opposed by theoretical models in which nerve cells were connected by contact but not continuity and in which the conducting pathway was always from an axon of one nerve cell to a dendrite or the cell body of another nerve cell. Those rival theoretical models are considered at length in Chapter 3. Cajal was the strongest

←

FIGURE 1.17. Lorente de Nó was one of the few neuroscientists with a firm grasp of both neurophysiology and neuroanatomy and he made valuable contributions to both. During a period in which neuroscientists were obsessed with mapping the cerebral cortex into distinct cytoarchitectonic regions based mainly on the spatial distribution of nerve cell bodies, Lorente de Nó drew attention to the importance of the circuit logic common to all regions, especially the significance of the arrangement of synapses on different parts of the dendritic trees. This is his diagram of the organization of neuronal circuits in the cerebral cortex, summarized in the small diagram on the left. In brief, he understood that in all parts of the cortex, "*what remains constant is the arrangement of the plexuses of dendritic and axonal branches, i.e., of the synaptic articulations. . . .*" and that "*the important data are those referring to the arrangement of neurons in synaptic chains. . . .*" and "*that the cortical chains are in no way different from the chains of internuncial neurons in any part of the central nervous system*" (Lorente de Nó, 1938c). He just failed to grasp the concept that direct inputs to pyramidal cell dendritic spines and to interneurons may be excitatory whereas indirect inputs via interneurons to pyramidal cell dendritic shafts may be inhibitory.

advocate of the discontinuous model. Cajal's evidence, which he interpreted as evidence of contact between cerebellar basket cell axons and Purkinje cell bodies, was published in 1888, and his demonstration that centrifugal retinal axons end by forming pericellular baskets on retinal amacrine cells was published in 1889 (*Anat. Anz.* 4:111–121), before his discovery of the dendritic spines in the cerebral cortex. Cajal understood that the dendritic spines were not essential for his model—he knew that some neurons lack spines but he included them in his model also. This was probably why Cajal hesitated, in his two 1891 papers on the cerebral cortex, to assign a function to the spines and did not insist dogmatically that dendritic spines are the elements on which axons terminate on cerebral cortical pyramidal cells.

People assumed that dendritic spines were the sites of contact between neurons, and some, like Tanzi (1893) and Monti (1895), even stated so explicitly, although the evidence was weak. Tanzi was the principal opponent of Golgi in his own country, but he lacked Golgi's authority and his interests were mainly in psychiatry and neuropathology. Similarly in the United States, neurologists like Barker (1899) and neuropathologists like Berkley were quick to see the possible relations of dendritic spines to their special interests. For example, H. L. Berkley (1897), who named the dendritic spines "gemmules," found that the number of spines was much reduced in cases of dementia, which he concluded to be caused by loss of contacts between neurons in the cerebral cortex. Circumstantial evidence of that sort piled up over the decades, but proof of the function of dendritic spines was not obtained. Sixty years passed before Fox and Barnard (1957) gave the first strong evidence of the postsynaptic nature of dendritic spines on Purkinje cells. The proof that dendritic spines are postsynaptic structures had to wait until the advent of the electron microscope (Gray, 1959).

It is very interesting to follow the conceptual logic and the technical advances that led to Cajal's inclusion of dendritic spines in his theoretical model. Cajal argued that the spines are not artifacts because they are demonstrable by more than one technique and always appear in the same place on the dendrites and not in other positions such as the cell body or axon. He arranged the conditions of his histological techniques to optimize staining of the dendritic spines, and he also stated that he took advantage of their greater abundance in the cerebral cortex of newborn mammals than in the adult. Cajal showed dendritic spines on a number of types of neurons in the cerebral cortex which later research showed to lack spines in the adult. Spines are found only on the dendrites of pyramidal cells and spiny stellate cells, but the latter were not described as such by Cajal, although they may have been included by him in the type of small pyramidal cells with arciform axons (DeFelipe and Jones, 1988). Cajal showed spines on dendrites of a variety of nonpyramidal cortical cells because he examined fetal brains in which there is a transient overproduction of dendritic spines on neurons that later lose most or all their dendritic spines (Marin-Padilla, 1969; Jones, 1975; Lund *et al.*, 1977).

It is also important to note Cajal's tenacity in perfecting the methylene blue technique until it clearly showed the dendritic spines. The technique was capricious and did not stain the spines under all conditions. S. Meyer (1895) had failed to stain the dendritic spines because his method of vascular perfusion with methylene blue solution did not provide adequate exposure of the brain to oxygen during the staining. Cajal applied methylene blue solution and crystals into cuts made in the fresh brain, thus greatly increasing the surface exposed to the air. He described his methods in more than usual detail in his 1896 paper on the spines of cerebral cortical cells stained with methylene blue. After 1896 Cajal took advantage of Bethe's new method of fixation of methylene blue preparations with ammonium molybdate (Bethe, 1896; Ramón y Cajal, 1896a) with the result that his demonstration of dendritic spines on cerebral cortical neurons was far more convincing than ever. Ironically, Bethe (1903) used the same method, and apparently saw the same things, but he argued that the dendritic spines connected the dendrites to an interstitial anastomotic network of fine fibers.

Even with good staining, Cajal could not have resolved the morphology of the dendritic spines without the high-power apochromatic microscope objectives which he possessed. Cajal notes that his observations on dendritic spines were made with a Zeiss 1.3 apochromatic objective in 1891 and a Zeiss 1.4 apochromatic in 1896. Cajal bought these objectives very soon after they were first made: apochromatic objectives became available from Zeiss only after 1886. Without those objectives he would have had great difficulty resolving the pedicle of the dendritic spine, which is much narrower and more faintly stained than the distal expansion of the spine. Cajal's 1.4 N.A. apochromatic objective, used with oil immersion and a 1.4 N.A. condenser, would have had a maximum possible resolution of 0.2 μm, which is close to the limit of resolution of the human eye. A person with very good eyes can use to advantage a magnification of 1000 times the N.A. of the objective. However, only the center of the field would have been in focus at that magnification with the objectives used by Cajal, and a very powerful light source would have been required to use a camera lucida to draw nerve cells at very high magnification.

In his 1891 paper in *Cellule*, describing the morphology of dendritic spines of cerebral cortical pyramidal cells, Cajal says that he used the camera lucida for the "*majority*" of figures using the Zeiss objective C, which was a low-power objective. Then he states that figure 5 showing the dendritic spines has "*been made with the very powerful objective E and Zeiss 1.30 apochromatic.*" For the reasons given above it would have been very difficult, probably impossible, for Cajal to have used the camera lucida to draw dendritic spines, using those high-power objectives. However, the matter can now be put to the test because Cajal's microscope, camera lucida, and Golgi preparations of the cerebral cortex have been preserved in Madrid (DeFelipe and Jones, 1988, 1992). Cajal's methodology is discussed further in Chapter 4. This example is given to show how important it can be to take the capabilities of the available instrumentation into consideration when evaluating the historical significance of a scientific discovery. In this and other case analyses I have tried to offer a rational explanation for why certain empirical data were selected or rejected, and I have attempted to reconstruct the logic by which the scientist reached certain conclusions.

The foregoing case studies support the arguments given in Section 1.3.4 that different observers may arrive at different explanations because they set different thresholds below which "noise" (artifacts) are cut off and above which the "signals" (facts) are detectable. Certainly, altering the conditions of a histological technique can result in quite different appearances. The level at which an individual scientist decides to set the threshold is governed by complex factors, not all of them objectively justifiable. Evaluation of the subjective and extrascientific factors as well as the objectively justifiable factors is possible only if the historian completely understands the techniques which were employed, which few historians are qualified to do. In this undertaking historians and scientists can cooperate profitably.

1.4.2. Case Study: Maturation of the Neurotrophic Theory

It is not unusual for claims to be made for originality where none are deserved. For example, we now read about *"The emerging generality of the neurotrophic hypothesis"* (Davies, 1988), after thousands of well-corroborated observations already support the hypothesis and more than two centuries after the nutritional functions of nerves on their innervating tissues were recognized as distinct from their roles in sensation and movement. What is worrisome about people whose historical perspective begins and ends with themselves, apart from what they say about the narcissism of our culture, is that they remove the incentive to read deeply in the original literature. There are few whose thoughts are so original that they can safely ignore their predecessors. The majority have more to gain from consulting their predecessors than to lose from being held back by the dead hand of the past.

Currently there are several auxiliary theories that have developed from the classical neurotrophic theory that emerged at the end of the 18th and beginning of the 19th century. The classical theory was that peripheral nerve fibers convey substances necessary for the vitality of the tissues they innervate. For example, Procháska (1784) wrote that *"Sylvius, Willis, Glisson and others considered that there were two fluids in the nerves, one thick and albuminous, subservient to nutrition, the other very thin and spiritous, intimately connected with the former, and subservient to sensation and movement. . . ."*

An auxiliary theory that developed after 1850 proposed that the vitality of nerve fibers depends on continuity with the nerve cell bodies which are the trophic centers (Waller, 1850). Another auxiliary theory, which developed at the end of the 19th century, proposed that trophic materials flow from the cell body to the nerve endings and perhaps from them to the tissues that they supply. About 1940 that auxiliary theory was accepted into a research program aimed at discovering the mechanism by which neurotrophic viruses spread, and into another research program aimed at promoting nerve regeneration. Both of those programs continue to be active (summarized in Jacobson, 1991, pp. 184–194).

Around the end of the 19th century the neurotrophic theory, which was concerned with the mechanisms by which the vitality of nerve fibers and their target cells are maintained, intersected with the neurotropic theory which is concerned with the way in which outgrowing and regenerating axons find their way to their peripheral targets. In both cases it was proposed that trophic and tropic molecules are involved: trophic molecules to maintain the cells, tropic molecules to promote nerve fiber growth to the targets. From then on the main aims of the research program were to identify molecules that have trophic and tropic effects and to elucidate their mechanisms of action in development, regeneration, and pathological conditions. An auxiliary program was started in the 1920s when it was discovered that the peripheral target tissues stimulate outgrowth of axons. That led to a period of ineffectual efforts, mainly carried out by amputation of limbs and grafting extra limbs, to try to discover how the nervous centers respond to increase or decrease of their peripheral fields (summarized in Jacobson, 1991, pp. 328–345). In one such experiment a sarcoma grafted to the body wall of a 3-day chick embryo was found to produce a large increase in the neurons in the spinal dorsal root ganglia supplying the sarcoma (Bueker, 1948). A similar effect occurred when the sarcoma was grafted on the chorio-allantoic membrane, without contacting the embryo directly, which indicated that the sarcoma produced a diffusible trophic factor (Levi-Montalcini, 1952). Finally the factor, called nerve growth factor (NGF), was isolated and characterized by Stanley Cohen from 1954 to 1960. Cohen and Levi-Montalcini shared the Nobel Prize in 1986 (Levi-Montalcini, 1987). A detailed account of the history of the discovery of NGF, its mode of action, and the discovery of other nerve growth factors can be found in Jacobson (1991, pp. 345–358).

In the 17th and 18th centuries the nerves were considered to be hollow tubes in which fluids flow from the CNS to the nerve endings (E. Clarke, 1968; Ochs, 1975), an "albuminous" fluid subserving nutrition and a "spiritous" fluid subserving sensation and muscle contraction. Thus, the concept of axoplasmic flow has been part of the neurotrophic theory from its inception. So has the concept of different materials flowing in the axon at different rates—the nutritional fluid was considered to flow slowly, the fluid subserving sensation and motion to flow very quickly.* The concept that nerves have nutritive functions was elaborated further by Bichat (1800). He conceived of life in two main divisions—"*vie de relation*" and "*vie de nutrition*," each with its own system of nerves. The nerves subserving "*vie de nutrition*" were thought to be what is now called the autonomic nervous system, then known as the vegetative or organic nervous system, and subsequently as the involuntary nervous system. I do not wish to digress further into the history of research on the autonomic nervous system because it had little further influence on the historical development of the concept of trophic functions of the somatic nerves.

There have been two conceptions of the nervous system's status in relationship to the rest of the organism. The first is that the nervous system exists primarily to receive sensory information and to control movement, the second is that it maintains the vitality of other systems. Cuvier developed the second concept that the nervous system determines the form of the entire organism. He thought of function determining structure. In his *Leçons d'anatomie comparée* (1800–1805) Cuvier compared the functional systems in different groups of animals, including invertebrates, and arrived at the unifying concept of the correlation of the parts. Cuvier conceived of the nervous system also determining the "type" of organism. The "type" represents an ensemble in which the parts are correlated in the organization of the whole (see E. S. Russell, 1916). The term "type" was introduced in 1816 by De Blainville, the successor of Lamarck and Cuvier in Paris. The theory of types, or archetypes, developed by Goethe emphasized the resemblances between objects rather than the differences. This led to the credo of the unity of plan as the key to understanding the organization of the nervous system, most

---

*A convenient review of the modern literature on axonal transport may be found in Jacobson (1991, pp. 184–194).

vigorously promoted by Kupffer (1906) in Germany, Gaskell (1889) in England, and J. B. Johnston (1906) and Kingsbury (1922) in the United States.

Is it at all surprising, then, that Cuvier should have thought of the nervous system, which passes from the centers into all the peripheral parts, as the unifying system which is responsible for correlation of the parts of the organism? That is the second of the conceptual origins of the neurotrophic theory, the first being the concept of proximo-distal flow of trophic factors in nerves. Cuvier thought that this correlation of the parts leading to development of the type can be traced back causally to the trophic actions of the nervous system and to use and disuse. Cuvier's theory has always had adherents, most recently Purves (1988), among many who have tried to explain causal relations between CNS and organs, for example in the cases of animals like ostriches and kangaroos with big differences between front and hind limbs.

This notion was also held by Berzelius (1813), who, in a lecture given in 1810, noted that *"the brain and the nerve determine altogether the chemical processes which occur within the body. . . ."* He applied a ligature to a nerve and noted that immediately *"the greatest disorders have arisen in the oeconomy of the animal, and continued as long as the ligature remained, although the nerve below the part tied always retained the same quantity of nervous substance as before. Again, if the ligature is loosened. . . . the disorders will cease. And why is this continuity so necessary in a channel, the contents of which remain on the same spot? It is clear that this indicates an effect by means of transmission, like that of electricity; although what we hitherto know about electricity, cannot here be applied in explanation."* Berzelius noted only the immediate effects of ligating the nerve, and although he did not study long-term effects, he seems to have recognized that something more than electricity must be transmitted to produce the observed effects, but as he failed to describe the effects in any detail, we can never know whether his conclusions followed from his observations.

These conceptions of trophic actions of the nervous system were well known at the beginning of the 19th century. As the century wore on, the interest in trophic functions of nerves diminished as interest focused on the localization of sensory and motor functions in the central nervous system. This concern was heightened in 1822 with publication of Magendie's conclusive experimental evidence showing that the anterior spinal roots are motor and the posterior roots are sensory. This is the main reason why neither Flourens (1824, 1842) nor Cuvier (1822) mentions the trophic functions in their reviews of experiments on the functions of the nervous system.

There are two other reasons for the neglect of the neurotrophic concept during the first half of the 19th century. At that time, other physiological problems predominated, especially the problem of the role of electricity in nervous conduction, and the problem of functional localization in the brain which gained interest from the popular phrenological movement. In addition, the prevailing understanding of the histological structure of the central nervous system did not encourage an experimental program to test theories of trophic actions of nerves. Until the 1840s it was thought that the brain was composed of separate ganglionic globules and fibers embedded in a semifluid ground substance (Ehrenberg, 1833, 1836; Valentin, 1836a,b; Purkinje, 1837a,b). The discovery that the globules and fibers belong to the same cell was made prematurely by Remak in 1838 but confirmation was delayed for reasons discussed in Chapter 3. In short, confirmation of

Remak's premature discovery of the unity of the ganglion cell and the nerve fiber could not be made with the existing technique of microdissection of single ganglion cells which left very short stumps of nerve fibers remaining attached. Remak's histological techniques were capricious and unreliable, with the result that both Johannes Müller (1840) and Albert Koelliker (1844) failed to confirm Remak's findings (see footnote 3.19). Consensus was delayed until the publication of additional observations by Rudolph Wagner in 1847, by Koelliker (1850), and by Remak himself (1853, 1855b). At first they all conceived of the fibers linking the cells in the central nervous system in the form of an anastomotic network as shown in Fig. 3.8B. With that concept it was inconceivable that any single nerve fiber could be entirely dependent on any single nerve cell for trophic support. That was why the significance of Waller's rediscovery in 1850 of peripheral nerve degeneration distal to the point of transection of the hypoglossal and glossopharyngeal nerves of frogs was not recognized immediately.

Waller observed that both sensory and motor nerve fibers degenerate distal to the point of transection, and he deduced that the phenomenon is not related to the sensory or motor functions or to the direction of conduction of the nerve impulse. He surmised that it was caused by loss of nutritional support of the fiber by the nerve cell body with which the fiber is directly continuous. In 1851 he first used the word "trophic" to describe the nutritional influence of the nerve cell body on its axon. He subscribed to the consensus that nerve cells in the CNS were connected internally by short cytoplasmic bridges, and could thus nourish one another. However, he thought that the longer peripheral nerve fibers are more vulnerable to disconnection from the nerve cells in the central nervous system or in the spinal ganglia. This concept was supported by his observation that degeneration of the peripheral nerve did not occur after cutting the posterior spinal roots proximal to their ganglia, but occurred after distal nerve section of both anterior and posterior spinal nerves (Waller, 1851a,b). He concluded that the *"central nutrition of the sensory spinal fibres is to be found in the vertebral ganglia, while that of the motor fibres is in the spinal cord."*

The significance of Waller's discovery was not self-evident at that time; for example, the Parisian professor E.-F.-A. Vulpian failed to recognize its significance in his excellent *Leçons sur la physiologie générale et comparé du système nerveux*, of 1866, because Vulpian held firmly to the conception of the histological structure of the CNS in terms of a network of cell bodies linked by cytoplasmic anastomoses.

Michael Foster in his *Text-Book of Physiology* (3rd ed., 1879, p. 614) concluded from Waller's findings that *"it would seem that the growth of the motor and sensory fibres takes place in opposite directions, and starts from different nutritive or 'trophic' centres. The sensory fibres grow away from the ganglion either towards the periphery or towards the spinal cord. The motor fibres grow outwards from the spinal cord towards the periphery."** Foster also cited evidence that long fiber tracts in the CNS are maintained

---

*This shows that by 1879, long before the neuron doctrine was promulgated by Waldeyer in 1891, two important components of that doctrine were sufficiently well established to be taught in a popular textbook: (1) that nerve fibers are outgrowths of nerve cells; (2) that nerve fiber growth is supported by the cell, which is a "trophic center." I trace the conceptual lineage of the neuron doctrine in Chapter 3, where I show that the doctrine was assembled over a period of about seventy years, starting in the 1830's, by convergence and coalescence of at least twelve separate theories.

by "trophic centres," and that fibers degenerate after they are disconnected from their trophic centers. The prevailing notion at that time was that most, if not all, nerve cells in the CNS were connected anastomotically but long nerve fibers in the CNS and long peripheral nerve fibers were maintained by "trophic centres." In the CNS some fibers were sustained by a single trophic center, and if cut off from that center the fibers degenerated. Other fibers survived section because they were sustained by two or more trophic centers. For example: "*Schiefferdecker [Virchows Arch. 67:542 (1876)] is himself struck by the fact that the great mass of the lateral column [of the spinal cord] is unaffected by section: this he explains by the hypothesis that the large number of the fibres of these columns, being connected at both ends with homologous nerve cells, conduct equally in both directions, and hold both their terminal cells as 'trophic centres,' so that when they are cut off from the one set they can still depend on the other*" (Foster, 1879, p. 764).

Different conceptions of neurotrophic functions depended on different views of how nerve fibers end in skin and muscles. Until the 1870s it was generally believed that peripheral nerve fibers anastomose with one another to form a fine plexus of nerve fibers in the skin and mucous membranes, or they form peripheral nerve loops, and that these recombine to form sensory nerves returning to the CNS as shown in Fig. 1.18 (Valentin, 1836; Beale, 1860, 1862). Those observations were made on tissues stained with carmine. Beale did not know that carmine does not stain fine nerve endings and so he

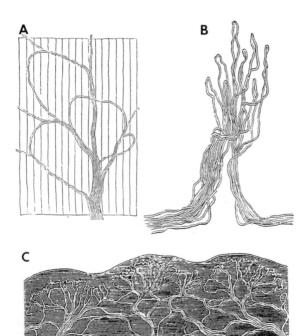

FIGURE 1.18. The concept that peripheral nerve fibers form loops in muscle (**A**), mucous membranes (**B**), fungiform taste papilla, and skin (**C**) was generally held from the 1830s through the 1850s. The concept was progressively corrected after about 1870 on the basis of more accurate microscopic observations made possible by improved histological methods discussed in Section 3.7. (From Carpenter, 1853.)

failed to observe the termination of motor nerves in motor end plates. Those were discovered by Trinchese in 1867. Nevertheless, Ciaccio confirmed Beale's observations in 1883, showing how tenaciously people cling to their preconceptions.

Freely ending peripheral nerve fibers were discovered in 1867 by Julius Cohnheim in tissues stained with gold chloride. Free termination of motor nerve fibers in muscle stained with gold chloride were demonstrated by Ewald in 1876. However, the gold chloride technique could not show whether nerve fibers merely contacted or were in cytoplasmic continuity with their peripheral targets. It was generally concluded that motor nerve fibers were in direct continuity with muscle fibers (Kühne, 1869), and that sensory nerve fibers entered peripheral cells, even hairs (Bonnet, 1878). Another interpretation was that the nerve fibers originate from sensory cells in the skin of the vertebrates as well as the invertebrates. For example, when Friedrich Merkel (1875, 1880) discovered the mammalian cutaneous sensory cells that now bear his name, he thought that they were ganglion cells from which the sensory nerve fibers originated.

That was the state of knowledge about the peripheral nerve endings and their relations to peripheral targets at the time when the discovery was made that taste buds degenerate after cutting the glossopharyngeal nerve (Vintschgau and Honigschmied, 1876; Vintschgau, 1880; Hermann, 1884). The observation was at first interpreted as evidence of the continuity of the nerve fiber with the peripheral sensory organs, not as evidence of a specific trophic action of the nerve on the sensory organ. Of course, the results were also regarded as corroboration of the work of Waller (1850, 1851a,b) showing that the nerve fiber depends for its nutrition on the nerve cell. Such findings of degeneration of sensory organs could not be explained by the concept of loss of functional stimulation of the nerve, but required a trophic stimulation of the sense organ by the sensory nerve in the direction opposite to the conduction flow of nerve impulses.

Not until the 1890s did it become evident that sensory nerve fibers end freely. The advance was made only after application of the Golgi technique showed that nerve fibers terminate on cells and organelles in the skin and mucous membranes by contact but not by direct cytoplasmic continuity (Retzius, 1892c–g, 1894c; van Gehuchten, 1892b). Lenhossék (1893) and Retzius (1892e) showed that sensory nerve fibers end freely among the cells of taste buds. Degeneration of taste buds after transsection of their sensory nerves could then be regarded as a trophic effect that extends across the contact zone between nerve ending and sense organ. After the contact zone between nerve cells was given the name of synapse by Sherrington in 1897, it became possible to conceive of trophic effects crossing the synapse from one neuron to another in the CNS. CNS degeneration, even including the cerebral cortex, long after amputation of a limb, could be understood in terms of retrograde transsynaptic trophic actions. That was one piece of evidence adduced by A. W. Campbell (1905) for localizing the sensorimotor regions of the human cerebral cortex.

Retrograde degeneration of the axon proximal to the point of axotomy had been described by Waller, and was called "indirect" Wallerian degeneration to distinguish it from the "direct" Wallerian degeneration of the distally disconnected axon. Degeneration of the nerve cell after injury to its axon was recognized only later, and the sequence of changes in the injured nerve cell was called "chromatolysis" by Marinesco in 1909 (see Jacobson, 1991, pp. 216–222).

The concept of flow of neurotrophic factors in peripheral nerves was not accepted into a research program until the 1920s, when J. M. D. Olmsted (1920a,b) rediscovered the degeneration of taste buds after denervation and observed that they regenerated and their structure was restored after reinnervation. Studies of the latent period between cutting the nerve and onset of degeneration in the lateral line organs of the fish showed that the more proximal organs start degenerating earliest (May, 1925; Torrey, 1934). It was shown that the onset of atrophic changes in the sense organs is proportional to the length of the distal segment of the nerve connected to the sense organ (Parker, 1932; Torrey, 1934). Those findings led to the rediscovery of the theory of Goldscheider (1898), which was restated by May (1925) as the *"flow of a hormone-like substance from the cell body of the neuron to its terminations."* That theory had been promoted 20 years earlier in the very influential book by Barker (1899, p. 307): *"The key to the whole problem [of the nutrition of the axon] undoubtedly lies, as the neurone concept teaches us, in the fact that the axone in all its parts, no matter how far removed from the cell body, is an integral part of a single cell. To explain the influence of the cell upon the fibre, Goldscheider has advanced a very ingenious hypothesis. He suggests that it is most probable that there is an actual transport of a material, perhaps a fermentlike substance, from the cell along the whole course of the axone to its extremity."*

The confluence of two premature theories—of trophic actions of nerves on peripheral tissues, and of the flow of materials in nerves—occurred at the end of the 19th century, but both languished for more than 20 years before acceptance into active research programs. Around 1900 there was a convergence of the neurotrophic theory and the theory of flow of materials in the nerve fiber. However, the latter entered a long period of neglect, mainly because it could not be tested by experimental methods then available. There were additional reasons for the neglect: experimental observations of the accumulation of materials above a ligature on a nerve were ignored or misinterpreted during the 1920s and 1930s. It was not until Weiss and his collaborators repeated the ligature experiments that the evidence was interpreted in favor of a theory of "axoplasmic flow" (Weiss *et al.*, 1945, 1962; Weiss and Hiscoe, 1948; reviewed by Bisby, 1982).

Cajal discussed the ligating of nerves at considerable length, and included several figures illustrating ballooning of the nerve proximal to a ligature and thinning of the nerve distal to a ligature on the sciatic nerve of the rabbit (Ramón y Cajal, 1928, pp. 290–304). But he arrived at the wrong conclusion, that the changes were the result of local factors at the site of the ligature. He concluded that *"the trophic influence exercised by the central neurones is of a dynamic and not a material nature."* He believed that trophic influence was related to the passage of nerve impulses which could pass in one direction only because the "dynamic polarization" of the neuron was based on a valvelike property of the neuron as a whole. Cajal concluded further that *"assimilation and growth of the axons are purely local processes . . . they are not influenced by materials or chemical reserves of the soma or the nucleus of the neurones"* (Ramón y Cajal, 1928, p. 302). Cajal failed to see the significance of the evidence, not because of technical limitations but because he viewed the evidence in terms of his theory of "dynamic polarization" of the neuron with respect to the direction of impulse traffic rather than in terms of the theory of proximodistal flow of materials in the axon. I argue in Section 4.3.4 that Cajal saw structure as a cause of function but not vice versa. In his view, the direction of impulse traffic was determined by the peculiar structural polarization of the nerve cell—that is the essence of his theory of "dynamic polarization" of the nerve cell. Therefore, Cajal resisted the idea that the structure of the axon is continuously renewed by flow of materials from the cell body. I suspect that another reason for the disbelief in axonal flow of considerable quantities of materials was its similarity to the discredited theory of flow of animal spirits in hollow nerve fibers that was the prevailing theory of the mechanism of nerve conduction during the 17th and 18th centuries (reviewed by E. Clarke, 1968).

The theory of transport of materials in the axon remained premature and was neglected after the 1920s. It emerged 15 to 20 years later as a mature theory after it was adopted by two different research programs aimed at discovering the mechanisms of spread of neurotrophic viruses in the nervous system (Koppisch, 1935; Bodian and Howe, 1941a,b; Bodian, 1948) and at elucidating the mechanisms of nerve growth and regeneration (Weiss, 1941; Weiss and Hiscoe, 1948; Weiss et al., 1962).

The concept of neurotropic substances that exert a chemotactic action on the ends of outgrowing and regenerating axons is different from the concept of neurotrophic factors, although the two concepts are related and have sometimes been conflated. When Cajal used the term "neurotropic" in his 1910 paper in which he discussed the evidence in support of the neurotropic hypothesis, he conceived of molecules originating from target cells that attract and guide outgrowing nerve fibers. "Trophic," however, connoted a nutritional function of nerves on target cells. The hypothesis that nerve growth is guided by a form of chemotaxis in response to an external agent was put forward by Ramón y Cajal (1892) on the evidence of reorientation of axons growing out of displaced retinal ganglion cells. The hypothesis was tested experimentally with inconclusive results by Forssman (1898, 1900) who invented the term "neurotropism" to describe the attractive influence of degenerating nerves on regenerating nerve fibers. Other experimental tests of the neurotropic hypothesis also gave negative results (for reviews see Dustin, 1910; Ramón y Cajal, 1910, 1928).

The theory that Schwann cells produce neurotropic and neurotrophic agents which attract axonal growth cones and promote axonal growth and regeneration originated from Cajal: *"The newly-formed fibres grow at first freely through the scar, but they soon become surrounded by a cellular sheath which probably emanates from the cells of Schwann. The penetration into the peripheral stump implies a neurotropic action, or the exercise of electric influences, by the latter"* (Ramón y Cajal, 1928, p. 196). Furthermore, *"the great influence that the proximity of the peripheral stump has on the growth and orientation of the outgrowing newly-formed fibres. . . . is exercised through ferments or stimulating substances formed by the rejuvenated cells of Schwann of the distal stump. . . . these substances have not only an orienting function, but they are also trophic in character, since the sprouts which have arrived at the peripheral stump are robust, show a great capacity for ramification"* Ramón y Cajal, 1928, p. 238). This was a premature theory, not able to be adopted by a research program because there were no techniques then available which could identify the tropic and trophic substances predicted by the theory.

Cajal seldom revealed any doubts about his concepts of structural organization of the nervous system but he expressed doubt whenever he had to consider the possibility of function determining structure. For example, he maintained a tentative position regarding the guidance of axons by neurotropic molecules: "*As for the theory of neurotropism, far from being for me a dogma, it is simply a working hypothesis which I am willing to correct or even abandon in the presence of better explanations*" (Ramón y Cajal, 1928, p. 195). Indeed, Weiss and Taylor (1944) claimed to have refuted the neurotropic theory by showing that the peripheral nerve stump does not have a greater effect than nonnervous substrates for attracting regenerating nerve fibers. But that obituary notice proved to be premature: Cajal's theory was resurrected in the 1980s when several reports showed that Schwann cells stimulate regeneration of axons *in vivo* and *in vitro* (see Aguayo *et al.*, 1983). This was followed by evidence that Schwann cells produce nerve growth factor (NGF) after nerve injury (Korsching *et al.*, 1986), and that Schwann cells express NGF receptors in culture (DiStefano and Johnson, 1988), and after nerve injury (Taniuchi *et al.*, 1986b), and that the Schwann cells may present NGF to the regenerating axons (Taniuchi *et al.*, 1988). Moreover, transplants of cultured Schwann cells promote axon regeneration in adult mammalian brain (Kromer and Cornbrooks, 1986). After languishing in a premature state for more than 50 years, the neurotropic theory finally matured in the 1980s, and has functioned as the guiding theory of a rapidly progressing research program.

1.4.3. Case Study: Myelination

The history of theories of myelination shows how competing theoretical models were made to explain the same set of observations. The myelin sheath was discovered in peripheral nerves by Remak in 1837 and 1838 and by Schwann in 1839. Four main problems were recognized: (1) Does myelin originate from the axon or the sheath cell of Schwann? (2) Is the myelin located intracellularly or extracellularly? (3) What is its composition and structure? (4) What are its functions? Theoretical models where constructed to include different answers to those four questions. A fifth question was whether myelin is the same in the peripheral and central nervous system.

Robert Remak (1837, 1838a,b) discovered the differences between myelinated and unmyelinated nerve fibers, notably the presence of a medullary sheath and the greater diameter of the former. He also observed that unmyelinated fibers, which he called "organic fibers," were characteristic of the sympathetic nervous system and could most easily be observed in the nerves to the spleen of ungulates. Remak (1838a,b) described and illustrated the nuclei attached to the organic fibers before they were again observed by Schwann (1839) in the sheath which bears his name. Nonmedullated nerve fibers in peripheral nerves were first described by Remak (1838a,b) in teased or dissociated peripheral nerves of the rabbit as elongated bands with an average diameter of 3-5 mm. These nonmedullated nerve fibers were henceforth called "fibers or bands of Remak."

The term "myelin" was introduced by Virchow (1858), to refer to the globules of fatty material, the so-called "myelin-figures," that appear in association with degenerating nerve fibers or after placing fresh myelinated nerve in water. The myelin sheath of normal nerves was generally referred to in the German literature as "*Markscheide*" or "*Nervenmark*," in the English literature as "medullary sheath," and in French it is usually called "*myéline.*" Ramón y Cajal took the term "*mielina*" and other terminology connected with the medullary sheath directly from Louis-Antoine Ranvier (1871, 1872, 1878–1879), who was one of the principal sources of Cajal's early knowledge of histology.

Summarizing what was then known about myelinated nerve fibers, Max Schultze (1870) states that "*the medullated fibres therefore consist essentially of two constituents, a cortex or sheath of medullary nerve substance, and an axial fibre or axis cylinder, which is either a primitive fibre or a bundle of fibrillae. The medullary sheath forms a more or less thick investment around the axis cylinder and consists of an oily substance containing protagon and capable of powerfully refracting light** . . . *The medullated fibres of the nerve centres are imbedded in an extremely delicate tenacious connective substance, the peculiar consistence of which preserves the fibres from injury. The medullated fibres of the peripheric nerves. . . . each possess in addition, and external to their medullary sheath, a special investment of connective tissue, constituting the so-called sheath of Schwann.*" It is significant that Schultze does not mention the "*cell of Schwann*" in connection with the medullary sheath because at that time the nuclei that Remak (1838) had found along the nerve fiber, both medullated and unmedullated, were not called Schwann cells. Ranvier (1871, 1872) appears to have been the first to designate the cell of Schwann as a separate entity and to have discovered that a single Schwann cell occupies each internodal segment. Vignal (1889) dissected individual Schwann cells off the axon and showed that they ensheath the axon (Fig. 1.19).

In 1839 Schwann himself saw only the nuclei of the medullary sheath, and thus confirmed Remak's description of them. But Schwann supposed that those nuclei had originally belonged to a chain of cells which coalesced to form the nerve fiber. This "*cell-chain*" theory of development of the nerve fiber continued to receive support for the following 60 years, even after His, Forel, Koelliker, and Cajal provided good evidence supporting the outgrowth theory of development of the axon (see Section 3.8). The sheath of Schwann continued to be thought of as a syncytium containing many nuclei. As late as 1928 Cajal was willing to admit that the Schwann cells may coalesce to form a syncytium during nerve degeneration (Ramón y Cajal, 1928, p. 130).

Observations of myelination of nerve fibers growing in the tail fin of living frog tadpoles were started by Hensen (1864) and Koelliker (1886) and later their observations were extended by Harrison (1904) and Speidel (1932). Hensen noted that the nerves are at first unmyelinated but became myelinated initially by separate droplets of myelin which later coalesced. Koelliker observed that myelination starts in the neighborhood of the nuclei of the sheath of Schwann and that "*the myelinated regions are at first separated from one another by long unmyelinated stretches.*" Koelliker (1886) then

---

*In 1864, Liebreich, a pupil of Hoppe-Seyler (1825–1895), one of the founders of biochemistry, proposed that protagon was the main chemical constituent of the brain, and that the other constituents, lecithin and cephalin, were its breakdown products. This was generally accepted for the next 30 years (reviewed by Wlassak, 1898), although in 1874 J. L. W. Thudichum (1829–1901) showed that protagon is not a chemical entity but a mixture of several chemicals, including phospholipids and cerebroside. In addition to discovering those chemicals, Thudichum also discovered and characterized brain glycolipids such as sphingomyelin. He is now justly regarded as the founder of neurochemistry, although in his time his findings were regarded as extremely controversial and were unjustly neglected (Drabkin, 1958; Tower, 1970).

concluded: "*I consider the primitive nerve fibers to be protoplasmic outgrowths of the central nerve cells, in which a central fiber separates from a thin protoplasmic mantle as the first rudiment of the axis cylinder. The nuclei of the sheath of Schwann possibly play a role, and under the influence of these nuclei the protoplasmic covering becomes transformed into real myelin by the deposition of fat. The myelinated fibers in the central nervous system show furthermore, that development of myelin can take place independently of external influences.*"

From the time of discovery of the myelin sheath by Remak there were two schools of thought about the origin of myelin, both of which had serious proponents until the 1950s: either it is a product of the sheath cells or it is formed by the axon itself. Proponents of the latter theory pointed out that they could not recognize sheath cells in the CNS, and therefore the central myelin must originate as a secretion from the axons. This theory was defended by Koelliker (1896) and by Ramón y Cajal (1909, p. 618), who states that all authorities agree that myelin is "*a product of secretion of the axis cylinder.*" Some continued to hold this opinion as late as 1958 (see Windle, 1958, p. 74). There were also differences in opinion about whether myelin is intracellular or extracellular. Ranvier (1871, 1872) believed that myelin is produced by the sheath cells and is formed inside them, whereas many others thought of myelin as an extracellular material. For example, Ramón y Cajal (1928, p. 41) states that "*the nerve tube contains a thick oleaginous sheath which is placed between the cell of Schwann and the neurite, and which is interrupted, at intervals, to let in more freely the nutritive plasmas.*"

By the beginning of the 20th century, the generally accepted concept of structure of the medullated nerve fiber, as described by Koelliker (1896) and Ramón y Cajal (1906), was that the myelin sheath is secreted by the axon and lies between it and the Schwann cells, which are surrounded by a clear "*membrane of Schwann.*" This is surprising when one considers the evidence, given below, that had been accumulating before 1900 indicating that myelin originates in nonneuronal cells supporting the axon. This is another example showing that mutually contradictory theories have coexisted for long periods in the history of neuroscience.

Vignal (1889) first proposed that myelin is produced by Schwann cells and by the "*cellule de revêtement*" which surrounds the myelinated fibers in the CNS. Vignal states that the central myelinating cells "*originate from embryonic cells of the grey substance*

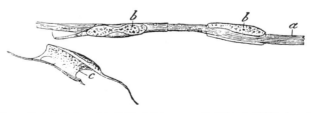

FIGURE 1.19. The earliest depiction of individual Schwann cells (Vignal, 1889). A peripheral nerve fiber (a) is ensheathed by two Schwann cells (b). One of these has a dumbbell-shaped nucleus, perhaps showing that it is about to divide. Vignal dissected single Schwann cells off the axon, showing that they do not form a syncytium, as was then commonly believed.

and have the same origin as the neuroglial cells." This concept of myelin formation by supporting, nonneural cells is shown in drawings on pp. 27 and 33 of Vignal's book *Développement des éléments du système nerveux cérébro-spinal* (1889), one of which is shown in Fig. 1.19. The origin of myelin from nonneural cells was also observed by Wlassak (1898), who showed that the first osmiophilic material appeared in the protoplasm of spongioblasts (i.e., the progenitors of neuroglial cells) in the CNS and in the protoplasm of the Schwann cells, and that these developed before the appearance of myelinated fibers. Wlassak (1898) concluded that *"the nervous supporting tissue at least in the embryonic organism does not merely have a mechanical but also a chemical function. It is a transport apparatus for certain materials which originate from the blood and are transferred to the nerve cells."* However, he was unable to show whether CNS myelin is made by the neuroglial cells or the nerve cells or both. This problem was not clarified further until the 1920s, when Rio-Hortega discovered the role of oligodendrocytes in formation of central myelin (see Section 1.4.4). The origin of the myelin sheath could only be revealed conclusively with the aid of the electron microscope.

There were also wide divergences of opinion about the structure of the nodes. Remak (1838a,b) and Koelliker (1852) had seen breaks in the medullary sheath but dismissed them as artifacts, and so they were regarded until 1871. Ranvier (1871, 1872) was the first to demonstrate that each internodal segment is a separate unit associated with a single Schwann cell. In nerves of the mouse stained with silver nitrate he showed that the nodes interrupted all the layers of the myelin sheath and exposed the naked axon, in which there is a disk or ring of material, possibly forming a constriction in the axon itself. In his publications in Spanish, Cajal also describes the nodes as *"interrupciones"* or *"estrangulaciones de Ranvier,"* terms which he borrowed from Ranvier, who used the terms *"anneaux constricteurs"* or *"étranglements annulaires"* for the nodes and *"segments interannulairs"* for the internodes. The nodes were termed *"Ranvierschen Schnürringe"* by Koelliker (1896) and other German writers. These terms show that the nodes were generally considered to be some form of constriction or interruption in the continuity of the axon, which were most easily seen in peripheral nerves. Bethe (1903) went so far as to propose that the nodes of Ranvier were septa completely interrupting the continuity of the axon except for the neurofibrils which penetrate the septa and which he thought are responsible for conduction of the nerve impulse. The presence of nodes interrupting the medullary sheath of nerve fibers in the CNS was first reported by Tourneux and Le Goff (1875) but denied by such authorities as Ranvier (1878–1879) and Koelliker (1896). Only after Cajal demonstrated nodes in the axons of neurons of the electric lobes of *Torpedo* in 1888 and in the granular layer of the cerebellum (1889) did Flechsig (1889) and Dogiel (1896) find nodes in other central myelinated fibers, using the methylene blue stain of Ehrlich (1885).

The incisures of Schmidt-Lanterman (Schmidt, 1874; Lanterman, 1876) were considered to be artifacts. Koelliker (1896, Vol. 2, p. 7–15) devotes eight pages to them and finally concludes that they are not present in normal living nerve fibers. An immense literature describing artifacts or debating the authenticity of structures in the axon and its medullary sheath was produced during the 19th century, and many terms had a transient currency. Consequently, the early descriptions of the myelin sheath are often impossible to reconcile with current knowledge. The most notorious of these fixation or precipitation artifacts are the cones of Golgi-Rezzonico, which were beautifully illustrated by Golgi (1880c) as periodically repeating, cone-shaped spirals of a threadlike material (Golgi's original drawings are reproduced in Zanobio, 1975). Various perinodal apparatuses were minutely described and illustrated (Frommann's lines, spiny bracelets of Nageotte, Ranvier's crosses). These were probably the result of selective penetration of stains into the axon at the nodes of Ranvier. Errors are universal, and they are noted here as the natural consequences of the limitations of the available techniques. Users are not to be condemned for the limitations of their methods, but the users themselves are made vulnerable by techniques.

The ectodermal origin of Schwann cells was first postulated by Nansen (1886–1887). This was supported by others (e.g., Froriep, 1907; Nageotte, 1907). Harrison (1906, 1924a) proved the neurectodermal origin of Schwann cells by showing their absence after removal of the neural crest in amphibian embryos. Nageotte (1918) observed that several nonmedullated nerve fibers are enclosed within separate tunnels in a single sheath cell, but Ramón y Cajal (1933a) claimed that each axon is individually enveloped by a special Schwann cell. Electron microscopy was to prove Nageotte in the right.

The history of this subject provides yet another example of a problem being hotly disputed by eminent neurohistologists for almost a century before it was finally resolved in a few years by the use of the electron microscope (Geren, 1954; J. D. Robertson, 1955). In retrospect it seems just as well that there were only a few investigators of myelin, because battalions of light microscopists armed with the best instruments and all the available techniques for staining myelin could never have resolved its structure and thus reached an understanding of its development. More progress was made in understanding myelin in the decade from 1954 to 1964 than had been achieved in the previous century. An excellent review of the early days of electron microscopy of myelin is given by J. D. Robertson (1987). The application of electron microscopy put an end to the debate and confusion that existed before 1953 by showing that in peripheral nerves the myelin sheath is a tongue of the Schwann cell wrapped around the axon like a scroll around a rod (Geren and Raskind, 1953; Geren, 1954; J. D. Robertson, 1955; A. Peters and Muir, 1959; A. Peters, 1960, 1964a,b). Ultrastructural evidence showing that myelination in the CNS is performed by oligodendrocytes wrapping cytoplasmic processes around axons was first provided by Farquhar and Hartmann (1957), Schultz *et al.* (1957), and Mugnaini and Walberg (1964). If the truth about the mode of myelination had been revealed to Schwann he would have found it utterly unbelievable.*

The role of glial cells in myelination of axons in the CNS also remained an area of uncertainty and disputation until the 1960s. Several investigators had produced circumstantial evidence that interfascicular oligodendrocytes are involved in myelination. The rapid increase in oligodendrocytes in central tracts prior to their myelination and the

---

*History provides many examples of the ingenuities of biological adaptations, for example as described by Charles Darwin (1862) in *"The Various Contrivances by which Orchids are Fertilized by Insects,"* which met with disbelief when they were first discovered. In a letter written in 1868, Darwin remarks. *"I carefully described to Huxley the shooting out of the Pollinia in Catasetum, and received for an answer, 'Do you really think I can believe that?'"*

invariable presence of oligodendrocytes during myelination in the CNS were very suggestive observations (Rio-Hortega, 1924, 1928; Penfield, 1924; Linell and Tom, 1931; Morrison, 1932). As the role of oligodendrocytes in myelination could not be proved, other hypotheses were entertained in which the role of myelin production was assigned to the axons themselves or to the astrocytes (Alpers and Haymaker, 1934; Scharf, 1951; Hild, 1957; Blunt et al., 1972). Finally, electron microscopy showed conclusively that there is continuity between the membrane of the oligodendrocyte and the myelin sheath and that the wrapping of the oligodendrocyte membrane around axons in the CNS is essentially the same as the process of myelination in peripheral nerves (Luse, 1956, 1960; Maturana, 1960; Peters, 1960, 1964a,b, 1966; Bunge et al., 1962; Kruger and Maxwell, 1966; Knobler and Stempak, 1973; Meier, 1976). Ironically, poor preparative procedures aided the initial studies of the layering of the myelin sheath: swelling resulted in slight separation of the layers and the scroll formation was easily seen. When better methods of intracardiac perfusion and fixation were used, the extracellular space between adjacent turns of the oligodendrocytes around axons in the CNS was absent and their interface was transformed into the intraperiod line (Peters, 1960; Maturana, 1960; Hirano et al., 1966; Hirano and Dembitzer, 1967; Hirano, 1968).

A powerful impetus to studies of development of myelinated fiber tracts was given by the discovery of degeneration (Wallerian degeneration) of the nerve distal to a transection, which was most easily traced in myelinated fibers (Waller, 1851a; see Clarke and O'Malley, 1968; Denny-Brown, 1970). Although this showed that survival of the distal nerve segment and its myelin sheath depends on connection with the nerve cell body, it was also interpreted by many, including Cajal, to mean that myelin is produced by the nerve fiber. In 1885 Marchi and Algeri showed that a solution of osmic acid and potassium bichromate stains myelin sheaths of degenerating axons but not normal axons. This proved to be the most useful method of tracing myelinated fiber pathways in the CNS until the introduction of radioactively labeled tracers, horseradish peroxidase, and fluorescent tracers more than 80 years later.

Because myelinated fibers were most easily traced with the available histological techniques, the 19th century literature is dominated by studies of myelogeny. Flechsig (1876; for review see Flechsig, 1920) was the first to recognize that myelination occurs at different times in different regions of the CNS, and to use this as a means of tracing fibers in the CNS. In 1876, Flechsig wrote: *"During certain periods of fetal life, fibers can be distinguished from each other in a very striking manner, which in the adult are of uniform consistency and differ little from one another. This is because some of them already have a complete myelin sheath, whereas others still exhibit their naked axis cylinders. Thus we are in a position, especially in the compact white matter, to follow for a considerable distance fibers and fiber bundles that later on, owing to the uniformity of their components, become masked in their course."* In this way he was able to demonstrate the course of the pyramidal tracts from the cerebral cortex to the spinal cord.

Numerous examples that support an evolutionary but not a revolutionary thesis of the nature of progress in neuroscience can be adduced from the history of myelin. Concepts did not spring into existence fully formed like Athena from the head of Zeus, nor were the concepts that were most consistent with the evidence at the beginning necessarily the ones that finally turned out to be true. Completely contradictory ideas were held by well-qualified people over many decades. For example, belief in the mesodermal origins of all glial cells was refuted very slowly, and the idea that myelin is a secretory product of the axons was held at the same time as the alternative theory that myelin is produced by nonneuronal cells. The contestants were either locked in a stalemate that could not be unlocked with the available techniques or they simply ignored their opponents. That is the main reason for persistence of the theory of axonal secretion of myelin even after the cumulative evidence was strongly, although not conclusively, against it. Crucial experiments that can solve a problem conclusively are very rare in the history of neuroscience, and that is one of the main reasons why conflicting theories can coexist for long periods. A theory that is ultimately accepted as true may languish unrecognized and unaccepted in the prevailing research programs: I call these premature theories.

The alternative theoretical concepts about myelination all flourished simultaneously from the 1850s until the 1950s when they were resolved by electron microscopy. That myelin is secreted by the axon was most widely upheld, although evidence advanced by Vignal (1889) showing that myelin is produced and stored by sheath cells in peripheral nerves now seems to be as compelling as the evidence supporting the alternatives. In the mind of Ramón y Cajal (1909, p. 267), the fact that there are no cell nuclei associated with myelinated axons in the CNS was sufficient to invalidate the concepts that the sheath cells produce myelin. Cajal also rejected the evidence (Vignal, 1889) that several peripheral axons share a single sheath cell, which Vignal suggested might also account for the apparent absence of sheath cells in the CNS. Cajal held the theory that the axons secrete myelin and was unconvinced by evidence to the contrary. He later believed that a special class of adendritic neuroglial cells (the third element) are involved in myelination in the CNS, but, as we shall see in the following section, he refused to believe that oligodendroglial cells are responsible for CNS myelination.

The history of myelination shows that neuroscientists rarely abandoned their concepts after counterevidence was produced. In the case of myelination, the decisive counterevidence could not be obtained until advent of the electron microscope, which could not have been imagined by anyone in the 19th century. Neuroscientists were unable or unwilling to define in advance the conditions for invalidation of their own concepts. Nor were they willing to specify the conditions under which they would have been prepared to abandon their research programs and adopt alternative programs. Popper (1959, 1962) has argued that progress in science is predicated on, among other things, the scientist's willingness to abandon a pet theory after certain conditions, agreed in advance, have been satisfied. Has the historian of science ever asked under what conditions Golgi would have agreed to abandon the concept of a nerve net (the so-called reticular theory)? The conditions that he would have required for refutation of the nerve net concept, if he was willing to grant them under any circumstances, would probably have been unattainable at that time, and thus of no practical significance in the process of selection between alternative concepts or theories. Golgi upheld the reticular theory until his death in 1926, and his disciples continued to safeguard the theory by means of ad hoc modifications (Held, 1929). This case is consistent with the view of Lakatos that no crucial experiment or logical refutation requires the scientist to reject his theory because "*with sufficient resourcefulness and some luck, any theory can be defended 'pro-*

gressively' for a long time even if it is false" (Lakatos, 1978, Vol. 1, p. 111). Outright refutation has rarely occurred in the history of neuroscience. A single contradictory fact never overturns a theory built on empiricist lines; to modify a phrase from T. H. Huxley (1894), it would indeed be a tragedy if a beautiful hypothesis was killed by an ugly fact.

If a certain theory is close to obtaining majority support, what is the significance of a recalcitrant minority who persist in supporting an opposing theory? My answer is that the recalcitrant minority can exert constructive effects on a research program. This is obviously true if the minority view turns out to be correct, but it may hold even if the minority view proves to be false. For instance, the mechanist-reductionist neuroscience research program has been opposed by vitalists, holists, and organicists whose theories have been largely refuted, but they have had the effect of introducing elements of doubt into the absolute certainty of the materialist-reductionist position (see Section 2.13). Such cases show that the tenacious dissent of the minority can have constructive effects. The recalcitrant minority can force the majority to test their theories more critically, and prevent sudden takeover of a research program by a majority. The dissenting minority can prolong the period of dialogue and critique necessary for progress of a research program. Here I emphasize the role of dissent in promoting dialogue and in enriching the research program before consensus is reached and the program comes to rest on authority. We can portray the history of science as a history of dissent from authority. The tension between authority and dissent is one of the forces driving science forward. New understanding of the meanings of any system of knowledge may also be arrived at by extraordinary individuals who break away from conventional modes of thought. The historian of scientific concepts must recognize and respect the roles of eccentric ideas, of dissent from authority and of inductive leaps.

1.4.4. Case Study: Neuroglia

The original concept of neuroglia meaning "nerve glue" was based on Rudolf Virchow's assumption that there must be a mesodermal connective tissue element of the nervous system (Virchow, 1846, 1858). Even if neuroglial cells did not exist Virchow would have had to invent them as a requirement for the prevalent theory, which he supported, that all cells of ectodermal origin must also be in direct contact with cells of mesodermal origin. But techniques were inadequate to either corroborate or refute that theory. The mesodermal origin of neuroglial cells continued to receive corroboration (Andriezen, 1893a,b; Weigert, 1895; W. Robertson, 1897, 1899, 1900) in the face of strong counterevidence showing that both neurons and glial cells originate from embryonic ectoderm (His, 1889a).

Virchow and his disciples were primarily interested in pathology and thus in neuroglial tumors and in the reaction of the neuroglia to disease and injury. Their theories guided practice in the direction of neuropathology, and away from normal development. Virchow's research program was aimed at showing that the causes of disease can be found in derangements of cells.

It is doubtful whether Virchow saw neuroglial cells in 1846 and there are no convincing pictures of them in Virchow's book, *Die Cellularpathologie* (1858), although he there discusses the theory of a neuroglial tissue. At that time the theory led and the facts followed. Progress was slow because techniques were inadequate in the 1840s and 1850s to provide reliable evidence. Virchow (1885) later described that period: *"The great upheavals that microscopy, chemistry and pathological anatomy had brought about were at first accompanied by the most dismal consequences. People found themselves helpless . . . filled with exaggerated expectations they seized on any fragment which a bold speculator might choose to cast out."** Virchow's theory of the neuroglia consisted of two such bold speculations: first, that neuroglial cells form a connective tissue, and second, that neuroglia develop from mesoderm in the embryo. These conjectures were not Virchow's original creations. He derived the concept of tissues from Bichat (1800), who first proposed that tissues are where the functions of life and the dysfunctions of disease occur—Virchow extrapolated, saying that cells are where life and disease occur. Virchow obtained the concept of mesoderm from Remak to whom he is also indebted for the idea that all cells originate from other cells (see Section 3.5).

The problem of the ectodermal or mesodermal origins of the neuroglial cells has a complex history. Wilhelm His (1889a) corrected Virchow's misconception that the neuroglia form the connective tissue elements of the nervous system by showing that neuroglial cells as well as neurons originate from neurectoderm. The problem was resolved in the last decade of the 19th century when numerous stains for connective tissue failed to stain neuroglial cells but only stained blood vessels in the CNS: the Unna-Taenzer orcein method (Unna, 1890) and the Weigert (1895) resorcinol-fuchsin method for elastic fibers; the van Gieson (1889) acid fuchsin-picric acid stain for collagen; and the Mallory (1900) aniline blue stain for connective tissue. That so many different connective tissue stains failed to stain neuroglial cells could not conclusively falsify Virchow's theory because it could be maintained that glial cells are a special form of connective tissue that is not stained by any other method. That was the argument used by Andriezen (1893a,b) and Weigert (1895). Contrariwise, the invention of specific neuroglial stains (Ramón y Cajal, 1913a,b; Rio-Hortega, 1919, 1921a,b, 1932) was not considered to be conclusive refutation of Virchow's theory. Cajal could still maintain in 1920 that the glial cells belonging to his "third element" are of mesodermal origin. The most compelling evidence against it was that neuroglial cells in the embryo originate from the neurectoderm (His, 1889a,b). However, the available evidence did not exclude the possibility of subsequent entry of mesodermal cells into the CNS. The gradual resolution of this problem is considered below, in connection with the history of the microglial cells.

His misidentified the progenitors of both the neurons and the neuroglial cells in the neural tube. He conjectured that neurons originate from germinal cells but the glial cells originate from a syncytial tissue named the *"myelospongium"* or *"neurospongium"* formed of spongioblasts. The theory that glial cells remain permanently anastomosed

---

*This state of affairs and of mind was not confined to science. In the mid-1850s Walter Bagehot in England was saying much the same thing about the social and political conditions in Europe. The period, he wrote, is an *"an age of confusion and tumult, when old habits are shaken, old views are overthrown, ancient assumptions rudely questioned, ancient inferences utterly denied. . . ."* (*Physics and Politics*, 1872; 1873 edition).

with one another continued to be corroborated by many authorities (Hardesty, 1904; Held, 1909; Streeter, 1912). In 1912, Streeter could confidently invert the truth by pronouncing that *"earlier conceptions of neuroglia cells were based on silver precipitation methods (Golgi) which failed to reveal the true wealth of their anastomosing branches, and there thus existed a false impression of neuroglia as consisting of scattered and independent cells."* At that time cells were believed to be as naked protoplasmic bodies, lacking a membrane, and connected by protoplasmic and fibrous bridges (see Section 3.10 for a more extensive exposition of this preconception and the effects it had on evolution of the neuron theory).

The neurospongium theory was based entirely on artifacts (Fig. 1.20), and could not be refuted conclusively until the 1950s when the electron microscope showed that all cells in the neurectoderm and neural tube are separated by narrow intercellular clefts from the beginning of development. In the 19th century the counterevidence to the neuroglial syncytium was already quite strong. Golgi staining of the neural tube always

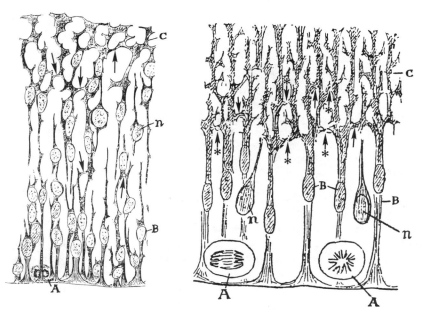

FIGURE 1.20. Early conceptual models of the structure of the neurectoderm of the vertebrate embryo neural tube. On the left from His (1887); on the right from Ramón y Cajal (1894b). Both scientists conceived of three different types of cells forming the neurectoderm: germinal cells (A) that undergo mitosis close to the inner (ventricular) surface; spongioblasts (B) forming a syncytium (neurospongium); neuroblasts (n) migrating as independent cells in the extensive intercellular spaces from inner to outer zones of the neurectoderm where they transformed into nerve cells. Both His and Cajal contributed to the construction of what came to be known as the neuron theory after 1891. Both believed in the autonomy of neuroblasts. His had no doubt about the authenticity of the intercellular cytoplasmic bridges which he depicted unambiguously (arrows). Cajal's doubts are shown by his ambiguous representation of some intercellular bridges, as if they were contacts by interdigitation (arrows with asterisks). He did not have evidence showing that they were histological artifacts, so they remained in his model of the neurectoderm.

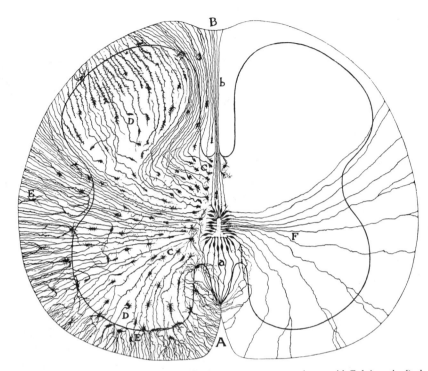

FIGURE 1.21. Spinal cord of a 14 cm human embryo (transverse section; rapid Golgi method) showing neuroglial cells at various stages of transformation from neurepithelial cells. A = anterior sulcus; B = posterior sulcus; a, b = bundles of slightly modified neurepithelial cells span the midline from the central canal to the pia mater; C, D = migrating glial cells at early and later stages of differentiation to neuroglial cells; E = almost mature glial cells; F = mature radial ependyma-glial cells. (From Lenhossék, 1893; labels added.)

showed separate glial cells forming a series of stages of development from the radially aligned spongioblasts to mature astrocytes, as shown in Fig. 1.21 (Koelliker, 1893; Lenhossék, 1893; Ramón y Cajal, 1894b). As evidence against the neuroglial syncytium Alzheimer (1910) showed that one neuroglial cell may undergo pathological change while its neighbors remain normal. By the 1920s the neurospongium theory had been abandoned, not because it could be falsified conclusively, but because it failed to predict new observations predicted by the alternative theory that glial cells are always separate (reviewed by Penfield, 1926).

The research program of His split into four separate research programs with different goals. The first dealt with the origins of neurons and glial cells from progenitors in the neural tube—His conjectured that neurons originate from germinal cells near the ventricle, and glial cells originate from spongioblasts which span the full thickness of the neural tube. The second theory dealt with the syncytial connections between cells in the neural tube. The third dealt with the enormous intercellular spaces which appeared as an artifact in sections of neural tube (Fig. 1.20). This space was supposed to contain a "central ground substance" in which the cells were embedded (Boeke, 1942; Bauer, 1953). A fourth theory dealt with the role of migration of neuroblasts in these spaces, guided by the radially aligned spongioblasts (Magini, 1888; His, 1889a, 1890). These show the important role of artifacts in construction of theories. They illustrate Cajal's dictum that *"in biology theories are fragile and ephemeral constructions . . . while hypotheses pass by facts remain"* (Ramón y Cajal, 1928). He was also well aware that the facts were often indistinguishable from the artifacts.

The Golgi research program on neuroglia started in 1875 with Golgi's studies of glial cell tumors. Golgi's earliest papers on neuroglia were published in 1870 and 1871 using hematoxylin and carmine staining which was incapable of showing either neurons or glial cells completely. By 1873 Golgi had developed his potassium dichromate-silver technique, the first of his methods for metallic impregnation which was capable of staining entire neurons and glial cells. In Golgi's 1875 paper on gliomas stained by means of his potassium dichromate-silver technique, he was able to give the first morphological definition of neuroglial cells as a class distinct from neurons. Golgi discovered the glial cell perivascular foot and conjectured that the glial cell "protoplasmic process" mediates transfer of nutrients between the blood and brain as shown in Fig. 1.22 (Golgi, 1875). This led ultimately to the modern theory of the blood-brain barrier. This is only one sign of the very progressive character of Golgi's research program, not only in terms of a great technical advance but also because of its theoretical boldness. By showing that the protoplasmic processes of neuroglial cells end on blood vessels and those of nerve cells end blindly, Golgi refuted Gerlach's theory that the protoplasmic processes (later called dendrites by His, 1890) end in a diffuse nerve net by which all nerve cells were believed to be interconnected. Having refuted Gerlach's theory, Golgi conjectured that the nerve net is really between afferent axons and the axon collaterals of efferent neurons. This reticular theory of nerve connections is discussed more fully in Section 3.11. Here it should be emphasized that it was a mature and progressive theory for its time, and that it became postmature and degenerative only near the very end of the 19th century. Guided by Golgi's conjectures about the relationship between neuroglial cells and blood vessels and about the active role of neuroglia in mediating exchange with the blood, progressive research programs have continued to the present time.

Golgi published his mercuric chloride method in 1879 but it went unnoticed until after he published his rapid method in 1887. The Golgi techniques were widely used after 1887 and were mainly responsible for the accelerated progress in neurocytology during the 1890s. One of the first advances was recognition of different types of glial cells. Astrocytes were named by Lenhossék (1891) who recognized them as a separate subclass although Golgi had identified them as early as 1873. Koelliker (1893) and Andriezen (1893a) subdivided them into fibrous and protoplasmic types according to the presence or absence of fibers in the cytoplasm. They noted that fibrous astrocytes predominate in white matter and protoplasmic astrocytes in gray matter.

There were two theories of astrocyte histogenesis. Schaper (1897a,b) conjectured that the germinal cells of His, positioned close to the ventricle, divide to give rise to neuroblasts, spongioblasts, and "indifferent cells." He speculated that spongioblasts give rise to glial cells only in the embryo but indifferent cells persist in the postnatal

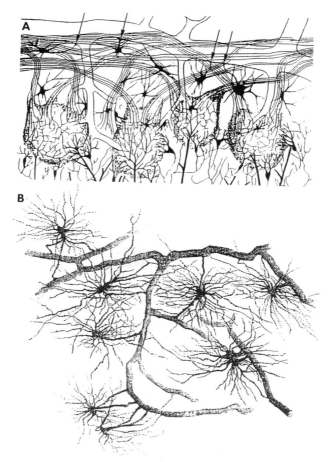

FIGURE 1.22. (**A**) Section of the olfactory bulb (Golgi, chrome silver preparation) showing four glomeruli and several glial cells with their processes apparently attached by means of a terminal expansion to blood vessels. (Arrows added; from Golgi, 1875.) (**B**) Glial cells (astrocytes; rapid Golgi preparation) showing the ends of some of their processes attached to blood vessels. (From Koelliker, 1896.)

period and give rise to both neurons and glial cells. A different theory was proposed by Koelliker (1890b) and supported by Ramón y Cajal (1909). They believed that spongioblasts persist into the postnatal period and are the only progenitors of all glial cells. Evidence for this was given by Lenhossék (1893), who showed a series of stages of astrocyte histogenesis, starting with detachment of radially aligned spongioblasts from the internal and external limiting membranes, followed by migration into the brain parenchyma where they continue to divide to give rise to astrocytes. Transformation of radial glial cells into astrocytes in the spinal cord was confirmed by Ramón y Cajal (1894b) in the chick embryo. Almost a century later those premature discoveries were rediscovered; one might say they matured (Schmechel and Rakic, 1979; Levitt and Rakic, 1980; Choi, 1981; Hajós and Bascó, 1984; Benjelloun-Touini *et al.*, 1985; Munoz-Garcia and Ludwin, 1986a,b; M. Hirano and Goldman, 1988). Both Lenhossék and Cajal observed that the radial glial cells are the first glial cells to differentiate and this has also been confirmed with modern techniques (Rakic, 1971a, 1981; Choi, 1981). Lenhossék and Cajal found that the peripheral expansions of some of the radial glial cells persist in the spinal cord to form the glia limitans exterior, and that too has recently been confirmed (Liuzzi and Miller, 1987).

Cajal's gold chloride-sublimate method stains astrocytes very well (see below), and this enabled him to confirm Lenhossék's theory of the origin of astrocytes from radial glial cells and to refute the theory that astrocytes originate from mesoderm (Ramón y Cajal, 1913a, 1916). He (1913b) also showed that astrocytes can divide in the normal brain; and F. Allen (1912) concluded that there is normally a turnover of glial cells in the adult mammalian CNS. This remained a premature theory until I. Smart and Leblond (1961), using autoradiography of tissue sections, confirmed that macroglial cells are produced in the adult mouse brain. These findings have more recently been corroborated (reviewed by Korr, 1980; Debbage, 1986; Jacobson, 1991, pp. 110–114). Mitosis of astrocytes after brain injury was demonstrated by Penfield and Rio-Hortega (1926). When those theories of glial cell origins were originally proposed, they could neither be corroborated nor falsified by means of the histological evidence because no techniques for tracing cell lineages existed at that time. Those theories of glial cell lineages were bypassed for a century until more reliable techniques for tracing neuroglial cell lineages recently became available (reviewed by Jacobson, 1991, pp. 105–109).

Oligodendroglial cells were discovered by W. Robertson (1899, 1900) using his platinum stain. He did not understand their significance in myelination and he conjectured that they originate from mesoderm (he called them mesoglia, not to be confused with the mesoglia of Rio-Hortega, which are brain macrophages). They were rediscovered by Rio-Hortega (1921b), who called them oligodendroglia because their processes are shorter and sparser than those of astrocytes. He made the distinction between perineuronal satellites in the gray matter, most of which he believed to be oligodendroglia, and interfascicular oligodendroglia situated in rows between the myelinated fibers of the white matter. From their anatomical position, and the fact that they appear only in late embryonic and early postnatal stages during the period of myelination, Rio-Hortega (1921b, 1922) conjectured that the oligodendroglia are involved in myelination in the CNS. This conjecture could not be tested with techniques available at the time, yet it remained a viable theory. This goes to show that it is not necessary for a theory to be refutable in order to be useful. Rio-Hortega's conjecture was finally corroborated in the 1960s, by means of electron microscopy (reviewed by Wood and Bunge, 1984). Rio-Hortega (1921b) and Penfield (1924) also conjectured that a common precursor migrates into the white matter and then divides to produce astrocytes or oligodendroglia. This conjecture has only recently been possible to corroborate (reviewed by Wood and Bunge, 1984; Federoff, 1985).

Before the introduction of good specific stains for glial cells (Ramón y Cajal, 1913a; Rio-Hortega, 1919) it was not possible to differentiate, with any degree of certainty, between processes of neurons and those of neuroglia. Neuroglial cell processes were identified by a process of exclusion. Carl Weigert said that *"one recognized neuroglia as the structure that one could not or would not call neuronal"* (Rieder, 1906). Weigert understood that there was an urgent need for a specific neuroglial stain, and he spent 30 years trying to perfect one (Rieder, 1906). He pioneered the use of hematoxylin for staining nervous tissue, developed a very good stain for myelinated nerve fibers, and introduced aniline dye stains (with the help of his cousin Paul Ehrlich). Weigert (1895) was the first to invent a stain (fluorochrome-methylviolet) that was specific for glial cells, but it only stained glial fibers intensely, the remainder of the glial cell weakly, did not stain neurons, and failed to stain glia in embryonic tissue. It showed glial fibers apparently outside as well as inside the glial cells. Those findings misled Weigert to conclude that glial fibers form a connective tissue in the CNS analogous to collagen.

Nineteenth century theories of neuroglial functions in the adult nervous system reviewed by Soury (1899) include their nutritional and supportive functions, formation of myelin, formation of a glial barrier between the nervous system and the blood and cerebrospinal fluid, their role in limiting the spread of nervous activity, their proliferation and other changes in response to degeneration of neurons, and their involvement in conscious experience, learning, and memory. These theories of neuroglial functions were sustained more by clever arguments than by the available evidence; indeed, they continued to flourish because the means to test them experimentally were not available until recent times.

Soury (1899, pp. 1615–1639) gives a masterful critique of the theories of glial function that were being debated at the end of the 19th century. The theories were weakest in dealing with the origins and early development of glial cells and with their functions in the embryo. This is not surprising because the specific methods required for identifying embryonic glial cells were not invented until much later [the glia-specific histological stains of Ramón y Cajal (1913a), and Rio-Hortega (1919); tissue culture of glial cells in the 1920s; identification of glial cells with the electron microscope after the 1950s; glial cell-specific antibodies after the 1970s]. The concept that glial cells help to guide migrating neuroblasts and outgrowing axons, first proposed by His (1887, 1889a,b) and Magini (1888), was not possible to test experimentally, for almost a century (Mugnaini and Forströnen, 1967; Rakic, 1971a,b). Their myelinating functions were suggested by a few but rejected by most authorities in the 19th century. It was thought that myelin in the CNS is produced as a secretion of the axon and in the peripheral nerves as a secretion by either the axon or the sheath of Schwann. This is considered in Section 1.4.3.

In 1913 Cajal introduced his gold chloride sublimate method for staining neuroglia. It stained astrocytes well but oligodendroglia were incompletely impregnated.* He (1920) mistook the latter for a new type of neuroglial cell lacking dendrites, which he called the "third element." He thought that these *celulas adendriticas* (adendroglia; Andrew and Ashworth, 1945) are responsible for myelination of fibers in the CNS, and

---

*Cajal's summary of his methods for staining the glial cells and his comments on the third element are given in the second volume of his autobiography, *Recuerdos de mi vida* (1917; 3rd ed., 1923), but these parts were omitted, together with most of Cajal's summaries of his methods, in the English translation of the 3rd Spanish edition by E. H. Craigie (*Recollections of My Life*, 1937, republished without correction, 1989).

that they are of mesodermal origin. Rio-Hortega's ammoniacal silver carbonate method (1919), which clearly stains oligodendroglial and microglial cells, showed that these are the authentic "third element" and that Cajal's conclusions were based on incompletely stained cells. Cajal opposed this explanation, and although he reluctantly, and with reservations, acknowledged the authenticity of microglial cells, he continued to deny the existence of oligodendrocytes long after they were clearly demonstrated and their role in central myelination had been revealed (Rio-Hortega, 1919, 1924, 1928, 1932; Penfield, 1924). However, it is only fair to say that the definitive proof that oligodendrocytes are solely responsible for CNS myelination had to await the ultrastructural evidence (Farquhar and Hartmann, 1957; Schultz et al., 1957; Mugnaini and Walberg, 1964; Schultz, 1964).

As Penfield (1924) put it: *"Oligodendroglia has received no confirmation as yet though accepted by several writers. This is probably due to two causes; first, the difficulty of staining this element, and second, the fact that Cajal, repeating the work of his disciple, was unable to stain these cells, and, although he confirmed microglia as a group, he cast considerable doubt upon the validity of del Rio-Hortega's description of the remaining portion of the cells previously termed by Cajal 'the third element.' "* Then comes the critical thrust: *"As Cajal, the great master of neurohistology, has himself so often pointed out, it is extremely dangerous to assign value to negative results."* Wilder Penfield (1977), who worked with Rio-Hortega in Madrid in 1924, describes how the disagreement resulted in an estrangement between the two great Spanish neurocytologists and may have been a factor which precipitated the older man into a state of depression.

Even Cajal did not escape being overtaken and corrected by his intellectual progeny, just as Emil du Bois-Reymond saw his theory of nerve conduction reduced to dust by the work of his student Ludimar Hermann (1838–1914). In historical perspective we see that what Cajal is to the neuron, Rio-Hortega is to the neuroglia. Rio-Hortega was the first to deduce the origin and functions of oligodendrocytes and microglial cells correctly, the first to show their structural transformations in relation to their functions and to emphasize the dynamic state of these cells in normal and pathological conditions. His artistic talents equaled those of his mentor, but while Cajal's drawings have the nervous vitality and intensity of vision of a Velázquez, Rio-Hortega's figures display the deliberately perfected beauty of a Murillo.

Two types of microglial cells in the mammalian CNS were first described by Rio-Hortega (1920, 1932): ameboid and ramified microglia. Ameboid microglia have short processes, appear to be motile and phagocytic, appear prenatally, and increase rapidly in the first few days after birth in the dog, cat, and rabbit. He concluded that these are macrophages originating from the blood, as Hatai (1902) had observed earlier. Marinesco (1909) showed that brain macrophages ingest India ink and thus behave like macrophages elsewhere. Rio-Hortega and Asua (1921) showed that microglial cells are morphologically very similar to macrophages in other parts of the body. They conjectured that microglia and macrophages both originate from the reticuloendothelial system, which at that time was being vigorously discussed (Aschoff, 1924). The ameboid microglia appeared to Rio-Hortega (1921a) to originate in what he called "fountains" of ameboid cells at places where the pia mater contacts the white matter:

beneath the pia of the cerebral peduncles, from the tela choroidea of the third ventricle, and from the dorsal and ventral sulci of the spinal cord. He identified another type, ramified microglia, with long processes, apparently sedentary and nonproliferative. These appear postnatally and persist in the adult. In his 1932 paper Rio-Hortega shows a series of transitional forms between ameboid and ramified microglia and concludes that these represent normal transformations between the two types of microglial cells, thus anticipating recent findings (Perry and Gordon, 1988). In the same paper Rio-Hortega shows that microglia migrate to sites of brain injury, where they proliferate and engulf cellular debris. These are the macrophages of the nervous system, whose roles in defense against infection and injury he was the first to recognize. Confirmation of most of Rio-Hortega's conclusions had to wait until the modern epoch, when the tools were forged that have made it possible to reveal the origins and functions of neuroglial cells.

The debate, started by Rio-Hortega (1932), about whether brain macrophages are derived from the blood or from the brain has continued for 50 years (reviewed by Boya et al., 1979, 1986; Adrian and Schelper, 1981; Schelper and Adrian, 1986). The presently available evidence shows that in adults both microglia and blood monocytes can contribute to brain macrophages, depending on whether the blood-brain barrier is intact or not. Present evidence shows that in the embryo the microglia originate from monocytes that enter the brain before development of the blood-brain barrier.

In modern times those who have concluded that brain macrophages are entirely hematogenous in origin include Konigsmark and Sidman (1963), S. Fujita and Kitamura (1975), E. A. Ling (1978, 1981), and Del Cerro and Monjan (1979). Those who have concluded that macrophages are derived from microglial cells include Maxwell and Kruger (1965), Mori and Leblond (1969), Vaughn and Pease (1970), Torvik (1975), and Boya (1976). The ultimate fate of the brain macrophages after repair of an injury is also controversial: they have been reported to degenerate (Fujita and Kitamura, 1975), or transform into microglial cells (Blakemore, 1975; Imamoto and Leblond, 1977; Ling, 1981; Kaur et al., 1987), but the latter possibility is denied by Schelper and Adrian (1986). The same techniques, in the hands of skillful workers, have led to diametrically opposite conclusions. The brain macrophages may indeed originate from more than one source and have multiple fates, but the neurocytologist tends to select his facts according to prevailing prejudices—in this he is no different from other scientists and nonscientists. The main difference between them is that the scientist, more often than the nonscientist, submits his prejudices for refutation.

1.5. DISCOVERY AND REDISCOVERY: DISPUTES ABOUT PRIORITY

The question of priority of discovery often arises in research programs—some would claim priority for the modifications of the theory necessary to bring it to maturity, others for the technical advances necessary to obtain the facts which they believe "prove" the theory. History often sanctions the award of priority to the rediscoverer, but rarely to the premature discoverer. Numerous examples prove that if one scientist does not make the important discovery, another will, and that is the reason for the race to establish priority (Merton, 1961). But priority is given to the person who succeeds in gaining a consensus more often than to the one who made the original discovery prematurely. As Richard Owen recognized: *"He becomes the true discoverer who establishes the truth, and the sign of the proof is general acceptance. Whoever, therefore, resumes the investigation of a neglected or repudiated doctrine, elicits its true demonstration, and discovers and explains the nature of the errors which have led to its tacit rejection, may calmly and confidently await the acknowledgements of his rights in its discovery"* (*On the Archetype and Homologies of the Vertebrate Skeleton*, p. 76, 1848).

The most notorious of all priority disputes in neuroscience was between Charles Bell and François Magendie. Both claimed priority for discovering sensory and motor functions of the posterior and anterior spinal nerve roots. The primary documents have been carefully analyzed by Cranefield (1974) and the essential facts are well known. It is not so well recognized that confusion about the functions of the spinal nerves persisted for decades after Magendie's definitive publication of 1822. In 1811 and 1821, Bell published evidence showing that the anterior spinal roots and spinal columns are both sensory and motor, and the posterior roots and columns have autonomic ("vital") functions. Magendie (1822) was the first to give conclusive experimental evidence identifying the sensory and motor functions of the posterior and anterior spinal nerve roots. The priority polemic started in 1824 when Bell published new versions of his 1811 and 1821 reports altered to conform with the results of Magendie's 1822 paper. On the strength of those claims, Bell continued to receive credit for the discovery, mainly in Britain, while Magendie's claim was upheld on the European continent (Olmsted, 1944; Gordon-Taylor and Walls, 1958). Here nationalism was an extrascientific factor that conditioned scientists' assessments of a scientific dispute. We see some interesting evidence of nationalism in the translations of Johannes Müller's *Handbuch der Physiologie* (1st German ed. 1833–1840; 4th ed. 1844). In the original German edition Müller awarded the priority to Magendie for the discovery and demonstration of the functions of the spinal nerves, but in the English edition of 1840, the translator added extensive footnotes in which priority was given to Bell. Needless to say, Magendie's priority stood unquestioned in the French editions of Müller's textbook (1st ed. 1845, 2nd ed. 1851).

I do not know who first had the notion of studying the rudimentary state of the embryonic nervous system rather than the complexity of the adult nervous system, but this ontogenetic method is now often credited to Cajal. Tiedemann (1816) and others had stated it before him, nevertheless he claimed priority for conceiving of the idea in 1888: *"Why did my work . . . suddenly acquire surprising originality and broad importance? . . . Two methods come to mind for investigating adequately the true form of the elements in this inextricable thicket. . . . The most difficult, consists of exploring the full-grown forest intrepidly, clearing the ground of shrubs and parasitic plants. . . . Such was the approach employed in neurology by most authors. . . . The second path open to reason is what, in biological terms, is designated the ontogenetic or embryological method. Since the full grown forest turns out to be impenetrable and indefinable, why not revert to the study of the young wood, in the nursery stage, as we might say? Such was the very simple idea which inspired my repeated trials of the silver method upon embryos of birds and mammals. . . . How is it, one may ask, that scientists did not hit upon so obvious a step? . . . Realizing that I had discovered a fertile direction, I proceeded to take advantage of it. . . ."* (Ramón y Cajal, *Recollections*, pp. 323–325).

Before he claimed to have hit on it, Cajal must have known of the ontogenetic method from the *Handbuch* of Jakob Henle (1871). Even earlier, in the introduction to his book on development of the human brain, Ignaz Döllinger (1814) remarked: *"It was also my intention . . . to show how the general features of brain structure were reproduced by comparison between the brain of the adult and the fetus."* The *raison d'être* of comparative neurobiology and developmental neurobiology was clearly stated by Ludwig Edinger in the preface to the second (1889) edition of his *Bau un Verrichtung des Nervensystems* (translated into English as *Twelve Lectures on the Structure of the Central Nervous System*, 1891). Edinger's main argument in favor of comparative and developmental studies was that they could reveal things present at earlier phylogenetic and ontogenetic stages which were hidden in the complexity of the latter stages: *"There must be a certain number of anatomical conditions which are common to all vertebrates— those which permit the simplest expressions of the activity of the central organ. It only remains to discover some animal or some stage of development in which this or that mechanism exists in such a simple form that it can be easily and clearly comprehended."*

The idea that to understand the adult brain it is advantageous to start with the embryo and fetus, advancing from simple to complex, had been proposed by Alexander Ecker in 1873 (p. 10): *"As long as attention was confined to the fully-developed brain, actual progress was not possible. Comparative anatomy and the history of development . . . first brought light into darkness . . . and thus tracing the development is certainly the way by which alone a correct insight can be gained."* However, the darkness was finally dispelled by the Golgi technique, which Cajal started using in 1887, not by his sudden invention of the method of working from the simpler stages of embryonic development to the more complex adult configuration. Cajal owed his early success during the period from 1888 to 1891 to the fact that the Golgi chrome silver technique works well on embryonic brain, rather than to the novelty of the idea of studying embryos.

It is now well known that a developing system might be easier to analyze than a fully developed one (e.g., Rosenblatt, 1962), and there is ample evidence showing that relations between genes and phenotypic characters are more clearly seen during early development than in the finished products (reviewed by Davidson, 1986, 1990). Working with prenatal or newborn animals has the advantage of giving a less complicated picture, but it can also be misleading in showing structures that are impermanent and later regress, such as transient axon collaterals and transient cell populations, and may also fail to show structures that appear at later stages or in the adult.

A claim as to priority has a greater probability of being successful if it is made with sufficient vehemence and well after the time at which the discovery is claimed to have occurred, when the originator can no longer defend the claim. This is one reason for the success of Cajal's claim to have invented the neuron theory, or at least his claim to have been the first to have shown that nerve cells connect to one another by contact, not continuity. Writing decades after the event, he says that in 1888 *"the **new truth** laboriously sought and so elusive during two years of vain efforts, rose up suddenly in my mind like a revelation. The laws governing the morphology and connections of the nerve cells in the gray matter, which became patent first in my studies of the cerebellum, were confirmed in all the organs which I successively explored. . . . The laws mentioned* *above, a purely inductive outcome of the structural analysis of the cerebellum was afterwards confirmed. . . . As happens with all legitimate conceptions, mine became more thoroughly established and gained progressively in dignity as the circle of confirmatory studies was extended*" (Ramón y Cajal, 1937, *Recollections*, pp. 322–323). If Cajal had made this claim at the time of his publications of 1888 the claim would no doubt have been challenged. But Cajal's priority claim was only made after the death of the real discoverers, His and Forel.

The first claim to establish priority for showing that every nerve cell is separate from the others was made by August Forel in his book *Der Hypnotismus* (2nd ed., 1891, p. 14): "*Independently, and in quite different ways, His (Zur Geschichte des menschlichen Rückenmarkes und der Nervenwurzeln, submitted in August, published in October 1886) and I (Hirnanatomische Betrachtungen und Ergebnisse, Archiv f. Psychiatrie, 1 January 1887, submitted August 1886), arrived at the view that anastomoses do not exist, that every nervous element is independent and anatomically separated from the others, and that every fiber is the outgrowth of a single cell.*" Near the end of his life Forel evidently thought that his priority was still not sufficiently well acknowledged, and he recounts the events, again without mentioning Cajal, in his autobiography (*Rückblick auf mein Leben*, published posthumously in 1935): "*I considered the findings of Gudden's atrophy method, and above all the fact that total atrophy is always confined to the processes of the same group of ganglion-cells, and does not extend to the remoter elements merely functionally connected with them. . . . All the data convinced me ever more clearly of simple contact. . . . I decided to write a paper on the subject and risk advancing a new theory . . . and sent it immediately to the Archiv für Psychiatrie in Berlin. However, this periodical was then appearing at long intervals, so my paper did not appear until January 1887. . . . Without my knowledge Professor His of Leipzig had arrived at similar results, and had published them in a periodical which was issued more promptly, in October 1886, so that formally speaking the priority was his.*"

The concept of connection of nerve cells by contact did not spring into existence from the mind of Cajal or anyone else, fully formed, like Athena from the head of Zeus. As I show in Chapter 3, there were at least 12 separate theoretical concepts that gradually converged over a 70-year period to form the neuron theory. Historical records show that priority for discovery of the neuron theory may be fairly assigned to at least eight different people. Priority of discovery of the unity of the nerve cell and fiber could be awarded to Remak for having shown the unity between fibers and sympathetic ganglion cells (1838); to Helmholtz (1842) for showing it in an invertebrate; to Koelliker (1844) for generalizing that concept to all nerve cells. Priority of discovery of outgrowth of fibers from nerve cells could be claimed by Bidder and Kupffer (1857), who first advanced the idea; by His (1886) for histological demonstration of fiber outgrowth from cells in chick and human embryos; by Cajal for discovery of the axonal growth cone, in 1890; and by Harrison (1910) for obtaining conclusive evidence showing nerve fiber outgrowth in tissue culture.

Is priority established by the one who pronounced the original theory, or who gave the first inconclusive evidence, or who finally proved the theory? Or should priority be given to those who invented the techniques that made it possible to obtain the facts? No doubt they all did well and all deserve praise, but I think that priority belongs principally to the one who planted the tree of knowledge, less to those who tended it and pruned it, and least to those who marketed the fruit.

Priority claims fail to recognize or confuse the relationship between technical, theoretical, and value goals in a scientific research program, for example, as shown in the recent dispute as to priority for discovery of the opiate receptor (Kanigel, 1986; Cozzens, 1989; Panem, 1989; Snyder, 1989). If we cease thinking in terms of priority claims for separate components of a scientific research program but start thinking of the conditions necessary for advancement of the whole program, we shall see that credit must be shared between the claimants, to the honor of all. Claimants might do well to consider the moving statement of Ludwig Wittgenstein (1889–1951; Vermischte Bemerkungen, 1977): "*I fully understand how someone may find it **hateful** for the priority of his invention, or discovery, to be disputed, and want to defend his priority 'with tooth and claw.'* . . . *Just **what** would Newton have lost, if he had acknowledged Leibniz's originality [for invention of the calculus]? Absolutely nothing! He would have gained a lot. And yet, how hard it is to admit this kind of thing, for one who tries it feels as though he were confessing his own incapacity. Only people who hold you in esteem and at the same time **love** you can make it easy for you to behave like this.*

*Naturally it's a matter of **envy**. And anyone who feels it ought to keep on telling himself: 'It's a mistake! It's a mistake!'* "*

1.6. CONCLUSIONS

A theory is a model formed in the mind of an observer to explain the meaning of some phenomena that have attracted the observer's interest. A theory begins as a concept which is a guess at what the truth might be. The initial concept is often formed by analogy with another domain, which has a resemblance to the phenomena under investigation. The analogous domain may be another nervous system, or another biological system, a physical system such as a communications network or a computer, or a mathematical model. For the nascent concept to develop further it must be generalized to include a larger set of observations and then to include related concepts in the same explanatory framework. Consilience is the process of unification of observations and lower-level concepts to form higher-level theories of increasing explanatory generalization. In the process, the explanatory generality of a theory increases and its predictive power may also increase. To achieve consilience, a methodology must be developed for making new observations, for determining the identities of objects and events in nervous systems, and for showing their correlations and causal connections.

Those fundamental relations (identities, correlations, and causal connections) have been most successfully revealed by reductive methods. Reductionism is a methodology, not a theory of the nature of reality, although in practice reductionism is the executive branch of a realist-materialist ideology. Reductionism is a methodology for finding relations between components of large systems, with the aim of explaining complex

---

*Leibniz's invention of the calculus was made in 1675–76 and published in 1684; Newton's was made much earlier but published in 1687. The consequent dispute about priority was discreditable to both parties.

systems in terms of simpler components. The aim is to achieve explanatory unification as well as simplification. Reduction can be carried on in either direction—from bottom-up, micro-to-macro, or from top-down, macro-to-micro. Reduction can proceed from molecules to nerve cells to nervous systems, as well as in the reverse order. In practice reduction often starts from some intermediate level of organization from which it is convenient to work in both directions. Reductionism, which includes molecular biology, does not have concepts of its own that can be tested by crucial experiments. Concepts are founded on observations of phenomena, not on methods. Indeed, for methods to be most effective they should be neutral with respect to rival concepts. However, methodological neutrality has rarely been attained in neuroscience.

Concepts can be formed about any phenomena, but when the phenomena excite the interest of many observers, they tend to form different concepts in conformity with their individual psychological and sociocultural prejudices. When different concepts are formed to explain the same phenomena, a process of competition begins that may end with one concept replacing its rival. This is rarely a sudden, all-or-nothing event in neuroscience. Many examples will be given to show that refutation of neuroscience theories usually occurs gradually, with rivals coexisting for decades. This is only one of the ways in which neuroscience theories differ from theories in physical science. Another is the relatively weak predictive power of neuroscience theories. The causes of those differences will be discussed in later chapters.

When observations, methods and theories reach a certain level of integration—when they mesh together effectively, and when they continue to raise the level of understanding of the phenomena—they form a progressing scientific research program. We can characterize the history of neuroscience as the development of research programs, and as the rivalry between different research programs that have been constructed in attempts to explain the same phenomena by different means. Research traditions are formed with respect to the internal logic of the scientific research, namely the progressive integration of scientific theories, increased coherence of data, and technical improvements. Research traditions are also influenced by psychological, sociocultural and other extrascientific factors. Those influences impose strains on the construction of neuroscience research programs, as we shall see in later chapters. But a program built in the form of a pyramid, with many least-inclusive theories at the base, and more-inclusive theories nearer the apex, is capable of withstanding the effects of many anomalies (Fig. 1.23). The least-inclusive theories stand closest to the substratum of phenomena which they attempt to explain, and they are most vulnerable to direct refutation. But revolutionary overthrow of a neuroscience theory by its rival occurs infrequently because crucial experiments are rarely possible, because counterevidence is rarely decisive, and because neuroscientists cling tenaciously to their theories in the face of countervailing evidence.

Explanation is the answer to the question "why?", as description is the answer to the question "how?" The connection of an empirical observation with a generalization may be either explanation or description, or both. In either case the explanations are contingent on the observations on which they are based: empirical science is like a pyramid standing on a base of observations from which deductions are made upwards. A theoretical system built on a broad base of many less-inclusive theories, closely related

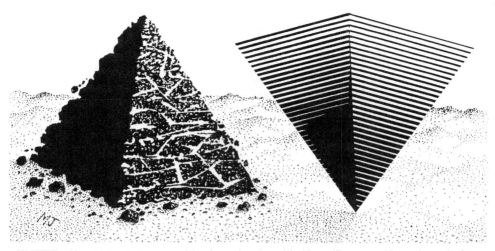

FIGURE 1.23. A theoretical system arising from a broad empirical base, like a pyramid standing on its feet, can withstand refutation of any of its higher-level theories, or of a few of its lower-level theories on which the higher-level theories are built. By contrast, a theoretical system based on axioms from which theories are deduced downwards, like a pyramid standing on its head, is liable to be toppled by evidence opposed to its basic axioms and thus refuting all theories deduced from them.

to empirical observations, can withstand demolition or replacement of many of its lower-level theories at the base of the pyramid without the collapse of the higher-level theories near the apex. In such a hierarchical system the higher-level, more-inclusive theory is connected to lower-level, less-inclusive theories with some level of plausibility decided by consensus. The alternative mode of construction of a theoretical pyramid, based on axioms, standing on its apex, is not seen in modern neuroscience. It can be found earlier in the history of neuroscience when *Naturphilosophie* was closely related to neuroscience in the early 19th century, and it can be seen in psychoanalytic theory: A single theory was deduced from a set of axioms to form a pyramid standing on its head, without any empirical basis (Fig. 1.23).

In my exposition of neuroscience theories I do not intend to describe and discuss all the lower-level concepts on which neuroscience now stands. Thousands of lower-level concepts now support a few higher-level concepts. My approach will be to show how, in general, lower-level concepts are related to the empirical data, and how lower-level concepts are unified in a hierarchy of neuroscience theories. I shall also show how concepts are modified or replaced, and how both scientific and extrascientific factors operate in the genesis and development of theoretical concepts and in the conflict that occurs when there are alternative ways of explaining the same things.

CHAPTER 2

Neuroreductionism

King Hui had a carver named Ting. When this carver Ting was carving a bull for the king, every touch of his carving knife was as carefully timed as the movements of a dancer. . . . "Wonderful," said the king, "I could never believe that the art of carving could reach such a point as this." "I am a lover of Tao," replied Ting, "and have succeeded in applying it to the art of carving. When I first began to carve I fixed my gaze on the animal in front of me. After three years I no longer saw it as a whole bull, but as a thing already divided into parts. . . . Unerringly my knife slips into the natural cleavages. . . . And so, by conforming my work to the structure with which I am dealing, I have arrived at a point where my knife never touches even the smallest ligament or tendon. . . . Where part meets part there is always space, and a knife-blade has no thickness. Insert an instrument that has no thickness into a structure and surely it cannot fail to have plenty of room."

Chuang Tzu (ca. 399–295 B.C.)

2.1. THE ARGUMENT

Neuroscience pivots on the answers to four connected questions: What are the relations between physical events in the brain and the mental events underlying perception, volition, memory, and consciousness? Those relations may be of three kinds: either identity, or correlation, or causal connection. Reductionism is a methodology for showing those relations. Reductionism is concerned with order, i.e., with information about relations, and not only with matter and energy. In addition to showing identities, and correlations, which are noncausal, reductionism aims at understanding causalities in terms of connecting, organizing, controlling, and stabilizing mechanisms or forces necessary for the functional relationships of parts to be maintained in the organized whole.

Reductionism is a way of regarding the whole nervous system "as a thing already divided into parts" and for understanding how "part meets part." It is a methodology for discovering relations between components of complex nervous systems by analysis of simpler subsystems, in terms of smaller components, and ultimately in terms of unreducible elements. For this it is necessary to know both how and where to make the cuts which reveal relations between parts. The aims of reduction are explanatory simplification and unification of different theoretical models of complex nervous systems.

According to the position that I am adopting here, we aim first for explanatory coherence and unification, even if that requires us to leave some complex organization unsimplified, as it often does. We aim for simplification only if that is consistant with the primary aim. Reduction is essentially the same as problem solving, namely finding relations between the whole and parts that explain the meanings of the parts (descending or top-down reduction), and finding relations between parts that explain the meaning of the whole (ascending or bottom-up reduction).

The reductivist program rarely advances systematically from the most complex to the most simple levels, but advances opportunistically and pragmatically, by trying whatever reductive methods are found to work. This requires faith in the methodology because research workers must believe that the effort has a reasonable probability of success, although they cannot predict in detail how the reduction will be done. Such procedures provide much data that are at first weakly coherent. The aim is to achieve more logical coherence of data within theoretical models of increasing inclusivity. The final goal is unification of all neuroscience theories, and unification of those with theories of chemistry and physics.

In reductivist terms, ontology (explanation of reality and of being) is carried on by experimental procedures designed to show relations between things in terms of higher and lower levels of organization. Inclusion of the observer in this program draws attention to the problem of knowing where and how to make the cut between the observer and the observed, and, in general, how and where to make the cuts between different levels. Ontological unification is achieved by showing relations between intermediate levels of organizational complexity, not only by reduction to the unreducible elements. One of the problems of reduction is to define the levels. For this it may not be necessary to make the cuts between conventional structural and functional levels, like cells and molecules, or like forebrain and brainstem, because the same structures can participate in different functions, and the same functions can be realized in different structures. This occurs because the hierarchical organization of nervous systems is formed of a combination of vertically stacked, branching, and nested patterns of organization.

Reductionism presupposes the validity of the principle of the uniformity of nature, without necessarily enquiring why nature should exhibit uniformity, or asking why biological systems, however complex they are, should conform to the laws of physics and chemistry. Although reductionism is not entailed by materialism— reductionism is a methodology, not an account of the nature of reality— reductionism amounts to a practical application of a materialist and realist philosophy.

Behind this program of macro-to-micro reduction of nervous systems are certain hidden assumptions that are so pervasive that they are hardly ever questioned. Some few dissenters on the fringes only make the common assent to those assumptions even more formidable. The consensus view is that all nervous systems are organized hierarchically in definable levels, so that each level includes everything in the levels below and is itself completely included in the levels above (Fig. 2.1). The conceptual level is not necessarily related to the physical hierarchy with which it deals (subatomic, atomic, molecular, tissue, organ, organism, popu-

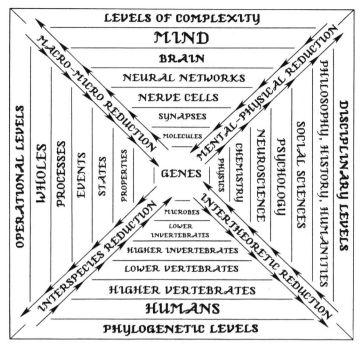

FIGURE 2.1. A neuroreductionist mandala.

lation). For example, at which level is the concept of mind? For theoretical physicists mental events may be conceived at the quantum level. For holists mental events are functions of the entire brain, while for cognitive neurobiologists they may be reduced to functions at some intermediate level of computational neural networks. Some psychophysical dualists conceive of the mind acting on the material brain, but not vice versa, while for other dualists the mind does not stand in any relationship to a physical hierarchy as we understand it.

Another assumption is that the most simple level is related to the most complex level by unbroken causal connections, without supervenience or emergence of entirely new properties. Another is that the same organizing principles exist for all nervous systems, from the most elementary to the most advanced phylogenetically, so that extrapolations can be freely made, for example, from flies to humans.

Another assumption that needs to be questioned is that all neural processes from the molecular to the expression of behavior are necessary and even purposeful. But, is it not obvious that much human behavior is gratuitous, and is not directed at any useful or adaptive goal? If so, unnecessary acts can be viewed as confirmation of free will that is unreducible in a strictly deterministic theory of behavior. Two related assumptions are that the element of chance is negligible and can be safely ignored and that spontaneous activity in nervous systems has little or no information content. Allied with this is the belief that there is no developmental noise, no persistent random variability of structure and function, and that "errors" and "exuberances" are rapidly eliminated from nervous systems.

Underlying all of those assumptions, and affecting them all, is the fundamental assumption, now popular and rarely questioned, that the mind is a macro-event which is nothing more than a large number of physicochemical microevents in the brain. It follows that mental events can be reduced completely into physical events, and it is assumed that the most relevant physical events are in neurons and their connections with one another. A currently fashionable assumption is that mental events are the same as computations in neuronal circuits, and that events in glial cells and metabolic events, which are not constrained within neuronal circuits, are of secondary importance.

The most extreme and uncompromising form of neuroreductionism demands that all things at the highest level (mental processes, events, and states) be completely identified with things at the lowest level at which the reductions are explanatory (neural events and states). In order to advance rapidly, some compromise form of reduction is more often practiced: reductions aim to find interlevel relations (identities, connections, correlations) that explain the coherence of the neural components in relatively high-level functional organizations.

No differences in meanings (including relevances, significances, and values) are implied by differences in conceptual level. Concepts about whole organisms and concepts about molecules are intrinsically equal in meanings but scientists with different biases may give them different meanings. Thus, molecular biologists tend to attribute "greater meaningfulness" to the molecular level, while population biologists may claim more for the organismal level. Reduction aims at unification and simplification, not at finding the "most basic" or "most meaningful" level. This aim of reduction will be emphasized wherever possible.

I do not intend to discuss whether all the goals of experimental reductionism are possible or even desirable. Experimental reductionists can justly claim to have achieved some important successes in a relatively short time. It seems to me that it is unreasonable to demand that experimental reductionists show in advance that their program is feasible. Whether experimental reductionism is capable of achieving all its goals can only be decided after the attempt has been exhaustively made. A more interesting criticism is that those goals are not the only ones worth aiming for, rather than that the goals cannot be reached.

Different people tend to feel most comfortable when conceiving of the organization of the nervous system in one of two ways—as either unity in diversity or diversity in unity. The former see mental and neural processes as unified although widely distributed in space-time, the latter see them as patterns of discrete events. The former have been called "lumpers," the latter "splitters." Reductionists, who are of course splitters, understand that a single component or event is recognized because it stands out from the rest, and because splitters are singularly interested in it, recognize its "relevance," and can thus identify it, and not because it actually exists and functions in isolation. Reductionists are, therefore, methodological typologists, as we see in Section 3.6: they recognize separate types of nerve cells because they can be identified by different sets of significant classifying attributes.

Reductionists often fail to see, or fail to admit, that in the end, to understand the brain, all the fractions have to be summed. This is not the fault of the reductionist program, but is a common failing of individual reductionists. The problem to which I have already alluded is how to account for the unity of our perceptions of outer reality and for the apparent irreducibility of consciousness. Our conscious perceptions are not broken up in jigsaw fashion but are formed as seamless, large-scale patterns. The problem is to discover how discrete neural events that are widely distributed in the brain are bound together to form comprehensive patterns.

2.2. DEFINITION, ASSUMPTIONS, AND GOALS OF NEUROREDUCTIONISM

The functions of nervous systems are smooth when observed macroscopically and consciousness appears to be seamless, and therefore not reducible. We know that nervous systems are formed of subsystems, and those are composed of cell complexes, which are constructed of molecules and still smaller components. How are all those parts joined together in functional relationships that appear to be indivisible at the macroscopic level of observation? What are the best places to make the cuts, and the best methods for reducing the nervous system to components, with the aim of understanding functional relations of parts to one another and to the whole system? Is anything left out or lost when reductive methods are used? Those are some of the questions that are raised whenever reductive methods are used in neuroscience.

Neuroreductionism is a methodology aimed at explanation of complex nervous systems by analysis of relations between levels of organization, and explanation of each level of organization in terms of simpler constituents. The methodology provides data for construction of theoretical models of nervous systems. Intertheoretic reduction is a logical method for relating higher-level, more-inclusive, theoretical models to lower-level, less-inclusive theoretical models, as we see in Sections 1.2 and 2.7. The final goal of intertheoretic reduction is the unification of all scientific theories. Before that final goal is reached, reductionism is committed to the achievement of logical coherence of less-inclusive theories within more-inclusive theories. I discuss this in relation to construction of the neuron theory in Section 3.5.

Neuroscientists have always been more interested in explaining the macro-processes of nervous systems by reducing them to micro-processes than in reduction of theories about neural processes. Also in physical sciences there is more interest in reduction of processes than theories (Brittan, 1970; Spector, 1970). By a neural process I mean a pattern of events (most simply a sequence), and events are defined as the occurrence of states, and states as instances of properties at locations in space-time (see C. Hooker, 1981, note 25). Thus, in mental-state to neural-state reduction, what is being reduced are states, not stuffs. We shall refer to this as M-N reduction, unless additional qualification becomes necessary for particular cases.

The goal of reduction is to explain macro-properties that the nervous system exhibits (e.g. sensation, perception, consciousness, imagination, memory, thought, will) either by explaining how they relate to micro-properties or by showing how the macro-properties do not appear in the individual micro-properties but appear in the organization of clusters, complexes, and hierarchies. Or conversely, by showing how micro-properties characterized by physical and chemical methods do not appear in macro-properties characterized by psychological methods.

Reduction of C to B and of B to A may be achieved via identity, connection, and correlation between A, B, and C. Reduction via identity requires that things included in C are the same as things included in B, and that things included in B are the same as things included in A. Reduction of C to B and of B to A also requires that a theory about C is explained by a theory about B, and a theory about B is explained by a theory about A. This follows from a definition of intertheoretic reduction: if theory B is about things in domain B, and theory A is about things in domain A (where all things in domain B are identical to some or all things in domain A), then theory B is explained by theory A, and B is reduced to A. The assumption is made that the boundaries of the different domains can be defined precisely, but as we shall see a little later the domain boundaries are rarely distinct.

In terms of relations between mind and brain, identity implies psychophysical monism, whereas connection implies dualism, and correlation implies parallelism.

The difference between correlation, connection, and identity can be illustrated by considering the relation between two clocks. (1) *Correlation*: two identical clocks that keep perfect time set to strike in unison, or set so that when the hands of one clock point to the hour the other clock strikes (the preestablished harmony of Leibniz).* (2) *Causal or nomic (lawlike) connection*: two identical clocks set so that when one clock points to the hour it causes the other clock to strike. (3) *Identity*: one clock seen reflected in two identical mirrors.

The clock example also shows that reduction via identities achieves ontological and nomological simplification that is not achieved by reduction via causal connection or reduction via correlation (Hooker, 1981). Reduction via identity achieves ontological simplification by requiring fewer numbers and fewer kinds of things, and nomological simplification by requiring fewer basic laws. In the case of a single clock reflected identically in two mirrors, the relations can be fully explained by the laws of optical reflection. But in the other cases the changes that are observed require further explanation. Thus, there are several different ways by which the indication of time by the clocks' hands and their strikes could be correlated or could be causally connected. The observation that one clock always strikes when the hands of the other clock point to the hour could mean that the clocks are connected causally or that they only seem to be connected but are really functioning synchronously but independently. Each of the clocks would have to be stopped alternately to find out whether its actions were connected or merely correlated with the other clock.

In practice, the use of reductive methods rarely leads to the construction of a research program by the logical procedure of reducing one theory to another; by

---

*Leibniz (1646–1716) regarded interaction between mind and matter as impossible because of the fundamental disparity between them. Therefore, he argued that mind and matter each exist in complete pre-established harmony, a term Leibniz first used in print in 1695.

someone deciding that the program has to be constructed as a straightforward logical sequence of experiments designed to test progressively more general theories. Rather, in practice, progress often occurs pragmatically and opportunistically, in the sense that scientists try whatever reductive methods are available and are found to work, regardless of the logical status that their results may eventually have in a research program. The opportunistic tactics rapidly lead to lots of independent advances, whose coherence is weak at first. It is a daunting task to achieve complete coherence of the data, even at the single level of nerve cells: for example mapping all the connections of all the types of neurons in a very simple organism such as a nematode worm (White *et al.*, 1974; Sulston *et al.*, 1983). It is even more difficult to achieve coherence between different levels, say of the genome and cellular phenotypes in the brain. To achieve complete coherence between the general functions of an organ such as the brain, and the molecular genetic mechanism is such an immense practical task that it cannot be embarked on by someone who lacks the faith that the goal will ultimately be reached. Assertions to the contrary by antireductionists are also based on faith, not scientific knowledge. For example, Ernst Mayr (1982, p. 60) has written that there are "*a number of severe limitations to such explanatory reduction. One is that the processes at the higher level are often largely independent of those at the lower levels. . . .*" What Mayr is really noting are the gaps in our understanding of the relation between levels, not the principle of such dependence.

Certain assumptions about the uniformity of nature and unity of science are held by all neuroreductionists, regardless of the ambition of their claims, and these deserve to be examined carefully. The first of these assumptions is that nervous systems are organized hierarchically so that complex parts in higher levels are composed of less complex parts from lower levels down to the lowest level at which elementary unreducible parts are found. Each level includes everything in all levels below it and is included in the content of the level above it. Leaving out the social level of relations between minds, such a hierarchy, in order of higher to lower level in the nervous system might be: the brain—brain parts—nerve cell collectives—nerve cells—nerve cell parts—macromolecules—molecules—atoms—subatomic particles. The nervous system thus appears to be amenable to experimental reductive analysis: higher-level processes may be investigated, explained, and understood in terms of lower-level processes. The assumption is often made that, in the final analysis, an unbroken chain of relations will be found to extend from the elementary particles to the mind, and that the mind will then be explained completely in terms of physical mechanisms.

This does not imply that hierarchical order in nervous systems is only structured vertically like a multistoried building or a segmented structure. Hierarchically ordered systems can also be in the form of structures boxed into one another, and recursive functions, as well as branching structures and functions, or any combination of those. A set of nested structures, like a set of Chinese boxes, is arranged in complete serial ordering, and so is a vertically layered or segmented structure. By contrast, a tree is a partial ordering: every asymmetric branch point is related to a unique sequence of descending and ascending relations. Therefore, it is often difficult to discover the cleavage points at which relations between levels can be found by means of reductive methods.

Another difficulty is to find the level that has most relevance to any neurobiological theory. It should be remembered that reductive methodology does not, by itself, test between alternative theories by means of crucial experiments. Reductive methods are, at best, neutral with respect to rival theories (methodological neutrality is discussed further in Chapter 1). Yet there are dialectical relations between theories and methods as we see in Section 3.4. Theories for explaining particle physics are different from theories for explaining neurophysiology, and the methods useful for investigating one are not useful for the other.

Reductionism is primarily committed to finding interlevel relations that are significant for explaining the functions of nervous systems. Finding a single, lowest, unreducible level is not the primary aim because different functions may have different lowest levels. For example, the lowest level necessary for explaining sensory-motor coordination and neural control of movement in general may be the cellular, whereas the lowest necessary for explaining mental processes may be the quantum level. However, the lowest level at which physical processes in the brain have significance in relation to mental processes is not self-evident. At the level of elementary particles all neural events are the same. A familiar analogy is that halftone pictures in the newspaper are all the same at the level of the minute ink spots. The physical events in the brain that are related to different mental events are evidently significantly different at the level of neuroanatomy and physiology, at the cellular and molecular levels. But perhaps that appears to be true only because those are the levels of most of our knowledge about the brain. However, there is another reason why I think that the molecular and cellular levels are the levels of explanatory significance: At those levels, form is more significant than content. To return to the previous analogy of the halftone picture—dots are the content and the picture's subject is the form. Random loss of a large fraction of the dots would reduce the content without equivalent loss of form. At what level can that occur in the brain?

Of course, all levels obey the laws of quantum physics but it would not be useful to think of neurophysiology as a baroque branch of quantum physics. The details of quantum behavior are irrelevant to relations between large components like macromolecules, cellular organelles and cells. The most relevant interactions between components at those levels are strong interactions at chemical bonds, or are directly determined by those interactions. Weak interactions between large components are relatively neutral with respect to dynamic behavior of the entire nervous system.

Another characteristic of the hierarchical order of biological systems is that evolutionary changes can occur at one level without necessarily affecting the other levels. A well-known example is the variation of the hemoglobin molecule in different species and even in the same species. The evolutionary changes of the hemoglobin molecule have occurred independently of other changes in the cardiovascular system. It is advantageous for evolutionary changes to occur in a component in one level without requiring coevolution of other components or other levels. Similar examples have not yet been found in nervous systems, but it is predictable that many molecular evolutionary changes in the nervous system have occurred without affecting the overall organization of the system. As I have suggested (Jacobson, 1974b), the gradual accumulation of such neutral changes in the nervous system may become advantageous rapidly under new conditions.

Another common assumption is that the processes found at a given level in one species of animals are the same as the processes found at the corresponding level in other species. It is also assumed that there are organizing principles of the nervous system that apply generally in all animals as well as in humans. Related assumptions are that all vertebrate nervous systems have a common plan and that it is possible to establish homologies of parts of the nervous systems of different taxa. Those assumptions were held by the founders of comparative neuroanatomy and have remained at the basis of application of the reductionist strategy to relatively simple nervous systems and then extrapolating to more complex nervous systems. In the following section we shall examine those assumptions.

2.3. REDUCTION BY ANALOGY, HOMOLOGY, AND COMPARATIVE EXTRAPOLATION

Reduction of components of the nervous system of one species to components of the nervous system of another species is interspecific reduction. Such reduction is either via analogy or via homology, and may be used as a means of extrapolation of observations made on one species to another species. Reduction via analogy is reduction by correlation. Homology implies evolutionary descent and conservation of genetic information, and therefore reduction via homology implies causal connection.

In nervous systems, reduction may proceed either via analogy or via homology, via correlation or via causal connection, respectively. Reduction via analogy proceeds on the assumption that because two things resemble one another in some respects they will probably resemble one another in other respects. Resemblance does not imply either identity or causality, but only correlation. Two things can be similar in all respects, in form and function, but made of different materials. Conversely, two things which have many of the same components can have different forms and functions. The term analogy is most often used to mean that different structural organizations have similar functions, like insect and vertebrate wings, or their sense organs, their nerve cells, or their brains. In many such cases there are no phylogenetic relations between the species, and reduction may proceed via analogy but not via homology.

Homologous structures in different species originate from a common ancestor. Reduction via homology is also reduction via causal connection, leading to explanations in terms of theories of genetics and evolution. Reduction by homology achieves ontological unification. By contrast, analogous structures may be correlated in very many ways that lead to explanations in terms of a diversity of theories of optimization of function, of correlation of structure and function, and many others that are only remotely connected with genetic and evolutionary theories. Reduction by analogy is rarely successful at achieving ontological unification.

It may be possible, and advantageous, to attempt a reduction of a brain of species C to a brain of species B and to a brain of species A. But the strategy has limits. For example, which is the simpler or the lower level: the vertebrate retina or the insect retina? There are structures, cells, and molecules in the insect retina not contained in the vertebrate retina. The vertebrate and insect retinae are based on different structural and functional organizational principles. True, there are analogies between them as wholes, and at the level of cell collectives, but authentic homologies are known only at the cellular and molecular levels. That is why it may be possible to reduce cellular functions of the vertebrate to molecular functions of the insect and vice versa. But that is not the same as reducing the vertebrate retina to the insect retina or vice versa.

Nerve cell complexes that are similar in both vertebrates and invertebrates may be analogous, performing similar functions, or homologous, derived from common evolutionary ancestral forms. A well known example is the form of complex known as the neural glomerulus of which the olfactory glomeruli of vertebrates and invertebrates have been well described (T. H. Bullock and Horridge, 1965). Each olfactory (chemosensory) glomerulus is formed of a bundle of afferent nerve fibers which converge on a glomerulus to form sparse terminal branches before forming excitatory synaptic connections with dendrites of the principal neurons (mitral and tufted cells). The complexity of the circuitry in the glomerulus is increased by interneurons and recurrent collaterals on the principal neurons. The basic circuitry and functional processing of the glomerulus are similar in general, although not in detail, in the vertebrates and invertebrates (J. Davis and Eichenbaum, 1991; Kauer, 1991; Shepherd, 1992). Yet, there is no evidence that olfactory glomeruli in the vertebrates are homologous with olfactory glomeruli of invertebrates. They are similar, that is, analogous, probably because there are restrictions on the number of ways that neural circuits can be organized to process information effectively. This is one of the reasons why many different neural circuits can be reduced via correlations to functionally isomorphic generalized networks (Hopfield and Tank, 1986; Rumelhart *et al.* 1986).

The related concepts of isomorphism and functional equivalence have been introduced before, in connection with building alternative models of nervous systems, and in connection with alternative ways by which the nervous system can function to reach the same goal. Sherrington spoke of the "luxury" of alternative pathways that enabled one component or subsystem to continue functioning after injury to the other. Functional equivalence is found at atomic and molecular as well as a cellular and systems level of organization. For example, at the atomic level different isotopes of the same element are functionally equivalent in their chemical reactions. At the molecular level there are alternative biosynthetic pathways.

The fact that different structures can be shown to be functionally isomorphic implies that they are analogous, not homologous. Actually, homologous structures can have different functions, as the many examples of vestigial structures show. Until there is evidence of true homologies between different neural circuits in the invertebrates and vertebrates, it will not be permissible to reduce vertebrate to invertebrate neural entities, like glomeruli, either via causal connections or via identities.

I am not arguing against the use of interspecies reduction under some circumstances, but I am pointing out some limits to the value of interspecies reduction that should introduce a note of caution in the bold claims to be able to reduce entities of the brain of one species to entities of the brain of another species. I am also asking what, after we have reduced the nervous system of the fly to its neural events and computations, we are going to do with that information. In the end, after the brain of the fly has been reduced fully, we may understand a lot about the brain of the fly, but little about the brain of the octopus or of the human.

Another argument that can be made against interspecies reduction has to do with reduction of mental to neural events. Animals with brains organized differently from human brains may also have minds. The mental events in them are probably quite different from mental events in humans. It is conceivable that in the human brain the same mental event will be realized in different neural events in different individuals. It is also probably true that the mental events in an octopus and a human, both perceiving the same object, such as a fish, are realized as different neural events in the two species. We cannot know what it is like for an octopus to think about anything, and its mental events are opaque to us and may always remain so. This is essentially the same as the conclusion reached by Thomas Nagel (1974) in his article "What is it like to be a bat?" But my skepticism goes deeper than his because I maintain not only that neural events can be different in different species thinking about the same thing, but also that mental events in different individuals of the same species thinking about the same thing can be realized in different neural events in the brains of different individuals. That may be one reason why one knows what it is like to be oneself, but one cannot know what it is like to be another person. And the individual variability of relations of mental to neural events greatly increases the difficulty of devising experiments aimed at reduction of mental events to neural events (M-N reduction). As I show in Section 2.12, M-N reduction via correlations and even via causal connections may be feasible, but there may be deep obstacles to achieving M-N reduction via identities.

Another reason why each person's mind is opaque to all others is that knowledge of other people's minds is based on inferences from behavior, to the extent that one person's behavior is similar to the behavior of others, as we shall see in Section 2.10. Empathy is founded on the same inference, that other people's emotions are the same as mine because their behavior resembles mine under the same conditions. We shall discuss empathy in more detail in Chapter 5 in relation to development of ethical behavior. Here it is relevant to say that we tend to develop empathy for animals whose behavior is similar to our behavior under the same conditions. The assumption is that the same behavior implies the same thoughts, feelings, and mental events in general. But that is to anthropomorphize animals that probably have different mental events from ours, although their behavior may resemble ours under similar conditions.

2.4. THE APPEAL OF SIMPLICITY

Do reductionist explanations appeal to us because we can understand simplicity more easily than complexity, or because the nervous system will eventually be found to be simple? Until we know the correct explanation we can only claim that reductionist methods will at least yield understandably simple explanations, but it remains to be proven that they are sufficient for arriving at understanding of complex systems. Many neuroscientists have invoked the concept of simplicity as a criterion of advantage for understanding "simple" nervous systems. They have never defined the meaning of the concept of simplicity because they assume that it is self-evident. Actually the concept is elusive.

If we define simplicity in terms of the ease with which we can reach understanding we are caught in circular arguments: an entity is simple because we understand it, and we understand it because it is simple. Or the question "What is a simple system?" can be answered only by recourse to another term: a simple system is small or has few components, or has some other property. The answer makes sense only relative to the uncritical acceptance of another property (Quine, 1969, p. 53). If we define simplicity in terms of the fewness of the variable parameters, we have to limit those parameters arbitrarily to avoid including random events which would introduce very large numbers of variable parameters. If we adopt the number of *significant* variable parameters as our criterion of simplicity and we eliminate random events entirely, we find that simplicity of nervous systems is not a linear function of the number of neurons involved. One reason is that control mechanisms in large systems greatly reduce the number of significant variables and parameters.

The progressive evolution of complexity of the nervous system is one concept that has gone virtually unchallenged, because it appears self-evident that evolution is from simple to complex. But that is disingenuous, because there is no *a priori* reason why complexity should have evolved continuously by natural selection unless complexity also confers greater advantages than disadvantages. But simplicity/complexity varies independently from advantage/disadvantage. In fact, organisms with nervous systems at many levels of complexity all coexist in a competitive equilibrium. Organization, not complexity on its own, confers functions that are adaptive and thus advantageous in the struggle for survival. Moreover, complexity and organization are not necessarily correlated. Thus, an animal like the octopus has a brain of the same level of complexity as the brain of the mouse, but with a different organization. The argument can be inverted: animals with differently organized brains can have similar perceptual and learning functions, e.g., octopus and mouse. This is relevant to the functionalist assertion that mental events are not identified with the number of material components, or with their complexity, but are identified with functional states, and with relations of components in organized systems.

In terms of significant variables, certain behavior patterns of a frog with a nervous system containing about ten million neurons may be of equivalent simplicity to the behavior of the nematode *Caenorhabditis elegans* with a nervous system composed of a few hundred cells in the adult worm. Although individual behaviors of the frog, e.g., locomotion, may be equivalent to some individual behaviors of the worm in terms of significant adjustable parameters, the frog far exceeds the worm in terms of the total possible number of behaviors. So, we can agree that the nervous system of the worm is simpler than the nervous system of the frog only in terms of all the possibilities of functions and behaviors but not in terms of single functions, and not in terms of single functional components like single synapses.

Simplicity is sometimes held up as an advantage in terms of the ease with which simple systems can be understood by comparison with the difficulty of understanding complex systems. But that is true only if understanding of the entire system is the aim. If the aim is to understand fundamental organizational principles, complex systems offer many advantages, both for top-down and for bottom-up reduction. The validity of this statement is shown by the overwhelming preponderance of discoveries of fundamental principles from research on complex neural systems like reflexes of the mammalian spinal cord which stand about midway between the most complex and most elementary functional organizations. They provide advantages both for top-down and for bottom-up reduction. In Chapter 4 we shall note many examples when reviewing the works of Cajal and Sherrington. Although work on the nervous systems of both the invertebrates and the vertebrates started at the same time and advanced together, many of the major discoveries were made by people working on the most complex nervous systems of mammals, and many of the major theories of organization of nervous systems were based largely on observations made on vertebrates.

2.5. REDUCTION REVEALS RELATIONS

Reductionism is generally defined in terms of asymmetrical *relations* between higher and lower levels. That definition contains several disguised assumptions: that causality works in one direction only, from lower to higher levels; that lower levels have priority over higher levels; and that lower levels have greater generality than higher levels. The reality seems to me to be that the concept of a hierarchy of physical levels is a heuristic device designed to facilitate macro-to-micro reduction. But one should not assume what one sets out to discover, namely whether a hierarchical system of levels of complexity from more to less is also a hierarchical system of large components standing at a higher level than small components.

In fact, molecules and other components of nerve cells have identities and reasons for existing that are context-dependent. They do not exist as independent entities but as synthetic products of cells, which are themselves composed of molecules. It may be convenient sometimes to think of molecules as the stuff of a lower physical level, and cells as the stuff of a higher physical level, but molecules and cells are really different aspects of the same thing or of the same process. Size is not a sufficient criterion for level. The absurdity of treating things as different levels based simply on size in a macro-to-micro reductionist scheme can be illustrated with familiar examples. Thus, DNA can be regarded as causally prior to RNA, protein, and to the phenotype (Schaffner, 1969). The molecule of DNA is larger and more complex than the messenger RNA molecules that are transcribed from the DNA, if only because much of the DNA consists of nucleotide sequences that do not code for RNA and of sequences that have no known functions. Are we to consider the DNA a lower or a higher level?

If we consider DNA and RNA to be at the same level, then how do we recognize a boundary between that level and the next higher level? Is the boundary located where different individual molecules interact, or is the boundary where different molecules assemble to form a visible structure, e.g., a chromosome, a ribosome, or a membrane, or is the boundary definable only where the molecules form an entire cell, or at all of those levels? If the concept of levels is to have explanatory value, then all levels and their boundaries must be identified, but if the boundaries do not exist in reality, then how can the levels exist except as convenient fictions? When talking of "the molecular level" and "the cellular level" we should understand that we talk of molecules and cells AS IF there is a boundary between two separate entities; AS IF all entities of one kind (molecules) belong in the same level, but all entities of another kind composed of molecules (cells) belong to a different level. In fact, the existence of molecules depends on the existence of cells and vice versa in complex relationships (identities, correlations, and causal connections). Relations are irreducible as Bertrand Russell has convincingly argued in *Problems of Philosophy* (1912). So it can be argued that the lowest level of reduction of nervous system is the level of relations in time-space. In other words, levels are convenient fictions which vanish when the nervous system is conceived as a continuous process in space-time.

Materialist reductionists say that meaning is no more than causal explanation of observations. However, things may be related by identity and correlation as well as by causality. Reduction of A to B via identities entails the identity of A and B but does not entail causal explanation. Demonstrating identity does not require causal explanation because identity is noncausal. For example, water is identical to the molecule H_2O, but hydrogen and oxygen are not causes of water or vice versa. Similarly, to claim that consciousness is no more than a brain process, asserts identity in the strictest sense, not just that they are correlated but that they correspond totally. Reduction of consciousness, perception, cognition, or other mental states via identities to physical events in the brain does not entail causal explanation. But it does require evidence showing that whatever is defined as "consciousness" or "perception" or "cognition" is nothing more than physical events in the brain. I shall argue shortly that it is unnecessary, and often not possible, to achieve reduction via identities for all entities at all levels.

Assertions that mental events are the same as neural events can imply similarity of structure and of function. Conceivably, mental and neural events may be the same physically but different functionally, like computer hardware and software. Or they may be the same functionally but different structurally, like mechanical and electronic calculators.

The identity theory, that mental events are identical with physical events in the brain is different from eliminative materialism (mental nihilism). Identity theory retains mind or mental events or states which can be reduced to physical events or states. By contrast, eliminative materialism accepts the existence of mind only as subjective knowledge, therefore partially or completely erroneous. Eliminative materialism aims at replacing all references to mind with references to physical processes. In short, eliminative materialism requires a study of brain processes completely outside the mentalistic theoretical framework.

What seems to me to be important is understanding the relations of things as components of a process, a system, an organization. Systematic relationships have meaning: cause-effect, part-whole, early-late, large-small, simple-complex are some of those relationships. To make that point about explanation requires that the significance of an entity such as a molecule or a neuron can be measured: it is the number and range of relationships of that entity in a multidimensional network of all other items of relevant knowledge. This is a measure of colligation, coherence, convergence, and consilience, as we discuss that unification of knowledge in Sections 1.2 and 3.4.

For realists, to explain an item of experience is to find a meaningful place for it in the multidimensional network of all other knowledge and to discern and understand the relationships between items in that network. The more relationships, and the greater their range in the network of interrelationships, the more the understanding. Interrelations of nerve cells to form hierarchies, complexes, and clusters appear to be a fundamental principle of organization of nervous systems. A problem is whether something new comes into existence as a result of such relationships, as a result of the process in which the relationships are realized, as Whitehead believed, and as some neuroscientists have claimed (e.g., Sperry, 1976, 1980). I shall argue in Section 2.8 that the notion of emergent properties has never been convincingly explained and that it is not only implausible but untenable.

The aims of reduction are to achieve explanatory and ontological unification (Hooker, 1981). In the process, conceptual simplification may also be achieved. Wimsatt (1976a,b) has argued that the aim of reduction is explanation of upper level phenomena like mind in terms of lower-level phenomena like cells and molecules, but it is not necessary to simplify or eliminate upper-level phenomena in the reduction. In that view the aim of reduction is not ontological simplicity, and identity claims are hardly ever possible to demonstrate for all properties of the reducing and reduced things, as Leibniz's law of identity requires.

Leibniz's Law requires that things which are identical share all properties possessed by either. Therefore, upper-level entities must be fully identified with and reducible to lower-level entities. For example, mental-neural identity requires that upper-level entities (e.g., consciousness, perception, thought, will, emotion) can be fully identified with and totally reduced to lower-level entities (e.g., cellular, biochemical, electrical). This is the most uncompromising form of neuroreductionism. But it seems to me to be arbitrary to demand complete identification of all entities of the highest level to entities at the lowest level, rather than to find interlevel reductions that are explanatory.

In practice, reductions can be made gradually, without making any assumptions about ultimate and complete reducibility. This can be achieved by making each claim for identity provisionally, standing only until it is refuted. The experimental reductionist will admit that cognitive states cannot, because of practical difficulties, be reduced in a single step to molecular states or even to states of neuronal circuits. Rather, the same result may be achieved by successive approximations via intermediate reductive levels. It has already been shown that the reduction of phenotypes to genes requires intermediate levels: the one-to-one relationship between a gene and its product is rarely sufficient to establish the relationship between genes and phenotypes. A significant percentage of genes have no overt function and have no overt effect on the phenotype. Molecular changes are not all reflected as changes at higher levels. A striking example is the unchanged morphology of living fossils such as the living coelocanth, *Latimeria*, and the horseshoe crab, *Limulus*, which have not altered morphologically over hundreds of millions of years, despite a normal rate of molecular evolution (Selander *et al.*, 1970). Genetic divergence without morphological change has also been reported in modern teleost fish (Sturmbauer and Meyer, 1992).

Those findings make it very probable that molecular changes can occur in the nervous system without related changes at higher levels. Thus there are neural states that are not identical with or in any way related to mental states. Some intralevel relations are independent of interlevel relations. The relations between levels do not form a linear chain but rather a treelike structure so that many entities at lower levels have no relation to entities at the highest level. In other words, many neural events are unrelated to mental events. This adds to the difficulties of mental-to-neural reduction. Other difficulties arise from redundancy in the nervous system, from heterogeneity of the neural entities, and lack of information about the initial conditions required for mental events.

Neuroscientists place great weight on macro-micro reductions that ensure coherence and maximize explanatory and ontological unification, as we shall see in Section 3.5. Therefore, neuroscientists tend to accept macro-micro reductions that make compromises with Leibniz's law, and that are a weaker version of the identity theory. Wimsatt (1976a,b) argues that for reasons having to do with the complexity of interlevel mapping, a weaker version of the identity theory is appropriate for biological systems. No doubt complexity increases the technical difficulties of reduction. But there is another good reason for practicing what I shall call moderate reductionism, namely that it aims not so much at ontological simplification as at explanation of the organizational principles of complex systems in terms of less complex systems, and to achieve that explanation requires less than mapping of *all* the properties which are possessed by *all* entities at either level. Explanation is the recognition of relations (identities, connections, correlations) that enable different things to cohere in a functional organization.

Philosophers make high claims for the identity theory, starting with LaMettrie, Cabanis and the later mechanistic materialists like Mach and Fechner, leading to modern neurophilosophers such as P. S. Churchland (1986a,b). LaMettrie argued that human beings are self-regulating machines, entirely self-sufficient and that the mind has no reality independent of the material structure of the brain (*Histoire naturelle de l'ame*, 1745; *L'homme machine*, 1748; see Vartanian, 1960). In his view, mental faculties such as consciousness, memory, will, emotion, and imagination are physiological functions that are based on irritability, which is a fundamental attribute of living matter, which is itself a function of motion which is a fundamental attribute of the elementary physical components of living matter. He implied that if we had the relevant knowledge we could reduce consciousness to motion of elementary physical entities.

Uncompromising insistence on seeking reduction via identity, at all costs, is characteristic of extreme reductionism. Moderate reductionism sidesteps the practical difficulties of reduction via identities, and accepts the explanatory value of reduction via connection or via correlation as sufficient for present purposes. In the following section we examine the different varieties of reductionism more fully.

2.6. ONTOLOGICAL STATUS OF NEUROREDUCTIONISM

Reductionist materialists understand ontology (explanation of reality and of being) not as discourse about words or about definitions but as a study of things that are observable. Ontology, in reductionist terms, is a scientific program of reduction aimed at explaining relations between things and events in the same level of organization, and explaining relations between different levels of organization. Reductionists also tend to be realists who think of the evolutionary origins of the nervous system resulting in an adaptation of nervous states to environmental states—as a process of adaptation and harmonization of brain processes with outer realities. Reductionists consider the theory of knowledge (epistemology) closely related to the explanation of reality (ontology), if not the same.

Most neuroscientists who use the reductionist methodology are realists: they believe in the possibility of knowing the real world (knowing that a molecule of water consists of one oxygen and two hydrogen atoms, and other knowledge of that sort), and they reject idealism and subjectivism (the view that objective reality is a mental construction), but they do not wish to be troubled much with epistemology. They admit that when matters are very complex, knowledge is uncertain, and proceed on the assumption that if there are limits they will be discovered empirically and not by the unprofitable business of speculating on them in advance.

For our purposes, realism is defined as the belief in the unity of all things, including all living things and mental processes, and acceptance only of observable phenomena—truth, meaning, and understanding are attainable only to the limits of observation of phenomena. The properties of things can only be discovered empirically; they cannot be fixed by definitions and logical arguments. The statement that a thing exists means only that it can be observed, and ideally, that it can be measured. Whether it has objective reality as part of an external world is a metaphysical question that is unanswerable. All that we can say is that there is a probability of observation of an object when we interact with it repeatedly, either directly or indirectly by means of instruments.

The status of unobserved and unobservable entities is controversial. Logical empiricists reject them, but realists admit unobservable entities into theoretical models when they have explanatory values. The synapse was such an unobservable entity from 1897, when it was invented by Sherrington to explain the difference between conduction in nerve fibers and conduction in reflex arcs, until the 1950s when the synapse was first seen with the electron microscope (see Sections 1.2 and 3.12).

Realists assert that our understanding of the truth can, under some contingencies, correspond with the external world, and that we can decide which of the present versions of our understanding corresponds with the real world. If there were no external world, there would not be any science. Human survival depends on the existence of an external world and on some correspondence between it and our knowledge of it, and our behavior in it. The human nervous system and mind has evolved as a part of the external world. Thus, the human mind is not separate and distinct from the external world but is an indivisible part of it. This relationship sets limits to our perceptions and understandings if we accept that we are products of evolution, without special exemptions.

From this definition of realism it follows that macro-to-micro reduction of neurons to molecules can actually be done. Clearly, that is different from reduction of a theory of neurons to a theory of molecules (i.e., reduction of a mental model of neurons to a mental model of molecules). I shall discuss intertheoretic reduction in the following section. For now, I can agree that the use of language to describe and explain objective reality in theoretical terms is limited, but there is no reason to suppose that the limitations reside in the real world itself. The stuff of the real world is impervious to intertheoretic reduction but not to scientific methods for reducing such things as cells to molecules. The distinction between intertheoretic reduction and experimental reduction does not lie merely in different roles they play in our thinking but rather in differences in the use of language to describe theories and to describe objective reality. Description and explanation of theories is bound to be imprecise because theories are themselves inherently so, but objective reality is not so, at least at cellular and molecular levels, although objective reality may be fictional if Heisenberg's interpretation of quantum theory is correct.

Neuroreductionists come in three varieties who make different claims for the ontological status of their form of reductionism—naive, moderate, and extreme. A significant proportion of neuroscientists practice reductionism without reflecting on it, without making claims for its ontological status, often without being able to give their method a name, and some would be surprised to be told that they were reductionists (Barnett, 1991). They remind one of the gentleman in Molière's comedy, who was astonished when he discovered that he was speaking prose. Since the mid-19th century extreme reductionists have also been extreme materialists, but moderate reductionists have generally been skeptical of the extreme materialist claim that animals and humans are only mechanisms that can be reduced entirely to molecular mechanics.

The extreme reductionist position is that the principles of construction and operation of any system, however complex, can be *completely* deduced from analysis of the structure and operations of its parts. From the materialist doctrine that mind is a state of matter, that mental processes and brain processes are identical, not separate things, it follows that mind can be investigated completely with the methodology of cell and molecular biology and biophysics. The extreme reductionist neuroscience program aims to elucidate *all* neuroscience on the foundations of neuroanatomy, neurophysiology, cell biology, molecular biology, and ultimately biophysics. Whether this is possible in practice, or even only in principle, is highly controversial. Saying that something is possible to do "in principle" implies that it is only temporariliy prohibited for technical reasons, but in fact it may really be impossible (Boyd, 1972).

Another reason why neuroscientists tend to be extreme reductionists is that they must have some belief in the successful outcome of a reduction (e.g. reduction of cellular processes to molecular and physical processes; reduction of reflex processes to synaptic processes) before they would be willing to undertake the difficult research program that such a reduction requires. Researchers must believe that such a reduction is possible and worth their efforts, even though they cannot know in detail how the reduction will be done and what the final results will be.

Extreme neuroreductionists view reduction of a theory of mind to a theory of neurophysiology as no different, in principle, from reduction of the wave theory of light to the electromagnetic theory, or reduction of thermodynamics to statistical mechanics. They do not recognize any fundamental differences between neurobiological and physical theories. Extreme neuroreductionists assume the uniformity of nature (Causey, 1977). In that view the uniformity is expressed in organization of neurons and glial cells in the same kinds of complexes under the same specified conditions, just as in chemistry the same elements always form compounds of the same kind under the same conditions.

Extreme neuroreductionists are also neuronal typologists as that term has been defined in Section 3.6. They argue that, regardless of the number of neuronal types, a nerve cell can be characterized completely and classified as a member of a unique type based on a set of significant classifying attributes (see Causey, 1972, for discussion of this general requirement for macro-to-micro reduction). For almost all practical purposes the classifying attributes are molecules to which the cell type can be reduced under certain conditions.

Extreme reductionists have resisted any recognition of the possible limitations of reductionism. They tend to equate the data with their meaning. They disagree with the proposition that we can possess data without ever, in principle, being capable of understanding their meaning. For instance, they say that we could completely understand the human brain if we had sufficient data. The extreme reductionist would not consider the possibility that the limits to understanding the human brain are in the human brain, not in the available data, as Emil du Bois-Reymond argued in 1877 and as Colin McGinn (1990) has argued again. This may prove to be no more than a technical obstacle that may be overcome if we could use computers as surrogate brains for recognizing patterns of relations in all the available data.

Moderate reductionists, having reflected on the limits of all methodologies that have proved useful in neuroscience research, do not attempt to give reductionism a monopolistic status. The modest reductionist makes no strong claims for the reductionist methodology beyond its immediate usefulness. The point of departure may be precisely specified, but there is no absolutely given point of arrival. This prudent form of reductionism generally goes with agnostic materialism and with realism, as we have defined them.

Moderate reductionists agree that the criterion of a good neurobiological theory is its explanatory power and not its predictive power, as I have argued in Section 1.2. However, extreme reductionists reject the widely held dictum that one of the fundamental differences between biological and physical theories is that the latter are the basis of strong prediction whereas the former are not. They hold that the differences are merely temporary, and will disappear when biological theories are eventually reduced completely to physical theories. It is said that Darwinian theory may allow retrodiction, but cannot predict future evolution of the species. But genetic theory does allow prediction of the phenotype, in general if not in all details, from adequate knowledge of the genotype. For example from knowledge of the genotype it is already possible to predict significant parts of the phenotype of a nematode worm or of a fly (Lawrence, 1992), and knowledge of the human genome may eventually allow prediction of substantial parts of the human phenotype. Knowledge of the genotype can also allow prediction of many kinds of behavior in worms, flies, and some kinds of behavior in mice and humans, but not of behavior under all contingencies in which the complexity of initial conditions defy definition, and in which there are many unknown variables. We shall consider those limitations shortly.

We can sum this up by saying that more accurate predictions can be made from low-level neurobiological theories than from high-level theories. Accurate predictions can be made from theories of nervous conduction, synaptic transmission, and muscle contraction, but not from theories of cognition, motivation, and mind. For example, accurate predictions can be made about whether activity of the motoneurons supplying the muscles of my finger will result in flexion or extension or no movement of that finger. But it is impossible to make accurate predictions, only estimates of probabilities, about whether I decide to flex that finger with the aim of scratching myself, tapping the table, or firing a gun.

This difficulty of prediction may arise for several reasons: because complexity of initial conditions may make prediction difficult; because new qualities emerge unpredictably at higher levels, and because neuroscience does not have fundamental theories from which accurate predictions of high-level function can be made. Predictions of voluntary behavior are inaccurate for the same reasons that predictions of complex physical events like earthquakes and the weather are inaccurate, and are limited to short-term forecasts: complexity, multiple variables, and inadequate knowledge of relationships between components of the system.

Neurobiology is not an exact science like physics, with fundamental and universal laws. Neurobiology is more like electronic engineering than like physics, as Smart (1963) has argued. Neither neurobiology nor electronic engineering has theories of its own that cannot be reduced to more general theories of chemistry and physics. Theories of neurobiology are generalizations from data obtained by means of many different methodologies ranging from those of biophysics and biochemistry to those of psychology and philosophy. In practice there are at present quite sharp divisions between the methods that are useful for dealing with such things as mitochondria and membrane potentials on one hand, and with such things as motivation and mind on the other hand. The methods that are now available for dealing with the former are ineffective for dealing with the latter, and vice versa. There is a similar division between the methods for studying reflex behavior and those for studying voluntary behavior. But that does not mean that, in principle, we could not discover all the intermediates between membrane potentials and mind, or between reflex and voluntary behavior, if we possessed the necessary methodologies and the intellectual powers necessary to understand the data.

The extreme reductionist sees neurobiology as a branch of biophysics and molecular biology, with special methodologies but not with special laws different from those of chemistry and physics. The aim is to find laws without which neuroscience could not be understood, but they would be general laws of physics. Reductionist materialists agree that neurobiology would have to include laws of physics that have not yet been discovered, but would never be required to include special fundamental laws of neuroscience. However, physics may be pushed by neuroscience to find new laws of physics that would explain mental processes. Therefore, one of the aims of the reductionist program is to achieve explanatory unification of neuroscience and physics. By contrast, dualists maintain that mind will always be left out after exhausting all physical explanations because there are mental and physical domains, both of which are fundamental, and the two domains are ontologically separate.

In addition to the admission of limits of the reductionist methodology, there is a more fundamental basis for the moderate reductionist's unwillingness to make claims for the finality of neuroreductionism. It is that in the final analysis science can be no more than is permitted by human desires, needs, and competencies. At the deepest level, meaning is what humans are capable of understanding. Moreover, those meanings are selected from all the many possible meanings of the objects of our conceptions, which are consistent with empirical observations. There may be limits to our abilities to understand the meanings of our observations of the human brain. Paradoxically, we may never know when we have reached those limits, even with the aid of computers. That paradox stands regardless of the best methodology we may ever be able to use.

For the reductionist the less fundamental problem, but a significant practical one, is whether the higher levels of brain organization are only approximately reducible to lower levels of organization. There are several reasons for us to be concerned about this limitation. First, the redundancy in the nervous system makes it possible to perform the same function in more than one way so that it may be difficult, or impossible in practice, to predict the future states from the initial conditions. Second, the nervous system is not in a steady state but changes irreversibly as it functions. Third, our understanding may be limited because we are forced to use some form of statistical treatment to replace the description of the microscopic state of each and every neuron and synapse. But the kind of statistical treatment appropriate for dealing with the nervous system would have to be different from that appropriate for a liquid or a gas whose homogeneity allows a macroscopic description to be given by a statistical averaging of the positions and velocities of the atoms or molecules. That kind of averaging is useful when dealing with events that are uniform, like nerve action potentials, but the nervous system is not composed of identical cells and synapses that can be treated by averaging. The information content of nervous systems is not only, or even mainly, in the molecular biology and electrical events of the individual nerve cells, but resides in the organization of neuron-glial complexes and of neural networks and circuits. Therefore, reduction of the whole to subsystems and of those to single units involve losses of information. These losses can only be estimated but, as far as we now know, they involve no occult entities such as those presupposed by vitalist and epiphenomenalist theories of the emergence of functions from complex structures.

2.7. EXPERIMENTAL AND INTERTHEORETIC REDUCTION

One has to distinguish between the principle of intertheoretic reduction and the goals of experimental reduction. The experimental reductionist treats reduction as a relation between things and not between theories. However, the philosopher treats reduction as a relation between theories in which the reduced theory is explained by the reducing theory (E. Nagel, 1961; Causey, 1972; Hooker, 1981; Rosenberg, 1985).

We have to be careful to say which form of reductionism we mean. For example, when biologists are setting out to reduce the phenomena of memory to the activities of cells and molecules, their program is different from that of the philosophers attempting to reduce a theory of memory to a cell and molecular theory. The philosopher is concerned only with the logical relationships between general statements and with the terms of the memory theory on the one side, and those of the molecular theory on the other side. For the reduction to be possible, the essential requirements are that the general statements of the reduced theory follow by deductive logic from those of the reducing theory and that the terms of the reduced theory be logically connected to the terms of the reducing theory.

Those requirements can be satisfied by theories of physical science in which general statements are logically related to one another, and which have precisely defined terms. But those requirements for intertheoretic reduction can rarely be satisfied for neurobiological theories, which are mostly analogical-inductive theories, in which general statements are conjectures, guesses at the truth, and are related to one another by analogy. Therefore, to reduce one neurobiological theory to another it is unnecessary to insist that the general statements of one are derived by logical deduction from those of the other. It is sufficient that the statements of the reduced theory are related to those of the reducing theory by analogy, and with some degree of plausibility. The degree of plausibility can only be established pragmatically for each case, not axiomatically for all cases. Another reason why reduction of neurobiological theories cannot satisfy rigorous logico-deductive requirements of intertheoretic reduction is that many neurobiological terms cannot be defined without recourse to other terms which are also imprecise. The terms that are precisely definable, like molecule and cell, belong to lower levels of neurobiological organization, whereas the terms that have controversial definitions or cannot be defined precisely, like consciousness and will, belong to higher levels of organization. It is therefore easier to reduce the lower-level neurobiological theories than the higher-level neurobiological theories to physical theories.

This is not only a difficulty of "cross-categorical reduction," as philosophers have said, but it is also the result of a real disparity between the terms of high-level and low-level neurobiological theories. There are no terms of cognitive psychology that are logically connected with terms of molecular neurobiology. Nevertheless there are ontologically valid relations between models of cognitive psychology and models of molecular neurobiology, via intermediate models of neurophysiology and neuro-anatomy. As we have seen in Section 1.2, the relations between thoeretical models need only satisfy a certain level of explanatory coherence to be plausible. Relations between theoretical models are justified pragmatically by their explanatory value, especially by their capability of unification of data previously believed to belong to different domains but now shown to belong to the same domain. I have argued in Chapter 3 that the neuron theory was successful because it achieved such explanatory unification.

The neurobiologist cannot afford to admit that the philosopher's requirements are necessary for what we shall call experimental reduction. In practice, the neurobiologist must ignore the philosopher's requirements for reduction of theories. The neurobiologist must insist on that because the philosopher's requirements can be satisfied so rarely when reduction of things is attempted, that they would hamper the experimental reductionist research program.

Experimental and philosophical reductionists join company in rejecting vitalism and teleology, but part company on the requirements for reduction. Most important, philosophical reduction has requirements which if satisfied in advance would virtually paralyze the experimental reductionist whose research tactics are sometimes opportunistic, trying one experimental method after another until some progress is made. In practice, even extreme neuroreductionists have to be satisfied with approximations to reductions. If they placed restrictions on the requirements for complete microreduction they would make very slow progress. For example, they could not require that the initial conditions be completely specified, and they could not forbid unobservables from the terms of the reduction.

The logical requirements for philosophical reduction established by logical positivist philosophers, namely logical deduction and connection of terms, are not necessary for experimental reduction and failure to satisfy those requirements is not an argument for abandoning the reductionist program. Cognitive psychology is riddled with subjective interpretation and differences of opinion about the meaning of its explanatory terms, which may make it impossible to deduce molecular theory from cognitive theory. Extreme complexity makes it difficult to reduce cognitive states to macromolecules, but we cannot assert that is an absolute limitation, and not merely a temporary practical problem. As it is actually accomplished in experimental neuroscience, reduction is achieved little by little, so that parts of cognitive states have been reduced to neural states and molecular states, while other parts have not yielded to experimental reduction.

For example, abnormalities of dopamine receptor types are known to play significant roles in the pathogenesis of neurological and psychiatric diseases such as parkinsonism, schizophrenia, psychosis, and cocaine addiction, which all result in cognitive disturbances. The connection between cognitive states and molecular states is also strengthened by the vast body of evidence that antipsychotic drugs bind specifically to particular dopamine receptors. For example, the D_2 receptor, which is known to be involved in schizophrenia, is the binding site of butyrophenones and related antipsychotic drugs. Thus, cognitive states known to be associated with dopaminergic neurotransmission can be reduced in part to molecular states. The reduction has not yet been completed through all intermediate levels, but that is not an adequate reason to say that reduction of cognitive to molecular states cannot be completed. This claim may not remain valid if something in addition to macromolecular states is found to be necessary for cognition. But, for the moment, that possibility is no more than an ad hoc hypothesis. Such hypotheses can be multiplied indefinitely to raise objections to any theory, but they are not serious rivals to the established theory unless they lead to predictions and explanations that cannot be made from the established theory, and/or lead to new counterevidence.

The practical difficulties of complete reduction of cognition to molecular biology appear to be insuperable at present, but is there any logical basis for concluding that the reduction is absolutely impossible? A practical basis for admitting that such complete reduction is impossible would arrive if it were found that there was more to cognition than just molecular interactions, say quantum uncertainty (Penrose, 1989; Lockwood, 1989).

2.8. ARE THERE EMERGENT PROPERTIES OF COMPLEX NEURAL SYSTEMS?

An emergent property of the whole is one that could not be deduced either from the properties of the parts studied in isolation, or from the properties of parts that could be predicted to come into play as a result of interactions between the parts; not because it is difficult to do so in practice, but because it is impossible to do so in principle. Some epiphenomenalists believe that the mind, the soul, is such an emergent property. Materialist reductionists say that animals and humans are only very complicated mechanisms and deny that there are any emergent biological properties (see E. Nagel, 1961, Chap. 6, Sect. 4; Smart, 1963, Chap. 3; Monod, 1971; Changeux, 1986).

Materialists, insofar as they admit the existence of mind, regard mind as fully deducible from the laws of physics and chemistry, if the necessary data were in our possession and if we were able to understand their meaning. I agree with that, but I think that it is an open question whether we could ever know the meaning of the data necessary to understand mind completely.

We are not exempt from the limitations imposed by the process of evolution of the human brain. Our mental capabilities are only what they now are by virtue of the contingencies of survival of the fittest of a multiplicity of possible forms. There is no basis for wishful thinking that the human brain is endowed with powers to understand everything that may ever be discovered. This is not the view of "the new mysterians" as Owen Flanagan (1991) calls those who deny the brain's ability to explain itself completely. Rather, it is the old notion of the limits of human understanding and the paradox that we can never know how close we have approached to the limits of human understanding. Even if we were in possession of all the facts we may never be able to understand them, or even worse, we may assent to a mass delusion. The belief that mind is an emergent property of brain processes seems to me to be the thin end of the wedge of that delusion. It explains nothing. It gains assent because of the perceived failure of other explanations, rather than because it offers a better explanation; namely an explanation that is founded on unification of a large number of facts from different sources.

Epiphenomenalism is the belief that consciousness is an epiphenomenon of the physical processes in the brain, not part of the physical processes themselves. If that is true, neural processes are the cause of consciousness and other mental phenomena, but the latter can have no causal effects on the neural processes. Exactly how mental phenomena may emerge from physical processes has never been explained, but epiphenomenalists maintain that emergent properties appear only when matter is organized in a certain way, e.g., the property of being solid, liquid, alive, dead, conscious, or unconscious. They maintain that emergent properties can be reducible, e.g., solidity, or irreducible, e.g., consciousness. However, the question is whether the irreducibility is in the nature of the emergent property, as epiphenomenalists maintain, or merely a temporary difficulty that may be overcome with the necessary techniques and knowledge. The history of vitalism parallels that of epiphenomenalism in that both theories have been gradually invalidated by improved knowledge of biochemistry and physiology. Life was regarded as irreducible until techniques made it possible to reduce life to a specific kind of self-replicating molecular organization. On that analogy, it has been argued that mind will remain irreducible only until the necessary techniques have been perfected.

Epiphenomenalists maintain that mental phenomena emerge when the brain becomes sufficiently complex. That was the opinion of T. H. Huxley in his paper "On the hypothesis that animals are automata" (1874) which was based on an earlier lecture, "Has a frog a soul?" Frogs are not complex enough to have a soul, was Huxley's answer, and most neurobiologists after him have reached the same conclusion. We now maintain that organization, not complexity, is the necessary condition for perception and learning. But, from either position, it is difficult to understand why goldfish and frogs, which have similar brain size and complexity, and with similar organization of their nervous systems, should have such markedly different abilities to learn in the laboratory situation. One possibility is that learning situations have been more natural or in other ways appropriate for fish than frogs in the laboratory.

It is also difficult to explain why animals with brains of similar complexity but different organization (e.g., octopus versus mouse) have perceptual and learning abilities that are similar in some important respects. The octopus, with a brain as large and complex as that of a small mammal, but organized differently, has observational learning abilities similar to those of small mammals such as mice (Young, 1964). The octopus can learn a task by watching another octopus learn that task in a separate tank (Fiorito and Scotto, 1992). I consider that to be good evidence that the octopus is conscious of other creatures, and possibly that it also has consciousness of itself. The only means of finding out whether octopus consciousness is similar to mouse consciousness is by subjecting both to operational tests. So far, the evidence indicates that the consciousness is functionally similar in those two species with differently organized brains.

The fact that a pattern of activity formed of many different neural elements is more than the sum of the elements is the result of the combinatorial relations of the elements and does not mean that there is an emergent property possessed by the whole pattern. Nor does it mean that there must be a pontifical programmer that receives all the relevant information at a central station, and then issues commands that ensure spatial and temporal integration of function. Perhaps there is not such an integrative control center, only a sequence of events entrained by local coupling combined with more widespread couplings between events resulting in a process which is a large-scale pattern of events in space-time. If so, reductionist methodology should, in principle, be able to discover the couplings between events, and to reduce the large-scale pattern of events to small-scale patterns. For more than a century neuroscientists have recognized that the problem is to understand how the distributed neural events are coordinated as functional patterns and are integrated to form a unified pattern of conscious perception. The modern consensus is that solutions to those problems will be found in terms of parallel distributed information processing in neuronal networks. That research program is advancing rapidly, as is shown by recent reviews (McClelland *et al.*, 1986; Bechtel and Abrahamsen, 1991; Dennett, 1991; P. S. Churchland and Sejnowski, 1992). Those reviews also show that much remains to be done before we can understand voluntary behavior and consciousness only in terms of neural events. Perhaps something has indeed been left out of those theories. I do not imply that there are emergent properties or occult properties that have been overlooked. Rather, it seems to me that some very important functions of the brain are carried on by glial cells, outside the neuronal circuitry, although having important functional effects on neuronal circuitry. That mental functions are widely distributed is now generally acknowledged, but the consensus is that distribution occurs exclusively via neuronal connections when it might also occur by diffusion and active transport of ions and molecules in brain space outside the confines of neuronal circuits. If so, we must define another level of organization of entities in brain space in addition to the level of synaptic connections and neuronal circuitry (see Fig. 2.1).

2.9. MENTAL-TO-NEURAL REDUCTION

Descartes posed the relation of mind to brain as a problem of two-way causal relations between a material brain and a nonmaterial soul (see especially *Meditations*, I and II, *Discourse on the Method*, 5). But there are several other ways to conceive of the relation, and they cannot all be correct. Later dualists allowed the brain to cause the mind but not vice versa, or they allowed no causal relation whatever (i.e., psychophysical parallelism). For Spinoza (*Ethics*, Part II, Prop. VII Note), mind and brain are different aspects of the same thing, and this double-aspect monism was developed by Bertrand Russell (1948) and Herbert Feigl (1958, 1960). Russell (1921, 1927) proposed a theory of "neutral monism" which asserts that mental and neural events only appear to be different but are really aspects of the same thing, which is neutral to both. This can be interpreted to mean that the mind is not caused by neural events or vice versa, but both are reflections of a deeper underlying reality.

A practical weakness of those theories is that there are no available methods, or even imaginable methods for testing them. That is probably why many neuroscientists prefer theories of mental-neural relations that can be tested with reductive methodology. For materialists who are single-aspect monists, starting with LaMettrie (1747), and including Bain (1855), Fechner (1860), Mach (1914), and Changeux (1986), the mind can be reduced completely to neural activities. As I have already pointed out, mental-to-neural reduction may be effected via correlation, causal connection, or identity. The notion that mental events are identical with neural events is called the identity theory. This has been summed up by Horace Barlow (1972) as the credo that "*the activity of each single cell . . . is related quite simply to our subjective experience.*" In the words of Pierre Cabanis (1805): "*Les nerfs—voilà tout l'homme!*"

Philosophers make a distinction between type-type identity and token-token identity. The former is identity as I have consistently used the term, and asserts complete physical identities, in all particulars, between mental macro-events and physical micro-events. Token-identity asserts only that there are functional, causal connections or correlations. Thus, type-type identity theory asserts that psychological states are no more than neurophysiological states, and the former can be reduced completely to the latter. But type-identity theory is not committed to saying which physical states are correlated with mental states or their cause. Although it is generally assumed that they are neural states, they may include glial states and any other physical states, including ones that have not yet been discovered

Type-identity theory maintains that the *qualities* of mental states are physical too, and many mechanists and all dualists depart from type-identity theory at that point. Another point of disagreement is the assertion of type-identity theory that identities are noncausal and therefore there can be no inherent purpose in relations between mental events and physical events. Obviously, different physical events may be causally connected and so may the relevant mental events, but type-identity theory asserts that $M = N$, which is a noncausal relation. We shall return to the problem of causality in Section 2.13. A related question is how free will, voluntary thought and behavior, and intentions can be accounted for by identity theory. Strict determinism asserts that the mental-physical domain is completely closed and self-sufficient. In that option, free will, which requres an open mental domain, is a delusion. In that view, human beings are physical automata, just like all other animals. Thus, the difference between voluntary and involuntary activity, or between pleasure and pain, is no more than differences in patterns of physical events in brain space-time.

"The mind" is an example of a reified concept in which an abstraction or an imaginary entity is treated as if it were a real thing. Disbelief in the existence of something called the mind is difficult because it is not easy to abolish reification. To avoid reification we refer to mental processes, events, and states and to reduction of those to neural processes, events, and states. It is widely recognized that this is an artificial division, a convenient fiction, required for communication. Whitehead regarded the mind-body problem as an example of "the fallacy of misplaced concreteness," and held the position that both mind and matter are different aspects of the same process (Whitehead, 1925, Chap. III; 1929, Chap. I). But it is hard to think of mind as a unified process in brain space-time. It is easier to conceive of a process as a pattern of discrete events, and events as the occurrence of discrete states. In fact, a mental event may be discretely localized or it may be widely distributed in brain space-time.

The mind-brain problem is made more difficult when the mind, an abstract concept, is treated as if it were a thinking substance. The mind is thus defined as an entity with attributes which enable its possessor to feel, perceive, will, think, remember—in short, to be conscious and self-conscious. This kind of definition was dismissed as a linguistic and conceptual error by logical behaviorists such as Gilbert Ryle (1949), who called it a *category* mistake, namely the transfer of a category from one domain to an unrelated one. It is as if one talks of "patriotism" as a thing separate from people, although related to them, when "patriotism" really describes people's attitudes and patterns of behavior. Having identified the mind-body problem as a "pseudoproblem," the logical behaviorists embarked on a program of logical analysis of the meanings of statements about mental states. They concluded that such statements are not at all about a person's inner states of mind but are really descriptions of what a person does or is disposed to do. Therefore, the mind-brain problem is really a neurobiological problem about the relation of attitudes and patterns of behavior to physical states in the nervous system.

Materialists can say that the existence of mental events, which are different from neural events, would have been demonstrated by the methods of introspection and philosophical argument, if the problem could ever be solved by those methods. Failure to demonstrate the nature of mental events means either that the methods are inappropriate or that there is not an authentic problem to solve, but only a misconception like the phlogiston theory of heat. From that skeptical viewpoint, the history of concepts of the mind is a history of deception and ignorance. Therefore, the aim of the historian of ideas about the mind is not so much a revelation of the truth, but, as seen from a different view, equally valid, as the attempt to discover the truth content of the lie, namely to discover how and why the misconception occurred.

I understand introspection to mean the ability to observe one's own mental events. But I am unable to do that. My mental events remain stubbornly hidden although I can infer that they are related to my thoughts, dreams, and conscious awareness. Since one is unable to report about one's own mental processes (see Nisbett and Wilson, 1977), knowledge of one's own mind is obtained, in the first instance, by inference from observations of one's own behavior or disposition to behave. That disposition may be conscious or unconscious, expressed after a short or longer delay. In many instances one does not know what one intends to do until after it is done. In other instances one thinks about intended behavior, but thinking is not the same as introspection.

Both the inner world of self-consciousness and the outer world are known by hypothetical reasoning from experience and the effects of behavior. My only way of authenticating my belief in the existence of outer reality is through consciousness of my own existence. From the knowledge of my own existence I infer that other beings also have self-consciousness. I become conscious of myself through sensory perception and by observing my own behavior, not by direct apprehension of my mental or neural events. I come to believe in the existence of other minds like my own, including the minds of other species, by observing their behavior and by inferring that it has the same validity as my own.

Some philosophers maintain that one can know one's own mind without inference (e.g., B. Russell, 1948, p. 224; F. Jackson, 1982; J. Foster, 1991). But even if the intuition of the existence of one's own mind was possible by introspection alone, further knowledge would require inference from observations of what we assume to be the effects of the mind, namely behavior. The concept of the mind as a process in the brain substance is a modern achievement: pre-neuroscientific man localized the mind in the brain ventricles or outside the brain in the liver, heart, or blood (Fig. 1.7). This shows the limits of introspection for gaining knowledge of mental events. In addition, one cannot obtain knowledge of neural events and physical processes in the brain by means of introspection.

If one does not have intuitive knowledge even of one's own mind, then I believe that knowledge of one's own mind is inferred from one's experience just as we infer that other people have minds. If introspection is nothing more than inference from sensation and perception, people denied some modality of experience of the world will show a corresponding lack of introspective knowledge. That is supported by the evidence showing that people who are born blind cannot have knowledge that is normally gained exclusively through visual experience—knowledge of colors and of visual space (Senden, 1960). The perception of space that blind people obtain by touch and kinesthetic sensation is quite different from space perception obtained by vision.

A congenitally blind person cannot have a brain state corresponding with the color red or with visual space as perceived by people with normal color vision and stereoscopic vision. An idea of anything is dependent on the sense impression of it, and that is different from the explanation of it. As Hume pointed out, *"We cannot form to ourselves a just idea of the taste of a pine-apple, without having actually tasted it"* (*A Treatise of Human Nature*). Evidence in support of this statement was obtained by Senden (1960) in his studies of congenitally blind people before and after restoration of vision. He showed that the perception of space derived from somatic sensation is not the same as the perception of space derived from binocular vision. The data of different modalities of sensation are dealt with by independent parallel processes in the nervous system, and we do not have the ability to mingle different sensory modalities.

We cannot add together olfactory sensations with sensations of color. It is because of this intrinsic inability to mix different sensations that some cross-categorical identifications and statements that also cross sensory modalities like "the color blue smells like a rose" are meaningless. For the same reason somatosensory perception of form, of a key for example, is different from visual perception of a key, and the statement that "the thing I touch looks like a key" is meaningless. A blind person who has learned to identify solid objects by handling them, and later has sight restored, still has to learn to recognize those objects by sight.

Senden concluded that a blind person has no visual spatial concepts at all, and that on recovery, visual space is at first incomprehensible. That conclusion is consistent with recent evidence, not available to him, that brain functions and structures fail to develop normally or may atrophy if they are not used during critical periods of development (reviewed in Jacobson, 1991, pp. 530–536). Therefore, failure of the congenitally blind to develop concepts of color and of space may occur because of failure to develop the neural organization required for visual perception of space, or because of disuse atrophy of that neural organization.

The advance of the reductionist neuroscience program can be illustrated by the changing status of introspection as a valid means of gaining access to cognitive states. Until the beginning of the 20th century cognitive states were believed to be accessible to no one but the experiencing individual, and could be studied only by introspection. Such mentalistic psychology was virtually replaced by behaviorism which held that psychology is not the study of mental events and cognitive states but of behavior and dispositions to behavior. Behaviorism aims at linking all psychological activities to experimental observations of patterns of behavior. The actions of the nervous system are to be inferred from behavior, precisely and objectively recorded by instruments of which the Skinner box is the classic example. *"One of the perhaps unforeseen consequences was that it now became feasible to generate large amounts of data with relatively little effort. Witness Skinner's last full-scale piece of empirical work, 'Schedules of Reinforcement' (1957)—700 large pages of experimental results, with many interesting nuggets and some acute observations, but mind-boggling in the ratio of undigested data to serious theoretical analysis"* (Mackintosh, 1990).

Experimental behaviorists threaten to sink themselves in seas of raw data. Few would now regard Skinner's work as an example of a successful reductionist strategy. It failed to achieve even its proximate goal which was to deduce the operational principles of the nervous system by controlling the input and recording every detail of the output. It became locked into a single functional level whose boundaries were demarcated with respect to the structure of the experimental paradigm rather than with respect to the structure of the nervous system. Therefore, Skinner was unable to advance the reduction to cellular and molecular levels. This example is given mainly to illustrate two of the difficulties of the reductionist methodology: how to carry the reduction through different neural levels, and how to deal with very large amounts of data, much of them incomprehensible.

These problems were not anticipated by philosophers such as Gilbert Ryle (1949) and Ludwig Wittgenstein (1953) when they argued that statements about cognition can be reduced into statements about behavior or a disposition to behave. In that case the only way to investigate states of mind is by complete description of the corresponding patterns of behavior. But that eliminates all the neurophysiological levels below overt behavior, and ignores the questions we have raised about branching patterns of relations between levels, and the importance of redundancy, the difficulty of defining initial conditions in the nervous system, and so on. The problems about neurophysiological levels have been misunderstood or ignored by philosophers such as Paul K. Feyerabend (1963a,b), Herbert Feigl (1958), J. J. C. Smart (1963), and Paul Churchland (1988), who have taken the next obvious step of identifying states of the mind with states of the brain. They beg the question of how M-N reduction via identities could be conceived, given the questions we have raised. A little later I shall show that mental-to-neural reduction via identities is not feasible, even in principle.

Philosophers, unlike neuroscientists, can choose to ignore the problem of the vast mass of data that it would be necessary to collect and analyze in order to show which brain states were identical with a mental state, which were causally connected, and which were only correlated or were not related. Assuming that all brain states (e.g., electrical, biochemical, metabolic) could be observed at the same time as a mental state such as happiness, the variability of the mental state and of all the physical parameters may make it impossible to identify the brain state that was uniquely related with the mental state. For the present, the program aimed at establishing identities of brain and mental states is a game philosophers play rather than a feasible research program for neuroscientists. There is something about research on the brain, especially on the mind, that both inspires and unhinges people. It inspires them to try to understand how the brain works and how it is related to the mind. It unhinges them to claim more understanding than they now have or are ever likely to attain.

The debate remains undecided on the matter of the validity of knowledge gained from introspection—the dualist accepts it as the most authentic knowledge, the logical behaviorist and the materialist reject it as a linguistic and conceptual error. Why should introspection give us truthful knowledge of the mind when introspection is utterly unable to provide knowledge of any physical brain processes? One of the reasons for this disability seems to be that we lack sensors inside the brain to provide information of the functional states of the brain. Without such a sensor I would know that I was in pain but not whether the central gray, or the hypothalamus, or any other neurons were firing in relation to my conscious perception. But, if I could provide an equivalent of a universal brain sensor to monitor the functional states of my brain continuously, would I be able to obtain objective knowledge of my mind? Would I be able to have a brain state of a brain state? Would I have an objective form of introspection? No, I could only have the equivalent of another sense organ, with all that implies. Even if the information from the brain sensor could be fed back into my brain in some way that would result in perception, I would only have a mental state (therefore subjective) of a brain state, not objective knowledge about my brain states. This leads me to think that objective knowledge is not the inverse of subjective knowledge but includes the latter.

What if there exists a brain state uniquely related to each and every mental state, but the relationship only occurs when permitted or instructed by another brain state (and additional permissive and instructive states could occupy a large part of the brain)? Then how could the philosopher or neuroscientist distinguish between the brain state that is identical to the mental state and the brain states that only permit or instruct, that are only correlated with the mental state? It is trivial to answer that mental states are identified with all the physical states in the brain.

That meanings must be intersubjectively communicable, and that there is no such thing as a private language known to only one person, were held by Wittgenstein and some of his followers. Alternatively, some philosophers, notably functionalists like Frank Jackson (1982) and dualists like John Foster (1991), argue that there are aspects of consciousness that are not communicable but which can be known privately by introspection. The argument that one has a privileged access to self-knowledge—that one is capable of knowing things that one is incapable of saying—can only be sustained by appealing to others who can confirm their inability to communicate aspects of their consciousness. But it is impossible to show that people share the *same* noncommunicable aspect of experience. It is impossible to refute the statement that people are deluded when they claim to have experiences that are incapable of being communicated to others. These limitations reduce the force of any argument that is supported by appeals to the existence of knowledge obtained by pure introspection.

As an example, consider the argument of Frank Jackson (1982) that mental states cannot be reduced to physical states, and that reductive materialism is inadequate, because there is private knowledge that is gained only from introspection which a person is incapable of communicating to others. Jackson argues that a scientist who has been born and raised in a black and white room, and who has been fully instructed in all the facts of color vision, would learn something new about color after visual experience of the normal world. From this difference he concludes that there is a private domain of knowledge entirely different from public knowledge. But the significant difference, I think, resides in the different input channels and nervous processes that are involved in direct experience of a thing compared with a thing learned indirectly. The fact that mental experiences are different when obtained through different sensory modalities has been demonstrated by psychophysical experiments. Obviously Mary could learn something from her own experiences that was different from anything she could learn from statements of the facts of physics and neuroscience, even if those facts were completely known. This difference occurs because visual sensation is different from auditory or tactile or any other modality of sensation. The differences between those sensory modalities can also be reduced to physical processes, and Mary may understand those differences fully, but nevertheless the understanding would not eliminate the differences. The argument fails to refute the identity theory. Similarly, the argument of Foster (1991) that a congenitally blind person could never obtain complete knowledge of vision from a complete neurophysiological description of vision is correct but it also fails to refute the theory of identity of mental states and neural states. The identity theory may be impossible to prove or disprove for reasons that go much deeper than the philosophical arguments, namely an uncertainty principle that I propose in the following section. It may turn out that reductionist methods may ultimately fail to give a complete physical explanation of visual experience, but that failure cannot be demonstrated in advance by philosophical arguments, only by further research in visual psychophysics and neurophysiology.

2.10. CAN IDENTITY OF MENTAL AND NEURAL EVENTS BE DEMONSTRATED?

Several methods for showing relations between structure and function, and relations between different events in the brain, are in current use in neuroscience: electrical stimulation of neurons, recording activities of neurons (and perhaps of associated glial cells), and removal or destruction of neurons (and of associated glial cells). Other methods are elimination, or blockage, or substitution of molecules that are known to participate in neural activities: for example, pharmacological blockage of release or uptake of synaptic transmitters, or introduction by transgenic technology of an enzyme in the neuron that alters its function in a known manner, knockout of a gene that is known to be involved in brain functions, and so forth. These methods can show that there are functional relations between mind and brain, but they pose difficult practical problems of establishing whether relations between neural events and mental events are identities, causal connections, or correlations.

Those tests of the relations between mental and neural events all rest on the assumption that the mind is a program of neural computations in neural networks. But what if the mind is really a separate physical process, not in the form of nerve action potentials or synaptic activities (although interacting with them) but in some other physical form that is distributed in the brain outside the recognized nerve pathways and connections? When we talk of neural events we think also of glial events, or of events in neuron-glial complexes. There is a correlation between the mental capabilities of a species and the ratio of glial to neuronal cells in its brain (see Jacobson, 1991, p. 82). What if the neurons were only the links between extraneuronal events and the inputs and outputs to memory circuits, sense organs, and effectors? If those conjectures receive empirical support, the current model of mental events as nothing more than computations in neuronal circuitry will have to be revised radically. And, of course, experiments to discover relations of mental events to neural events will have to be founded on rather different assumptions to those now in vogue. To prove identity, both the mental events and the neural events must be observable. But the mental events are private and can only be inferred, never observed directly, while observation of all the relevant neural events is beyond the powers of the present methodology, and may always remain so.

We can now consider how, given the assumptions and objections to them stated above, neuroscientists might go about obtaining evidence to show whether mental and neural events are identical. Would it be sufficient to show that a specific pattern of sensory input was always related to a specific pattern of cerebral cortical activity? The answer is no. This notion is based on the misconception that all or most cerebral cortical activity is driven by sensory input. In fact, it is estimated that about 1% of the input to the cortex is extracortical (Braitenberg, 1978; Eggermont, 1990). The cortex works to a large extent on its own output.

Would it be enough to show that destruction of certain neurons, or of their interconnections, abolishes perception, consciousness, memory, and so on? The answer is no, because it is necessary to show that the lesion did not merely interrupt a link in the perceptual process (e.g., as a lesion in the striate cortex causes a visual scotoma).

What about showing that stimulation of certain populations of neurons, or even of a single neuron, produces certain perceptions in the absence of the sensory stimuli that normally result in those perceptions? For example, the problem of the relations between perceptual processes and neural processes could be investigated by stimulating the brain directly by noninvasive methods, e.g., by magnetic field pulses (Belliveau *et al.*, 1991; Mills, 1991). Such stimuli would appear to make it possible to apply patterned stimulation to evoke complex perceptual processes, behavior patterns, emotions, thoughts, and memories. We do not now have the knowledge required to evoke anything of interest. But even if that practical limitation could be surmounted, that demonstration would fail to prove the identity of neural and perceptual activity unless it could also be shown that the neural activity related to a given perception was always confined to a given population of neurons. In practice, however, focal brain stimulation, capable of evoking perception, always causes widespread brain activity. This occurs because the number of synaptic connections between neurons that are involved in a perceptual event far exceeds the number that are active during that event. The result of direct stimulation of neurons may show that they are somehow involved in that perception, but not that they are both necessary and sufficient for that perception, and certainly not that they are identical with that perception.

Effective patterns of stimulation could be designed only if we knew how to localize stimuli to particular populations of nerve cells and if we understood how the stimuli affect those structures. However, the obstacles to that program have become apparent recently with the knowledge that a single neuron can have multiple functional domains, and that small parts of a single neuron can act as more or less independent units (Bullock, 1959; Y. Fujita, 1968; Ross *et al.*, 1990). Furthermore, the redundancy of neurons and of the synaptic connections between them indicates that the same function could be performed by different combinations and permutations of neural events. It seems to me to be probable that the same mental event can be realized in different neural events in different people, and also in the same person at different times. That would not necessarily prohibit the finding of a particular combination of neural and mental events that were identical but the probability of finding that combination would be much reduced.

If a perception is evoked by a stimulus aimed at a single neuron or a cluster of neurons and glial cells, it does not necessarily mean that those cells cause the perception, or that they may code for and represent the perception in a single locus where they are completely identified with the perception. A version of such a view sees one neuron or a local cluster coding for an element of perception, many of which can be related to each other by logical operations to form a higher-level percept like a face or grandmother's umbrella. Each neuron or cluster may be necessary but not sufficient. They may participate in a distributed coding or representation, and they may permit other events in neuronal circuits and in brain space as a whole, that are identical with the perception—they may be the trigger, not the bullet. If direct focal stimulation of a part of the brain invariably resulted in a perception of a moving object, the result would not prove that the perception was identical to the neural activity, as some people have claimed (e.g., Barlow, 1972). That demonstration falls short of establishing the identity of perceptual and neural processes. It only shows that they are related in some way. Relations can be scalar, having a certain weight or numerical value, and vectorial, having numerical order showing direction, in brain space-time. The relations can be represented by coordinates, namely by a network or a matrix. Here I wish only to emphasize that the relations may be widely distributed. Details can be obtained from many recent works on distributed neural processing (Hornick *et al.*, 1989; P. M. Churchland, 1989; Bechtel and Abrahamsen, 1991; P. S. Churchland and Sejnowski, 1992).

Now, if one agrees with Hume that cause and effect are no more than habitual coincidence, one can also claim that the neural events cause the perceptions, but not that they are identical. To show that they are identical one would have to show that no other events preceded, coincided with, and followed the neural events that did not also precede, coincide with, and follow the perceptual events. Anything short of that demonstration could show only that neural and perceptual events coincide at some stage of the process. In any case, to show that a neural event was the true cause of a perceptual event would be very difficult, perhaps impossible, if a true cause is defined as an event for which no prior event was necessary.

It should be said that this uncertainty enters only with proving identity, not with proving correlations or causal connections. Of course, it has already been found possible to show correlations and causal connections between mental states and physical states in the brain—electrical activities of single neurons in the cerebral cortex can be correlated with perception; electrical stimulation of the brain can be correlated with perception in awake human subjects; electroencephalographic patterns can be correlated accurately with sleep, alertness, and so forth. Pharmacological and biochemical state can be causally connected with mental states. But such successful demonstrations of correlations and of causal connections do not prove identity.

It may be possible for an observer to record all the activity in my brain and to recognize from the patterns of physically observable neural events that I am thinking about a mathematical problem or about my next vacation. However, the observer will only be recognizing neural events, not my thoughts and intentions. Although the observed neural events may in some ways be correlated with my mental events or even causally connected, they are not identical with them. The strong artificial intelligence theory is that if a computer of sufficient power could be hooked up to the human brain and could monitor all the activity there, the computer could eventually learn everything the brain can do and would do it better (Moravec, 1988). But such an experiment has yet to be tried, even on a mouse. I find no reason to believe that the computer could be programmed so that it could distinguish the neural events that are relevant to particular mental events from neural events that are not relevant to those mental events but may be relevant to others or to none. Assuming that every physical event in the brain could be monitored in the necessary temporospatial detail, they could only be correlated with overt behavior, not with intentions, or with thinking that was not followed by behavior. Even if a computer is hooked up to a human subject willing to report his intentions and thoughts, the subject's verbal behavior is not the same as his related mental events. If, as I have argued in Chapter 1, we form mental representations of outer reality, then a computer hooked up to a human brain could, at its very best, only form a representation of a representation. That may be the only view that we can have of the frontier separating our minds from objective reality—an infinite regress of representations of representations.

At the beginning of this work I raised the problem of how we can compare our internal representations with external objects, including our own brains, to see if they match. Here I shall outline a method for comparing the internal and external representations, and for comparing two internal representations with one another, to see if they match. I arrive at the conclusion that if they matched we could not know it.

Mind-brain relations can take three possible forms—identity, causal connection, or correlation. An individual mental event can be related to a single neural event or events in only one of those ways at any one time. However, it is conceivable that an individual mental event can be related to a neural event in more than one way successively, but not simultaneously. Also, individual mental events may be related to different neural events in more than one of those ways. In addition, the relations may appear to be different depending on how we observe them, and our observation may change the M-N relations. For the purposes of this argument, and to keep the argument simple, let us assume that neural events are always related to mental events in one and only one way, either via identity, or connection, or correlation. Let us also assume that our methods of observation of mental and neural events do not alter the relations between them. Then, how could we discover how mental events and neural events are related? I shall argue in the following pages that we could never be certain of the kind of M-N relationship even if we were able to record all mental and all neural events simultaneously and continuously.

Even if I were the observer of my own brain processes I believe that I would not recognize that the objective observations of my neural events were identical with my subjective mental states. I would recognize that the neural and mental states coincided or were correlated in other ways, but I could not experience them as identical. That nonidentity could arise in two ways: (1) if the mental and neural events are not identical in reality, and (2) even if they are really identical they would seem to be nonidentical because the subjective experience and the objective observations of those experiences would necessarily involve noncongruent sets of events in my brain. If probes in my brain were able to record mental events (or the neural events identical to the mental events) and if the record derived in that way could be returned to my brain in another way (by sensory stimulation or by direct brain stimulation), I predict that I would always recognize the record to be in some way different from the mental events from which it was derived. Therefore, I believe that subjective mental states will elude such attempts to identify them with neural states.

We have already discussed some of the difficulties, if not the impossibility, of intersubjective communication of mental processes, states and events. For this argument let us assume that I am able to report my mental events and that my neural events could also be observed and recorded with instruments attached to my brain. Let us assume that all my neural events could be recorded accurately. Let us assume that the record of my neural events could be returned to my brain directly or represented by a computer for others to observe.

Given the complete records of neural events, a computer can be programmed to determine how certain neural events are related to my reports of certain mental events— a certain mental event can appear with a certain probability with respect to a certain neural event. Such habitual concomitance may mean that mental events cause neural events or vice versa or that they are correlated but not causally related. By that method we could not prove that any neural event was identical with any mental event.

Now let us imagine what might occur if the record of neural events from my brain, related to certain mental events, could be reentered in my brain. We shall call the original mental events and neural events primary, and the reentered neural events and the mental events that they evoke, or with which they are associated or are identical in my brain, we shall call secondary. We assume that when the recorded neural events are reentered in my brain I experience mental events. The problem is whether I experience those secondary mental events as identical with, or causally connected with, or correlated with the primary mental events. This is when the uncertainty enters: if I perceive the primary and secondary events as one and the same I cannot assert identity because I cannot be certain that I am only perceiving one or the other, not both. I could not know if the reentered stimuli had failed to produce secondary mental events. In other words, I could only recognize when the primary mental events were different from the secondary mental events, but could not know if they are the same. Programming a delay between the primary recording of neural events and their reentry would not circumvent that difficulty. Even if the primary and secondary mental events could be temporally separated from one another they could not be analytically distinguishable. I mean that the subject could not then know whether they were both the same or causally connected, or only correlated.

This experiment avoids the difficulties associated with naming or describing mental events. The difficulty arises because different people may ascribe different sensory qualities to the same physical events in the nervous system. For example, one could not prove that the subjective sensations of color experienced by different people were the same or whether they were based on an "inverted spectrum," as Sidney Shoemaker (1984) has argued. In the proposed experiment that problem does not arise. The subject's task is to compare two events, a primary mental event with a secondary mental event, without requiring that the events be given names or be described.

This experiment also avoids the problem of viewing the mind-brain dichotomy as a dichotomy between the observer and the observed. The proposal is for the same subject to be both the observer and the observed. That seems to me to be the most favorable arrangement to see if mental and neural events are the same, but the result appears to me to mean that the subject would not be able to tell whether his mental and neural events were identical. Observation of mental states gives misleading information about the physical activities of the brain, whereas physical measurements of brain activities give misleading information about the mind, and physical measurements may perturb consciousness and may even cause the collapse of consciousness just like the quantum wave function collapses into a particle when it is observed. In other words, consciousness is a purely theoretical, probabilistic function that includes all the brain's possible states—until it is observed. Then consciousness collapses into electrochemical events with which we are all familiar. In this view, consciousness is a probabilistic function that has the same relation to any discrete physical event in the brain as weather prediction has to rain at a particular time and place.

A complementarity of mind-brain was suggested by Bohr (1958) by analogy with his complementarity theory in physics which regards atoms and smaller micro-entities as having dual wave-particle natures. This is related to Heisenberg's uncertainty principle—as particles the entities have precise positions but not precise momentum, whereas as waves they have precise momentum but not precise position. Bohr suggested that the mind may have precise properties such as free will but not precise mechanistic-reductionist properties, whereas the brain has precise mechanistic-reductionist properties but not precise free-will properties. If so, observation of the properties of mind would preclude simultaneous observation of the properties of brain. This is consistent with maintaining that there is a radical bifurcation and discontinuity between the data of sense perception and the objects in the physical world with which those data are correlated or connected. This position of epistemological dualism differs from psychophysical dualism which asserts the radical discontinuity of the mental and physical domains in the brain.

2.11. PSYCHOPHYSICAL DUALISM: ITS PLACE IN NEUROSCIENCE

It is necessary to make distinctions between two kinds of dualism (and the two kinds of monism that oppose them): namely epistemological dualism and psychophysical dualism, as Lovejoy (1960) has pointed out. Epistemological dualism maintains that there is a discontinuity between the data of sensory perception and the objects of the physical world from which those perceptions originate. This bifurcation arises both from uncertainties inherent in making observations, and from inherent limitations of the cognitive capabilities of the human nervous system. Those limitations make it impossible to prove the identity of any perceptual data with any objects of the physical world, although it may be possible to find correlations and even causal connections between them. Psychophysical dualism goes even further in denying any identities of mental events with physical events in the brain.

Epistemological monism recognizes no radical break between the subjective domain (including perceptual and cognitive) and the objective physical domain. Psychophysical monism maintains that the mental domain is the same in all respects as the physical domain in the nervous system. Most neuroscientists are both epistemological and psychophysical monists, although it is logically possible to affirm either one without the other.

There is a strong tendency for neuroscientists to dismiss dualism as a quaint relic, at best, and as unscientific nonsense, at worst. This is unfortunate because it eliminates dualism from the dialogue about how mental events are related to physical events in the brain. My position is that neither dualism nor monism is subject to empirical refutation or to final verification. My skepticism comes from a conviction that an uncertainty principle thwarts all attempts to demonstrate the relations between mental events and neural events, and prohibits any conclusive, or even persuasive demonstration of the identity of neural with mental events.

I do not need to discuss dualism as it is implied by the religious belief in an immaterial and immortal soul. An enormous literature, from ancient times to the present, has been devoted to arguing this theme from every conceivable position (Arnett, 1904; Révész, 1917; Bailey, 1959; Swinburne, 1986). The modern theory of psychophysical dualism begins with René Descartes. For Descartes the only indubitable, authentic knowledge was that the knower thinks that he exists—*cogito ergo sum* (I think, therefore I am). In reply to one of his critics Descartes wrote, "*One thing is certain: I know myself as a thought, and I positively do not know myself as a brain.*" This is borne out by the inability of prescientific man to locate the mind in the brain. Instead, it was located in the heart, the blood, the liver, or the diaphragm or some other part of the body. Experimental demonstrations that consciousness, memory, thought, will, and emotion are functions of the brain are among the principal achievements of neuroscience. Descartes conceived of the mind as rational by its very nature, and as a nonmaterial force capable of causing effects in the physical substance of the body.

The Cartesian definition of "self" is the realization of brain functions that are sufficient for awareness of the existence of one's own mind, from which the existence of other minds can be inferred. The Cartesian "I" is as indivisible and as separate from other individual minds as Newtonian atoms. Like billiard balls, minds may interact but never exchange or merge. The Cartesian theory of the indivisibility of the mind stood against opposing theories of some kind of subdivision of the mind, like the Freudian theory of levels of mind, or the Jacksonian theory of a functional mental hierarchy because those theories have gained little or no empirical support. The concept of separate minds has also been upheld against all the spurious claims to have demonstrated mental telepathy.

Recently, the evidence of the split brain experiments (Sperry, 1968, 1969, 1970a,b, 1973, 1982) has been said to have overturned the notion that the mind is irreducibly holistic. The reductive cut was made in the corpus callosum, thus largely disconnecting the two cerebral hemispheres from one another. As a result, functions were abolished that required cooperation of both hemispheres, but each hemisphere could function independently (Fig. 2.2). The surprising observation was not only that functions of the two hemispheres are different (that has been long established) but that the functions that are regarded as "consciousness" are predominantly exercised by the left cerebral hemisphere in right-handed subjects (the dominant or major hemisphere), whereas the subject is not conscious of activities exercised by the other (subdominant or minor) hemisphere. Note that the terminology is explicitly reductive, based on the assumption that the major is at a higher level than the minor hemisphere. An equally reasonable alternative assumption is that the two hemispheres are not different levels in a hierarchy, but that they are functionally specialized to perform different roles in a unified process.

The split-brain patients showed more clearly than before that memory does not require consciousness. This had long been inferred from clinical observations, as had the converse, that consciousness does not require memory. The split-brain subjects experienced consciousness by use of the dominant cerebral hemisphere. By contrast the use of the minor cerebral hemisphere did not enable the subject to have conscious experience although it did subserve memory, nonverbal intelligence, and concepts of spatial relations. Therefore, if it is true that the dominant hemisphere is sufficient for consciousness, then the Cartesian theory of the unity of conscious experience and of the relation of the mind to a localized part of the brain, has been upheld. Unity and continuity of consciousness are preserved in the split-brain subject. The subject is not aware of any change in his own consciousness, such as a reduction or a separation into two

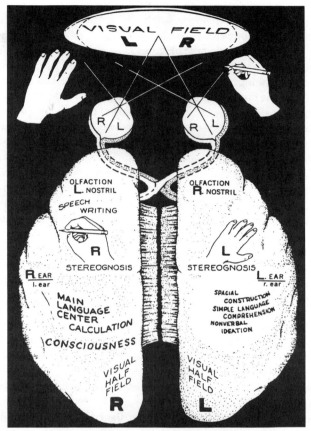

FIGURE 2.2. Cutting the midline commissures connecting left and right cerebral hemispheres results in loss of functions that require cooperation of both hemispheres. Functions are saved that each hemisphere can perform independently, including "consciousness" subserved by the dominant, left hemisphere. (Modified from Sperry, 1970b.)

consciousnesses. Therefore, the results do not overturn the Cartesian notion that one knows oneself as a unique indivisible mind, not as a unique brain. The results show that consciousness can be uncoupled from memory and from some aspects of cognition. The two hemispheres are prevented from coordinating their activities, but consciousness is not divided by cutting the corpus callosum. Nor does the operation result in a discontinuity of the subject's consciousness so that the preoperative state of consciousness is perceived to be discontinuous from the postoperative state of consciousness. Similarly, in a person suffering from a multiple personality disorder, the person's consciousness is not duplicated. That person believes that she is a different person at different times, but the person's consciousness is not fragmented but remains continuously unified. The person admits that "yesterday I was Mary and today I am Joan," but the personalities that may be expressed at different times are all related to a unique and indivisible consciousness. Similarly, the split-brain subject has only one consciousness, even after learning that there has been a loss of perception and of memory of behavior mediated by the functions of the minor cerebral hemisphere.

Descartes conceived of all animals, except for man, as pure automata. It is not necessary to conceive of a Cartesian mind in the brain of the bee switching the input signals to the appropriate output channels. Nor do we have to think of the evolution of the bee's behavior, its dance, for example, as an evolution of the mind of the bee. When a bee performs a dance which communicates to other bees that there is food at a certain distance and direction from the hive, it does not know or understand that its behavior is purposeful. Similarly, a parrot that can utter meaningful sentences can never understand their meaning. But a person is conscious of purpose of many forms of behavior, and can design behavior to achieve preconceived goals. It can be argued that motivation and intention (and therefore choice of goals which implies an ability to make value judgments) are some of the necessary abilities that distinguish organisms that have mind from those that lack mind (e.g., Dennett, 1987).

The distinction between reducible brain processes and irreducible mind is one of the main roots of Cartesian dualism. According to that theory, mechanistic reductionist methods are necessary and probably sufficient for arriving at an understanding of *how* but not of *why* human conduct is related to brain processes. The question "why?" has been dismissed from neuroscience as unnecessary teleology and as mischievous metaphysics. I argue, in this and the final chapter, that a neuroscience research program is incomplete unless it addresses the question "why?" Why people want different things, why we select different goals, and why we consider some things true and others false, some good and others bad, are fundamental problems of neurobiology. Whether neuroscience is ever likely to be equipped to answer such questions is another problem that we shall address in the final chapter.

All theories of psychophysical dualism, to the effect that the mind can control matter, have to violate some of the laws of classical physics. Descartes's theory of the action of the soul on the animal spirits in the brain had to be modified, or some say abandoned for that reason, and so do modern theories of the same sort, for example, the theory of Sherrington (1941) and of Eccles (1953, 1974, 1976, 1989). Dualists cannot explain how a mind or a soul is nonspatial yet it can occupy a space in the brain, and how a nonmaterial mind can interact with matter.

Dualist theory has had to be modified by ad hoc hypotheses as the opposing scientific evidence accumulated. It is worth considering this process of modification of the dualist theory because it has been interpreted differently by protagonists and antagonists of dualism. Antagonists say that as the mind is reduced to molecular mechanisms, dualism will be refuted, if it has not already been refuted. Dualists reply that one of the signs of strength of their theory is its capacity for constructive modifications, and that dualism is entirely consistent with what is know about mental process (J. Foster, 1991). Dualists do not deny that the brain operates by physical processes, nor that very complex brain processes can be reduced to simpler physical processes, but they hold that the mind is not a physical process as we now understand physics. Arguments against dualism are trapped in the paradox that if dualism is true then all the things we know about the brain as a physical system are known only because the brain is in liaison with the mind, and if dualism is false, we shall not be able to refute it until the mind can be explained fully in physical terms.

There are a number of philosophical arguments for dualism. The evidence of introspection is the most frequently given: dualists admit that introspection cannot lead to knowledge of such things as the atomic constitution of matter and of the brain itself, but the argument goes on to claim that only the mind can have direct knowledge of such things as pleasure, pain, happiness, and emotions generally, and of sensory qualia. The empiricist doctrine that *"whatever is in the intellect must previously have existed in the senses,"* attributed to Aristotle and adopted by Adrian as the motto of *The Basis of Sensation* (1928), is explicitly denied by Descartes (*Meditation* II, p. 21, and *Meditation* V, p. 44) but upheld by Gassendi in his fifth set of *Objections to Descartes' Meditations*. The modern argument against dualism goes something like this: the color red can be reduced to photons of a certain quantum energy reflected from objects and hitting the red cones in the retina, resulting in biophysical and biochemical changes that have been identified and can be specified as a causal chain from the retina to various neurons in the cerebral cortex. Similarly, all sensation, including those which originate from inside the body, can be reduced completely to physical events that can be measured without leaving a qualitative residue. If we really understood all that goes on in the brain we would not need to refer to sensory qualia. The dualist replies that even if all the physical events were to be known, they would not account for the qualities of the mental events because the mind is outside the domain of physics.

Some of the arguments that have been used against dualism are naive—that computers and machines with artificial intelligence can outperform any person at arithmetical calculation is not a good argument. Dualists counter that argument with the objection that machines and computers are products of human minds and could not perform functions for which they had not been designed—they have no creative powers of their own. In the parlance of experts on artificial intelligence, computers have "derived intent" while only humans have "original intent." Descartes in his *Discourse on the Method* (Part 5, pp. 56–57) was the first to criticize the notion that automatons have a mind. His first objection was that machines may be able to utter words but could not create meaningful sentences in answer to any question. *"Secondly, even though such machines might do some things as well as we do them, or perhaps even better, they would inevitably fail in others, which would reveal that they were acting not through understanding but only from the disposition of their organs. For whereas reason is a universal instrument which can be used in all kinds of situations, these organs need some particular disposition for each particular action. . . ."* In other words Descartes implied that a machine that could think could never be made. Turing (1950) proposed a test of whether machines could think but he did not arrive at an answer. Turing's test was to put the same questions, whose solutions require thought, to a human and an intelligent machine, and if an observer could not distinguish between the behavior of the two, then there were no grounds for denying that the machine can think like a human. As is now well known, multipurpose machines have been designed that can perform calculations or individual tasks more efficiently than humans but machines are unable to perform the simplest of tasks, like writing this sentence or doing the grocery shopping, without having received detailed instructions from a human operator.

Making physical models of neural activities is not required by reductionism but is a useful adjunct of reductionist methodology, as a collateral method of testing theories—not on the nervous system itself but on a simulacrum. For some years now it has been possible to make electronic simulacra or analogues of individual neurons and to link them to form analogues of neuronal circuits and complex networks. Those electronic analogues are models of theories of brain functions, as we have seen in Section 1.2. But they are subject to all of Descartes's criticisms of automata. They mimic what is already known from direct observations of nervous systems. For example, they can mimic the sodium and potassium currents and behave in conformity with Hodgkin-Huxley equations, or they can mimic neuronal activities such as long-term potentiation (Mead, 1989; Andreou, 1991; Mahowald and Douglas, 1991).

Such neuromimes tell us much more about how machines may be able to mimic brains than how brains function like machines. In other words neuromimes and neurons are not really functionally isomorphic in practice because the functions of one cannot be completely translated into the functions of the other. The feasibility of designing a mechanism that is completely functionally isomorphic with mental processes such as conscious feeling, thinking, and willing is seriously compromised by our inability to define mental functions completely, and may ultimately be defeated by an uncertainty principle. In Section 2.13 I shall explain what I mean by such an uncertainty principle.

Difficulties with localization of the mind in the brain beset Descartes and continue to inconvenience his successors (see Riese, 1958; Jefferson, 1949, Riese and Hoff, 1950). Descartes selected the pineal gland as the seat of the soul and the mind (Descartes says, "*I consider the mind not a part of the soul but as the thinking soul in its entirety.*" Reply to Gassedi's objection 356). He chose the pineal because it satisfied his prerequirements for a site at which impressions from paired sense organs could be united and because he could conceive of movements of the pineal on its stalk regulating the passage of animal spirits into different nerve endings.

Note that the tendency to locate the soul strategically interposed between input and output has persisted since Descartes's original concept of the soul directing the traffic of nervous activity. The early 19th century concept of separation of nerve globules and fibers was fully consistent with the Cartesian model (Fig. 3.4). However, the latter was flatly contradicted by models in which the input nerve fibers and nerve cells were in direct cytoplasmic continuity with the output cells and fibers (Fig. 3.9). The soul or mind was later to find another place (its final refuge?) in the synapses (Eccles, 1989). Very soon after the synapses were defined they were regarded "*as the seats of the psychophysical processes*" (McDougall, 1901). This quaint notion has persisted in the numerous writings of J. C. Eccles on the mind-brain relation. Most recently he has argued "*that mind-brain interaction is analogous to a probability field of quantum mechanics, which has neither mass nor energy yet can cause effective action at microsites*" (Eccles, 1989). He believes that these sites are the points of release of synaptic vesicles into the synaptic cleft. It is remarkable that occult forces are usually conceived in terms of action between bodies at a distance—a distance as narrow as the synaptic cleft is apparently regarded by some as sufficient for the action of a nonmaterial force.

Descartes conjectured that the pineal is the "*seat of the imagination and sensorium commune*" as well as the main seat of the soul. This was necessary in his scheme in which the soul was in command of the nerve endings, regulating the passage of animal spirits into them from the ventricles. The first difficulty met by Descartes was that the soul has a localized seat and yet it is in the whole brain or even the whole body. Gassendi raised the objection that however small the seat of the soul, it nevertheless occupies space, so how could the soul be in the whole body and yet be localized at one place? Descartes's answer to Gassendi's objection was that the soul extends to all parts of the body but operates only through the pineal—this is an essential point of functional interaction where the soul exercises its functions "immediately" or "more particularly."

Gassendi addressed Descartes playfully in the vocative case—"O Mind," and Descartes replied—"O Flesh," as an indication of their fundamentally opposed positions—what we might now term mentalistic and materialistic. Although Descartes ends his response with the usual pleasantries, declaring himself the winner, it is clear that Gassendi's main objections stand. Indeed, Kant raised anew Gassendi's objection that the soul cannot be diffused throughout the body and simultaneously lack extension.

Descartes accepted the first law of motion according to which a body left to itself will move with a constant velocity in a given direction. He also accepted the law of conservation of motion according to which the total quantity of motion in the universe is conserved. His theory that the soul, located in the pineal, alters the direction but not the quantity of motion of animal spirits, was in accord with those physical laws. However, his theory had to be abandoned in its original form when it was found to violate the law of conservation of momentum. According to that law the total quantity of motion in any given direction is constant. To save Descartes's theory one of his followers, Arnold Geulincx (1624–1669), invented the "two-clock" analogy. According to that analogy the relation of the mind to the body is like that of two clocks that both keep perfect time so that when the hands of one clock point to the hour the other clock strikes, making it appear as if one had caused the other to strike. He asserted that the mind and the body are both set going synchronously by God and the mind does not move the body nor the body the mind. This is rather like the preestablished harmony between body and mind as conceived by Leibniz (see footnote on p. 102). The two-clock analogy makes both mind and matter strictly deterministic, and makes man an automaton with a mind but without free will. It was only a short step to LaMettrie asserting in 1747 that man is a machine without mind.

A theory that mind is able to influence neurons because of the unstable equilibrium of some chemical process was considered by Bertrand Russell (1948, p. 41). The objection to this theory is that it fails to explain how mind could select the precise times and places at which to influence chemical equilibria in myriads of neurons and then continue the control of the activities of neurons that are necessary to complete any voluntary action. Alternatively the completion of a voluntary action may be automatic—the mind only being required to control a small set of "command neurons" from which the physical effects of voluntary actions begin.

Eccles (1953, 1973, 1974, 1976, 1989) has developed a dualist neurophysiological theory of the action of mind on the brain. According to that theory, the mind first determines the instantaneous states of millions of neurons in the cerebral cortex many of which are at a state just below the threshold for excitation. Then the mind exerts a minimal additional excitation sufficient to fire a subpopulation of neurons that perform the functions determined by the mind. Eccles conjectured that the influence of the mind on matter could be exercised within the range of indeterminacy of quantum theory or by changing the behavior of elementary particles. This theory is untenable, as Rosenblueth (1970) has shown, if only because the energy required to fire a neuron is on the order of 10^{17} quanta. Even if the neuron is close to threshold, the quantum energy required to fire the neuron would be outside the range of quantum indeterminacy. The macromolecules and organelles, like synaptic vesicles, that are involved in firing neurons, are relatively large bodies which behave more like footballs than like subatomic particles. One cannot conceive of them swerving from their paths or changing their forms and functions at the command of the mind. Eddington (1935, 1939, pp. 182–183) considered the possibility of the mind influencing neurons within the range of quantum indeterminacy. In 1935 he considered that it ought to be possible but in 1939 he rejected the possibility because he could see no grounds for violating the indeterminacy principle— *"there is no half-way house between random and correlated behaviour. Either the behavior is wholly a matter of chance, in which case the precise behaviour within the Heisenberg limits of uncertainty depends on a chance and not on volition. Or it is not wholly a matter of chance, in which case the Heisenberg limits, which are calculated on the assumption of non-correlation, are irrelevant."* Schrödinger (1951, pp. 61, 67) also arrived at the conclusion *"that quantum physics has nothing to do with the free-will problem."* This problem is considered at the end of Section 2.10 and again in Section 2.13 in relation to causality in neuroscience.

2.12. OPPOSITION TO REDUCTIONISM

There has always been a significant minority who oppose the reductionist program. In the 19th century the dissenting minority were dualists who affirmed that the brain is a physical system which is completely reducible, but denied that mind is reducible. Vitalists opposed materialism of any sort, especially materialist neurobiology (Lange, 1877–1892; Gregory, 1977). The dissent was particularly strong against neuroreductionism because consciousness, the soul, and the artistic and ethical faculties of the mind appeared to be threatened by reduction to physical mechanisms. In the 20th century the dissent has taken the forms of neovitalism, organicism and holism (Driesch, 1908, 1914; E. S. Russell, 1946; Koestler and Smythies, 1969). Those schools of dissent from materialistic reductionism have been relatively ineffectual by comparison with the general success of reductionism. Nevertheless, the recalcitrant minority, even if it is ultimately refuted, has had constructive effects—by prolonging the period of dialogue and critique and by delaying consensus. Thus, the reductionist program continues to be enriched by fresh concepts that might not have been admitted into the program if it had been taken over prematurely by extreme reductionist materialists.

The main objection against reductionism, that analysis of the organism down to its individual parts leaves out some important unifying principle, stimulated efforts to try to explain how the emergence of the whole is different from the sum of its parts. These objections to what is often called "crude" reductionism to distinguish it from "refined" alternatives, have taken many different approaches: the holism and organicism of philosophers such as Bergson and Whitehead; the vitalism of biologists such as Hans Driesch; the Gestalten of psychologists such as Wolfgang Kohler, the general systems theory of Ludwig von Bertalanffy.

The claim that the progress of biology has resulted from the use of teleological explanation, not materialistic reductionism, has been made by E. S. Russell in his book *Form and Function* (1916). Russell gives as his prime example the progress of embryology in the late 19th century and first decades of the 20th. But it is now quite obvious that the teleological explanations of embryological mechanisms were erroneous, and that the progress of embryology has coincided with elimination of teleological explanations and with use of reductionist methodology. Russell's claims, that teleological thinking and antireductionism have been associated with some of the greatest advances in biology, have persuaded such historians of science as Michael Polanyi (1968) and Timothy Lenoir (1982). But in my opinion, which I share with Jacques Monod (1971, 1974), Russell's arguments amount to nothing more than special pleading for the privileged status of living systems that render them impervious to reduction into irreducible elements. I do not accept all of Monod's arguments that living systems can be understood completely by complete reduction. I remain agnostic with regard to the possibility of our understanding all the data that will result from such a complete reduction, not because I believe that some emergent properties will be lost, but because I doubt whether human beings have the necessary intellectual equipment to comprehend the data, and because an uncertainty principle may frustrate complete understanding.

Holism is now only admitted in neuroscience research programs on probation until such time as a complete reductionist program can be mounted. This is a reasonable compromise, unequal but fair, considering the great success of reductionism and very limited explanatory power of holism. But it is important not to banish holism completely while the possibility remains that in the future the holistic philosophy may have some important explanatory power. Now that reductionists threaten to sink themselves in oceans of data obtained analytically, there is likely to be a renewed interest in synthesis. A move in that direction is seen in the growing interest in the concept of multineuronal units of functional activity in the cerebral cortex. From the old concept of the functioning of the cortex as a whole, the parcellation into ever smaller functional cortical areas has occurred more recently, until the functions of the cortex were conceived in terms of single neurons with unique operational characteristics. At its most extreme this took the form of the notion of "command neurons," or of the "pontifical neuron." The trend is now being reversed—understanding neural computation in terms of groups of neurons forming functional modules is again in vogue (see Section 1.2). The significance of the integrated operations of large regions, perhaps of the entire brain, are being reestablished by theories of computation in neurons connected in distributed networks. But those theories do not yet take into account physical events which are not action potentials in neuronal circuits and are not events at synaptic connections.

Another source of opposition to reductionism comes from taxonomists and population biologists engaged in a debate about the definition of species and types (see Section 3.6). Transferred to the question of whether neuronal populations can be reduced to separate neuronal types, this debate polarized into those who deal with the statistics of variation of any parameter (e.g., structural, functional, operational) in populations of nerve cells versus those who regard individual nerve cell types, defined by means of type-specific molecular markers, as real and highly significant.

The orthodox Marxist dialectical materialism also supports the notion of emergent qualities: Mind is not directly determined by matter but results from the dialectical interaction of the nervous system with social, economic, and political conditions. Those interactions are supposed to result in the gradual accumulation of quantitative differences which quite suddenly produce a significant qualitative change. I have never succeeded in understanding how this magical change occurs, but it is obviously not amenable to reductive analysis.

Because in many cases reductionism has not succeeded in completely explaining the workings of complex systems, its critics assert that it is inherently unable to deal with interactions between components of complex systems: "*Our reductionist arguments from the logical requirements of function may lead us very far astray here. Such an approach works very well when we are dealing with the mechanical or hydraulic aspects of life. . . . But in the case of the nervous system we are . . . . dealing with a quite different **kind** of system. . . . The difficulties and inadequacies of the logico-reductionist position are clearly illustrated in the work on regeneration within the visual system*" (Gaze, 1970, p. 263). In my view, the opposite is illustrated by that example, namely that understanding the regeneration of a complex system can only be attained by reducing it to simpler cellular and molecular processes. What else does such a complex system contain that makes it a "*different kind of system*" from a simpler one?

Here is another example of the antireductionist claim that something is left out of the reduction: "*Extreme analytical reductionism is a failure because it cannot give proper weight to the interactions of the components of a complex system. And an isolated component almost invariably has characteristics that are different from those of the same component when it is part of its ensemble, and does not reveal, when isolated, its contribution to the interaction . . .*" (Mayr, 1982, p. 61). Here Mayr is simply ignoring the facts when he says that reductionism fails to take interactions into account; does he regard the reduction of cellular functions to complex biochemical networks or reduction of some nervous functions to neural networks as failure to "give proper weight to the interactions"? Mayr is evidently ignorant of the successes of analytical reduction—neuroscience has advanced rapidly because of the successful experimental analysis of single nerve cells and subcellular components. It has been possible to advance a synthetic theory such as Sherrington's theory of integrative action precisely because of the success of analytical methods for elucidation of the physical mechanisms of single neurons and single synapses.

For those opposed to it, reductionism is at best an interesting analytical research program and at worst an exercise in futility. Reductionists react to this kind of criticism by dismissing holism, vitalism, and dialectical materialism as antiquated nonsense. As a consequence of this mutual incomprehension there has been little dialogue between the opposing sides that might lead to a creative synthesis. The gulf between the two sides is shown by their different views of the nature of emergent or supervenient properties, defined as properties that appear at higher levels of organization but are not predictable in terms of lower levels. Weaknesses in the notion of emergent properties have been discussed in Section 2.8. In brief, emergent properties that appear when matter is organized in different ways, e.g., the property of being solid or liquid, dead or alive, can be explained by reducing the differences entirely into differences of physical states. While it is true that the phenomena that we identify with mind are not now predictable from the properties of molecules or of nerve cells, extreme reductionism holds that if we knew enough about the complex interactions at all the levels below mind we would be able to predict it from the lower levels. Mind would then be regarded as a resultant property of brain organization and not an emergent property.

Vitalists and other opponents of reductionism place their philosophical account of nature first and their methodology second. This is the reverse of the order assigned by reductionists to ideology and methodology. The primary place of methodology cannot be understood by philosophers who have not spent time in the laboratory. It is important to see how much of the experimenter's efforts are devoted to assembling, maintaining, repairing, modifying, and improving experimental techniques rather than to theoretical speculation or even to obtaining and analyzing the data. Personal experience teaches me that the more strongly a research program is committed to reductionism the more time its proponents devote to tinkering with experimental techniques. This is only to be expected from research working at the limits of accuracy of its current techniques. The impossibility of complete exclusion of artifacts arising from the limitations of the techniques is well known to the experimentalist, and I have discussed some of the consequences in the previous chapter. However, the problem of the limits of accuracy of techniques may become the central problem for reductionists attempting to deal with mind-brain relations. If I am correct in what I have said about the limits to obtaining knowledge of mental events by means of any techniques, there is an uncertainty principle that will come into operation to prevent conclusive demonstration of the identity between mental and neural events.

2.13. ON CAUSAL EXPLANATIONS IN NEUROSCIENCE

The most generally accepted measure of progress of neuroscience is the accumulation of observed effects for which causes can be assigned and understood. Classical mechanics eliminates the concept of purpose and describes the course of all events as the automatic consequence of given initial conditions. But modern developments of atomic physics expose the limitations of mechanical causality and a new interpretation has been required.

Cause has been defined in many ways so that the notion of cause can vary depending on context. Causality implies that an event can influence the future but it cannot influence the past. Cause has traditionally been implied by (1) concomitance between events and habitual association between bits of empirical knowledge; (2) the success of prediction; (3) the notion of purpose or goal-directed action; (4) temporal succession; (5) process—the development of states over time. The strongest sense of causation is strict determinism. A weaker sense is that there is a certain probability of one event following another. In very complex systems the calculation of the probability of one event following another may be beyond the powers of any human being, but that does not alter the principle of causal connection.

All of those implications of causality are meaningless in the quantum world in which subatomic particles appear, change states, and disappear, without known causes. Causality is conceived mechanistically in terms of events that are either coincident or are connected through a causal process consisting of a continuous series of connected events. This view of causation assumes connection in space-time. But quantum theory allows events to be causally related although they are separated in time and space. Mind too seems to be unconstrained by time and space—thought is not apprehended as if it occupies a space, one can remember events that occurred in the past and imagine events in the future. Voluntary thought and action appear to deny mechanistic causality. Such features of memory, imagination and volition, as well as the apparent interaction of an observer's consciousness with quanta (Wigner, 1967; Wheeler, 1980), have led to much speculation about mind in terms of quantum theory (Eddington, 1935, 1939; Heisenberg, 1958; Lockwood, 1989; Penrose, 1989; Hodgson, 1991). My sympathy extends strongly to such efforts, marred as they are by a good deal of neuroscientific unsoundness. Lest I be accused of unsoundness in my understanding of quantum physics I shall not attempt to argue about the validity of the basic notion that the quantum universe underlies and unifies mental events and physical events.

We claim to understand the meanings of neural events and mental events from their cause-and-effect relations. This argument depends on the form of one's belief in the mental domain: *Mental nihilism* denies the existence of a mental domain; *mental realism* claims that mind exists in a fundamentally different domain from the physical; *mental reductionism* claims that the mind can be reduced to and can be understood completely in terms of physical processes. Cartesian dualism grants causal efficacy to mind, whereas monism denies that mind can have causal effects on the brain. Mind can be conceived as entirely physical in terms of bottom-up causation, from molecular states through intermediate states to mental states. But the converse, top-down causation, from mind to molecules is not conceivable in terms of physical mechanisms. Another difficulty arises from the argument for top-down causality: if mental states cause molecular states, and molecular states cause mental states, we are caught in an empty tautology, a vicious circle. This may be resolved by maintaining the (noncausal) identity of mental and physical events, but if I am correct in arguments given in Section 2.10, we shall never be able to show that a mental and a physical event are identical. If so, we are permanently limited to finding correlations and causal connections, but never identities, between neural and mental events.

When correspondence between different events are one-to-one we may assert cause-and-effect if the events are habitually associated in the same temporospatial relations, where those relations can be predicted, and when we can understand their functional significance. For example, the release of a synaptic transmitter molecule at the presynaptic ending and binding of the molecule with a receptor of the postsynaptic membrane are one-to-one cause-and-effect events. Their meanings are increased when they can be related to a long chain of cause-and-effect relations, for example when they are components of reflex withdrawal of the hand after noxious stimulation. At least since Descartes identified such behavior as purely mechanical, there has been no question that reflex behavior will eventually be reduced completely to mechanical cause-and-effect relations. We discuss this further in Chapter 4, in relation to Sherrington's research program. That research program is progressing rapidly. The problem now is to define whether there are limits to that progress. I have argued that there is a limit to showing that neural and mental events are identical. We should also question the assumption that a reductionist program will be able to show an unbroken causal chain from molecules to mind. Reductionism has to make compromises, namely to accept correlations rather than causal explanations, to achieve explanatory simplification and ontological unification.

Consider the problem of many-to-one causal relations between neural states and behavior. The vertebrate brain is known to be able to deliver the same output although it has received different inputs, and to use different structures to perform the same task: a motor task, like opening a box, can be completed by a variety of motor effects, using hands, feet, or teeth, and the serial order of these acts can be varied without affecting the final goal. This concept of "motor equivalence" (Hebb, 1949, p. 335) raises profound questions about the variability of cause-and-effect relations between structures and functions. An animal trained to perform a function in one way, if prevented from exercising that one way, can often complete the function in another way. For example, monkeys trained to open a problem box by hand will do so with a foot or mouth if the hand has been paralyzed by destruction of part of the motor cortex (Lashley, 1924; Jacobsen, 1932; Glees and Cole, 1950).

There are several ways in which many-to-one causal relations between brain states and behavior may be achieved. Alternative pathways and structures can be used to accomplish the same behavior. Activity in large populations of neurons may contribute to a single behavior in such a way that there is no correlation between the activity of any one neuron and any one behavior. Behavior can be controlled with respect to body representations in the brain. In the brain, a central representation of the body, the "body image," is an operational model which is constructed during embryonic development and tested and modified on the basis of signals from many peripheral sources. The body image is the model with respect to which voluntary movements of parts of the body are coordinated to achieve relevant effects (reviewed by Olson and Hanson, 1990).

The number and heterogeneity of causes that determine mental processes and behavior are very large, varied, and mostly unknown. We have no general theory of the neurobiology of mental processes or of behavior with which to explain their causes. Obviously one of the higher-level theories that stand behind neurobiology is the theory of evolution, and especially the theory of fitness of the individual in the competition for survival of the species. This implies that the individual has the anatomical, physiological, and biochemical mechanisms necessary to survive, and function in the given environment so as to increase the individual's reproductive rate. This effect can be measured. Fitness is a property of the relation between the organism and its environment measured by the advantage for survival of the organism, especially differential reproduction. The causes of fitness are all the conditions in the organism and in the environment that promote differential reproduction. Of course we can point out such causes of fitness—sensory-motor coordination, perceptual abilities, communication between individuals culminating in language and its sociocultural effects But this notion has weaknesses, not the least of which is that much human behavior is not adaptive. Another criticism is that the units of selection must include nerve cells, but we are unable to show that the function of a single neuron or of neuronal groups, clusters, or modules in the mammalian brain have evolved because they enhance the survival and reproductive advantages of the species.

In the human brain the cause of a mental process is not only the present neural process but certainly includes the entire past history of the individual from the beginning of development of the brain, and may even be conceived in terms of more remote evolutionary antecedents. Even if we include only the proximate cause within an individual life span, the initial conditions from which a succession of neural states starts can never be specified completely. Because of the unknowability of the initial conditions it is impossible to specify the next neural state in the succession of neural states that occur during a person's everyday life. Since the initial conditions in the nervous system are not known, we can explain neural events only insofar as they are regularly associated with and are covariant with other neural events, and in relation to the laws of neuroscience as they are presently known.

Because of the limitations of our methods for observing all or many parts of the brain simultaneously for long periods, and in comprehensive cellular and even molecular detail, we are forced to observe functional states of small neuronal populations or single neurons independently of others. When a given event is observed repeatedly as one of a series of events, we may conclude that those events belong to a causal chain when they should be understood more realistically in the context of functional networks. Finding such separable causal chains is necessary before we can in some degree know how different causal chains are inseparably interconnected to form the whole system. It is well known that the interconnections include negative feedback and feedforward loops with the result that effects become part of their own causes. Feedback is a well-known example of downward causation (Campbell, 1974), of higher-level events that regulate lower-level events in hierarchical systems. Sperry (1969, 1973) has suggested that the action of the mind on the brain is an example of downward causation, but that falls far short of an explanation, even in the most cursory terms, of the mind-brain relation.

We recognize that one of our problems is to understand how the cause-effect logic of individual components and of small subsystems is related to the functional logic of the entire system. Thus, electronic components that operate by digital logic can be assembled to form a system that has analog output, and in any system, including the nervous system, the functional logic of individual components need not be identical with the functional logic of the entire system. This principle is certainly true of the nervous system in which the cause-effect logic of individual neurons and of subsystems composed of many neurons are not identical with the functional logic of the entire system.

For example, the circuit logic of the intrinsic neurons of the cerebellar cortex is well known but it is not the same as the functional logic of all the subsystems that together control body movements. The overall effect of the entire system is to produce smooth and coordinated movements which have evolved so as to ensure survival and reproduction. But that effect could not be predicted only from knowledge of the circuit logic of the cerebellar cortex, or even from understanding the meaning of every physical event in the cerebellar cortex. This is one reason why functionalist theory is opposed to the identity theory that says that for every mental state there is an identical neural state. Functionalist theory is materialistic in holding that mental states are physical states, but not that mental and neural states are necessarily identical. Rather, the mind cannot be understood only in terms of the physical composition of the brain, but by its operational programs and functional states (Putnam, 1975, 1980; Shoemaker, 1984; Dennett, 1991). One of the problems of both behaviorists and functionalists is that when they say that a person's mental state is ultimately to be construed from his behavior and behavioral dispositions, they imply at least one causal condition in the brain, and probably a large number of them, and these must include beliefs and intentions (or their physical identities). Neither theory can escape from the difficulty of defining behavior in terms of causes in the brain.

Failure to relate the function of the whole system to the functions of its components is a problem of context-dependence. Simple voluntary acts such as moving a finger cannot be initiated as if there were no prior conditions determining them, and the more complex the act, e.g. movement of a finger in the context of all the movements necessary to eat a meal or play a musical instrument, the more important it becomes to take context into the ontology of the act. Failure to do so is one of the main weaknesses of behaviorism. Another is the inability to give an account of beliefs and intentions that may be expressed as behavior. The complexity of behavior is accepted by functionalists but at the cost of denial of the possibility of ever reducing behavior or of arriving at a complete causal explanation of behavior.

The difficulty of defining causes of human conduct can be demonstrated by a single familiar example—what are the causes of a simple action such as contraction of the muscles that flex the second finger? Many physiological events are known to occur between the conscious initiation of a voluntary movement and the muscle contractions. Let us confine our attention only to the well-known sequence of events leading from depolarization of the motor nerve terminal to the well-known molecular changes involving muscle actin and myosin. It can easily be shown that no single one of those events, e.g., release of acetylcholine from the motor nerve ending, is either necessary or sufficient for the motor effect (not sufficient because many other events are necessary, not necessary because they may be substituted, e.g., direct electrical stimulation of muscle may cause the motor effect). Hence, no single event is the necessary and sufficient cause of the behavior of the whole system, although for simplicity it may be regarded as one of many causes of the motor effect. Now, we may choose to ignore all but a few conditions that are known to be necessary for the observed effect, and we may decide to call those the significant causes of the effect, knowing that they are not all the causes. Alternatively, we may attempt to include all the conditions that can possibly precede the effect. But that attempt will lead to a very long process of reduction to find the first cause or to an expansion of relations by inclusion of all conditions that regularly precede the effect, many without being causes. In practice, neither of those strategies will lead to understanding of meanings of neural effects and behavior in terms of mental causes, or of intentions and motivations (Flew and Vesey, 1987).

I am also referring here to the psychological perception of causal action, for example, to the feeling that one's intentions and motivations are the "cause" of one's actions. When we initiate a voluntary action we seem to be "causing" our bodies to act, and to be "causing" the effects of those actions. The meaning of the intentional act is commonly perceived in terms of a neurophysiological process—a development of a succession of neural states over time. As we have seen, many of the same succession of states can be part of different processes. By different processes I mean processes that have different effects and are presumed to have some different causes, although they can also share many of the same causes. Consider the example of the voluntary flexion of the second finger. Consider the following three cases in which all the physiological conditions (e.g., nerve conduction, synaptic transmission, muscle contraction) are the same but the motivations are different. In one case flexion of the finger scratches the skin, in the second case flexion of the same finger depresses the key of a piano, and in the third case flexion of the finger pulls the trigger of a gun. It is clear that these three different effects share a large number of the same causal conditions (e.g., electrochemical events in the same sets of neurons, release of synaptic transmitters). Therefore, those common conditions cannot be significant causes of the differences in the final effects—the significant causes are the different motivations.

People give meaning to observations of things and events by recognizing their relations and connections to other observations, to form theoretical constructs, and by observation of their effects. Returning to the example of the effects of flexion of the second finger, it is obvious that the effects can be very complex—why did the finger scratch? In what musical context did the finger strike the piano key? At what object was the gun aimed? In short, what were the goals of the action and what was the context in which it occurred? It is a truism to declare that a single piece of empirical evidence has no meaning. It acquires meaning in the context.

We arrive at the conclusion that different meanings are attached to different effects in the three cases of flexion of the second finger. Different meanings are also attached to the significant different causes, namely motivations, but not to all the known causes of the effects. Now, motivations are themselves effects of an unknown number of causes, some of which may be held to be not significant because they are common to all motivations. We can continue this regression in search of significant causes until we find a "first cause" or a "true cause." But how would that be recognized as such? Psychophysical dualists, from Descartes on, answer that the initial cause is the soul or mind, that is a sufficient cause in itself. For Descartes the mind-body problem could be solved by the analytic method, by starting from the beginning and showing how effects depend on causes. He rejected the synthetic method of examining causes by means of their effects because he thought that synthesis can never show the method by which the truth has been discovered. Descartes chose the soul as the origin for his analysis of the causes of behavior, the modern bottom-up reductionist chooses DNA as the origin. Finding an origin also has the advantage of preventing the descent into tautology of the kind that argues that the brain is the cause of behavior which is the cause of evolutionary selection of fitter brains, and in general of the kind of argument that maintains that it works because it is fit, and it is fit because it works.

The great disadvantage of beginning the reduction either from the mind or from the genome is the enormous distance separating those logical starting points from the middle ground that is of greatest practical interest to the neuroscientist, the relatively well-understood domain of mammalian, or at least vertebrate, nervous structure, function, and dysfunction. Taking the pragmatist approach, it is often advantageous to work on a part of a nervous system that appears to be amenable to reduction in both the descending and the ascending directions. As we shall see in Chapter 4, that is what Sherrington did when he decided to begin his research on the knee jerk, and that is what Cajal did when he decided to start his research on the vertebrate retina and cerebellar cortex. Sherrington saw the knee jerk as a fraction of many complex behaviors, and as itself a mechanism that could be reduced to elementary components. Cajal saw the retina and the cerebellar cortex as parts of more complex systems, and as themselves reducible to more elementary modular structures. Part of their success can be attributed to their selection of levels of organization that offered them opportunities for reduction in both directions, upwards and downwards.

CHAPTER 3

Struggle for Synthesis of the Neuron Theory

> *The independence of science from philosophy means at the same time its irresponsibility—not in the moral sense of the word but in the sense of its incapacity and its lack of any perceived need to give an account of what it itself means within the totality of human existence, or especially in its application to nature and society. That of course is not what the so-called theory of science regards as its task. But if it really intends to be a giving of accounts, should it not go beyond the task of an imminent justification of the doing of science?*
> Hans-Georg Gadamer, *Reason in the Age of Science*, 1982.

3.1. THE ARGUMENT

The neuron theory is the most inclusive theory that has yet been constructed in neuroscience. The mode of its construction, historically and conceptually, can be understood in terms of a realist conception of the history and philosophy of neuroscience that I have laid open in previous chapters. I might emphasize that I have not attempted to find laws of historical development that can be used to explain the past or predict the future of neuroscience.

My aim is to enrich discourse about neurobiological research with relevant historical and philosophical examples and arguments. Neurobiological theories cannot be elaborated in an unhistorical framework, limited to the concerns and methodologies of the present. We must avoid the exclusive concern with the most recent empirical data and most modern techniques because that drives out the necessary concern with the ongoing historical process of constructing explanatory models. To be most useful such models must be seen in their historical context. They must never be viewed as final truths given authenticity by the superiority of modern scientific research methodology. I believe that construction of explanatory models must include immanent criticism which we bring to bear when making decisions about changes to our conceptual models. This critical apparatus must be applied vigorously by scientists themselves. Critique of science cannot be projected effectively into scientific research by philosophers and historians of science working outside the structure of scientific research programs.

Historians and philosophers of science have themselves to blame if their voices find no echo among working scientists. Philosophers and historians of science write at a vast distance from the practical concerns of scientists working in laboratories. Hans-Georg Gadamer is correct in his estimate of the harmful effects of separation of scientists from the philosophy and history of science. Unfortunately his critique has been elaborated at so great a distance from our practical concerns that it hardly touches the real problems of science. Thus, it is incapable of evoking a strong response in us.

Scientists cannot work with abstract philosophical, ethical, and historical ideas denuded of relevance to practical problems. We should not blame philosophers for that. We have ourselves to blame for failing to accept the responsibility of enriching our explanatory models with historical and philosophical components.

The neuron theory is an explanatory model of the organization of nervous systems. It is a special case of the cell theory, constructed to explain the evolution, development, and functional organization of nervous systems. The nerve cell is seen as the unit of histogenesis, morphogenesis, and function of the nervous system. Historically and conceptually the neuron theory was constructed, during the 19th century, by convergence and combination of several more-restricted, less-inclusive theories, of which 12 are identified and considered in some detail.

The neuron theory is part of a research program aimed at constructing explanatory models of the organization of nervous systems, and at understanding the workings of nervous systems by reducing them to their elementary cellular and molecular structural components and to elementary functional events. Historical evolution of this macro-to-micro reductionist neuroscience research program is closely related to philosophical materialism and realism, and to the development of relevant experimental techniques. In addition to laboratory techniques, the invention and improvement of printing technology was of great importance for the progress of neuroscience, and especially for the progress of the neuron theory, which was critically dependent on pictorial communication of empirical evidence and on graphic representation of concepts. In the 19th century, inexpensive production of well-illustrated scientific periodicals made it possible to communicate detailed information rapidly and widely for the first time in history.

The mode of construction of a more-inclusive theory (such as the neuron theory) on several less-inclusive theories (such as theories of development of nerve cells, and theories of the forms and functions of connections between nerve cells) has very important implications. Less-inclusive theories are less abstract, more restricted in the range of phenomena that they explain, and more directly related to the relevant body of observations than are more-inclusive theories. One important consequence is that by contrast with more-inclusive theories, less-inclusive theories are often founded on conjectures based on minimal evidence and are more vulnerable to refutation by counterevidence.

The dynamics of refutation of theories and corroboration of their opponent theories is discussed in this chapter. For example, several theories of the mode of formation of the axon were proposed, but the outgrowth theory was the only one to have withstood refutation and to have been repeatedly corroborated. Several theories of the mode of connection between two or more nerve cells were proposed, but only the contact theory was progressively corroborated, while the rival theories of continuity between different nerve cells and theories of formation of nerve networks were gradually refuted.

While two opposing theories are each based solely on different sets of empirical observations, there is no way to choose between them, or the choice is a matter of taste. At that stage there is as much to be said for preferring one theory to the other as for preferring onions to oysters. The contact theory became the accepted theory only when it was shown to be consistent with a much wider range of observations than the network theory, and when the contact theory explained the phenomena explained by the network theory but not vice versa. The contact theory was also accepted because it predicted discoveries not predicted by the network theory or forbidden by it. Only when those advances had occurred was the choice of one theory no longer a matter of personal preference but of empirical necessity. It was a matter of which theory could explain the meaning of a progressively larger set of well-corroborated observations. One does not accept observations because they lead to a preferred theory, but one accepts a theory because it explains a growing number of observations and because it correctly predicts new observations. One abandons a theory that has been refuted by a crucial experiment, but such cases of outright refutation are rare in neuroscience. More often a theory is abandoned because many of its supporting observations cannot be corroborated, because it fails to explain observations or to predict new ones, or because it does so less effectively than a rival theory. A theory is considered to be true when it explains, simplifies, and unifies knowledge from different sources, and when it achieves increasing explanatory generalization. It is considered to be false when it has to be modified, and becomes increasingly complex, in order to square with the facts.

The lower-level theories on which the neuron theory was founded were of an essentially descriptive, imaginative, conjectural nature, in which analogical-correlative thinking was very important. This class of neuroscience theories is based on models in which complex nervous systems were represented by analogy with simpler nervous systems and with mechanical, electrical, or other systems.

Histological drawings were an important form of analogical problem-solving. A neurohistological representation of spinal cord or cerebral cortex is a conceptual model that may be a more or less realistic description of the complexity of nervous organization. Such a model may represent an accepted theory of nervous organization or it may go beyond the accepted theory to include unobservable entities and to make more or less accurate predictions from theory. It may extend accepted theory by importing features from a well-understood domain (like the spinal cord) into a model of a less well-understood domain (like the cerebral cortex). In such cases the two domains must share at least one causally relevant condition. The concept of the neuron as the unit of all nervous organizations provided that condition. That concept was the convertible currency required for enriching the poorly understood domains of the nervous system by importing features of the well-understood domains.

Some personal characteristics that have enabled scientists to construct the neuron theory are conceptual innovativeness, imaginative interpretation of observations, and tenacity in holding hypotheses with little supporting evidence or in the face of counterevidence. Yet, scientists are ultimately ruled by the logic of integration of the components of their research programs—values, theories, techniques, and observations. In this process, powerful techniques are as important as powerful intellects and imaginations capable of making novel conjectures and of giving meanings to observations.

The concepts of nerve cell typology, neuronal specificity, and functional division of labor are outgrowths from the neuron theory. Typologism developed during the 19th century as better methods were invented for identification of specific nerve cell types, morphologically and physiologically. The methodology for identification of essential intrinsic properties of specific types of nerve cells is termed methodological essentialism. Although condemned as an error by some historians and philosophers of science, methodological essentialism has flourished in neuroscience. Typologism has been firmly established in neuroscience as the result of the successful application of molecular biological methods for identification of molecules that are uniquely possessed by specific types of neurons and glial cells.

Molecular biology is a research methodology, not a complete research program in which methods are fully integrated with theories. Molecular biology does not test between theories of molecular biology by means of crucial experiments. It is a set of techniques for making observations and measurements and for making correlations between those data. Of course those data may be useful for helping to decide between theories in neuroscience. That is why the name "molecular neuroscience" has been given to the business of using molecular biological techniques to obtain data relevant to neurobiological theories, and even to theories that go beyond the molecular level of explanation. An advantage of using molecular biological techniques is that they are usually neutral with respect to rival theories in neuroscience. Methodological neutrality can, in principle, ensure that observations are not slanted toward one of the alternative theories. This is necessary in order to decide between rival theories of nerve cell lineages, nerve cell differentiation, and correlations between structures and functions.

Useful theories are conceptual models that are continuously under construction and revision, and may be regarded as permanently incomplete. An indication of its incompleteness is that the neuron theory does not give an explanation of the workings of neurons in conscious thought and will. We may envisage a most-inclusive theory relating mental events to neural events. Although many books have been written on that subject, what we really understand about mind-neuron relations is as yet inadequate for the construction of a general theory of mind standing higher than the neuron theory in the hierarchy of neuroscience theories. Failure to construct a theory that explains how mental events are related to neural events shows that the neuron theory is essentially incomplete.

3.2. REMARKS ON HISTORIOGRAPHY OF NEUROSCIENCE

There is no single "correct" theory of history, just as there is no perfect historian.*
I do not pretend to be a philosopher or a historian, but I still need theories of the philosophy and history of science to enable me to understand the history and philosophy of science. Therefore, we must define our theories explicitly so that they can be criticized, modified, and replaced if necessary. It is obviously best to follow a correct theory, but that is not always possible under the given circumstances, and "*it is probably true that a false theory is better than none*" (N. R. Campbell, 1920, p. 152).

I do not propose to recount the historical development of the neuron doctrine as if it were a linear progression of ideas and a sequence of entrances and exits of the well-known contenders for scientific truth. That approach has already been taken by many others whose names are well known to us. In spite of much that can be admired in their works, their view of history contains deep flaws.

It is easy to write the history of neuroscience as if it was the progressive discovery of truth and elimination of error. It is nice to read the biographies of the heroes who did "good" neuroscience and to be told how the antiheroes did it badly. The history of neuroscience has been written as if it was a succession of theories that have approached closer and closer to the correct explanation of how nervous systems work. But such naive views of the cumulative nature of scientific progress and the objective character of scientific explanation have been under suspicion for decades—at least since 1931 when Butterfield called them "Whiggish."

The problem for the historian is to explain how the false was replaced by the true explanation of the phenomena. Was it by replacement of the old by new observations, by replacement of one answer by another, or by replacement of the old by a new question? In many examples that will be given, replacement of one answer by another answer to the original question did not lead to progress—for that to occur it was necessary to replace the old question by another that could lead to a correct answer. That change of the question has frequently led to progress in neuroscience. For example, understanding reflex action could not progress while the problem was conceived in terms of nervous excitation caused by flow of electricity from nerve cell to nerve cell. The confusion of thinking about inhibition prior to Sherrington's definitive clarification can be seen from Verworn's (1900) paper on nervous inhibition, in which inhibition is portrayed as a passive process or as a drainage of an active process. Progress occurred only after the old question was replaced by new questions about the causes of delay in reflex arc conduction which was greater than could be accounted for by conduction in nerve fibers, and new questions about reflex inhibition which could not be accounted for by simple reduction of excitation, but which required inhibition to be an active process. The historian of science has to explain how and why the old question was replaced by the new.

---

*"*The perfect historian must possess an imagination sufficiently powerful to make his narrative affecting and picturesque. Yet he must control it so absolutely as to content himself with the materials which he finds, and to refrain from supplying deficiencies by additions of his own. He must be a profound and ingenious reasoner. Yet he must possess sufficient self-command to refrain from casting his facts in the mould of his hypothesis. Those who can justly estimate these almost insuperable difficulties will not think it strange that every writer should have failed. . . .*" Thomas Babington Macaulay (1800–1859). "History," *Edinburgh Review*, 1828.

To understand the wider meaning of neuroscience it is necessary to disclose its historical causalities and to comprehend its values. These are interdependent because historical causalities cannot be understood fully without comprehension of values, and values cannot be identified without reference to their historical origins and effects. Knowledge of history is necessary to have as a reference standard against which to measure the outcome of our reflections on contemporary events. Without an attempt at objective reference to history, self-reflection tends to validate our wishes and prejudices. The process of iterative reflective judgment, discussed in Section 5.4, leads to a more objective understanding of the meaning of contemporary science in its historical context.

Most histories of neuroscience have been attempts to show that science is a progressive accumulation and verification of facts and elimination of errors. These belong to the justificationist and Whig varieties of historical writing. Histories of that sort can be useful when they return to the original texts to find the sources of ideas, as Paul Cranefield (1974) has traced the concept of separate nervous input and output pathways, and republished the original documents in the dispute between Charles Bell and François Magendie over priority for discovering the functions of the spinal nerve roots. Cranefield accurately appraised the effects of nationalistic values on the dispute about priority. However, he failed to consider the fact that many well-qualified neuroscientists continued to ignore Magendie's experimental demonstration, made in 1822, that the posterior spinal roots are sensory, and the anterior are motor. For example, G. H. Lewes in his *Physiology of Common Life* (1860, Vol. II) held that anterior and posterior spinal roots contain both sensory and motor nerve fibers. Wilhelm Wundt in his *Menschen-und Thier-Seele* (1863, Vol. I, p. 222) believed that the anterior spinal root nerve fibers transmit both the motor impulse to the muscles and the sensory information about "muscle sense" from the muscles to the CNS. Alexander Bain held a similar theory (*The Senses and the Intellect*, 3rd ed. 1868, p. 76). There were others who believed that muscles lack special sensory nerves entering the posterior nerve roots, but both the motor and the sensory impulses from muscles are conveyed by the nerve fibers of anterior spinal nerve. I raise these cases here to support my thesis that neurobiological theories are never refuted rapidly, even by such compelling evidence as Magendie's, which is now regarded as conclusive.*

---

*There were two opposing theoretical models: either sensory and motor impulses are both conducted in the same nerve fiber or they are conducted in separate nerve fibers. The first model was described clearly by Descartes in his posthumously published *De homine* (1662). Descartes conceived of nerve fibers as tubes containing many delicate fibers extending from pores in the cerebral ventricles to all parts of the body. Sensations were produced by physical stimuli moving the peripheral ends of the fibers in the nerves, whereas motor action was produced by animal spirits, in the form of very small particles, flowing from the cerebral ventricles in the nerve fibers to the muscles. Thus, both sensory and motor actions were conducted by the same nerve fibers. Albrecht von Haller (1708–1777) argued for this concept. It was first seriously challenged by Alexander Walker (1779–1852) with his notion that the anterior spinal nerve roots were sensory and the posterior roots motor, published in his *New Anatomy and Physiology of the Brain in Particular, and of the Nervous System in General* (1809). In spite of Walker's misattribution of functions to the spinal nerve roots, he deserves credit for originating the concept of their functional specificity. This concept was developed further by Bell (1811), correctly demonstrated by Magendie (1822), and elevated to a general principle by Johannes Müller (1826, 1833–1840). This opened the way for studies of the anatomical basis of sensory and motor functions of the spinal cord, and the discovery of separate motor and sensory tracts (Stilling, 1842; Schiff, 1858–1859; Brown-Séquard, 1860, 1866; Edinger, 1889; Bechterew, 1894).

Some historians have given useful evaluations of the strengths and weaknesses of experimental techniques. Histories have also been valuable in giving English translations of some of the classics of neuroscience, and in recounting the biographies of significant scientific discoverers, even if they have rarely been successful in explaining causalities and if they uniformly fail to recognize the significance of values. In addition there are picturebook histories of neuroscience about which it is fair to say that "*one need only read a few lines and look at a few pictures, in order to know them entirely; the rest is there only to fill up the pages*" (Descartes, *Cogitationes Privatae* X, 214).

One cannot understand a scientific theory or discovery as if there is nothing outside it or preliminary to it, nor can one understand the history of ideas as if they have neither antecedents nor successors. The models of neural organization that have been constructed at different times have been strongly informed by the prevailing concepts of physical communications systems—railway, telegraph, telephone, and now computers. This is only one indication that neuroscience is an affair largely influenced by culture if not created entirely by it as constructivists like Kuhn (1962, 1970) have argued. Nevertheless, structuralists such as Popper (1959) and Lakatos (1978) deny the validity of the belief that great works of science and of literature are produced in response to stimuli in the nonscientific world. I argue in Chapter 1 that structuralists are incorrect in that denial: one cannot understand a scientific theory or discovery entirely from within its own structure without references to external causes and consequences.

Structuralists attempt to think from within the system that is described. I believe that to be limited to systems in one's own tradition. Since Collingwood wrote *The Idea of History* (1946), some historians have taken their task to be to enter into the minds of their historical subjects and to recover their subjects' own sense of their life and times. This sort of empathy is possible only when both the historical subjects and the historian belong to the same intellectual tradition. It seems to me to be the preferred method whenever possible, but it is often not possible. There are obvious advantages to thinking from inside the system one actually inhabits rather than presuming to be able to think from inside a system one can only pretend to inhabit. For example, I cannot pretend to think from within such systems of belief as remote from my experience as the Ayurvedic system of ancient India or the Taoist system of ancient China, although they contain some fascinating concepts of the functions of the nervous system and of the relations of the mind to the body (Zimmer, 1935; Wong and Wu, 1936; Veith, 1950, 1957, 1966). We can analyze the beliefs of those or any other systems only from our experience, using our conventions and methods, limited as they may be for the task.

More to our point, why do historians of neuroscience continue to write in the Whiggish tradition? Why have all historians of the neuron theory failed to understand that their premises about the nature of scientific explanation and scientific progress have been completely undermined by recent advances in the philosophy and historiography of science? The reasons are easily found. First, most neuroscientists work in research institutes and medical schools, separated from philosophers and historians of science. This is one reason why philosophers' histories of science are different from scientists' histories of their limited fields of expertise. Another is the persistent element of 19th-century naive positivism in 20th-century neuroscience, and that includes the uncritical acceptance of empiricist and positivist views of the nature of scientific explanation, and adoption of a theory of the history of science based on a correspondence theory of truth.

According to that theory, scientists and historians of science have only to hold better mirrors up to nature to obtain more truthful reflections of events in the real world. The correspondence theory makes arbitrary distinctions between the human mind and the outside world, and it treats the mind as a mirror of outer reality rather than as an active constructor of models of the world and even of the mind itself.

An important part of my explanation of the construction of the neuron theory, and of neuroscience theories in general, takes into account the personal characters and professional styles of the neuroscientists. This is worked out in some detail in Chapter 4. Scientists have different degrees of imagination and boldness in asserting their claims, and different degrees of tenacity in holding and defending them. Scientists do not withhold their claims to support or refute a theory *until* the evidence is sufficient and conclusive. Nor do scientists abandon a theory because of claims to have refuted it, but they show tenacity in holding a theory in the face of counterevidence. This enabled the nascent neuron theory to survive and eventually to gain a consensus. Claims accumulate gradually until a theory either has gained a consensus in its favor or is demonstrably false. The process of gaining a consensus occurs in relation to advancing techniques that expose anomalous data, and against a changing social and cultural background.

3.3. FACTS ARE NECESSARY, EXPLANATIONS ESSENTIAL

In the historical process of construction of theories, the neuron theory among others, it frequently happens that more than one theory is proposed to explain the same group of phenomena. In this chapter we consider how different theories are erected to explain the same facts, how their rivalry may result in progress, and how one theory may absorb or eliminate the others. It is worth emphasizing here that resolution of scientific conflicts can occur only when opposing theories occupy the same theoretical framework. When rival theories attempt to give different explanations of the same set of facts, and when they share the same theoretical and practical terms and aims, they may engage in some form of intertheoretic discourse that may lead to a resolution of their differences. Of course, there are additional conditions, discussed in Chapter 1, that are necessary for theories to be rivals and for their conflicts to be resolved. But if theories do not relate to the same set of facts, they will either ignore one another or come into conflict only with respect to a subset of shared facts. In that case one theory may modify parts of the other or exclude parts of the other, but one theory cannot eliminate the opposing theory totally.

Alternative theories concerned with the principles of cellular organization of the nervous system were able to engage in discourse which eventually led to the synthesis of the more-inclusive theory we call the neuron theory. More-restricted theories, each giving a different explanation of the way in which the components of the nervous system are organized, all used the same theoretical terms, mainly the terms of the cell theory. They also aimed for the same goals, principally to reduce complex nervous systems to simpler components, and to elucidate functions and dysfunctions of nervous systems in terms of that reductionist research program, as we consider it in Chapter 2. Reductionism is a methodology for obtaining data by systematically cutting large systems into smaller components. It is not the data, but the explanation of them, that matters. Theories are explanations that relate to observations of phenomena at different reductive levels. Theories relate to each other in a hierarchy. Modern neuroscience is supported by gene theory, cell theory, and evolution theory, and those stand in relation to more general theories of chemistry and physics. In the hierarchy of theories, the neuron theory is most closely related to the cell theory. Ranked in terms of generality, from less to more inclusive, the neuron theory stands at a level below the cell theory, as a special case of it, stated in terms of nerve cells.

The modern neuron theory is more than a single proposition or observational statement about one aspect of the nature of nerve cells (e.g. that nerve fibers are outgrowths of nerve cells or that all nerve cells are separated from one another by synapses). Because it is mostly descriptive and not strongly predictive, some prefer to call it a doctrine rather than a theory. To rank as a theory at a higher level of generality than any one of many possible propositions about the nature of nerve cells or of nervous systems, the neuron theory must explain many or all such propositions. Exactly how many is a good question to ask because it draws attention to the ongoing process of construction of the theory. That process has continued for about two centuries to the present.

At the same time as the neuron theory has changed, the related theories have also progressed. The precursors of the cell, genetic, and evolutionary theories, just like the precursors of the neuron theory, were continuously evolving from their early forms at the beginning of the 19th century to their maturer forms at the beginning of the 20th century (Watermann, 1964; T. S. Hall, 1969; Stubbe, 1963). We can relate the neuron theory, as it existed at some stage in its construction, to the other theories only at the same historical stage. Yet, in terms of the generality of theories, we can discern that the neuron theory and its precursors have always been less inclusive relative to the cell, genetic, and evolutionary theories.

The level of one theory, such as the neuron theory, in relation to other theories in a hierarchy, is given by the generality of its explanations of the meanings of observed phenomena. A high-level theory, like the neuron theory, includes, explains, and unifies a wide variety of phenomena that would require several lower-level theories to explain. Simply stated, a more-inclusive theory can explain everything included in less-inclusive theories but not vice versa. Stated in terms of intertheoretic reduction, namely the notion of reduction of general theories to more-restricted ones, the reduced theory is explained by the reducing theory, but not vice versa. You will no doubt have recognized the difference between reduction of more general theories to less general theories, done by philosophers of science, and macro-to-micro reduction of things at higher levels of organization to lower-level constituents, done by scientists. This also shows that the explanatory value of natural science is different from that of the philosophy of science. In short, scientific explanation shows how observations are organized into theories of increasing generalization from which we can make reliable predictions. By contrast, philosophical explanation of scientific theories shows how different levels of scientific explanation are related. These distinctions are more fully considered in Section 2.7.

The neuron theory evolved historically as a higher-level, more-inclusive theory formed by convergence and synthesis of many lower-level, less-inclusive theories. The lower-level theories, which we shall soon consider in detail, were directly related to relevant empirical observations, but the higher-level theory was related to those observations only indirectly. One of the common errors of virtually all previous accounts is to pose a conflict between the continuity theory of nerve connections, a lower-level theory, and the neuron theory, a higher-level theory. The former was in conflict only with the contact theory of nerve cell connections, one of several lower-level theories on which the neuron theory was constructed.

The higher-level status of the neuron theory in relation to lower-level theories was understood in principle by those who gave the neuron theory its definitive form, in the decade after 1890. For example, when Cajal reviewed the field in 1894 in his monograph *Les nouvelles idées sur la structure du système nerveux*, he claimed that his own work, which started at the end of 1888, had demonstrated the following: 1. *"The nerve cell is an independent unit which does not form anastomoses. . . ; 2. Every axis cylinder ends by means of varicose and flexuous branches in the manner of the nerve branchings at the motor plate of muscles; 3. The branches are applied either on the body or on the protoplasmic expansion of the nerve cells, establishing connections by **contiguity**, by **contact**, as efficiently for the transmission of currents as the connections would be by **continuity**; 4. The cell body, just like the protoplasmic expansions, plays the role of conductors and not just a nutritional role"* (Ramón y Cajal, 1894a, p. 9). Cajal saw the neuron theory standing on those four lower-level theories, which he claimed to have placed there. But the neuron theory was actually supported on several more lower-level theories in addition to those four. If Cajal's four had not been in place the neuron theory would have been less secure but would not have collapsed. And if the great microscopist had not provided his support some other microscopists would quickly have done so, as Merton (1961) has argued, and as I argue in Section 4.3.

Restriction of the neuron theory, in the manner of Cajal, led early critics of the neuron theory to call it a nontheory, and to denigrate its explanatory value. As one early critic asserted, the neuron theory *"is not, properly speaking, a theory at all. It explains nothing. It is a statement, supposed to be true, that every nerve-cell, with all the parts belonging thereto (axon, collaterals, terminal arborizations, dendrites) is anatomically distinct from every other nerve cell. The conclusions with regard to physiological autonomy and the restrictions of pathological processes which are often regarded as deductions from the 'Neuron Theory' are really deductions from the Cell Theory. . . ."* (Hill, 1900). This position has been defended in more detail by Spatz (1952).

To many during the 1890s the neuron theory was the cell theory applied to the nervous system. For example, that was what Barker declared in his textbook *The Nervous System and Its Constituent Neurones* (1899): *"The attempt has been made to apply the neurone conception—that is the cell doctrine—as consistently as possible to the explanation and description of the complex architectonics of the nervous system."* Barker (1898, 1899) made it clear, as Waldeyer had already done in 1891 when he introduced the term *neurone*, that Cajal's contributions were mainly limited to observations that axons end freely, and observations that dendrites are in the nervous conducting pathways, not the entire neuron theory, which was the product of many minds. A similar generous appraisal of the credit for construction of the neuron theory was given by Beevor (1901), in his review of Barker's book.

You will have asked how the construction of the cell theory was related to the construction of the neuron theory. You will have recognized that those who made statements about those theories during the course of their construction were not at that time able to envisage the theoretical edifice that would eventually be built on their foundations. By the 1890s the cell theory could be stated in the form of the following propositions (adapted from Baker, 1948–1955):

1. Most organisms consist of microscopic bodies called cells.
2. Cells are all of essentially the same nature and are the units of structure of multicellular organisms.
3. All cells originate from preexisting cells, usually by mitosis.
4. Organisms consist of nothing but cells and their products which form the tissues between cells.
5. Cells are to some extent individuals, and there are therefore two levels of individuality: That of the cells and that of the multicellular organism. A related proposition is that evolution occurs by selection and survival of the fittest at the level of the cell and of the organism as a whole.

Evolution of nervous systems was conceived in terms of the cell theory in the 1880s, before Waldeyer's pronouncement of the neuron theory in 1891. During the 1880s, the idea took shape of evolution of the central nervous system of higher invertebrates, starting in the annelids and culminating in the vertebrates, by inward migration of nerve cells from the surface and by grouping of nerve cells to form ganglia and the central nervous organs. The concept of migration of neuroblasts had by then become accepted after the work of Wilhelm His (1868, 1879) showing the migration of neural crest cells and their grouping to form the ganglia of the peripheral nervous system of vertebrates. A scheme of evolution of vertebrate neurons from independent neurons and from independent effectors of invertebrates was sketched by Retzius (1892g) as shown in Fig. 3.1. A scheme of the early evolution of neurons was worked out in more detail by Parker (1919) and Ariëns Kappers (1929). The first stage was thought to be muscle that contracted independently of nerves, as in the oscular and pore sphincters of sponges. The second stage was seen in the receptor-effector system of coelenterates, in which a patch of sensory cells, linked together functionally, is attached to a group of muscles. This was recognized to be a primitive reflex system by G. J. Romanes (1877, 1878, 1893b), who was one of the first to understand that research on the nervous systems of coelenterates and other invertebrates might throw light on the evolution of the nervous systems of vertebrates. The third stage of evolution was seen in some coelenterates, like the actinians in which ganglion cells are interposed between the sensory cells and the muscle cells (Hertwig and Hertwig, 1879, 1880). It was seriously proposed that the presence of those neurons was sufficient to justify a science of psychology of the actinians (Piéron, 1906a,b). In a few decades a theory of evolution of nervous systems was constructed that was consistent with the cell theory (which had been ignored by Darwin), and which showed how evolution of complexity of organization of nervous systems might be related to evolution of more complex behavior.

By the 1890s the main propositions of the cell theory would have been accepted by almost all biologists although there was some controversy, especially about the proposition that the individual cell is the unit of structure of all multicellular organisms. There

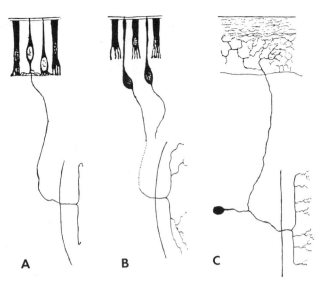

FIGURE 3.1. Diagrams showing the differences in morphology of the primary somatic sensory neuron, especially change in relative position of the cell body, in (**A**) an oligochaete worm (*Lumbricus*); (**B**) a polychaete worm (*Nereis*); (**C**) vertebrates. These diagrams were made by Gustaf Retzius (1892g) to represent his theory of a phylogenetic progression of neuron morphology. He implied that the nerve cell (or neuron type) may be the unit of natural selection during evolution of nervous systems. Although he recognized that these neurons are analogous, and not proven to be homologous, his proposal had the great merit of bringing the neuron theory and Darwinian evolutionary theory into consilience.

was a related dispute about whether the individual nerve cell is the unit of structure of all nervous systems, or whether groups of nerve cells are coupled together to form a functional unit. This controversy has been reopened recently by discovery of gap junctions linking adjoining cells, allowing passage of ions and small molecules between cells (Loewenstein, 1981). In the 19th century, the discussion was focused on the definition of the limiting boundary of a single cell and therefore on the nature of a surface membrane, and on the extent of protoplasmic continuity between different cells. The main objectors to the cell theory, especially Whitman (1893) and Sedgwick (1895), believed that the cell lacks a surface membrane and that cells are joined together by protoplasmic connections. Therefore, they held the unit of structure and function to be a group of interconnected cells, and in some cases the whole organism. The effects that their notion had on the concept of the neuron as the unit of structure and function of nervous systems will be discussed later in this section and more fully in Section 3.10.

I must emphasize again that the neuron theory was constructed at a level of generalization above the level of either the network or the contact theory of connections between nerve cells. Although the contact theory eventually replaced the network theory and other theories of cytoplasmic continuity between neurons, the change occurred gradually over a period of about 70 years. During that period the nerve network theory was regarded by many competent people as compatible with the neuron theory, as

Waldeyer-Hartz (1891) observed when he reviewed the evidence showing that the neuron was the functional and structural unit of the nervous system: "*If . . . we accept the existence of nerve networks our interpretation is somewhat altered, but nevertheless we can retain the nerve-units. Then the boundary between two nerve-units would always lie in a nerve network. . . .*" The same conclusion was reached by Hill (1900), who subscribed to both the network theory and the neuron theory. When Waldeyer reviewed the evidence in 1891, both less-inclusive theories could still be accommodated in the more-inclusive neuron theory, although the consensus was growing in favor of the contact theory, and in favor of the individual nerve cell as the unit. But Waldeyer was careful not to commit himself to equating the single nerve cell with the nerve-unit (*Nerveneinheit*) that he called a neuron. The term "neuron" was invented by Waldeyer (1891) in the first of seven papers with the title "Ueber einige neuere Forschungen im Gebiete der Anatomie des Centralnervensystems": "*Thus a nerve-element exists (a 'nerve-unit' or 'neuron,' as I have decided to name it). . . .*" He was able to conceive of several nerve cells linked together to form a functional unit.*

Formulation of the neuron theory occurred so gradually that when it was announced publicly by Waldeyer in 1891, Cajal was taken by surprise and complained that "*all Waldeyer did was to publish in a weekly newspaper a resumé of my research and to invent the term neurone.*" Indeed, the idea to which Waldeyer gave the name "neurone" was not radically new—it was widely known and was accepted by many others in addition to Cajal. Sherrington (1949) notes that Waldeyer's review was published "*without consulting Cajal*" and "*somewhat to Cajal's surprise.*" This goes to show that Cajal thought of the neuron theory as his personal achievement. He did not understand that Waldeyer was not presenting himself as a claimant for priority, but only as a reviewer of the work of many men and as the spokesman of a growing consensus. Waldeyer accomplished those tasks very ably, showing that the "neurone doctrine" was a part of the cell theory and giving the evidence for the concept of a nerve unit (he hesitated to define its limits dogmatically because he accepted the evidence that some nerve units were connected via networks).

When the eminent American neuroanatomist J. B. Johnston reviewed the neuron theory in 1906 (p. 84), he did not hesitate to declare that "*in 1891 the cell theory was stated in a special form as it applied to the nervous system. This statement of the cell theory of the nervous system has since been known as the neurone theory of Waldeyer, who formulated it.*" Johnston accepted the neuron theory as a special case of the cell theory but he concluded that the contact theory was inadequate because "*it has been clearly shown that in many cases nerve cells are in continuity with one another . . . that part of the neurone theory which states that nerve cells make connections with one another only by contact is definitely disproved*" (J. B. Johnston, 1906, p. 87–88). That

---

*The idea of functional and anatomical coupling of groups of neurons to form a multineuronal functional unit would be revived repeatedly. Electrically coupled synapses were found that enable synchronization of activity of adjoining neurons (Bennett *et al.*, 1963). Discovery of intercellular communication through gap junctions has again brought the cell theory into question (Loewenstein, 1981). But in fact those intercellular junctions permit the passage of ions and small molecules but maintain the separation of macromolecules. Macromolecules that determine genetic differences between cells normally remain locked up within the boundary of the individual cell.

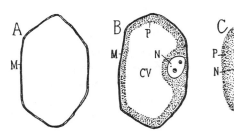

FIGURE 3.2. The concept of the nerve cell lacking a surface membrane originated with Schwann (1839), and was supported by many able neurocytologists (Schultze, 1870; Duval, 1897). This diagram, from Duval (1897), shows: **(A)** a plant cell with a surface membrane; **(B)** an animal cell with a surface membrane (M) and a fluid-filled vacuole (CV); and **(C)** a cell lacking a membrane, formed of a mass of protoplasm (P) and a nucleus (N).

conclusion was consistent with a modified version of the cell theory founded on the evidence of intercellular continuity (Whitman, 1893; Sedgwick, 1895), in vogue from before 1890 to as recently as the 1920s.* As might be expected from the more-inclusive relation of the cell theory to the neuron theory, this modification of the cell theory was soon adopted by many neuroscientists as a modification of the neuron theory. They proposed that *"cells are not always bounded by cell walls and in many cases adjacent cells are directly continuous with one another by means of strands of protoplasm"* (J. B. Johnston, 1906, p. 86). One of the effects of that concept, which will be discussed in Section 3.12, was to prolong the dispute between the rival theories of nerve connections (Jacobson, 1991, p. 163–166).

It is unfortunate that the concept of the membraneless cell has been overlooked by all previous historians of the neuron theory (e.g., van der Loos, 1967; E. Clarke and O'Malley, 1968; A. Meyer, 1971; E. Clarke and Jacyna, 1987; Shepherd, 1991). But without recognizing the ubiquity of that concept it is hard to understand the changing relations between the cell theory and the neuron theory in the late 19th century, or to understand why so many able neurohistologists of that period were able to embrace the continuity theory of nerve connections (Fig. 3.2).

Of all the less-inclusive theories that together formed the neuron theory, the contact theory was no doubt the most significant from the point of view of physiological functions. That is perhaps why the contact theory is often equated with the entire neuron theory. To avoid that distortion of historical perspective we should rather see the neuron theory maturing in relation to more general theoretical constructs. First it matured in relation to the cell theory, then in relation to an epigenetic theory of development, third in relation to a theory of integration of physiological functions, and fourth in relation to the Darwinian theory of evolution.

I shall explain in the following section how it was possible for the neuron theory to be constructed by synthesis from several lower-level theories, some of which were refuted while their rivals were progressively corroborated. The continuity theory of

---

*As late as 1923 Emil Rohde proposed that all cells, including neurons, originate from a multinucleated syncytium in the embryo rather than by cell division. I mention this only because it was accepted by holists such as E. S. Russell (1930, p. 221) as evidence for the "wholeness" of the organism, part of which was the apparent indivisible nature of consciousness. That was why holists like Russell supported network theories of the organization of the nervous system.

nerve connections was not reduced to the rival contact theory, nor did the two theories merge. The continuity theory was eliminated because many of its supporting data could not be corroborated, and because all its well-corroborated supporting data could be assimilated in the contact theory, but not vice versa.

From this position, many matters that were previously obscure now become more clearly understandable. We can now understand why the neuron theory has survived numerous anomalies in one or another of the less-inclusive theories which formed the more-inclusive neuron theory.* It becomes possible to understand why some less-inclusive theories dealing with origins of individual nerve cells, their migration, and outgrowth of their processes were consistent with both the neuron theory and the continuity theory. Therefore, it was possible for Golgi and other supporters of nerve network theories to assent to several less-inclusive theories that eventually became assimilated in the neuron theory. For example Golgi (1880a,b, 1883b,c) discovered the outgrowth of axon collaterals in the chick spinal cord and he never disputed the validity of other observations showing that the axon and dendrites are outgrowths of the nerve cell body. He conjectured, of course, that axon collaterals form a continuous network connecting nerve cell to nerve cell as shown in Figs. 1.2 and 3.8C, but that concept was not inconsistent with the outgrowth theory of development of nerve fibers if the network developed only after the initial outgrowth of free nerve endings. Now we know better, but then there was no inconsistency in believing both the outgrowth theory and the nerve net theory—it was a normal part of the process of construction of more-inclusive theories from whatever less-inclusive theories happen to be consistent with one another at any time. Of course, inconsistencies may be present but not understood at the time, or anomalies in theories may accumulate so gradually that their presence is not immediately apparent, although they may be glaringly obvious at a later time. The historian using what may be termed a "retrospectroscope" can easily get a distorted view of the past, seeing the errors and anomalies that could not be seen at the times when they actually happened.

3.4. CONSTRUCTION OF THE NEURON THEORY BY CONSILIENCE FROM LOWER-LEVEL THEORIES

The term "consilience" was introduced in 1858 by William Whewell (1794–1866) to mean that lower-level theories can be unified under the terms of higher-level theories. For example, consilience occurs when the neuron theory effectively reinterprets data originally thought to belong to different domains as now belonging to the same domain.

---

*Another possible anomaly in the neuron theory has been the finding that small parts of one nerve cell can function as quasi-independent units (Bullock, 1959; Fujita, 1968; Ross et al., 1990). In a more-inclusive theory that includes many less-inclusive theories, the concept of the regional functional specialization of the nerve cell is easily accommodated. In fact, the concept of regional specialization was introduced in the mid-19th century when the different structures and functions of the dendrites and axon were recognized. In the ensuing century the evidence of regional specialization has grown so that there are now many examples of individual neurons that can function as a number of functional units. We can then regard the integrative action in a single cell as in some ways comparable to integrative action in a multicellular system.

Whewell viewed theory construction as an inductive process in which knowledge from different sources is combined to form a unified explanatory concept. The further unification of inductions he called consilience. Consilience is explanatory unification, and in Whewell's view unification and simplification are the main measures of a true theory. In a false theory the opposite occurs—convergence of independent evidence adds to the complexity. In that view, false theories are abandoned because they have to be made increasingly complex to square with the facts rather than because they are refuted outright.

Consilience also occurs when induction from the facts of one domain supports induction from the facts of a different domain; for example, when induction from the facts of the histological domain supports induction from the facts of the neurophysiological, or biochemical, or psychological domains. We have seen in Chapter 2 that consilience is a kind of intertheoretic reduction, when lower-level theories can be included under the terms of a higher-level theory.

There are many reasons why it is advantageous to consider the hierarchical order of theories and to rank the neuron theory as a more-inclusive theory in relation to many less-inclusive, more-restricted theories, such as theories about the genesis, lineages, differentiation, growth, death, maintenance, and functions of nerve cells. Ranking theories in terms of their inclusivity can show logical relations between theories, and the historical process of their construction. A more-inclusive theory, like the neuron theory, is formed by the combination of several less-inclusive theories. The neuron theory is ranked as a more-inclusive theory in relation to less-inclusive theories (listed below) because it succeeded those theories historically, because it includes the explanatory terms of those theories, and because the more-general neuron theory can be reduced to the more-restricted theories, and thus those theories can be explained by the neuron theory.

Mere historical succession of several related theories is not sufficient to rank them in a theoretical hierarchy. At least three additional conditions must apply: There must be a progressively more abstract relationship between empirical observations and progressively more-inclusive theories; the explanatory terms of less-inclusive theories must correspond with those of more-inclusive theories, and the more-inclusive theories must be reducible to the less-inclusive theories as we saw in Chapter 2 (for various views of how theories are related to data and to one another see Chapter 1 and also Duhem, 1906, Campbell, 1920; Quine, 1953, 1960, 1969; Braithwaite, 1953; Popper, 1959, 1974; E. Nagel, 1961; Kuhn, 1962, 1968, 1970; Hesse, 1966, 1980; Rescher, 1970; Cupples, 1977; Lakatos, 1978; Thagard, 1978).

The closeness of the relation between a theory and a group of relevant observations is one important way of defining the level of a theory. There are direct relations between a restricted group of empirical observations and a less-inclusive theory, but more-inclusive theories are generalizations made from less-inclusive theories and do not relate directly to the observations primarily supporting a less-inclusive theory. The terms of the more-inclusive theory are abstractions, far removed from observations (see Section 1.2.2). Terms such as "reflex," "functional integration," and "neuronal specificity" are such abstract terms that have been adopted into the modern neuron theory.

I have perhaps labored this point unnecessarily, but it is one that is habitually ignored by many historians of the neuron theory who relate empirical observations directly to the neuron theory rather than to the relevant less-inclusive theory. This is important because that practice diminishes the historical and explanatory significance of the less-inclusive theories. Placing the primary data in direct one-to-one relations with the more-inclusive theory eliminates the one-to-many relations between more- and less-inclusive theories. The practice of relating the original observations directly to the neuron theory rather than to relevant less-inclusive theories also distorts the corroborative or refutative effects that the observations had on the less-inclusive theories. That practice also distorts the historical relations between observations and theory—the neuron theory had not yet come into existence when most of the observations were made that were directly related to relevant less-inclusive theories.

As I argue in Chapter 1, neurobiological theories are rarely refuted outright. It is obvious that the refutative effects of counter-evidence will be greatest on the theories constructed directly on the available evidence, and least on a more-inclusive theory like the neuron theory that is the consequence of several less-inclusive theories. Before discussing the implications of this view, it is necessary to define the 12 less-inclusive theories that entered into construction of the more-inclusive neuron theory.

A theoretical model goes beyond description of phenomena to attempt unification of knowledge from different sources, explanatory coherence of different observations, and explanation of the phenomena in terms of cause and effect, function, and purpose. Whether such an explanation is justified, or is even possible, is not the question that we are now asking. The question is about the difference between description and explanation that relates to the difference between a report of observations and an attempt to understand their meaning with reference to a theoretical model. Construction of a theoretical model proceeds by inclusion of more observations, and by assimilation of less-inclusive into more-inclusive models. The 12 theories summarized below are regarded as less inclusive because of their relations to the more-inclusive neuron theory, and because they are consistent with a restricted range of relevant empirical observations. These less-inclusive theories are listed below in historical order, allowing for the considerable temporal overlaps between them. A fuller historical account of these theories will be given in the following sections of this chapter.

Theory 1a (Fig. 3.3). The CNS is composed of microscopic subunits: fibers (also found in peripheral nerves) in the form of hollow or fluid-filled tubes (Leeuwenhoek, 1685; Fontana, 1781; Treviranus, 1835) and globules confined to the CNS and ganglia (Leeuwenhoek, 1685; Soemmerring, 1791–1792; Ehrenberg, 1833, 1836; Valentin, 1836b; Purkinje, 1837a,b).

Theory 1b (Fig. 3.4). The ganglionic globules (nerve cells) and nerve fibers are anatomically separate structures, although they may interact functionally (Prévost and Dumas, 1823; Valentin, 1836a,b; Beale, 1860). When the connection between globules and fibers could be resolved with achromatic microscopes, theory 1b was replaced by theory 2.

Theory 2 (Fig. 3.5). Ganglionic globules (nerve cells) and nerve fibers are parts of the same nerve cell (Remak, 1838a,b, 1853, 1855b; Hannover, 1844; Koelliker, 1844, 1848, 1852; Bidder, 1847; Wagner, 1847a,b).

Theory 3 (Fig. 3.5). The ganglionic globules each contain one or more nuclei and

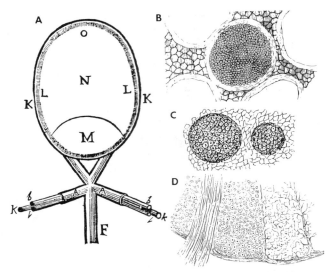

FIGURE 3.3. The Cartesian model of the nerves cast a long shadow into the 19th century. (**A**) Descartes (*L'homme*, 1664) depicted the *"nerve A like a pipe with inner pipes b, c, k—the outer coats of these nerves are continuous with the meninges, K, L, which envelop the brain M, N, O."* He conceived of motor impulses conveyed in the space between the pipes and outer sheath while sensory impulses were conveyed in the inner pipes. This model was adapted by the early microscopists like Leeuwenhoek (1685) as shown in (**B**), and reproduced by Purkinje in 1837 (**C**), and by Sharpey *et al.* in 1878 (**D**). Depictions of peripheral nerve structure remained virtually unchanged even after the concept of animal spirits, supposed to be conveyed in hollow nerve fibers, was replaced by the concept of animal electricity (Galvani, 1791).

nucleoli (Ehrenberg, 1833, 1836; Valentin, 1836b; Purkinje, 1837a,b). After Schwann (1839) saw the nucleolus and nucleus as the essential components of the cell, he recognized that nervous systems are formed of ganglion cells. They originate, migrate, and differentiate like cells in other tissues (His, 1879, 1889a,b, 1890, 1904a; Schaper, 1894a,b, 1895, 1897a,b).

Theory 4 (Fig. 3.5). There are many types of nerve cells which are structurally and functionally specialized (Koelliker, 1852; Berlin, 1858). The two main types are the ganglion cells and the glial cells (Virchow, 1846, 1858; Golgi, 1873, 1875; His, 1889a). A number of auxiliary theories were formed to account for the modes of histogenesis and differentiation of diverse types of nerve cells and for their various functions.

Theory 5 (Fig. 3.6). Dendrites (protoplasmic processes) and axons (axis cylinders) are different prolongations belonging to the same nerve cell (Wagner, 1851 in 1842–1853; Remak, 1854; Deiters, 1865; Golgi, 1873, 1880a,b, 1883a,b,c). *Corollary to theory 5*: Like many other cell types, the nerve cell has regionally specialized parts. This specialization is the basis of nerve cell typology. For example, Golgi (1873) conjectured that protoplasmic processes are specialized for exchange of nutrients between nerve cells and the blood, while the axon and its collaterals are specialized for reception, conduction, and transmission of nerve activity. Alternatively, Ramón y Cajal (1891b, 1894b,c)

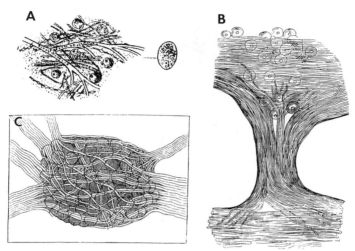

FIGURE 3.4. The concept of anatomically separate ganglionic globules and nerve fibers persisted into the 19th century. There was no significant progress from the representation by Purkinje (1837a,b) (**A**) to later representations in major textbooks: (**B**) from Laycock (1860) shows ganglion cells in the spinal cord and fibers of the third dorsal spinal root of a frog; (**C**) from Sharpey et al. (1878) shows a sympathetic ganglion of a mouse. They show complete anatomical separation of ganglion cells and nerve fibers, although after about 1850 there was increasing evidence showing that they were parts of the same nerve cell. Old neuroscience theories are not eliminated rapidly by counterevidence but gradually fade away—last of all from textbooks.

and van Gehuchten (1891c) proposed that dendrites are specialized for reception of nerve activity from other nerve cells and for conduction to the cell body, whereas the axon is specialized for conduction away from the cell body and transmission of nerve activity to other nerve cells. Sherrington (1897b) conjectured that the synapse is formed of transverse membranes that are specialized to allow transmission in one direction only.

Theory 6. The part of the nerve cell containing the nucleus is the trophic center of the entire cell. When the nerve cell is cut in two, only the part containing the nucleus is capable of long survival (Waller, 1850, 1852a,b; Gudden, 1870; Forel, 1887). *Corollary to theory 6*: After a nerve cell is cut in two parts, regeneration can occur only from the part containing the nucleus (Waller, 1852c). However, many reports of reunion of the cut ends of the vertebrate nerve continued to be made throughout the last half of the 19th century (reviewed by Ochs, 1977).

Theory 7a (Fig. 3.7). Nerve fibers develop from fusion of chains of cells (Schwann, 1839; Balfour, 1881)

Theory 7b (Fig. 3.7). Nerve fibers develop from preformed protoplasmic filaments connecting cells (Sedgwick, 1895; Held, 1897, 1905).

Theory 7c (Fig. 3.7). Nerve fibers are direct outgrowths from nerve cells (Bidder and Kupffer, 1857; His, 1886; Ramón y Cajal, 1890b; R. G. Harrison, 1910). This theory was in conflict with theories 7a and 7b and eventually replaced them.

Theory 8a. Nerve cells are generated in the form of a syncytium in which the

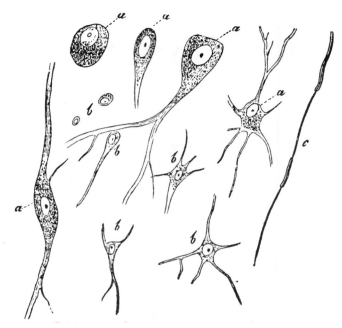

FIGURE 3.5. By 1854 Koelliker could confidently depict nerve fibers and globules as parts of the same nerve cell.

nucleated cell bodies are connected by protoplasmic bridges of various degrees of fineness (Schwann, 1839; Koelliker, 1852; Whitman, 1893; Sedgwick, 1895; J. B. Johnston, 1906). This theory was in conflict with theory 8b, and was finally replaced by it.

Theory 8b. Nerve cells originate by cell division (Remak, 1852; Virchow, 1858) and migrate as free cells from germinal zones to their final positions, where they differentiate and grow as self-contained cells in surface contact but not in cytoplasmic continuity with other cells (His, 1879, 1886, 1887, 1888a,b; Ramón y Cajal, 1890a).

Theory 9a (Fig. 3.8). Conduction of nervous activity occurs via cytoplasmic connections linking nerve cells, either via extensive intercellular cytoplasmic bridges (Müller, 1833–1840; Wagner, 1842–1853; Schroeder van der Kolk, 1859, 1863); or via fine networks connecting the protoplasmic processes (dendrites) of different nerve cells (Deiters, 1865; Koelliker, 1867; Gerlach, 1872), as shown in Fig. 3.8C; or by fine networks connecting axon collaterals of different nerve cells (Golgi, 1882–1885, 1907), as shown in Fig. 3.8C; or by fine fibrils connecting axons of one nerve cell to dendrites of another (Fig. 3.8D and 3.8E; Apáthy, 1897; Bethe, 1900, 1903; Nissl, 1903). All these variants of the theory of intercellular cytoplasmic continuity were in conflict with theory 9b and were eventually replaced by it.

Theory 9b. Neighboring nerve cells are connected by surface contact only and not

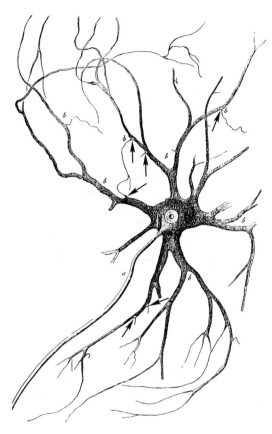

FIGURE 3.6. The concept that protoplasmic processes (b) and the axis cylinder (a) are different prolongations of the same cell was represented by Deiters in 1865 in a neuron dissected from the spinal cord of an ox (from Max Schultze, 1870). Deiters thought that input to the protoplasmic processes (dendrites) was via fine fibers connecting by means of a trumpet-like expansion (arrows added). Thus, he narrowly missed finding the presynaptic endings discovered by Auerbach in 1898 (see Fig. 3.16.)

by cytoplasmic continuity (His, 1886, 1887; Forel, 1887; Ramón y Cajal, 1888a,b, 1890a,b, 1894a,b; Koelliker, 1890a–d; Lenhossék, 1891; van Gehuchten, 1891; Retzius, 1891a,b, 1892b).

Theory 10. Nerve conduction can occur in either direction in the axon (du Bois-Reymond, 1849), but in the CNS the nervous impulse is conducted in a preferred direction from nerve cell to nerve cell. This concept was generally held after about 1855, and defined by Koelliker (1867) for nerve cells connected by cytoplasmic anastomoses (Fig. 3.9). In terms of the contact theory, one-way conduction was conceived as an inherent property of polarized cell structure—from dendrite to axon in the same nerve cell and from axon of one cell to dendrite or cell body of another ("law of dynamic polarization" of the nerve cell: Ramón y Cajal, 1891b; van Gehuchten, 1891a; Lenhossék, 1893).

Theory 11C. An intercellular barrier, the synapse, between neurons in contact with one another, allows nervous transmission between neurons in one direction only (Sherrington, 1897b).

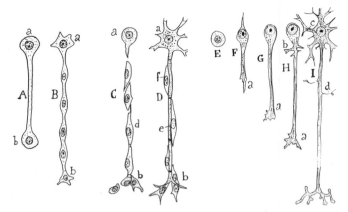

FIGURE 3.7. Three theories of the mode of formation of the axon. **A**, **B**, showing the axon developing from a preformed cytoplasmic bridge connecting neuroblasts a and b. **C**, **D**, showing axon e developing by fusion of a chain of neuroblasts, d, which become Schwann cells, f. **E–I**, showing the axon, with growth cone, a, growing out of the neuroblast: **E**, germinal cell; **F**, bipolar phase with initiation of growth cone, a; **G**, neuroblast phase; **H**, appearance of basal dendrites, b; **I**, formation of apical dendrites c, axon collaterals d, and axon terminals. (Redrawn from Ramón y Cajal, 1901–1917, Plate XXVIII.)

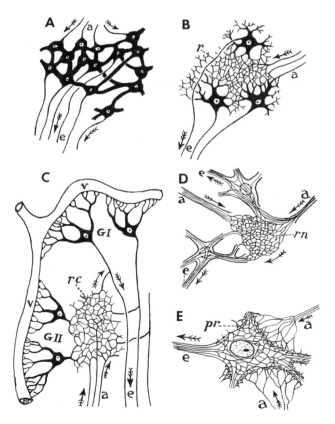

FIGURE 3.8. Theoretical models made to explain how nerve cells might be connected together by cytoplasmic continuity. Theories made over a 50-year period by the following neuroscientists are represented: (**A**) Schroeder van der Kolk (1859); (**B**) Gerlach (1865) and Deiters (1865); (**C**) Golgi (1891a, 1907); (**D**) Apáthy (1897); (**E**) Bethe (1900, 1904) and Nissl (1903). a, afferent axon; e, efferent axon; GI, Golgi type I neuron; GII, Golgi type II neuron; r, network of branches of dendrites; rc, network of axon collaterals; rn, extracellular network continuous with intracellular neurofibrils; pr, pericellular net connected with intracellular neurofibrillary net; v, blood vessel.

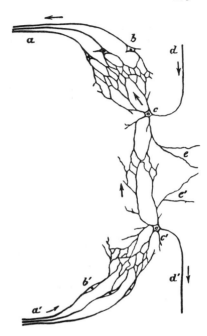

FIGURE 3.9. The idea that nervous impulses travel in a preferred direction—in dendrites toward the cell body, and away from it in the axon—is shown in this representation of Gerlach's nerve net theory (from Koelliker, 1867). For the contact theory of nerve connections the idea was presented as the "law of dynamic polarization" of the neuron (van Gehuchten, 1891b,c; Ramón y Cajal, 1891b).

Theory 12. If synaptic transmission from one (presynaptic) neuron to one or more (postsynaptic) neurons has any effect on the latter, it always has the same effect—either excitation or inhibition. But when more than one presynaptic neuron makes synaptic connections with different parts of one postsynaptic neuron, there are spatial and temporal summations of the inhibitory and excitatory synaptic actions of individual presynaptic endings on the postsynaptic nerve cell. These synaptic events are the basis of the integrative actions of the nervous system, as represented in Fig. 4.8 (Sherrington, 1906). *Corollary to theory 12*: all presynaptic endings belonging to a single neuron release the same chemical neurotransmitter (Dale, 1935).

It is only for convenience that these theories have been listed in this manner. Their interrelationships should be represented more realistically at four levels simultaneously. First is the level of the logical relationship within and between theories. Neuroscience advances by flowing together of knowledge from many sources, by convergence of knowledge from different domains, and by cohesion of observations made by different methods. Second, progress can be viewed as the construction of various conceptual models. These are representations which approach closer to real nervous systems as more facts are unified in the same theoretical model and as different models are unified under a more-inclusive model. Third, those neurobiological theories have to be viewed in relation to more-inclusive, higher-level, physical theories that are being constructed at the same time. Fourth, we must comprehend those theories as part of the sociocultural fabric into which they were woven. It is hardly necessary for me to disclaim the knowledge of science, history, sociology, and philosophy necessary for this task. I shall have succeeded if I point in the correct directions without myself traveling all the way to the indicated destinations.

3.5. NEURON THEORY IN RELATION TO MICROREDUCTION OF NERVOUS SYSTEMS

The modern era of neuroscience may be said to have started in the late 18th century when the concept of the molecular organization of nerve cells was related to the concept of the molecular constitution of all living matter. This great advance was associated with the philosophy of materialism and the practice of reductionism (F. A. Lange, 1877–1892; Merz, 1904–1912). We have considered reductionism at some length in Chapter 2. Here it is sufficient to define reductionism as a methodology aimed at discovering the identities of higher-level components of the nervous system in terms of lower-level components, and explaining everything in the higher level in relation to components at lower levels. Because higher and lower are often the same as large and small, this method is often called macro-to-micro reduction, or simply microreduction.

In practice the historical order of discovery is often from the macroscopic to the microscopic, but the historical order of conceptualization is often unrelated to the empirical demonstration of a physical hierarchy. Just consider that some pre-Socratic philosophers, especially Democritus, conceived of atoms before they could be understood in terms of Dalton's atomic theory and before the nuclear theory of the atom was proposed by Rutherford in 1911. Likewise, the concept of the molecular structure of living organisms was advanced by Gassendi and Borelli in the 17th century, but it was not until the end of the 19th that biological theories began to be constructed in which molecules were an essential part. Theories were constructed in which synapses (Sherrington, 1897b) or viruses (Lwoff, 1957) were essential parts long before those entities could be seen by means of microscopes. The historian who chooses to portray the history of neuroscience concepts as a linear progression of reduction from larger to smaller components of nervous systems does so at considerable risk of misrepresentation of the history of ideas. Theories in neuroscience have sometimes evolved in hierarchical order, starting with theories about the brain as a whole, then about gross anatomical subdivisions, then about microscopic anatomy. However, it would be vulgarization of history to portray the progress of science in that way exclusively. As I have shown in Chapter 1, progress was driven by many forces, especially the need to fit the ever-increasing amounts of new data, obtained by rapidly improving research techniques, into theories and models of greater inclusiveness and of increasing explanatory generalization. Progress was not an orderly advance in a straight line toward an agreed-upon goal. Nor was progress only driven by the accumulation of data. Extrascientific and psychological factors, for which there are no simple explanations, impelled different people to prefer different data, theories, and goals.

To say that observations "acquire" meaning is the same as saying that meanings are formed in the minds of people who think about how new observations can be fitted into their private mental models. I argue in Sections 1.2 and 5.4 that mental models are formed by the process of imaginative reflection, and by induction in which knowledge from different sources is unified to form an explanatory model, and by communicating models in the public arena for constructive critique. This is the basis for my understanding of the role of the process of reflection-communication-critique in forming theories of increasing explanatory generalization which also gain increasing consensus.

René Descartes (1596–1650) and later the iatromechanists Pierre Gassendi (1592–1655) and Giovanni Alfonso Borelli (1608–1679) taught that if we knew enough we could reduce biology to mechanics.* They believed that the processes by which organisms develop, life is sustained, and animals move, are purely mechanical and strictly determined by physical law. The fibrous structure of the brain was known to Descartes, who based his theory of the flow of animal spirits on that foundation. The concept that living organisms are reducible to elementary components, invisible atoms, corpuscles, and fibers, which were already in the 17th century termed molecules, was promoted by Pierre Gassendi and Robert Boyle (1627–1691) among others (reviewed by Hall, 1969).

The atomistic and corpuscular theory, according to which all natural phenomena can be reduced to the size, shape, and motion of invisible atoms and molecules was reconciled with the Christian religion, for the Bible taught that "*God makes all things by weight, measure and number*" (Ecclesiastes 11:21). The ideal of complete reduction of the universe to the fundamental mathematical order was regarded as the goal of science and philosophy by the rationalists. Although rarely reached in practice, they believed that the goal was attainable in principle because "*the world is an epistle written by God to mankind (as I believe Plato has said); he could have added, written in mathematical characters*" (Robert Boyle, *The Excellence of Theology*, p. 36–37, 1674). Revelation could thus be seen as identical to intuition and reason, but it could be attained only by those who knew the language of mathematics. The *laboratorium* was the new temple with new instruments and techniques through which intuition and reason could reveal the mathematical order of the world. This was the very beginning of the belief that science is all that is needed to be done to attain knowledge. That belief coupled with the realization that scientists have the ability to make history are the sources of two failings of many scientists of our times—intellectual pride and the belief in the self-sufficiency of science.†

We suspect that reality is underpinned by mathematics. Otherwise, how could the axioms of mathematics be absolutely certain, and how could we explain the universality of mathematics in all our attempts to explain nature? But the workings of the brain are not absolutely certain and indisputable, like mathematics. One neuron plus another neuron does not always add up to the same sum—they violate what Rudolf Carnap called "correspondence rules." Because no two neurons are identical, and because they are all in process of change, adding up two neurons is like adding two people chosen at random

---

*The analytical method was defined by *Descartes in Discourse on the Method* (1637, Part 2, p. 18): one should start "*with the simplest and most easily known objects in order to ascend little by little, step by step, to knowledge of the most complex, and by supposing some order even among objects that have no natural order of precedence.*" In practice, reduction has been done in both directions—top-down as well as bottom-up (see Chapter 2).

†This is what Kant refers to as "the luxury of science as affording food for pride, through the insatiable number of inclinations thus aroused" (*Critique of Judgement*, Sect. 83). He also refers to the beauty of the arts and sciences which make us more civilized to the extent that they lead us to reflect on the purposes of nature.

in a crowd or chosen because they are of the same height or age. If we add only the essential, universal, timeless attributes of two neurons we denude them of the attributes which are most relevant to our interests in their roles in organized nervous systems. This is one of the problems of reductionism that we confronted in Chapter 2.

Atomistic and molecular theories of living organisms were conceived in the 17th and 18th centuries, long before the cell theory was advanced in the 19th. Those conjectures about molecular mechanisms were often close to the truth as we now understand it. For example, the modern concept of "treadmilling" whereby actin filaments grow by addition of actin monomers at the plus end and an equal loss of monomers occurs from the negative end of the actin filament (L. G. Tilney *et al.*, 1981) was first defined more than 300 years ago by Descartes. He advanced a theoretical concept of the movement of animal spirits in the form of small particles along the fibrils which form nerves: "*As fast as any particle is detached at the extremity of each fibril, another is attached at its root*" (*La description du corps humain*, first published in 1664). This was a premature theory which could not be accepted into a neuroscience research program because it was too far in advance of any research program at that time.* It was also an immense inductive leap from negligible knowledge to a molecular theory of movement of molecules in nerve fibers. Many conjectures about molecular mechanisms were made in the 17th, 18th, and 19th centuries which could not be tested experimentally until the 20th century. Nevertheless, although the methods were not available to implement the reductionist program, that was clearly seen as the desired goal.

As new facts emerged, the neuroscientists' image of the brain changed from gross macroscopic to low-power microscopic to higher powers of microscopic analysis. The word "molecular" was used with increasing frequency in statements about nervous activities, both by neurophysiologists like Michael Foster and by neuroanatomists like Koelliker. In the first edition of his *Manual of Human Histology* (1853) Koelliker wrote "*If it be possible that the molecules which constitute cell membranes, muscular fibrils, axile fibres of nerves should be discovered, and the laws . . . of the origin, growth, and activity of the present so-called elementary parts, should be made out, then a new era will commence for Histology, and the discoverer of the law of cell genesis, or of a molecular theory, will be as much or more celebrated than the originator of the doctrine of the composition of all animal tissues out of cells.*" Michael Foster in the third edition of his *Text-Book of Physiology*, published in 1879 wrote that "*the central nerve-cells concerned in reflex action are to be regarded as constituting a sort of molecular machinery.*" These are indications that the research program of molecular neurobiology had begun and that neurobiology had entered the modern age.

Marcello Malpighi was the first to claim to have seen microscopic bodies in the brain, and he believed that they were glands that secrete animal spirits (Malpighi, 1686, 1697). However what Malpighi saw were almost certainly artifacts. That Malpighi could

---

*The concept of maturity of a theory deals with the role of that theory in construction of a research program and is discussed in Section 1.4. Briefly stated, a theory is mature when it can function as an explanatory model around which a research program can be constructed, using various methods to make observations that may be used to modify the theory or to corroborate it or to refute it. A premature theory is unable to function maturely because the techniques are not available to make observations that can test the theory and because of a variety of extrascientific factors that are discussed in Section 1.4.

not have seen nerve cell bodies has been demonstrated recently by using a microscope and histological methods like those used by him (Belloni, 1966, 1968; Clarke and Bearn, 1968). The globules that Leeuwenhoek and other 17th and 18th century microscopists observed in the brain were probably lipid droplets and particles surrounded by halos which were the effects of optical aberrations (Baker, 1948–1955). Some of the globules may have been cells but it is virtually impossible for us to sort out which were really cells and which were artifacts. Joseph Jackson Lister, who developed the achromatic objective in 1830, concluded that virtually all previous microscopic observations of histological structure were invalid because of the gross optical aberrations produced by the available lenses (Lister, 1870). The poor methods of preparation of specimens for microscopic examination also set severe limitations on the accuracy of histological observations. Construction of the research program concerning nerve cells and fibers and their connections was closely linked to the progress of microscopic and histological techniques, which is considered in Section 2.7.

Rapid progress in understanding the cellular organization of the CNS occurred in the decades from 1830 to 1860. Several conditions favored the rapid progress of neurobiology, especially in Germany: the improved cell theory of Schwann, published in 1839 and developed while he was a student of Johannes Müller at Berlin (Baker, 1948–1955; Hall, 1969); favorable conditions for research by professional scientists, especially at German universities (Mendelsohn, 1964; Ben-David, 1971; McClelland, 1980); improved microscopic and histological techniques; and the improved techniques for making paper, setting type, and printing illustrated books that led to the mass production of scientific periodicals.*

The concept that the brain is composed of separate microscopic globules and fibers suspended in a fluid or ground substance was formulated in the 1830s on the basis of more reliable observations that were possible because of improved histological techniques and use of achromatic microscope lenses (Ehrenberg, 1833, 1836; Purkinje, 1837a,b; Valentin, 1835, 1836a,b, 1838; among many others).

The discovery of nerve cell bodies, then called ganglionic globules, was made by Christian Gottfried Ehrenberg (1833, 1836), not only in the vertebrates but also in the much larger nerve cells of the leech and the snail. The first engraving of a nerve cell nucleus and nucleolus are found in the papers by Ehrenberg (1833) and Valentin (1836b). The nucleus had already been discovered in 1702 by Leeuwenhoek in the red blood cells of the salmon. According to Baker (1949), Fontana (1781) first discovered the nucleus and nucleolus in animal cells other than red blood cells. After 1830, intracellular structures could be resolved in nerve cells by the use of the recently invented achromatic microscope objectives (Lister, 1870). That technical revolution, more than anything else, resulted in the discoveries of cell structures that were made rapidly in the period from 1840 to 1890.

---

*Apparently the social conditions were favorable in Germany. However, the economic conditions were much more favorable in Britain which was the world's richest country from about 1850 to about 1910. From 1870 to 1900 the purchasing power per person in Britain was almost double that in Germany (*The Economist*, 20 June, 1992). Yet neuroscience stagnated in Britain during the 19th century until well after 1860, mainly because of disadvantageous social conditions that are discussed in Section 4.4.

The relations between the cell theory in its successively modified versions and the components of the nascent neuron theory can be reviewed briefly. Schwann conceived of five modes of relationships between cells in tissues, ranging from complete separation of autonomous cells (as in the blood) to complete coalescence (as in muscle and nerve). He conceived of various degrees of fusion of cells ranging from coalescence of the cell walls to complete mingling of the cell contents (Schwann, 1839, p. 177). He thought that nerve fibers form as a result of fusion of chains of cells. Schwann's concept of cell formation, as he originally described it in 1839 in his *Mikroskopische Untersuchungen*, begins with formation of a nucleolus inside a mother cell by a process analogous to crystallization. Then the nucleus emerges from the nucleolus and surrounds it. This sequence is repeated until the mother cell bursts, allowing the daughter cells to escape. Johannes Müller added the possibility that nuclei can also form in the extracellular fluid. He maintained that notion in the 1844 edition of his *Handbuch der Physiologie*.

In spite of Remak's demonstration in 1841 of cell division of leukocytes, there was no reason to bring that evidence into the cell theory as it was then conceived. It was not until 1852 that Remak was able to propose that all cells originate by binary division of mother cells and that *"pathological cells and normal cells in general are . . . products or descendants of normal tissue cells."* In his book *Untersuchungen über die Entwickelung der Wirbelthiere* (1855a, pp. 164–170), Remak developed the concept that all cells originate from other cells in the embryo, and not from a cytoblastema as Schwann and Müller had conjectured. Virchow's celebrated pronouncement *"Omnis cellula e cellula,"* made in 1858, is directly derived from Remak, although Virchow did not acknowledge this. Remak's priority for this correct version of the cell theory is now well established (Pagel, 1945; Ackerknecht, 1953; Kisch, 1954; Hall, 1969). Both Remak's and Schwann's theories of cell formation were considered valid by Koelliker in the first editions of his *Mikroskopische Anatomie* (1850) and his *Handbuch der Gewebelehre* (1852).

It was not until 1873 that the stages of the mitotic cycle were correctly described by Schneider (in the embryo of a flatworm). Two years later Strassburger described cell division in plants. A much more accurate description of the stages of cell division was given by Flemming in 1882. He named the entire cycle *"Karyomitosis"* and the individual nuclear changes *"Mitosis."* Before Flemming's work it was believed that the nucleus dissolves during each cell division and reforms anew in each daughter cell. The publication, in 1882, of Flemming's *Zellsubstanz, Kern und Zelltheilung* (*Cell Substance, Nucleus and Cell Division*) was the main impetus given to studies of mitosis in the neural tube and the discovery of the neuroepithelial germinal cells by Wilhelm His (1887, 1888a,b, 1889a, 1890). Flemming's book showed the first clear depiction of duplication and division of the chromosomes (named by Waldeyer in 1888). Fleming recognized that chromatin (the name he gave to the material in the nucleus which he stained with azo dyes) is probably the same as the nucleic acid which Miescher (1871) had isolated from the nuclei of leukocytes and had called *"nuclein."* The nucleus was identified as the cell's genetic center by Koelliker, Oscar Hertwig, Strassburger, and Weismann, almost simultaneously in 1884 to 1885. By about 1885 it was generally thought that chromatin, the material of the chromosomes, is the basis of heredity. Those discoveries, made in the short period from 1840 to 1890, linked together the domains of cytology, embryology, biochemistry, and evolution.

Understanding of the functions of the nucleus increased slowly during the 1840s. Nevertheless, depiction of ganglionic globules containing one or more nuclei and nucleoli became conventional after 1833. Gustav Gabriel Valentin (1836a,b) confirmed Ehrenberg's findings in the abdominal ganglion of the leech He introduced the word "nucleus" (in 1836) and the word "nucleolus" (in 1839) into the terminology of animal cytology. Jan Evangelista Purkinje (1837a,b) corroborated the presence of the nucleus and nucleolus in the large nerve cells in the cerebellar cortex that now bear his name. At that time he showed only the Purkinje cell body and failed to recognize the relationship of the protoplasmic processes (i.e., dendrites) to the nucleated part of the cell.*

Of course, before the publication of Schleiden and Schwann's cell theory, these early workers had no correct ideas about the functions of the ganglionic globules and their nuclei and nucleoli. I agree entirely with Baker (1949) that the great accomplishments of Schleiden and Schwann were to gain recognition for the ubiquity of the nucleus in development and function of cells, and for the nucleated cell as the unit of most organisms. However, they were only starting to form theoretical concepts explaining why the nucleus is ubiquitous. Ehrenberg (1833) called the nerve cell bodies "*grössere drüsenartige Kugeln*" (large glandlike corpuscles), thus perpetuating Malpighi's misconception of their secretory functions. Another misconception was that the nucleus and nucleolus in the ganglionic globules were in some way analogous to the germ in the egg. This idea came from Prévost and Dumas (1823) and was fostered by the romantic ideas of *Naturphilosophie*, especially the idea that the macrocosm contains an infinite series of nested microcosms.

Both Valentin and Purkinje were vitalists, influenced by *Naturphilosophie*, and as a result they came into conflict with Schwann and other materialists (or vital-materialists, as defined by Temkin, 1946) about the "correct" interpretation of the cell theory. I agree with Clarke and Jacyna (1987) that *Naturphilosophie* did not have entirely negative effects on neuroscience in the 19th century. It encouraged the use of analogy and metaphor in thinking about the meaning of phenomena (see Purkinje, 1839). But it gave rise to some totally misconceived models such as Oken's vertebral theory of the morphology of the skull, Goethe's color theory, and Carus's notion of the fundamental globular structure of everything from the planets to cells.

Purkinje and Valentin were guided correctly to discover the universality of ciliary motion, but they were also misguided by their beliefs into seeing false analogies. For example, Carl Gustav Carus (1814, 1835) proposed that the primary form of all organization is the spherical, for instance the globule of mercury, of blood, the planets and stars. Hence, the earliest stages of embryos are spherical and are composed of spherical globules. Likewise, he argued that the elementary units of the nervous system must be spherical globules, and that is what his followers believed they saw in the CNS. They saw analogies between the ganglionic globules in the nervous system and the germinal vesicle of the avian egg, which Purkinje discovered. They saw both those structures as the same arrangement of concentrically nested spheres (the *Urform*), and saw both as sources of generative forces (the *Urphänomen*). Above all they recognized the teleological principle, an irreducible causal principle, the "natural purpose" (*Naturzweck*).

---

*Purkinje's name was given to the cell by Obersteiner (1869) in his doctoral thesis.

Powerful microscopic and histological techniques were developed just as the influence of *Naturphilosophie* diminished, and I am tempted to find a causal relationship there. From about 1840 the new observations made possible by the general use of achromatic microscopes and more reliable histological techniques were inconsistent with the basic notions of *Naturphilosophie*. If that philosophical inclination had any effects, they were to guide the new powerful techniques in the wrong direction, to seek analogies between microscopic components of nervous systems and the structure of the psyche in the manner of Freud and his followers, as we have seen in Section 1.2.6.

The notion of the morphological type designed for fulfilling a natural purpose was held by many in addition to the *Naturphilosophen* who were influenced by the philosophy of Hegel and Schelling. Timothy Lenoir (1980, 1981, 1982) has discussed the influence of Kant on the Göttingen school which started with Blumenbach and his students such as Reil, Treviranus, Tiedemann, and Wagner, to mention only some who participated in the construction of the neuron theory in the early period from about 1830 to 1850. Lenoir calls those people "vital-materialists" because, following Kant (*Critique of Judgement*, 1790), they conceived of living systems as mechanisms that were reducible but which included an irreducible and ultimately unknowable teleological principle. They were all vigorously opposed to mechanistic reductionists like Helmholtz and du Bois-Reymond who believed that all biological systems could be reduced completely to chemistry and physics. This brief digression has been made with the purpose of showing that theoretical models made by neuroscientists were strongly affected by their ideologies.

None of the discoverers of the nucleus and nucleolus in ganglionic globules had the faintest idea of their real functions. Adequate analysis of the significance of the early discoveries of the structure and functions of nerve cells requires much more detailed attention than I have the means to give here, to the history of biology and especially to the cell theory. Some of the relevant information can be obtained elsewhere (Baker, 1948–1955; Hall, 1969). Here it is relevant, in connection with Purkinje's contributions both to the cell theory and to the neuron theory, to say that in 1825 Purkinje discovered the germinal vesicle of the egg of the bird, and he conceived of the entire egg (ovum) as a cell. In 1838 Schwann agreed that the ovum is a cell but he thought that the germinal vesicle was the nucleus and the yolk the cell body (see Schwann, 1847, pp. 152–153). It was not until 1861 that Gegenbauer showed that the entire ovum is a cell.

The concept that the nervous system consists of separate globules and fibers prevailed until the 1840s. Purkinje and his student and collaborator Valentin conceived of the ganglionic globules and the neighboring nerve fibers as anatomically separate, although in some way functionally related. They conjectured that the ganglionic globules were "nervous energy generators" and the nerve fibers were "nervous energy conductors." They recognized that, in some unknown way, the nervous energy must be transmitted from the ganglionic globules to the nerve fibers. It was believed that globules and fibers originate separately and that when they join, which occurs only rarely, it was a secondary and perhaps an impermanent union.

In the aftermath of Magendie's demonstration in 1822 of the functions of the spinal nerves, the concept of the CNS as an input output system became generally accepted. The outstanding problem was how inputs were related to outputs in the brain and spinal cord. The concepts of Purkinje and Valentin must be viewed in the framework of the problem of how nervous inputs are transferred to nervous outputs. Note that as long as it was conceivable that the input fibers were physically separated from the output cells, there remained room for the immaterial soul and for vital forces to affect the connections. Such ideas would have been congenial to Purkinje and Valentin, who were vitalists, and were adherents of *Naturphilosophie*.

The state of understanding at that time of the relations between nerve cells and fibers was summarized by Johannes Müller: "*It would be important to know whether the large globules of the grey substance in the brain and in the ganglia have, or have not, any connection with other parts or with one another. Certain processes, which are, under favourable circumstances, seen issuing here and there from the globules, suggest the probability that they are connected with each other by fibres. I saw these tooth-like processes first on the club-like bodies in the medulla oblongata of the petromyzon. Remak observed them soon afterwards on the globules of the grey substance of the brain and ganglia. He not only saw fibres coming off from the surface of the globules of the ganglia, but succeeded even in isolating them to the extent of several or many times the diameter of the globules. These fibres of the ganglionic globules have some similarity with the delicate grey filaments which Remak has observed in ganglionic nerves; and if the latter filaments which form the grey fasciculi of the sympathetic are organic fibres, it becomes in some degree probable, or at least possible, that this is the mode of origin of those fibres*" (Müller, 1840, p. 657).

I suggest that Müller struggled with Remak's discovery of the continuity of the nerve fiber and the nervous globules because the implication was that the inputs to the nervous system feed directly into the input fibers of the nerve cells and the output fibers of those cells continue as the outputs of the nervous system as a whole. That model was soon pictured in the textbooks of the time (Fig. 3.8A,B). There was no room for the soul between input and output in that model, and although not explicitly stated, it was implied that the soul was not required for routing the nervous activity in the brain and spinal cord. The nucleated part of the nerve cell then occupied the place between input and output formerly occupied by the Cartesian soul. The great advance made possible by Remak's discovery of the unity of nerve cell body and nerve fibers was the ability to conceive of the input-output functions of the nervous system determined by the polarity of its constituent cells.

These preliminary remarks may help to give a more general significance to the discovery of the continuity of nerve globules and fibers. Increasingly more reliable observations indicating that the globule and fiber belong to the same cell were made by Robert Remak (1838a,b, 1853, 1855b), Albert Koelliker (1844, 1850), and Rudolph Wagner (1847 a,b). The theory of the unity of nerve cells and fibers was generally accepted before the 1850s and it was clearly stated by Koelliker in the first edition of his *Handbuch* published in 1852. That theory gained additional empirical support from histological evidence indicating that nerve fibers grow out of the nerve cells (Bidder and Kupffer, 1857; His, 1886). These data were obtained from histological sections, and were open to criticism regarding their authenticity as evidence of the condition of the living nerve cell. Therefore, the converging evidence that nerve fibers degenerate after they are disconnected from the nerve cell was strong evidence for the unity of the nerve cell. This evidence led to the concept of the nerve cells as a trophic center (Waller, 1850, 1852a,b), which is discussed at some length in Section 1.4.2.

These advances in understanding the relations of parts of the nerve cell gain significance when seen in relation to steps in the promulgation of the cell theory. The conceptual leap from the solidist concept, that each tissue is formed of its own type of some relatively homogeneous solid material, to Schwann's cellular concept was made in two steps. First, Schwann accepted that the initial condition of any tissue was a uniform matrix, which he called the cytoblastema, and second, he thought that cells emerged in some way analogous to crystallization out of the matrix. This goes to show that one cannot leap across a chasm in two steps. The successful conceptual leap was made by understanding that all cells arise from other cells by cell division. This leap was first made by Robert Remak in 1852, although Rudolf Virchow, who made it in 1858, is conventionally awarded the priority. The idea that all nerve cells originate from other cells by a process of binary division entailed that the nerve cell originates without fibers for it was inconceivable that a cell with long fibers could divide or that retraction or detachment of long nerve fibers could occur before cell division.

The discovery of the difference between dendrites and axons was the next big conceptual advance toward understanding the physical basis of nerve cell polarity (Wagner, 1851; Remak, 1853; Deiters, 1865). The genealogy of concepts about regional specialization of nerve cell processes will be traced in Section 3.9. Increasing understanding of the differences between axons and dendrites intensified efforts to discover how nerve cells are connected together. The earliest concepts formed in the 1850s and 1860s were based on the assumption that protoplasmic processes (called dendrites by His, 1890) form the link between neurons as shown in Fig. 3.8A (e.g., Bostock 1825–1828; Müller, 1839; Schroeder van der Kolk, 1859, 1863). Gerlach (1865, 1872) and Deiters (1865) thought this linkage was made by fine fibrils connecting dendrites of different neurons (Fig. 3.8B). Camillo Golgi (1882–1885) conjectured that the linkage was formed by a fiber network interposed between afferent axons and collaterals of efferent axons (Fig. 3.8C). These theories will be considered more fully in Section 3.11.

It is significant that the first good evidence in support of the contact theory was experimental and not merely histological, because the histological methods in common use at that time were not capable of resolving the problem. August Forel (1887) showed that, after eye enucleation or lesions of the visual cortex, degeneration was confined to the injured neurons and did not extend to those in contact with them. Only later did the histological evidence accumulate to support the theory of nerve cell contact (Ramón y Cajal, 1888a,b, 1890a,b; van Gehuchten, 1891a–e; Lenhossék, 1891). But that evidence remained relatively indirect, therefore weak, because the contact zone could not be resolved with light microscopic histological techniques. In Section 3.12 we discuss the construction of different models of the modes of contacts between neurons, between sense organs and neurons, and at the neuromuscular junction.

Theories, techniques, and values meshed together effectively during the second half of the 19th century to facilitate construction of what we can now call the neuron theory. We have already indicated how the techniques were developed from about 1830 to 1885. We have also summarized the main events leading to construction of the neuron theory from several less-inclusive theories. These will be discussed in more detail in the following sections. Values too had changed during that period. The positivist conviction that science has the power to explain the structure and functions of the nervous system and to apply that knowledge for the benefit of mankind gave added value to research on the nervous system. The conviction that the human brain is a product of organic evolution (Darwin, 1854, 1871; Huxley, 1863; Haeckel, 1874, 1883; Romanes, 1893a) gave added value to research on comparative neurology. The mechanical materialistic conception of nature gave added value to reductionist methods for studying the nervous system. Reductionist materialism proved itself immensely successful during the 19th century, even in its power to show relations between the human mind and the brain.

The concept that the size of an animal's brain is proportional to its intelligence originated near the end of the 18th century and is explicitly stated in the influential work, *Vom Hirn und Rückenmark* (1788), by S. T. Soemmerring. This was an advanced concept, and it was soon accepted into a research program aimed at discovering the anatomical correlates of mental faculties (reviewed by Bastian, 1896). Study of the brains of geniuses was thought to provide evidence supporting correlations of mental abilities with cerebral convolutions and brain size (Spitzka, 1907; Economo, 1929). The evidence showing extraordinary features of the brains of geniuses started with the collection of brains assembled at Göttingen by Rudolph Wagner (1805–1864). In 1816 Friedrich Tiedemann directed that after his death his brain be removed for study. Tiedemann had been one of the principal opponents of phrenology, and one of the first to demonstrate the regularity of the pattern of cerebral convolutions in humans and mammals (Tiedemann, 1816).

Attempts to find unique structures of the human brain, not found in subhuman primates and other vertebrates, were stimulated by the phrenological theories of Gall and Spurzheim. Thomas Laycock (1860), Paul Broca (1861a,b,c, 1877, 1878; see Schiller, 1979), and John Hughlings Jackson (1884), under the influence of the phrenological movement, continued this task after Gall's death in 1828 with the aim of finding cerebral localization of human intellectual faculties. In *The Origin of Species* (1859), and especially in the second edition of *The Descent of Man* (1874, with a "note on the resemblances and differences in the structure and development of the brain in man and apes"), Charles Darwin pointed out that the comparative study of the brains of primates might give some indications of human origins and the evolution of human behavior. However, all attempts failed to find unique structures of the human brain (Huxley, 1863). For example, the claim by Richard Owen (1858) that the distinctive feature of the human brain is the possession of a structure he called the "hippocampus minor" was shown to be false when the same structure was found in apes and seals (Huxley and Flower, 1862).*

---

*This led Charles Kingsley to write a parody for children called the *Water-babies* (1863), in which he wrote: "*The professor had even got up at the British Association and declared that apes had hippopotamus majors in their brains, just as men have. Which was a shocking thing to say; for if it were so, what would become of the faith, hope and charity of immortal millions? You may think that there are other more important differences between you and an ape, such as being able to speak, and make machines, and know right from wrong, and say your prayers, and other little matters of that kind; but that is only a child's fancy, my dear. Nothing is to be depended upon but the great hippopotamus test. If you have a hippopotamus major in your brain you are no ape. . . .* "

Search for uniquely human brain structures was abandoned soon after the beginning of the 20th century. By then the comparative neuroanatomists had demonstrated the homologies of all the main parts of the human nervous system. They could confidently claim to have demonstrated a continuous evolution of the brain from fishes of the Devonian to modern man. A pointed example of this certainty is found in Cajal's study of the motor cortex of the human brain. He concluded that *"there are only quantitative differences, not qualitative differences, between the brain of a man and that of a mouse. Accordingly, all cortical regions which are vested with a specific structure and a specific function and are differentiated in humans are also represented—with the corresponding simplification and reduction—in the mammals and probably even in the lower vertebrates"* (Ramón y Cajal, 1890e). Then where do the differences in functional capabilities reside? Are they to be found in the structural organization, complexity, size, or in none of those, or are they emergent properties of the whole brain? Those questions are discussed in Section 2.8.

During the period from 1830 to 1885, gross anatomical studies gave way to microscopical, biophysical, and chemical investigations of the structure and functions of the nervous system. A tremendous conceptual leap was made from descriptive gross anatomy and understanding of global functions of the brain to understanding its microscopic organization and the corresponding physiological activities. The old macroscopic view was expressed fully for the last time by Burdach in his masterpiece, *Vom Baue und Leben des Gehirns*, three volumes published from 1819 to 1826 (see Edinger, 1904; K. F. Meyer, 1970). The new view of the structural organization of the nervous system began with observations of Stilling, made possible by his invention of the technique of serial section reconstruction in 1842. Prior to Stilling, individual cells and fibers had been teased out of nervous tissue, but observations of isolated cells and cell fragments could not show how they fitted into the complete organization. That understanding began with the histological investigations of the CNS by the 19th century giants of microscopic neuroanatomy. Concurrently, neurophysiologists used that knowledge of microscopic neuroanatomy to interpret their experimental observations. This led to the debate about the primacy of structure or of function that enlivened neuroscience during the 19th century. We shall consider it in Section 4.3.4.

After this brief summary of the initial steps in forming some of the lower-level theories on which the neuron theory was eventually constructed, we can now look at some of them in greater detail.

3.6. NEURONAL TYPOLOGY

Typologism with respect to nerve cells is the belief that for each type of nerve cell there is a nontrivial set of properties, genotypic and phenotypic, that is essential for an individual cell to be a member of that nerve cell type. The advantages of this claim are that it provides a basis for classification and that it can lead to further reduction of each cellular property to its molecular properties. Typologism is also essentialism, both of which have been decried by some biologists.

In the 19th century the concept of neuronal types was undoubtedly related to the type-concept in classification of animals and plants (Farber, 1976), that is, as examples selected for convenience of illustration and comparison, not because it was thought that ideal types exist in nature. The latter method of identification of taxa on the basis of some essential, indispensable, property, attribute, or characteristic (the essence) is what has been called *methodological essentialism* (Popper, 1962, p. 206; also see D. L. Hull, 1964–65). The concept of neuronal typology was at first based on the former assumption (convenience for purposes of description and comparison) and has only recently been based on the latter (definition of some essential intrinsic quality or qualities).

Methodological essentialism has been embraced by molecular neurobiology, as one of its fundamental laws, that cellular differentiation is based on differential gene expression and development of nontrivial molecular properties that are characteristics of different types of cells. Molecular neurobiologists have adopted methodological essentialism because of their success in identification of nerve cell types by one or more molecular markers, for example the possesion of unique kinds of messenger RNA molecules and of antigens that bind specifically with monoclonal antibodies. Each of these methods recognizes a unique *molecular phenotype* (reviewed in Jacobson, 1991).

The idea that different cell types exist because they fulfill some teleological need—that they each have a special purpose in the total economy of the organism, and that their diversity is based on a utilitarian division of labor—was implied by all histologists from Malpighi to Bichat. But the idea was neither examined critically nor worked out systematically until it was adopted as an explanatory theory by Henri Milne-Edwards (1826, 1857–1881). The concept of cell typology could not be developed without the related concepts of the autonomy of the cell and the origin of all cells from other cells, which were the major proposition of the cell theory. The concepts of division of cellular labor and of cellular differentiation were also related to theories of evolution: first to the Lamarckian theory, which emphasized the graded effects of the environment on the differentiation of types and the gradual transformation of one type into another; then to the Darwinian theory (which originally retained some elements of Lamarckism) which emphasized the random variation of types and selection of the fittest, leading to discontinuous variation of types. These theories dealt with the types of individuals, but they were soon extrapolated to cell types, first by Wilhelm Roux in his seminal work, *Der Kampf der Theile im Organismus*, published in 1881. Roux made the distinction between genetic inheritance of cell type from ancestral cells, and ultimately from the fertilized egg, and functional adaptation of cells in response to use and disuse, and to environmental influences in general.

That there are many types of cells in the nervous system was well established by 1852. In the first edition of Koelliker's *Handbuch der Gewebelehre*, published in that year, he classified nerve cells according to shape (e.g., pyriform, fusiform), and according to the number of processes emerging from the cell body (apolar, unipolar, or bipolar). The regulation of cell polarity is now well understood (Nelson, 1992), but it was Koelliker who first explicitly defined the concept of cell polarity in terms of different functions performed by the basal and apical parts of the epithelial cell. At that time he could not say how nerve cells were related to one another in terms of their polarity. The reason was that his cellular typology was originally based on the appearance of nerve cells dissociated from the tissue (Fig. 3.5). By the time he published the fifth edition of his *Handbuch* in 1867, he could conceive of nerve cell polarity in terms of inputs and outputs of nerve cells connected in networks, and the flow of nervous energy in one direction through the network, as shown in Fig. 3.9.

The first evidence showing that nerve cells are arranged in a regular layered order in the cerebral cortex, with recognizable cell types occupying stereotypical positions, was reported by Rudolph Berlin (1858). In addition to layers, Berlin observed various types of cells which he called pyramidal cells, spindle cells, and granule cells. He was able to see them because they were stained with a carmine dye introduced by Gerlach in the same year. This was the beginning of the general concept of neuronal architectonics, namely that different structural and functional types of nerve cells are organized in a regular spatial pattern that is characteristic of each part of the nervous system as a whole. The concept of hierarchical organization is implied: the nerve cell is regionally specified and polarized (dendrites and axon; inputs and outputs), different nerve cells (including receptor cells and glial cells) are connected in specified relations in each region, and different regions are connected together in specified patterns to form larger systems. That concept was closely related to the concepts of nerve cell typology, neuronal specificity, and functional division of labor among different types of cells. Once more the definition of a functional unit was raised: it could be a small part of a single nerve cell, an entire cell, a group of cells, or an entire system. Once again different views are taken by lumpers, who regard the group or the system as the significant unit, and splitters, who regard the individual synaptic connection as the significant functional unit.

The question that was asked by those neurohistologists like Koelliker, who thought about cell differentiation, was whether nerve cells all started life with the same potentialities or with restricted potentialities. Koelliker (1896, p. 810; see footnote 10 in Chapter 4) concluded that neurons start as multipotential cells and change their functions and structures in response to stimulation. Only after the cell theory was widely accepted was it possible to try to explain the diversity of cell types in terms of the origins of cells from qualitatively different progenitors, and the differentiation of different cells under different external influences. In other words, the concept of cell typology became associated with the concepts of genetic and epigenetic mechanisms of cell differentiation. Those concepts reached their apotheosis with publication of the great textbooks relating cytology to development and evolution (Hertwig, 1893–1898; Wilson, 1896).

The way was now opened for progress in three directions: first, to find out how the different cell types develop; second, to relate cell structure to function; third, to discover the underlying molecular mechanisms. Those goals were clearly conceived by the 1850s but attaining them was slow, mainly because of the limitations of the available techniques which are discussed in the following section.

The rate of progress can be seen by comparison between the depictions of histological sections of retina, cerebral cortex, or cerebellar cortex, or the depictions of nerve cells, in various authoritative texts published between about 1850 and 1900.*

---

*The six editions of Koelliker's *Handbuch der Gewebelehre des Menschen* are indispensable sources. They appeared in 1852, 1855, 1859, 1863, 1867, 1889 (Vol. 1), 1896 (Vol. 2, devoted entirely to the nervous system). The *Handbuch* was preceded by his *Mikroskopische Anatomie oder Gewebelehre des Menschen*, published as separate parts of Vol. 2 in 1850, 1852, and 1854 (Vol. 1 was never published).

These show that neurocytologists were struggling with the histological techniques on the one hand, and on the other with the conceptual problem of depicting the "normal" or "ideal" cell types in the same figure or model showing the variability seen in large populations of nerve cells. The problems of constructing models of nervous organization are discussed at length in Section 1.2. As an example, Fig. 1.6, showing cerebellar Purkinje cells depicted by different observers over the relatively short period of 50 years, is a pointed illustration of the problems scientists had with the rapid change of the concept of a single type of nerve cell. For purposes of reduction to the cellular level neurohistologists were forced to simplify and to idealize. The danger of reversion to some form of idealistic morphology was not explicitly noted by any of the neurohistologists, nor did they discuss the problem of essentialism as it related to nerve cell typology.*

The criteria for ranking objects into sets under common denominators should enable us to identify an object, such as a cerebellar Purkinje cell, and to pick it out again if so required. Otherwise we could not identify it in the first instance, could not relate it to other objects in the same or different set, and could not reidentify it if we chose to do so. Thus, typology is useful, even indispensable. If we could not identify the neuron that we were describing as, say, a motoneuron in the third lumbar segment of the spinal cord of a particular animal species, or if we could not identify a certain type of nerve cell after it had been changed by disease, someone else could not confirm or refute our observations. Cell-type-specific antibodies have been produced which enable nerve cells to be identified even under conditions like tissue culture or disease states that severely perturb normal cell morphology. Typology is necessary for reidentification, which is necessary for scientific communication and understanding. This kind of characterization necessary for identification makes no claims about "essences" in the Platonic or Aristotelian meaning of the word, namely the immutable and timeless nature of particular material objects, nor in any of the meanings used by later philosophers.

Nerve cell typology is concerned only with establishing criteria, the possession of which are necessary and sufficient for ranking neurons and glial cells into types, for the purposes of unambiguous identification and reidentification. The criteria by which nerve cell types can be identified must be founded on experimental evidence. Criteria that can be defined logically but not empirically should be avoided or eliminated. Philosophers may claim that this program can never be entirely successful—that we shall never succeed in eliminating all terms that cannot be defined empirically. But, without defining neuronal types in terms only of empirical procedures, there can be no meaningful scientific discourse about neuronal types.

Essentialism has been roundly condemned as a conceptual error, notably by population biologists and taxonomists who would prefer to replace essentialism with "population thinking" as one of the conceptual foundations of the synthetic theory of

---

*Cajal sometimes used the word "neuron" to mean any of the different types of cells of which the nervous system is composed: nerve cells, glial cells and receptor cells. When Cajal wrote *La rétine des vertébrés*, published in *La Cellule* in 1893, he had not yet fully recognized the distinction between neurons and glial cells, and the first of his "General Conclusions" was that "*the nerve cells, the glial cells, the cones, and the rods of all vertebrate retinas are discrete cells, the true neurones of Waldeyer.*" That sentence remained unchanged in the revision of the monograph made by Cajal near the end of his life and published as an appendix to the *Travaux du laboratoire du recherches biologiques de l'Université de Madrid* in 1933.

evolution. Ernst Mayr (1976) has been one of the most articulate critics of essentialism (typologism) and a strong advocate of population thinking: *"The assumptions of population thinking are diametrically opposed to those of the typologist. . . . Individuals, or any kind of organic entities, form populations of which we can determine the arithmetic mean and statistics of variation. Averages are merely statistical abstractions; only the individuals of which the populations are composed have reality. The ultimate conclusion of the population thinker and of the typologist are precisely the opposite. For the typologist the type . . . is real and the variation an illusion, while for the populationist the type (average) is an abstraction and only the variation is real. No two ways of looking at nature could be more different"* (Mayr, 1976, pp. 28–29).

Can this sort of either-or thinking be applied usefully to neuroscience? I do not think it can because both the population and typological views of nerve cells have their own validities. They are complementary rather than conflicting. Whether one sees the discontinuities between different types of nerve cells or sees the continuously variable properties will depend on the focus of one's attention and the methods used. The typologist is interested in finding qualitative differences, and the unique possession of any property, even of a single kind of molecule or of an epitope detected by a monoclonal antibody, is a sufficient mark of the unique cell type. That will be true even if that cell type has other properties, like size, that are continuously variable in a population that includes cells that have different molecular properties. There are also many alternative relationships possible between morphological and functional variability. If the type is based on functional criteria, e.g., serotonergic neurons are regarded as a type, then morphological differences must be excluded altogether. However, if the type is based on morphological criteria, e.g., local circuit neurons, then functional differences must be excluded. The artificiality of this extreme form of typology is obvious, even to those who use it for the purposes of easy discourse.

The question of the number of types of nerve cells that exist can be answered in several ways: that there are no types, meaning that the concept of typology is itself erroneous, or that there is a finite number of distinctly identifiable types, or that there is an infinitely long series of types. The problem is not unlike that of taxonomy in which there are difficulties in definition of the species and subspecies within a genus. Thus, G. G. Simpson says, *"Certainly the lineage must be chopped into segments for the purpose of classification, and this must be done arbitrarily . . . because there is no nonarbitrary way to subdivide a continuous line"* (*Principles of Animal Taxonomy*, 1961). But are there such continuous series of nerve cell forms, in which an infinite series of intermediate forms relate the extreme ends of the typological spectrum? Or are neuronal types quite discrete as defined by a small number of necessary and sufficient characteristics? The latter has been demonstrated by means of cell-type-specific monoclonal antibodies. In some cases the cell type can be decisively defined by a single antibody, in other cases, two or at most three antibodies are necessary and sufficient.

One of the unforeseen difficulties encountered by neurotypology is that as more molecular differences between nerve cells are discovered, the possible number of types may increase to the extent that virtually every nerve cell is different from all others. For example, when only one type of dopamine receptor was known, dopaminergic neurons could be regarded as a single neurophenotype. However, in the 1970s two types of dopamine receptors were discovered, D_1 and D_2, a third type, D_3, was discovered in 1990, and two other types, D_4 and D_5, were reported in 1991. Several more types of dopamine receptors are likely to be discovered. The different types of dopamine receptors have quite different physiological roles, and the functional differences that are already known are not trivial. Even so, it is conceivable that neurons could be fully characterized on the basis of a large number of cell-type-specific molecules, each known to possess significant structural and functional properties. The neurophenotype thus defined in terms of macromolecular structure and function may then be reduced further to the molecular genetic level.

Some promising starts have already been made in reducing different psychological states to different neurotransmitters. There is a large body of evidence showing that antipsychotic drugs bind specifically to particular dopamine receptors, and that is correlated with a change in psychological states. Thus, the D_2 receptor, which is known to be involved in schizophrenia, is the binding site of butyrophenones and related antipsychotic drugs, and the resulting change in dopaminergic neurotransmission is associated with a change in cognitive state. But that does not reduce the cognitive state to molecular states because the reduction has not been carried out through all intermediate levels. Something in addition to macromolecular states may be necessary for cognition. This and other objections to claims to be able to reduce mental events to neural events is discussed more fully in Chapter 2.

Showing a causal relationship between abnormalities of dopamine receptors and abnormal psychological states is only showing some correspondence, not the *identity* of a specified neural state with a specified mental state. The relationship between cognitive states and molecular states is not in question, but the identity theory demands more than that—it requires demonstration that cognitive states are identical in all respects with neural states.

In the experimental program to reduce the nervous system to neurocytology, neuroscientists have to deal with the problem of the relation of neural events to mental events (see Section 2.10). Dualists treat the nervous system as a mechanism in liaison with a nonmaterial mind. They see the mind as invulnerable to reductionism. Therefore, they do not have to consider the problem of elimination of mind. Those who believe that the mind is an epiphenomenon or emergent property of complex neural organization recognize the limitation of reduction of that organization to cellular and molecular components. Whether they perceive the problem correctly is considered more fully in Chapter 2. The reductionist materialists simply equate the mind with the neuron—the "cell of the mind." They see nothing lost by reduction of the brain to its cells and molecules; indeed, they see much gained in explanatory simplification. The aims of reductionist materialism are either to eliminate the mind completely from neuroscience or to identify the mind with physical entities in the brain.

The neuron theory was seen by some of its founders to be a way of either eliminating the mind or identifying it with nerve cells. Sherrington, who held to the dualist theory of mind, was not pleased when he attended Waldeyer's class in histology in 1886 and Waldeyer portrayed the nerve cell as *"das Denk-Organ."* Later Sherrington wrote, *"for a classroom to exhibit an isolated brain cell and label it large 'The organ of thought' may be dramatic pedagogy; it is certainly pedagogical overstatement"* (*Man on his Nature*, 1941, p. 268). What he would have said about modern psychopharmacological diagrams labeled "molecules of the mind" I leave to the reader's imagination.

3.7. THE DICTATORSHIP OF TECHNIQUES

Techniques make scientific research possible and scientific research now makes techniques indispensable. Scientists are vulnerable to the limitations of their techniques. Only our critical vision keeps us the masters of our methodology and guards us from becoming slaves of our techniques. These remarks apply particularly to the research program aimed at reducing complex nervous systems to their cellular constituents. Reductionism is a methodology and is therefore critically dependent on the successful application of techniques that are capable of isolating the structural and functional components of complex neuronal systems. Development of microscopy, histological techniques, electrophysiological techniques for recording activities of single neurons and single synapses, and biochemical and molecular biological techniques have played critical roles in construction of the neuron theory.

There are scientists who are so fully employed in the techniques of science that they have become unemployed as builders of theories. Enslavement to techniques starts whenever we think of a research problem primarily as a methodological problem rather than as a problem of finding meanings of observable phenomena, first and foremost. Theories are required for that purpose, and molecular biology has no theory of its own. The relevant theory, from our position, is the cell theory and its less-inclusive subsidiary, the neuron theory. When we deliberate first about means, and not about meanings, we become more vulnerable to the limitations of techniques. That is a hazard of this era of molecular biology, which is a methodology for obtaining data, not for explaining them. One would have to be silly to believe any claim to have either proved or disproved a neuroscience theory by a single crucial experiment. On critical scrutiny, virtually all methods are biased in favor of one of the alternative theories. Methodological neutrality is hard to achieve, as numerous examples have shown. Revolutionary techniques can produce results that make it possible to choose between alternative theories that claim to explain the same group of phenomena. Rarely in neuroscience has the choice between theories been based on observations made by means of a single technique, or has a revolutionary technique enabled a crucial experiment to be performed. Rather, the choice is made as the result of convergence of evidence obtained by means of several different techniques. For example, resolution of the conflict between different theories of the modes of connections between neurons was entirely a function of the use of better techniques—light microscopic histological techniques and, later, the electron microscope settled the dispute about the mode of connection between neurons. Microelectrode recording from single neurons and muscle cells settled the dispute about functional connections. Advances in electrophysiological techniques will be considered where relevant, especially in Chapter 4 in connection with Sherrington's work, and in Section 1.3.3 in connection with the effects of revolutionary techniques on the progress of neuroscience. The following is confined to considering the effects of advances of histological techniques on construction of the neuron theory.

To appreciate the achievements of the early neurocytologists it is necessary to recognize the technical problems they had to deal with and overcome in order to make progress. A summary of the technical methods they had to work with will show that after 1830 when the achromatic compound microscope became available, and apochromats after 1886, the limits to what could be observed were set by methods of tissue processing more than by the resolving power of available microscopes. Alcohol was the only means of fixation until Adolph Hannover started using chromic acid for fixation in 1840. Until formalin was used as a fixative (Ferdinand Blum, 1893), it was not possible to fix large pieces of nervous tissue without producing gross artifacts. Frozen serial sectioning, done freehand, was introduced by Benedikt Stilling in 1842. Stilling's magnificent atlas of low-power ($\sim 10\times$) serial sections published in 1859 "*laid the foundations for the modern anatomical study of the spinal cord*" (Clarke and O'Malley, 1968). Actually Stilling's method did more than that: in the estimation of Edinger (1891, 1909), understanding the spatial relationships of cell assemblies in the brain (first termed "nuclei" by Stilling), and the main fiber pathways connecting nucleus to nucleus, was made possible by three-dimensional serial section reconstruction. This was done by freehand sectioning until several models of freezing microtome were invented in the 1870s and 1880s. Embedding in paraffin wax was introduced in 1869 and in celloidin in 1882. However, a microtome capable of cutting serial section ribbons was not invented until 1884 (Threlfall, 1930; Bracegirdle, 1978).

The complexity of interlacing of fibers in the CNS led to attempts to loosen the tissue by various methods of maceration. Nerve cells teased out of macerated tissue were the first animal cells to be seen in isolation. The well-known illustration of a single motor neuron dissected by Otto Deiters (1865) from the macerated spinal cord is a good example of what could be accomplished by a very skillful operator (Fig. 3.6). Cajal (*Recollections*, p. 305) says: "*What a delight it was when, by dint of much patience we succeeded in isolating completely a neuroglial element etc. . . . The worst of it was that such a rather childish display of technical virtuosity was incapable of satisfying our eagerness to elucidate the inscrutable arcanum of the organization of the brain.*"

The method of dissecting out individual neurons led to serious artifacts. One was the attachment of fine axon terminals to dendrites which led Deiters and Gerlach to the wrong conclusion that the fine fibers originated from the dendrites. Another was the depiction of multinucleated nerve cells dissected out of the brains of vertebrates (Fig. 3.10). This particular form of syncytium exists in the vertebrate liver and in many invertebrates but not in nerve cells of vertebrates. I suppose that Valentin, Purkinje, and others who depicted multinucleated nerve cells probably saw a cluster of cells, including glial cells, and mistook them for a single nerve cell. There is another side to their drawings of multinucleated nerve cells: they were ways of depicting their theoretical models of the mode of formation of cells. Schwann and his followers conceived of cell formation inside a maternal cell by a process akin to crystallization of the nucleolus, followed by emergence of the nucleus from the nucleolus, and the repetition of those processes within a maternal cell until the latter burst to release several daughter nuclei each surrounded by some maternal cytoplasm.

Partial separation of cells was more often used in tissue sections, for example by Max Schultze in his studies of the nerve endings and structure of the inner ear (1858a),

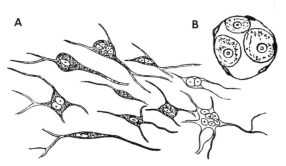

FIGURE 3.10. Schwann's theory of cell genesis by division of nucleoli and nuclei inside a mother cell was adopted by neurocytologists during the 1840s and 1850s. This enabled them to "see" multinucleated nerve cells when none actually exist. Two examples are given: (**A**) From Koelliker (1854), showing multinucleated nerve cells in the human spinal cord; (**B**) from Laycock (1860), showing "three corpuscles included in one cell" in the Gasserian ganglion of a calf embryo.

olfactory organ (1862), and retina (1866). Schultze fixed the tissues in osmic acid and mounted them, with light compression, in iodinated serum, which caused slight separation of individual cells and tissue layers.

Little progress in staining was made until the 1860s (Baker, 1945). Carmine was first applied to animal tissues by Joseph Gerlach in 1858, but although it stained cell bodies quite distinctly, it stained nerve fibers faintly, and this led to errors in tracing nerve fibers to their terminations. Peripheral nerve fibers could be stained crudely with silver nitrate (Wilhelm His, 1856), or with gold chloride (Julius Cohnheim, 1866), but those stains produced serious artifacts when applied to the CNS. For instance, the gold stain, much improved by Apáthy (1897), led him to see neurofilaments crossing the synaptic junctions and extending without interruption through several neurons (for other examples of artifacts mistaken for facts see Sections 1.3.4 and 1.3.6).

Golgi published his mercuric chloride method of impregnation in 1879 and his rapid method in 1886, which was later used so profitably by Cajal. Golgi was also the first to use the method of injection of fixative directly into an artery (Golgi, 1886, p. 30). Why did these revolutionary technical advances go unheeded until their importance was recognized by Koelliker in 1887? There are several reasons that are worth considering because they illustrate some of the peculiarities of communication of scientific information at that time.* Publication of his 1873 and 1879 papers in local Italian journals consigned Golgi's results to virtual oblivion for a decade until he published his 1886 paper in a journal that had a wide circulation for its time. Cajal contended with similar difficulties of very limited circulation of his papers published before 1889.† The

---

*Golgi's works were collected in his *Opera omnia* (three volumes), printed in an edition of 325 copies in 1903 by the publisher U. Hoepli in Milan. A fourth volume was issued by the same publisher in 1930. All the plates were lost when the premises of the publisher were destroyed during the Second World War.

†This could result in disputes about priority. For example, in 1890 Cajal (*Anat. Anz. 5*, 1890) disputed Golgi's claim to have discovered the axon collaterals in the spinal cord in 1880, published in what Cajal characterized as a "*modest and unknown local medical Bulletin*" (Ramón y Cajal, *Recollections*, p. 379). But Cajal based many of his own claims to priority on publications in just such unknown local Bulletins. The recurrent collaterals were to become understood 60 years later in terms of inhibition and negative feedback via interneurons, the Renshaw cells in the spinal cord, for example (Eccles *et al.*, 1954), but neither Cajal nor Golgi was in any position to understand their functions. They saw the axon collaterals only in terms of their theoretical preconceptions—Golgi saw them as the origin of nerve networks, Cajal saw them as evidence for the free termination of nerve fibers.

German-speaking states of central Europe were then the centers of research on microscopic anatomy in general, and especially on neurocytology. Until the mid-1880s there were more neuroscientists working in those central European lands than in all other countries. Communication of new ideas and techniques was fastest within the central countries, slower from the central to the peripheral countries, and very much slower in the other direction.

Koelliker made several visits to Golgi at Pavia (Fig. 3.11), the first in 1887, and after that he started using the Golgi technique very profitably. He was largely responsible for renewing interest in the Golgi technique after its long neglect (Koelliker, 1887a,b).* In the same year Cajal first learned of the Golgi technique and used it so effectively that he was able to demonstrate his Golgi preparations, showing evidence in support of the neuron theory, at the congress of the German Anatomical Society, at Berlin in 1889. But the facts are that the Golgi technique had been in use by Koelliker and others for some time before Cajal's first experiences with it in 1888. Obersteiner (1883, 1888) refers to Golgi preparations approvingly. Koelliker had started using the Golgi technique in 1887, Nansen in 1886, and Edinger (1891, p. 4) says that he learned of the Golgi technique in 1883. Many of the savants at the meeting with Cajal were already convinced of the validity of the theory of nerve cell connections by contact. I think that Koelliker was particularly interested in Cajal's preparations of spinal cord because they provided more evidence invalidating Golgi's reticular theory of nerve connections, and also showed, contrary to Golgi's theory, that the dendrites are part of the nervous conduction pathways. That the spinal cord was one of the main battlefields on which those theories were contested can be seen from the flurry of papers dealing with the histological structure of the spinal cord that appeared between 1887 and 1895 (reviewed in Lenhossék, 1893, revised edition 1894; Barker, 1899).

The Golgi techniques were mainly responsible for the rapid advances made in neurocytology after 1887, and in the resulting change in the image of the brain. Before Golgi the cellular composition of the brain was half-hidden and the globular cells were seen floating in an amorphous ground substance which occupied more than 50% of the volume of the gray matter. The Golgi techniques revealed the dendrites fully for the first time and gave an image of cellular diversity that could not have been imagined before. The picture of the brain that was revealed by the Golgi technique provoked urgent questions about how to reduce the multiplicity of neuronal types to some unifying principle and how to give functional meanings to the diversity of neuronal structures. Another question was whether neurons are connected together randomly, or in networks, or in circuits. Those questions could not have been asked before the Golgi era. Their answers now appear self-evident but we should remember that they only became so because of the adventurous minds of our 19th century predecessors.

Other techniques served different purposes, for example the myelin stain of Weigert (1882) and the methods of anterograde and retrograde tracing of fiber pathways

---

*In his autobiography, Koelliker (1899, p. 169) described his journey to Pavia to visit Golgi in 1887: *"This visit was very meaningful for me, because as a result of it I became familiar with this very exceptional savant and with his new method of staining nerves, which I was the first to introduce into Germany. Since that time the friendship with this pioneering researcher has become ever closer and has been sustained by frequent reunions."*

(Gudden, 1870, 1889; Marchi and Algeri, 1885–1886; Forel, 1887; Nissl, 1894a,b). In 1904 Max Bielschowsky published his silver impregnation method which stains both myelinated and unmyelinated nerve fibers, and which shows neurofibrils with great clarity. This led Bielschowsky (1908, 1928) to support the theory of Bethe (1898, 1900), which proposed that neurofibrils are the conducting elements of nerve cells. Silver staining was supplemented by the pyridine silver technique of Ranson (1911) and the silver protargol method of Bodian (1936) which stained unmyelinated fibers better than any previous methods. Ranson (1911, 1912, 1915) first showed that the majority of unmyelinated fibers in peripheral nerves are afferents and traced them to the small neurons of the dorsal root ganglia. Selective silver staining of degenerating nerve fibers and nerve terminals was greatly improved by the methods of Nauta (1950, 1957) and of Fink and Heimer (1967).

An important technical advance was Ehrlich's discovery in 1885 that methylene blue injected intravenously in living animals could stain their nerves (Ehrlich, 1886). One advantage of this technique was as a complement to the Golgi technique. For example, showing that dendritic spines have identical appearances when stained with methylene blue and with his double-impregnation modification of the Golgi method, Ramón y Cajal (1891a,b, 1896a) demonstrated that dendritic spines are unlikely to be artifacts caused by metallic precipitation as Golgi, Koelliker, and others at first suggested (see Section 1.4.1 for a case study of the discovery of dendritic spines).

We have briefly considered those neurohistological techniques with which data were obtained for construction of the neuron theory and its less-inclusive associated theories. The importance of some biochemical and molecular biological techniques is emphasized again in Chapter 2 where we consider reductionism. Electrophysiological techniques are at least equal to the others in importance and they are discussed in Section 4.4 in connection with Sherrington's research program, and in Section 1.3.3 in relation to the importance of revolutionary methods for promoting the progress of neuroscience.

The techniques of scientists are conventionally limited to laboratory methods but there are excellent reasons for including the methods for communication of scientific information, of which the printed word is still the most important. Both neurophysiology and neuroanatomy depend on graphic representation of empirical observations and on pictorial representation of concepts (see Tigerstedt, 1911–1912, for a detailed account of the history of physiological recording methods during the preceding century). Dependence on accurate pictorial representation of data became critical with the use of histological staining techniques. The earliest textbooks of histology showing nervous tissue (Koelliker, 1850) used black-and-white illustrations printed from woodblock or copper plate engravings. However, color lithography was used increasingly for neurohistological figures as the century progressed.

←

FIGURE 3.11. Camillo Golgi (1842–1926) on the right, Albert Koelliker (1817–1905), center, and Giulio Bizzozero (1846–1901). The latter collaborated with Golgi on studies of the life cycle of malaria parasites in human red blood cells. Koelliker was one of the main founders of neurocytology. He learned the Golgi technique during a visit to Pavia in 1887. This photograph was taken at Pavia, about 1900, by Aldo Perroncito, Golgi's adopted son and professorial successor. (Courtesy of P. P. C. Graziadei.)

Progress in printing scientific periodicals and textbooks was one of the important factors promoting rapid progress of neuroscience in the 19th century. Book production became much cheaper after 1880 (Chappell, 1970). This was the result of a combination of several conditions: increased literacy and demand for books for popular entertainment as well as for education; mass production of paper (with decline in quality as wood pulp replaced cotton as the raw material); greatly improved typesetting machinery and power printing presses (Steinberg, 1961; Moran, 1973). Those advances in printing technology coincided with a rapid expansion of scientific research in the European universities, and with the founding of periodicals for reporting the results of scientific research.

The invention of lithography, by Senefelder in about 1798, made it possible to make multiple copies of drawings on the lithographic stone, made in subtle gradations of tones and colors. That had been impossible with the previously used methods of printing from woodblocks. Those techniques continued to be used for black-and-white illustrations, but the blocks wore down after printing a few hundred copies. After about 1820 those techniques were replaced by etching and engraving on copper plates for reproduction of black-and-white scientific figures. Thousands of perfect copies could be printed from the metal plates.

The process of lithography was adapted for printing from metal plates in 1804, resulting in a considerable reduction of the cost of printing colored lithographic illustrations. As a result there was a steady increase in the number and quality of lithographic illustrations in the neurobiological books and periodicals after 1804. That technique was principally used for colored illustrations until replaced after 1890 by photoengraving.

Photoengraving is the technique of breaking up the halftones into dots by photographing through fine screens and then etching the image on zinc or copper plates from which the illustrations could be printed in black or in several colors, using a separate plate for each color.

The progress in printing techniques can readily be studied in the successive editions of the great textbooks of histology published during the 19th century. By midcentury copper plate engraving had replaced wood-block engravings as the method of printing black-and-white figures. This can be seen in the first edition of Koelliker's *Handbuch der Gewebelehre* (1853), and Stricker's *Manual of Histology* (English translation 1870), which do not have colored figures. The beautiful lithographic figures in the early textbooks of histology such as Hassall (1849) and Morel (1864) were printed by the lithographic process from metal plates, which was the preferred method of lithography for book illustrations. Near the end of the century the change from steel engraving and etching techniques to photoengraving can be seen by comparing the illustrations in the first (1894), second (1896), third (1900), and fourth (1906) editions of van Gehuchten's magnificent *Anatomie du système nerveux de l'homme*. In the first and second editions all the figures were printed from steel or copper plates engraved and etched by hand (and often signed by the engraver). In the third edition, halftone figures made by the photoengraving technique (produced anonymously) have replaced most of the illustrations made by the older, individually signed, manual engraving and etching techniques.

3.8. THEORIES OF NERVE FIBER DEVELOPMENT

After 1839 there were three main theories of development of the axon that seemed to be consistent with Schwann's cell theory (Fig. 3.7): the cell-chain theory, the plasmodesm theory, and the outgrowth theory. According to the cell-chain theory, originated by Schwann (1839) and supported by F. M. Balfour (1881) among other excellent embryologists, the axon is formed by fusion of the cells that form the neurilemmal sheath. This theory of development was an extrapolation from interpretations of regeneration of peripheral nerves, in which Schwann cells and fibroblasts were mistaken for the precursors of the regenerating axons. Ramón y Cajal (1928, pp. 7–16) heaps scorn and derision on this theory and on its proponents who are *"little acquainted with the severity and rigour of micrographic observations, and with the secrets of histological interpretation,"* but his rhetoric cannot conceal the fact that histological observations alone, without experimental evidence, were insufficient to disprove the cell-chain theory. The cell-chain theory was finally refuted by the elegant experiments of Harrison (1904, 1906, 1924a), when he showed that removal of the neural crest, from which the Schwann cells originate, results in the development of normal nerve fibers in the absence of Schwann cells. He also demonstrated that removal of the neural tube, which contains the developing neurons, prevents the formation of nerves, although the Schwann cells are left intact.

The plasmodesm or syncytial theory originated with Viktor Hensen (1864) and was supported by Hans Held (1897, 1909). According to this theory, the nerve fiber differentiates from preestablished filaments that connect all the cells of the nervous system. This theory was founded on the fundamental assumption that cells originate as a syncytium in the embryo and retain protoplasmic bridges throughout life. This theory was consistent with much of the evidence available at that time and was widely supported. For example, as late as 1925 in the third edition of *The Cell in Development and Heredity*, E. B. Wilson was still defining plasmodesms as *"the cytoplasmic filaments or bridges by which in many tissues adjoining cells are connected."* Held conjectured that the plasmodesms became incorporated into the developing nerve fibers.

Herman Braus (1905) devised an experimental test of the plasmodesm theory: he grafted limb buds containing no nerves to the body of frog tadpoles and observed the development of nerves in the grafted limb. He found that the pattern of nerves in the limb is determined by local conditions in the limb and not by the spinal or cranial nerve origins of the limb nerves and he interpreted this in favor of the plasmodesm theory. This conclusion has since been largely confirmed (Harrison, 1907a; Piatt, 1940, 1957a,b; A. C. Taylor, 1943, 1944; Pettigrew *et al.*, 1979), but has been interpreted in terms of the outgrowth theory of formation of nerve fibers. Such experiments could not, however, provide crucial evidence refuting the plasmodesm theory because ad hoc hypotheses could be invented to save the theory, for example, that the plasmodesms were invisible and that the axon developed by outgrowth as well as by incorporation of plasmodesm material. Ross Harrison understood this when he set out to test the theory by means of his celebrated tissue culture experiments. Harrison (1910, p. 790) understood *"that in all of the first experiments the nerve fibers had developed in surroundings composed of living organized tissues, and that the possibility of the latter contributing organized material to the nerve elements stood in the way of rigorous proof of the view that the nerve fiber was entirely the product of the nerve center. The really crucial experiment remained to be performed, and that was to test the power of the nerve centers to form nerve fibers within some foreign medium which could not by any possibility be suspected of contributing organized protoplasma to them."*

The correct theory that the nerve fiber is an outgrowth of the nerve cell was originally proposed by Bidder and Kupffer (1857, p. 116): *"It can be stated with the greatest degree of certainty, that the nerve cell is endowed with the conditions for allowing the fiber to grow as a direct extension out of itself. . . . every fiber must thereafter until its peripheral termination, regarded morphologically, be conceived merely as a colossal 'outgrowth' of the nerve cell."* This was an imaginative theoretical concept based on very flimsy evidence, but it was a mature theory in the sense that it could be accepted into a research program as I define those terms in Section 1.3.1. Its heuristic power was tremendous, and it has continued to guide research even today. It was also a progressive theory in that it was ahead of the facts. It continued to lead the facts for the following 50 years or more, as they were slowly accumulated, culminating in Harrison's (1910) demonstration of nerve fibers growing in tissue culture.

Wilhelm His was the first to obtain histological evidence showing that the axon grows out of the nerve cells in the spinal cord of the chick embryo. In 1886 His made this pregnant statement: *"As a firm principle I advocate the following law: that every nerve fiber extends as an outgrowth of a single cell. That is its genetic, its nutritive and its functional center; all other connections of the fiber are either merely collaterals or are formed secondarily."* From the deduction of Forel that nerve fibers end by contacting other nerve cells in the nerve centers, and those of His that the fiber is an outgrowth of the nerve cell, the neuron theory began to be constructed. His could not see the full extent of outgrowing nerve fibers in his preparations stained with carmine, silver, gold or hematoxylin, and he did not have the advantage of the Golgi technique, which Golgi had used for his studies of the spinal cord of the chick embryo (Golgi, 1880a,b, 1883b,c).

Neither Golgi nor His had the advantages of apochromatic lenses, which became generally available after 1887. Therefore, His was unable to see the growth cones, and could not at that time deal explicitly with the question of how one neuron connects with another. In 1856 His had obtained the earliest evidence that nerve fibers end freely in preparations of cornea stained with silver nitrate and blackened by exposure to light. His suggested that nerve fibers in the centers might also end freely but could not obtain evidence with the techniques at his command. In his papers which definitively showed that all peripheral nerve fibers end freely, Retzius (1892c,d,e,f) reviewed the observations of the previous 50 years showing anastomoses between peripheral nerve endings. He concluded that they were all based on optical aberrations and histological artifacts. Microscopy has probably produced more spurious data than any other method used in neuroscience research. Those spurious observations were frequently accepted in research programs for decades before they were exposed as anomalies. We consider their effects in Sections 1.3.4 and 1.3.5.

Using his metallic impregnation technique, Golgi (1880a,b, 1883b,c) obtained the first good evidence showing that axons and dendrites grow out of the nerve cell bodies in the spinal cord of the chick embryo. He discovered that the axons put out collateral branches and he showed that the dendrites grew out after the axons. From this it is clear that Golgi did not regard the evidence of outgrowth of nerve processes from the nerve cell to be in conflict with his theory of a reticular connection between axons and axon collaterals.

In 1889 Cajal began his studies on the development of nerve cells in the chick spinal cord. He came to the conclusion that the nerve fibers and their collaterals (a term that he introduced although Golgi had priority for their discovery) grow out freely, tipped by growth cones (Fig. 3.12). He concluded that the axon collaterals of one nerve cell make contact with dendrites of other cells (Ramón y Cajal, 1890c,d), and not with other axon collaterals, as Golgi had concluded. In both Cajal's and Golgi's view, the crucial fact was the mode of connection of the axon collaterals, not their mode of outgrowth. At the same time as Cajal, Lenhossék (1890a,b) published observations of the mode of outgrowth of nerve cell processes in the embryonic spinal cord, and also arrived at the conclusion that the axons and axonal collaterals of one nerve cell end freely in relation to the cell bodies and dendrites of other nerve cells.

An indirect approach to this problem was taken by August Forel in 1887 which led him to obtain the first experimental evidence showing that nerve cells connect by contact and not by continuity. Forel's work was based on the discovery by Bernard von Gudden (1870) that removal of the eye of the newborn rabbit results, after a short survival period, in atrophy of the visual centers. He thus provided the first method for tracing pathways in the CNS. He extended this method in 1881 to show that lesions in the cerebral cortex result in degeneration in the corresponding subcortical structures. It was clear that

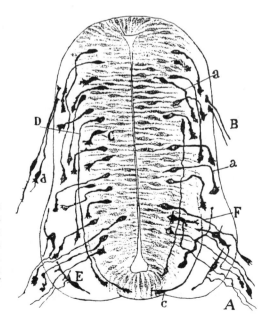

FIGURE 3.12. Outgrowth of axons, tipped with growth cones, from neurons in the spinal cord of a 3-day chick embryo. A, ventral spinal nerve fibers; a, axon; B, dorsal root fibers; C, growth cone on a very short axon; c, growth cone on a commissural fiber; D, commissural fiber; d, dorsal root ganglion cell; E, ventral horn motor nerve cell; F, fiber of lateral motor nerve cell. (From Ramón y Cajal, 1890c.)

acute degeneration is confined to the injured nerve cells but Gudden did not understand the general significance of his results. That was accomplished by Forel, who in 1887 showed that removal of one eye of a rabbit results in degeneration restricted to the optic nerve fibers without extending to nerve cells of the lateral geniculate nucleus, whereas removal of the visual cortex results in loss of the lateral geniculate neurons without apparently affecting the optic nerve fibers. From this Forel made the brilliant deduction that there are two separate neurons linking the retina to the cerebral cortex and that they make contact but do not form direct connections in the lateral geniculate nucleus.

Cajal, in Chapter 5 of his autobiography, deals with the contributions of His and Forel in the following manner: "*Two main hypotheses disputed the battlefield of science: that of the network, defended by nearly all histologists; and that of free endings, which had been timidly suggested by two lone workers, His and Forel, without rousing any echo in the schools. . . . My work consisted just in providing an objective basis for the brilliant but vague suggestions of His and Forel.*" Their statements were certainly neither timid nor vague, but neither were they dogmatic. Cajal seems to have misunderstood their unwillingness to assert their conclusions dogmatically. I suggest in Chapter 4 that Cajal's perception of the contributions of His and Forel reflects his lack of understanding of their classical or Apollonian style in contrast to his own romantic or Dionysian style.

The outgrowth of nerve fibers was eventually demonstrated in tissue culture by Ross Harrison (1907b, 1910).* Harrison excised pieces of neural tube from early tailbud frog embryos, at a stage before any nerve fibers are present, and explanted the tissue into a drop of frog lymph suspended from a coverslip. Nerve fibers grew out of the explant, in some cases from single isolated cells, for distances up to 1.15 mm at rates ranging from 15.6 to 56 μm per hour. Harrison's observations were very rapidly confirmed by other reports of outgrowth or regeneration of axons in tissue culture of nervous tissue of amphibians (Hertwig, 1911–1912; Legendre, 1912; Oppel, 1913), chick (Burrows, 1911; Lewis and Lewis, 1911, 1912), and mammals (Marinesco and Minea, 1912a–d).

The locomotion and growth of epithelial cells, young neurons and nerve fibers in tissue culture were shown to occur only when they are in contact with a surface such as fibrin fibers in a fluid medium, or are at the interface between the solid substratum and liquid medium, or are at the liquid-air interface (Loeb, 1902; Harrison, 1910, 1912; Lewis and Lewis, 1912). This phenomenon was called *stereotropism* by Loeb (1902) and Harrison (1910, 1912), *contact sensibility* by Dustin (1910), and *tactile adhesion* by Ramón y Cajal (1910, 1928). Wilhelm His was the first to recognize the importance of mechanical factors in embryonic development. This and many other important contribu-

---

*Harrison is usually credited with having invented tissue culture but he had several forerunners. Nevertheless he deserves the credit for having seized the opportunity to use a new technique to solve an old problem. The earliest report of culture of fragments of chick embryo medullary plate in warm saline was made by Wilhelm Roux in 1893, and in the following year Gustav Born was the first to culture fragments of amphibian embryos (reviewed by Loeb, 1923; J. Oppenheimer, 1971). After 1910, Harrison abandoned tissue culture in favor of surgical alteration of whole embryos because he recognized the limitations of tissue culture for answering questions about specification of structures in relation to position in the embryo. For Harrison's contributions to experimental embryology see J. Oppenheimer (1966).

tions of His to developmental neurobiology are reviewed by Picken (1956). His clearly understood and described cases of axonal guidance by the tissue substratum and his 1894 review of the mechanical basis of animal morphogenesis contains numerous apercus of the concepts and mechanism of nerve growth later promoted by Ross Harrison and by Paul Weiss.

Harrison (1910) observed the growth of the axons from the Rohon-Beard cells, which are the primary sensory cells and which can be seen in the dorsal part of the neural tube just beneath the epidermis in living frog embryos. Beard (1896), in his original description of the development of Rohon-Beard neurons, had accurately depicted the outgrowth of the axon from the cell body, but he failed to draw the general conclusion. Harrison observed that as the axon grows out of the Rohon-Beard cell into the subepidermal tissue, it slowly increases in length and gives rise to many branches. The growth of the axons of Rohon-Beard cells occurs in the same way as the growth of axons in tissue culture. The tip of the initial outgrowth, as well as the end of each branch, consists of an enlargement from which ameboid terminal filaments are constantly emitted and retracted. These were the first observations of the activities of the growth cone during normal development in a living animal. They fully confirmed Cajal's descriptions of growth cones in fixed specimens, and they provided a standard by which to assess whether growth cones in histological preparations are normal or artifactual. Harrison's observations on the growth of living nerve fibers agreed with descriptions of the growth of axons in histological preparations of the developing nervous system. This established at one stroke the validity of histological observations of Golgi (1880a,b, 1883a,b), His (1886, 1887, 1889a), Ramón y Cajal (1888a, 1890a,b,c), and Retzius (1893b), which were far more detailed and diverse than any that could be obtained *in vitro* at that time. Harrison's experiments had, in his own estimation, taken the mode of formation of the axon *"out of the realm of inference and placed it upon the secure foundation of direct observation."*

3.9. THEORIES OF NERVE CELL MICROSPECIALIZATION

The concept of regional functional differentiation of the cell was stated in Schwann's original version of the cell theory in 1838–1839. As we have noted earlier, the discovery of ciliary action by Valentin and Purkinje in 1835 was a dramatic example of regional cellular specialization. Koelliker expressed the concept of cell polarity in the first edition of his *Handbuch* in 1853. He pointed out that in epithelium the basal and polar ends of the cells are functionally and structurally different. Yet the concept of cell polarity excited little interest in general cytology. It is barely mentioned in the main textbooks of cytology that appeared at the end of the 19th century (Hertwig, 1893–1898; Wilson, 1896). The concept had by then been well developed by neurocytologists who had begun to understand the differences between axons and dendrites in terms of nerve cell polarity.

Nowadays it is taken for granted that the individual nerve cell is a mosaic of molecular domains each specialized for its own functions. That concept of the morphological differentiation of nerve cells began to develop in the mid-19th century in the attempts to understand the relations between different parts of the same cell, and the relations between different cells in the same tissue. The following three sections summarize the development of concepts of regional microspecialization of nerve cells.

Let us first consider the history of concepts of dendritic form and function and the origins of the concept that dendrites are fundamentally different from axons. That concept began to develop after Rudolph Wagner described the large nerve cells in the electric lobe of the brain of *Torpedo*, and noted that usually only one of its several processes is continuous with the nerve fiber (Wagner, 1847a,b, in Wagner, 1842–1853, Vol. 3, p. 377). Before that time nerve cells had been described as ganglionic globules, notably by Ehrenberg (1833, 1836), Valentin (1836a,b; 1839), and Purkinje (1837a,b), as we have noted earlier, but the relationship of globules to nerve fibers was incorrectly understood, and the dendrites had not been identified. In 1837 Purkinje described the large nerve cells in the cerebellar cortex and he depicted the cell body and proximal part of the dendrites without identifying the latter (Fig. 1.6A). In 1844 Koelliker did not make any distinction between different kinds of fibers from spinal ganglion cells, although he recognized that the fibers were prolongations of the ganglion-globules, as Remak had observed eight years earlier. Koelliker stated in 1844: "*the fine fibers arise in the ganglia . . . as simple continuations of the processes of the ganglion-globules. In other words, the processes of the ganglion-globules are the beginnings of these fibers.*"*
In the first edition of Koelliker's *Mikroskopische Anatomie* (1850–1852) the dendrites are not identified as such.

Wagner's identification of two different types of nerve cell processes was confirmed in the multipolar nerve cells of the spinal cord of the ox by Robert Remak (1854), who also confirmed that the axon is in direct continuity with the nerve cell body. Those observations were made by Remak on tissue sections sent to him by Stilling, who was the master of the freehand technique for cutting thin frozen sections.

The relationship of the cell body to two different types of elongated processes was described more clearly by Otto Deiters who dissected single motor neurons from the spinal cord of the ox, after macerating the cord in a weak solution of potassium dichromate (Fig. 3.6). He showed that the dendrites, which he named protoplasmic processes, are different from the axon, and he generously gave priority to Remak for discovery of two different nerve cell processes: "*I find the fundamental features of a theory of the central ganglion cells in the observation of Remak, that every cell makes connections exclusively with only one motor nerve cell root, and that this is a fiber chemically and physiologically different from all other central processes. . . . The body of the cell is continuous, without interruption, with a more or less large number of*

---

*In 1838 Remak had made the discovery that nerve fibers come off the ganglion cells, but Johannes Müller failed to confirm Remak's observations and pronounced them erroneous (Müller, 1840, p. 657, footnote). Therefore, Koelliker could claim priority for that discovery in 1844. In his argument against Remak's claim to have discovered that nerve fibers were continuous with the nerve cell body in the fish dorsal root ganglion, Koelliker started a long-lasting misconception that the majority of fibers pass through the mammalian dorsal root ganglion without making direct contact with cells there. In other words, Koelliker failed to recognize that the cells are T-shaped (pseudo-unipolar) in mammalian dorsal root ganglia (first shown by Ranvier in 1871), but are bipolar in fish dorsal root ganglia, as Remak had correctly observed in 1838.

processes which branch frequently. . . . These processes which. . . . must not be considered as the source of axis cylinders, or as having a nerve fiber growing from them. . . . will hereafter be called protoplasmic processes" (*Untersuchungen über Gehirn und Rückenmark des Menschen und der Saugethiere,* 1865).

Deiters also described a separate system of fine fibers originating from the dendrites, which he believed to run into the ground substance in which he thought nerve cells are embedded. Short segments of those fine fibers, apparently emerging from the large dendrites, are shown in the famous figure of the isolated nerve cell dissected out of the spinal cord of an ox (Deiters, 1865). As they are difficult to show in a reproduction of the original lithograph, the same cell is shown in Fig. 3.6 of a woodblock engraving from an article by Max Schultze (1870). Of course, we cannot be certain of what Deiters observed, but I agree with Schultze's conjecture that Deiters saw axon endings attached to the dendrites and mistook them for dendritic branches (Schultze, Introduction, in Deiters, 1865).

This artifact was the source of the idea that there are two systems of fibers connecting nerve cells. Deiters, and later Gerlach, thought that axons connected with one another form one system and that another system is formed by the dendrites of different nerve cells, also connected with one another. Deiters's observations, which were published posthumously in 1865 by Max Schultze, elevated the difference between axon and dendrites to a general theory of nerve cell morphology. This theory was characterized by Henle (1871, p. 26) in his historical review of the progress of anatomy, as the single most important advance made up to that time in understanding the nervous system.

Only the proximal segments of large dendrites could be seen until Golgi, in several works published between 1873 and 1886, showed the complete dendritic trees of neurons in the spinal cord, olfactory bulb, cerebral and cerebellar cortex. However, it was only in 1890 that they were named "dendrites" by His (1890). Another important distinction was made by Nissl (1894a,b), who showed that the basophilic granules which now bear his name extend from the cell body into dendrites but never into axons.*

Golgi made significant advances in understanding the morphology of dendrites, starting with his 1873 paper in which he introduced the technique of metallic impregnation of nerve cells, using potassium dichromate and silver nitrate. In that paper Golgi gave the first essentially correct description of the morphology of the major classes of cells in the cerebellar cortex. It was not surpassed, except by Golgi's paper of 1886, until

---

*We owe to Ramón y Cajal (1889c) the suggestion that the peripheral fibers of spinal dorsal root ganglion cells are dendrites. It is one of his less fortunate conjectures, and was soon corrected by Retzius (1891b), although the mistake was subsequently repeated by many others until the electron microscope resolved the problem by showing that both the central and peripheral processes of dorsal root ganglion cells have the ultrastructural characteristics of axons. Retzius (1891b) showed that in *Amphioxus*, which lacks spinal dorsal root ganglia, the sensory neurons, which are located in the spinal cord, have dendrites which branch in the spinal cord and axons which run in the dorsal roots to the periphery. He also pointed out that analogous conditions exist in annelids and crustaceans (Retzius, 1890, 1891a). This evidence was ignored by Cajal when he promoted the theory of "dynamic polarization of the neuron," according to which dendrites conduct nerve impulses toward the cell body and axons conduct away from it (Ramón y Cajal, 1891b, 1895).

the early work on the cerebellar cortex by Ramón y Cajal (1888d, 1890a,b, 1892), Koelliker (1890c), van Gehuchten (1891a,b), and Retzius (1892b), all of whom relied on the Golgi technique.

In the century after the invention of the Golgi technique, the study of dendritic growth and form had barely advanced to the stage that the plant and animal taxonomists had surpassed two centuries earlier. Classification of the forms of dendritic trees were based on the concept of the neurophenotype and on the conviction that the form of each neuron, especially the form of its dendritic tree, is a characteristic and invariant feature of a distinct neuronal type. Individual neurons belonging to the same type may exhibit a limited variability, but they are assumed to share certain invariant features that make it possible to assign that cell to its proper type. Such a system of classification of neurons can easily become an artifice, based merely on differences or on similarities of form that are useful as an aid to identification but that have no other functional or developmental significance. The danger of dwelling on externals alone is well shown in the history of taxonomy (Goerke, 1973, pp. 89–105). The danger is greatest when the classification is based on a single technique such as the Golgi method. This has resulted in classifications of dendritic branching patterns which are reminiscent of the efforts of the taxonomists before Linnaeus to classify plants into trees, bushes, and herbs (Ramón-Moliner, 1962, 1968; Percheron, 1979a,b). More advanced methods of topological analysis of dendritic branching patterns have recently been introduced (Van Pelt and Verwer, 1982, 1984; Verwer and Van Pelt, 1983; Uylings et al., 1983; Berry and Flinn, 1984).

Evidence showing that dendrites are in the direct conducting pathway accumulated rapidly to refute Golgi's theory that dendrites only have nutritive functions. First, many bipolar neurons had been described, for example in the spinal ganglia and cranial ganglia of fish (Wagner, 1847b), and in the cochlear and vestibular ganglia. In those neurons the dendrites had to be in the conducting pathway. Max Schultze (1870, p. 174) says "*It is obvious that such a ganglion cell is only a nucleated swelling of the axis cylinder.*" Second, many neurons were found in which the axon comes off a dendrite rather than the cell body, for example, in cerebellar granule cells (Ramón y Cajal, 1888d,e, 1890a,b,c). In the invertebrate nervous system, where the cell body is outside the line of conduction, the beautiful methylene blue and Golgi preparations of Retzius (1890, 1891a,b, 1892a) clearly showed dendrites as a necessary part of the conduction pathways and failed to show any signs of Golgi's conjectured network.

The problem of continuity of nerve fiber and ganglion cell also engaged the young Sigmund Freud. In his earliest neurobiological research, published in 1877 and 1878, Freud traced peripheral nerve fibers to their origins in the giant ganglion cells (Reissner's cells) in the spinal cord of *Petromyzon*, a cyclostome fish. The origins of nerve fibers from the ganglion cells had already been well established before Freud began his research, as we have noted. Freud correctly concluded that Reissner's cells are homologous with the dorsal spinal ganglion cells of higher vertebrates. In his third and final neurocytological paper published in 1882, he noted that the unipolar ganglion cells of the crayfish may be homologous with the multipolar motor neuron of the mammalian spinal cord depicted by Deiters in 1865 (see Fig. 3.6).

Freud's contributions to the neuron doctrine are minor but they have been overestimated as part of the blatant hero worship of Freud (Brun, 1936; Jelliffe, 1937; Bernfeld, 1950; Jones, 1953; Shepherd, 1991). Freud came to accept the neuron theory in the form proposed by Waldeyer in 1891. Thus, in Freud's *"Project for a Scientific Psychology"* written in 1895 he included the neuronal circuit shown in Fig. 1.10 in which neurons are depicted in contact, not in continuity. Freud called the connections between neurons "contact-barriers" and assumed that they acted as resistances to the quantity but not the quality of nervous energy transmitted. He held the quaint old notion that the quality of nervous energy is encoded as differences in the periodic motion of the neurons (Freud, 1895, p. 371). This idea, originating in the final paragraph of Isaac Newton's *Principia*, had been developed in David Hartley's *Observations on Man* (1749). Freud's understanding of neuronal function was characterized in a letter to Wilhelm Fliess written in 1896: *"There are, as it were, three ways in which neurons can affect one another: (1) They can transfer quanitity to one another; (2) They can transfer quality to one another; (3) They can, in accordance with certain rules, have an exciting effect on one another."* Freud conceived of the quantity of nervous energy ("the amount of cathexis") as equivalent to facilitation. He thought that "repression is removal of cathexis" whereas "facilitation is storage of cathexis" (Freud, 1895, pp. 380–384). Thus, nervous energy could not be generated in the nervous system or lost from it, but could only be facilitated, repressed, or rerouted (e.g. from conscious to subconscious paths).* These concepts were to become the cornerstones of Freud's theory of psychological structure. Insofar as psychoanalysis has any neurobiological roots, it is rooted in the neurobiology of the 1890s and earlier. The roots of psychoanalysis can also be found in *Naturphilosophie*, in which metaphor diverted attention from reality, and in which the morphological type was seen to be designed for fulfilling a natural purpose. Thus, Freud spoke of different functional classes of neurons, for example, "perceptual neurons" (Freud, 1895, p. 372–374). He conceived of "the ego" as a collection of neurons (Freud, 1895, p. 384–389) in which he imagined that facilitation and repression of nervous energy occurred (see Fig. 1.10).

3.10. THEORIES OF THE NERVE CELL MEMBRANE

The history of ideas about the forms and functions of neurons shows that the conditions which permit different scientists to uphold totally opposed theories are, first that the evidence is contradictory and inconclusive; second, that one or both theories are based on erroneous assumptions; and, third, that the opponents ignore counterevidence. One of the major assumptions of the late 19th century was that animal cells lack a cell membrane (Fig. 3.2). The notion of the membraneless nerve cell persisted in the face of experimental evidence showing that electrical potentials in nerve and muscle are generated at the cell surface (du Bois-Reymond, 1849; Hermann, 1870; Bernstein, 1871).

---

*Freud did not know that nerve impulses are conducted in an all-or-none manner. The all-or-none law for nerve was discovered by Gotch (1902; see Liddell, 1960, p. 89). The all-or-none law for muscle, namely that the contraction of the muscle fiber is always maximal, was first shown for heart muscle by Bowditch (1871), and for skeletal muscle by Fick (1882).

The cell surface was believed to be a transition between two phases, with special functions but without special structure. Therefore, it was assumed that cytoplasmic bridges between cells could freely appear and disappear. To recognize the significance of this fundamental assumption is to gain an entirely fresh view of the history of rival theories of formation of nerve connections. Proponents of the *contact theory* believed that nerve cells only come into close contact and are never in direct cytoplasmic continuity, whereas proponents of the *continuity theory* believed that nerve cells are directly connected by cytoplasmic bridges or networks. The continuity theory is consistent with the fundamental assumption that cells lack membranes; the contact theory is in conflict with that assumption.

The concept of the cell without a surface membrane is as old as the cell theory itself. Schwann (1839, p. 177) wrote: "*many cells do not seem to exhibit any appearance of the formation of a cell membrane, but seem to be solid, and all that can be remarked is that the external portion of the layer is somewhat more compact.*" Schwann conceived of five classes of cellular organization: (1) isolated, autonomous cells (blood cells); (2) connected but autonomous cells (pigment cells, epithelial cells, crystalline lens cells); (3) cells connected by anastomoses (cartilage, bone); (4) cells forming fibers (connective tissue); (5) cells whose walls and contents have coalesced (muscle, nerves, capillary walls). Thus, from the inception of the cell theory, nerve cells were placed in a class of cells whose walls coalesce.

The concept that the cell lacks a membrane was supported by Carl Gegenbauer in his influential essay on the evolution of the egg, published in 1861. Max Schultze (1870) defined cells as "*membraneless little lumps of protoplasm with a nucleus.*" Sedgwick (1895) regarded the embryo as a giant protoplasmic mass in which numerous cell nuclei are embedded. In discussing the structure of nerve cells, Koelliker (1896, p. 48) believed that the "*central cells lack a definite membrane and possess as boundaries only the tissues of the grey substance, which consists in varied proportions of nerve fibers, glial cells and blood vessels.*" In the first edition of E. B. Wilson's very influential book *The Cell in Development and Inheritance* (1896, p. 38), I came across the statement that "*the cell-membrane or intercellular substance is of relatively minor importance, since it is not of constant occurrence, belongs to the lifeless products of the cell, and hence plays no direct part in the active cell-life.*" Wilson maintains the same opinion in the second edition (1902, p. 53), but in the third edition (renamed *The Cell in Development and Heredity*, 1925, p. 54) he provides some evidence for the existence of a plasma membrane.

Cajal was the most vigorous advocate of the contact theory of nerve cell connections, and it is significant that he never doubted the objective reality of the cell membrane. The sources of his conviction are difficult to trace because he does not discuss the evidence or defend his belief in the existence of the cell membrane. Already in the first edition of *Elementos de Histologia Normal* (1895, p. 303) he defines a "*fundamental membrane*" which is "*a living organ of the cell which is a continuation of the protoplasm.*" In the fourth edition of his *Manual de histologia normal* (1909, pp. 150, 154) he says: "*All the cells of the central nervous system and sensory organs as well as the sympathetic possess a membrane of extreme thinness, a fundamental membrane. . . . This membrane is not a peculiarity of certain neurons, it is a general property without exceptions."* This statement is made *ex cathedra*, without supporting evidence, as if it were a self-evident truth, at a time when most authorities, even some who supported the contact theory, held the opposite opinion, namely that the cell membrane either is a histological artifact or is a lifeless structure without significant function.

I have been unable to determine whether Cajal's belief in the existence of a cell membrane preceded or followed his adoption of the contact theory. The two beliefs are now seen to be so obviously interdependent that it is not easy to understand how, at that time, it was possible to affirm the one while denying the other. Yet none of the other supporters of the contact theory shared Cajal's deep conviction in the authenticity of the cell-surface membrane. Nor can one find any discussion of the authenticity of the nerve cell membrane in the 19 volumes of *Biologische Untersuchungen* (1881–1921) of Gustaf Retzius, a consistent supporter of the contact theory. They either did not understand the importance of the cell membrane or denied its very existence. For example, Koelliker (1896, p. 48) stated that *"with reference to the envelope of the nerve cell, it can be shown with certainty that the latter apparently at all times lacks a cell membrane, and possesses the nature of protoplasts."* Another defender of the contact theory, Mathias Duval in his *Précis d'Histologie* (1897, p. 774) said of the nerve cell that *"formerly one described it as having an envelope, by reason of artifacts produced by coagulating reagents; nowadays it is recognized that it is a naked protoplasmic body."* All those who opposed the contact theory were at least logically consistent in also denying the existence of a nerve cell membrane. Thus, after reviewing the evidence, Sterzi (1914, p. 19) concluded that *"a cellular membrane does not exist. . . . The nervous cytoplasm is in direct relationship, through fine reticular fibrils that constitute the interstitial part of the nervous tissue."*

In his final statement on the evidence for the contact theory, Cajal (1933b) still found it necessary to ask: *"Do the terminal nerve arborizations actually touch the nude protoplasm of the cell or do limiting membranes exist between the two synaptic factors?"* Then comes the prescient conclusion: *"I definitely favor this latter opinion, although with the reservation that the limiting films are occasionally so extremely thin that their thickness escapes the resolution power of the strongest apochromatic objectives."* He then admits that *"Neuronal discontinuity, extremely evident in innumerable examples, could sustain exceptions. . . . for example those existing in the glands, vessels and intestines."* The cell membrane was one of those unobservable entities like "the synapse," "the gene," "the virus," "the atom," which were given a name before they could be identified physically. Cajal's reservations show that even by 1933 the cell membrane remained a hypothetical construct rather than a physically identified structure.

3.11. THEORIES OF NERVE CONNECTIONS: CONTINUITY THEORIES

The intellectual climate of the 19th century nurtured the concept of cytoplasmic anastomoses between different nerve cells. When the inadequate histological methods of those times failed to show membranes between cells, it was quite reasonable to assume that the cells are connected to form a syncytium. This false assumption, as much as the histological artifacts, formed the basis for reticular theories of connections between neurons.

All the reticular theories of neuronal connectivity portrayed the nerve cells in direct cytoplasmic continuity to form various types of networks (Gerlach, 1865, 1872; Golgi, 1883a–c, 1884, 1891a; Apáthy, 1897). The outlines of these theories have often been reviewed (Stieda, 1899; Soury, 1899; Barker, 1899; Bielschowsky, 1928; Spatz, 1929, 1952; Ramón y Cajal, 1933b; Scharf, 1958; Andreoli, 1961; Buess, 1964; van der Loos, 1967; Clarke and O'Malley, 1968; Meyer, 1971; Clarke and Jacyna, 1987; Shepherd, 1991). However, none of those authors seem to have recognized that the unstated assumption beneath all the variant theories was that cells normally lack a cell membrane. In the preceding section we have discussed the causes and effects of that false assumption.

Earlier microscopists saw large cytoplasmic bridges connecting nerve cells to form a network, as shown in Fig. 3.8A. Deiters was the first to surmise that a system of fine fibers connected the protoplasmic processes (dendrites) of different nerve cells as shown in Fig. 3.6. Gerlach was the first to claim to see a very fine reticulum forming the nexus. In sections of the cerebral cortex, cerebellar cortex, and spinal cord stained with carmine or gold chloride, Gerlach (1865, 1872) depicted a fine feltwork of fibers in the gray matter. He interpreted this to be a genuine network formed by anastomosis between branches of the protoplasmic processes (Figs. 1.6B and 3.8B). Remak (1854) and Deiters (1865) had shown that the branched protoplasmic processes are different from the single, unbranched axis cylinder, but they had not been able to show how they end or form connections in either the central or peripheral nervous systems. Gerlach (1872) depicted the sensory fibers of the dorsal roots originating indirectly by branching from a diffuse nerve network formed by interconnected branches of the protoplasmic processes. Gerlach's concept was accepted by all neuroanatomists at that time because it was consistent with the prevailing belief in cytoplasmic bridges connecting cells in general.

Gerlach (1872, p. 353) says: *"The cells of the grey substance provided with nerve and protoplasmic processes are therefore doubly connected with the nerve-fiber elements of the spinal cord, on the one hand by means of the nerve process which becomes the axis-fiber of the tubules of the anterior roots, and secondly through the finest branches of the protoplasmic processes which constitute a part of the fine plexus of nerve-fibers of the grey substance."* The functional consequences of such a system did not seem to have occurred to Gerlach. Gerlach's well-known figure showing the dendritic terminal branches of two nerve cells connected by a fiber with a lateral branch does not indicate the direction of flow of nerve activity, but from his description it is clear that he believed that the input was through the fine fiber running into the nerve net, and the output was from the network via the cell bodies to the axis cylinder. Koelliker (1867) had made an attempt to show the direction of the nervous inputs and outputs in his scheme of connections between nerve cells in the spinal cord (Fig. 3.9). Gerlach simply did not think in such terms.

Gerlach failed to understand the Bell-Magendie law when he wrote (1872, p. 354): *"Jacubowitsch (1857). . . . transferred the law of Bell, which had been applied more or less successfully to the white columns of the spinal cord, to the grey substance, and maintained that the larger cells of the anterior cornua were motor, and the smaller cells of the posterior horns the sensory elements, notwithstanding that every tyro was aware that neither the conditions requisite for voluntary movements, nor those for sensation, are present in the spinal cord. . . ."* This statement was a reflection of Gerlach's belief, then prevalent, that the sensory fibers passed without interruption, from the dorsal spinal roots to the brain. While Gerlach was venting his scorn on Jacubowitsch's essentially correct concept he was unable to see the defects in his own theory.

In the first edition of Michael Foster's *Text-Book of Physiology*, published in 1877, which was generally used in Britain, the nervous system is described as a *"protoplasmic network"* which is *"mapped out into nervous mechanisms by the establishment of lines of greater or lesser resistance, so that the disturbances in it generated by certain afferent impulses are directed into certain efferent channels. But the arrangement of these mechanisms is not a fixed and rigid one."* Foster did not bring the reticular theory into question until publication of the fifth edition of 1888–1890 of his *Text-Book*, and the term "synapse" was used for the first time in the seventh edition of 1897, in which Sherrington was assistant author. Sherrington later defined the term more explicitly, but still with qualification, as a junction *"presumably without actual continuity of substance,"* in E. A. Schäfer's *Text-Book of Physiology* (Vol. 2, 1900).

Adequate counterevidence to Gerlach's reticular theory could not be obtained with the existing techniques and was delayed until invention of the Golgi technique, which revealed the entire neuron for the first time. Golgi (1880a,b, 1882–1885) showed protoplasmic processes (dendrites), fully stained for the first time, in the spinal cord, cerebellar cortex, cerebral cortex, and olfactory bulb. He showed that the protoplasmic processes end freely, without any connections to one another and thus totally demolished Gerlach's theoretical construct. We should recognize the significance of Golgi's discoveries in relation to his progress ahead of his predecessors and contemporaries and not only in relation to the later advances made by his followers. Golgi's view of the cellular structure of the CNS was as far in advance of his predecessors as the views of Cajal were in advance of those of Golgi. Cajal could see farther not only because he had sharp vision but because he stood on Golgi's shoulders (Fig. 3.13).

Golgi proposed that dendrites end on or close to blood vessels, and he believed that they have nutritional functions and are not in the main conducting pathways. As Koelliker (1896, p. 53) observed, these connections between dendrites and glial cells and blood vessels were no more than hypothetical, and *"In his main work Golgi does not show a single illustration of such a connection, and his descriptions also ring less than true."* Because Golgi thought that the protoplasmic processes must have nutritive functions, he looked elsewhere for the conducting pathways between nerve cells. He believed that he had found them in the axon collaterals, which he discovered. Golgi then made the distinction between two main types of neurons: motor, or type I with a long axon, and sensory, or type II with a short axon. He suggested that the axon collaterals of sensory neurons are connected directly by a nerve network (*"reticola nervosa diffusa"*) to branches of the axons of motor neurons, thus excluding the dendrites from the conducting pathways as shown in Fig. 3.8C. By "sensory" and "motor," Golgi meant afferent and efferent neurons, for example, as he conceived of them in relation to the nerve network in the hippocampus and the cerebellar cortex (Fig. 3.14).

FIGURE 3.13. *"Though there were many giants of old in Physick and Philosophy, yet I say with Didacus Stella: A dwarf standing on the shoulders of a giant may see farther than a giant himself. . . .* Robert Burton, *The Anatomy of Melancholy*, 6th ed., 1651. *"If I have seen farther, it is by standing on the shoulders of giants."* Isaac Newton, letter to Robert Hooke, Feb. 5, 1675.

Gerlach had thought of nerve fibers arising exclusively from networks, but Golgi and his followers such as Bellonci, Bela Haller, and Nansen conceived of two modes of origin of the nerve fibers: directly as an extension of the nerve cell body, and indirectly from a nerve fiber network. There were several versions of that model. In an early version the protoplasmic processes were connected by large anastomoses while the nerve fibers were connected with one another via a network of fine fibers as shown in Figs. 1.6B and 3.8B. In a later version proposed by Golgi the protoplasmic processes (dendrites) terminated freely, and the axons were connected together via collaterals to the fine fiber network as shown in Figure 3.8C.

Golgi did not show a picture of his conjectured network or reticulum but in 1886 he described it in very guarded terms: "*Out of all these branchings of the different nerve processes there arises, of course, an extremely complicated texture which extends throughout the whole grey substance. It is very probable that out of the innumerable further subdivisions there arises a network, by means of complicated anastomoses, and not merely a feltwork; indeed one would be inclined to believe in it from some of my*

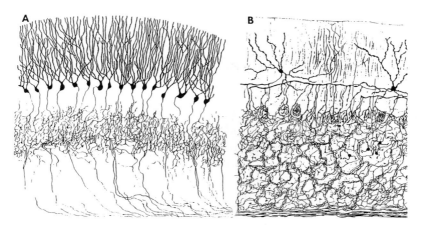

FIGURE 3.14. Golgi's representations of his hypothetical diffuse nerve network. Golgi (1907) states that these give an "idea" of the network: in other words, they were his mental models. In that model the network was formed of the terminal axonal branches of Golgi type II neurons and branches of axon collaterals of efferent (Golgi type I) neurons. Dendrites and neuron cell bodies were supposed to have trophic functions but were excluded from the conduction pathways. **A** shows the network formed of axon collaterals of granule cells of the hippocampal dentate gyrus; **B** shows the network formed of axon collaterals of cerebellar cortical basket cells. (From Golgi, 1907.)

preparations, but the extraordinary complexity of the texture does not permit this to be stated for certain."

Golgi at first treated his model as a speculative hypothesis and the diffuse nerve network is merely discussed but never shown in any of the numerous figures in his 1882–1885 papers, nor in his long polemical paper of 1891 which deals specifically with the question of the functional significance of the nerve net. Golgi was quick to understand that his theory precludes strict functional localization in the CNS. This theoretical issue is a significant component of Golgi's thinking about structure-function correlation in the CNS which merits far more attention than can be given to it here. One can merely comment that supporters of reticular theories of neuronal connections were unanimously opposed to the concept of strict functional localization while those in favor of the contact theory generally supported the idea of localization of functions in the CNS. Reticularists also underestimated the significance of specificity of neuronal connections and therefore overestimated the role of plasticity in development and in recovery from injury of the CNS (e.g., Bethe and Fischer, 1932a,b). This is considered further near the end of this section.

Pictures of the "*reticola nervosa diffusa*" are shown for the first time in two figures in a communication from Golgi which appears in the *Trattato di fisiologia* of Luigi Luciani (1901), and the same figures are reproduced in Golgi's *Opera omnia* (1903, Figs. 41 and 42) and his Nobel lecture (1906). One figure shows a network formed by axon collaterals of granule cells of the hippocampal dentate gyrus, the other depicts a network in the granular layer of the cerebellar cortex formed by axonal branches of the basket cells (Fig. 3.14). The text also reveals Golgi's reluctance to commit himself to specific details, saying that these figures only "*give an idea of the network*" (Golgi, 1907, p. 15).

There followed the well-known dispute between the supporters of Golgi's reticular theory (notably Apáthy, 1897; Bethe, 1898, 1900; Held, 1905, 1909, 1929; Nissl, 1903) and the supporters of the alternative theory of neuronal connections by contact (Forel, 1887; His, 1886, 1889a; Ramón y Cajal, 1890a,b,c; Koelliker, 1890a,b,c; Lenhossék, 1890a,b, 1893; van Gehuchten, 1891a,b; Retzius, 1892b,c; Vignal, 1893). None of the supporters of the neuron theory actually saw the synaptic contact zone, only the free nerve endings. Both theories were based on unobservable entities. The difference between them was that the synapse was eventually observed, whereas the diffuse nerve network was not. The diffuse nerve net and all the other incredibly fine interneuronal fibrils and networks were unobservable entities that existed only as convenient fictions (see Section 1.3.8). It is difficult to refute a theory supported by unobservable entities because those entities can be reinvented as required to save the theory. That strategy saved the reticular theory for over 80 years until the electron microscopic evidence conclusively refuted it.

It should be noted that all these proponents of the theory of neuronal contact at first also admitted that there were some neurons linked by protoplasmic anastomoses. The difference between the opposing factions was that while one side asserted that contacts between nerve cells are the rule and anastomoses the exception, the reverse was asserted by the other side. This is most clearly seen in Koelliker's theoretical position which shifted progressively from the reticular to the contact theory. The confrontation between these opposing theories, unpleasant as it often was, had great heuristic value, resulting in efforts to obtain corroborative and refutative evidence.

The reticular theory, increasingly destitute of empirical and intellectual support, found its final refuge in structures of incredible subtlety, at the limits of resolution of light microscopy. These are aptly described by Golgi (1901) as "*di organizzazione di meravigliosa finezza.*" In their final incarnation these structures took the form of tenuous fibrillar networks that were depicted as direct extensions of intracellular neurofibrils as shown in Fig. 3.8D and 3.8E (Apáthy, 1897; Bethe, 1900, 1903, 1904; Bielschowsky, 1904, 1908; Held, 1905). The presence of fibrils connecting neuron to neuron was regarded by many as perfectly consistent with the neuron theory. For example, the psychologist William McDougall (1901) believed that "*these observations of Held and Apathy seem to be not only not incompatible with the neurone-theory but rather to prove the existence of just such a state of affairs as is demanded by that theory when conceived in its only rational form*". By that McDougall meant that the neurons are entirely separate at the beginning of development, but make contact later via synapses which are the plastic elements. He thought that some synapses become fixed by intercellular fibrils, the "fibrillar substance" which Nissl (1903) believed to be the material responsible for higher mental functions. Such fancies had to be abandoned when it was shown that neurons are separated at some large synapses by a membrane which is not crossed by neurofibrils and that the neurofibrils do not enter all synaptic terminals (Bartelmez and Hoerr, 1933; Hoerr, 1936; Bodian, 1937). The neurofibrils that are stained with silver in light microscopic preparations of mammalian CNS tissue were eventually shown to be clumps of neurofilaments when examined with the electron microscope (Peters, 1959; Gray and Guillery, 1961).

3.12. THEORIES OF NERVE CONNECTIONS: SURFACE CONTACTS AND SYNAPSES

In the first half of the 19th century it was generally believed that there were three possible modes of termination of peripheral nerve fibers—by anastomosis, by formation of loops, and by free ending of fibers (Fig. 1.18). The problems raised by anastomosis of nerve fibers in peripheral nerves were discussed and solved in principle by Robert Whytt in 1751 and by Johannes Müller in 1840. Whytt (1768, p. 505) argued that the peripheral nervous filaments must be insulated from one another in order to mediate sensory discrimination and motor specificity. Müller's law of specific nerve energies (1826)—that stimulation of a sensory nerve, by any means and at any position along the length of the nerve, always results in the sensation that is specifically that of the sense organ which it serves—was predicated on the belief that nerves do not anastomose. He argued against the many observations which indicated that nerves anastomose (*Elements of Physiology*, 1840, pp. 651–653). He understood that private pathways are necessary to ensure sensory discrimination and motor specificity: "*A knowledge of the course of the primitive fibres in the nerves is of the utmost importance, for. . . . we must at last come to the question, where do the primitive fibres contained in a nervous fasciculus arise, and where do they terminate? . . . . If the primitive fibres never anastomose, it must follow that the cerebral extremity of each fibre is connected with the peripheral extremity of a single nervous fibre only, and this peripheral extremity is in relation with only one point of the brain or spinal cord; so that, corresponding to the many millions of primitive fibres which are given off to peripheral parts of the body, there are the same number of peripheral points of the body represented in the brain. If, on the contrary, the primitive fibres anastomose with each other in their course within the small fasciculi, ... then the cerebral extremity of a nervous fibril will be in relation with very many peripheral points. Now. . . . the irritation of a primitive fibre in a single point of the skin would necessarily be propagated through all the anastomoses,—in other words, no local impression on a definite point would be perceived by the brain. . . . The possibility of our establishing an accurate theory of the action of the nerves consequently rests wholly on the question, whether or not the primitive nervous fibers anastomose.*"

Because neuroscientists recognized the importance of the problem of how sharp the point-to-point connections between the periphery and nervous centers really are, the modes of termination of nerve fibers were a topic of heated controversy during the first half of the 19th century. Microscopic examination showed the nerve fibers terminating in three different ways: by the formation of a plexus, by formation of loops, and by free ending of isolated fibers (Müller, 1840, p. 653). The plexiform mode of termination was reported by Schwann in the tail and mesentery of the frog; the mode of termination by recurrent loops joining neighboring fibers was first reported by Prévost and Dumas (1823), and corroborated by Emmert (1836) and Valentin (1836a,b). Looping of nerve fibers was also reported to be the mode of termination of cutaneous sensory nerves and sensory nerves of the inner ear (Breschet, 1836; Valentin, 1836a,b).

These observers actually saw fascicles of nerve fibers, rarely single nerve endings. Free ending of nerve fibers was discovered in the retina and inner ear and olfactory nerves by Treviranus (1835) and confirmed by Schultze (1858b, 1860, 1862, 1866, 1870). Schultze (1862) discovered that each olfactory nerve fiber originates from a single olfactory cell in the nasal mucous membrane. I mention this to show that such a discovery was premature in the sense that it was not ready to be explained by a theory. The discovery was confirmed by Ramón y Cajal (1890f), van Gehuchten and Martin (1891), Retzius (1892c), and Koelliker (1892), all using the Golgi technique which stained the entire olfactory neuron. Those authors showed that the axon of the olfactory nerve cell ends freely in the glomeruli of the olfactory bulb whereas the short distal process of the olfactory nerve cell, seen by Schultze, is really a dendrite.

Valentin (1836a,b, 1839) extended the idea of nerve fiber looping to include the central ends of the nerve fibers as well as the peripheral ends. He conceived of the nerve fibers looping around the nerve cells, but he regarded the nerve cells and fibers as separate structures (Fig. 3.4). The idea was that the cells generated the nervous energy, which was transferred to the fibers, which conducted the energy to the muscles, and from the periphery back to the CNS. This concept required transfer of nervous energy by physical contact between nerve cell and fiber, and between nerve fiber and muscle, rather like transmission of electricity between conductors in contact. It would be a mistake to regard that concept as a preliminary to the notion of synaptic transmission because it was really totally misconceived both in terms of function and structure. Yet, the idea of transmission of nervous energy, like electrical conduction, between nerve elements in contact was persistently held, for example by Cajal, until the correct idea of unidirectional conduction at the synapse was stated explicitly by Sherrington (1897b, 1900b). Even so, the idea of "ephaptic" transmission between nerve fibers in contact in the CNS was a model proposed as late as the 1950s to account for the very short delay of about 3 msec for transmission between neurons in the CNS (reviewed by Grundfest, 1959). The idea of ephaptic transmission has recently been revived as an explanation of a mechanism of synchronization of groups of functionally-coupled neurons such as hippocampal pyramidal cells (Taylor and Dudek, 1984).

The modes of connections in the CNS were at first thought to be quite different from those of peripheral nerves. The possibility that central and peripheral nerve endings might be the same or homologous was first studied by Willy Kühne in the 1860s. Kühne (1862) observed that the motor nerve endings penetrate the sarcolemma and end in close relation to the contractile tissue of the striped muscle in the frog. In his Croonian Lecture of 1862, Koelliker reported that all the motor nerves to frog skeletal muscle terminate in free ends outside the sarcolemma. Koelliker expanded these observations in the fourth edition of his *Handbuch der Gewebelehre des Menschen* (1862, Figs. 111, 157, 158). Both Koelliker and Kühne agreed that the motor nerve fibers terminate on a small region of the muscle fiber. However, Beale (1860, 1862) had reported seeing an anastomotic network of fine nerve fibers covering the muscle fibers. We cannot be certain of what Beale actually observed, but most probably he mistook connective tissue fibers for nerve fibers.

By 1869 Kühne had recognized that motor nerves may either end on muscle fibers by contact or by cytoplasmic continuity. He asked the correct question: *"In what way do nerves terminate in muscle,"* but he arrived at the wrong conclusion: *"We now believe that we are able to perceive the direct continuity of the contractile with the nervous substance."* Then he expresses some doubt: *"Yet it may still happen that, in consequence of further improvements in our means of observation, that which we regard as certain may be shown to be illusory."* Only after 1886, when Kühne obtained apochromatic objectives, was he able to see the cleft between the motor end plate and the muscle fiber. In his Croonian Lecture to the Royal Society in 1888, with Sherrington in the audience, Kühne declared that *"nerves end blindly in the muscles. . . . Contact of the muscle substance with the non-medullated nerve fiber suffices to allow transfer of the excitation from the latter to the former."* Kuhne had discovered the neuromuscular synapse, but it was Sherrington who generalized the concept and gave the synapse its name in 1897. We shall discuss the steps leading to his discovery in relation to Sherrington's research program in Chapter 4.

Evidence showing a discontinuity from neuron to neuron came from four quarters: embryology, histology, physiology, and pathological anatomy. The first embryological evidence of the individuality of neurons was reported by Wilhelm His (1886, 1887, 1889a). He demonstrated that neuroblasts originate and migrate as separate cells and that nerve fibers grow out of individual neuroblasts and have free endings before they form connections. In 1887 His described with remarkable accuracy the outgrowth of the nerve fiber: *"The fibers which grow out from the nerve cells advance by growing into existing interstitial spaces between other tissue elements. In the spinal cord and in the brain, the medullary stroma, already formed, provides pathways for expansion and its structure undoubtedly determines the course of the process of extension. . . ."* These observations were later confirmed by Ramón y Cajal (1888a,d,e, 1890a,b,c, 1907, 1908). This evidence showing free outgrowth of axons did not shake the faith of the reticularists because they could point out that it did not exclude the later development of protoplasmic continuity between neurons after they had made contact (Nissl, 1903; Bethe, 1904; Held, 1905). Of course, the fact that axons and dendrites develop as independent extensions of the neuron did not prove that the entire neuron remains an independent cell, but evidence for that gradually accumulated. Ramón y Cajal (1888a,d,e, 1890a,b,c, 1906) showed that neurons stained by means of the Golgi technique always appeared to be completely isolated from others, and he followed Golgi in assuming that they were revealed in their entirety. This assumption was only possible to confirm empirically more than fifty years later (Blackstad, 1965).

The most compelling pathological-anatomical evidence of discontinuity between one neuron and the others came from the experiments of Gudden (1870, 1889) and Forel (1887). After axons were cut, they observed that degeneration is confined to the corresponding neurons. Reactive changes occurred in neighboring glial cells (Nissl, 1894a; Weigert, 1895), but the neighboring, uninjured neurons remained unaffected. After the discovery of specialized nerve terminals by Held in 1897 it appeared from the light microscope histological evidence that these endings are not in direct continuity with the neurons they contact. This could be deduced with greater certainty from the fact that injury to one neuron in a chain of neurons resulted in rapid degeneration of its own nerve terminal structures but not of the neurons that they contact downstream (Hoff, 1932; Foerster et al., 1933). After nerve injury resulting in retrograde degeneration there was no immediate effect on the nerve endings belonging to the upstream neurons in contact with the degenerating neurons (Barr, 1940; Schadewald, 1941, 1942). That the acute degenerative changes were confined to the injured neuron was much later confirmed with the electron microscope (De Robertis, 1956; reviewed by Gray and Guillery, 1966).

Confrontation between the continuity theory and contact theory did not end with these advances—the continuity theory merely changed from a progressive to a retrogressive theory, meaning that it was overtaken by the facts and continued to be supported only by histological artifacts. For example, in support of the continuity theory, there were several reports which claimed that fine filaments cross between neurons at the synapse and that those filaments persist even after degeneration of the presynaptic terminals (Tiegs, 1927; Boeke, 1932; Stöhre, 1935). In support of the contact theory it was shown that the end bulbs of degenerating axons swell and disappear completely within 6 days after axotomy, without apparently affecting the cell on which they terminated (Hoff, 1932). This became an effective method of tracing fibers to their terminations, and it generated a vast literature from the 1930s to the 1950s. The continuity theory was by then only a historical relic. It was finally refuted when the synapse could be studied with the electron microscope (Robertson, 1953; Palade and Palay, 1954; Palay and Palade, 1955; Palay, 1956; Gray, 1959) and the narrow synaptic cleft of about 20 nm between pre- and postsynaptic cell membranes was revealed.

Many examples of axosomatic junctions were also compelling evidence that a fine network does not form the link between neurons. For example, Cajal showed the contacts between cerebellar basket cell axonal terminals and Purkinje cell bodies in 1888, the contacts between centrifugal optic nerve fibers and retinal cells in birds in 1889 and 1892, and the terminations of cochlear nerve fibers on cell bodies in the ventral cochlear nucleus of mammals in 1896. This evidence was not sufficient to refute Golgi's theory conclusively, but it was strong counter-evidence. Nevertheless, Golgi continued to adhere to his reticular theory despite the counterevidence—an outstanding case of tenacity.

The original demonstration of a specialized presynaptic ending at an identified synapse was made by Held (1897), who showed that during development of the giant presynaptic terminals (the calyces of Held) on the cells of the trapezoid nucleus there appeared to be a clear line of demarcation between the axon terminals and the nerve cells on which they ended (Fig. 3.15). However, Held incorrectly concluded that the two neurons fuse later in development. The axon terminal expansions were named *"Endfüsse"* (end feet) by Held (1897), and *"Endknöpfchen"* (terminal buttons) by Auerbach (1898), who understood the significance of these specializations as a means of increasing the contact area, and he was also the first to show that presynaptic terminal expansions occur as a rule and to demonstrate them clearly by means of his reduced silver stain (Fig. 3.16). In 1903 Cajal refers to these axon endings as *"mozas," "anillos," "varicosidades," "bulbos"* and *"botones"* (knobs, rings, varicosities, bulbs, and buttons). They were called *"boutons terminaux"* by van Gehuchten (1904). The terminology was simplified when Sherrington (1897b) introduced the term "synapse" which he soon

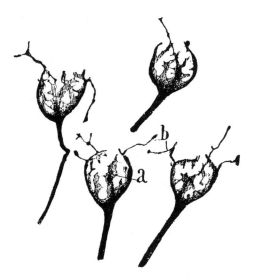

FIGURE 3.15. The chalices (or calyces) of Held in the nucleus of the trapezoid body in a kitten a few days old. Golgi method. From Ramón y Cajal (1933b): "*one receives the impression that the chalice is something external to the cell although very intimately applied to it.*" Even at that late date there was some uncertainty about the relations between axon terminal and postsynaptic cell. a, principal axonal terminal branches; b, thin appendices, probably growth cones.

changed to "synapse," and it became conventional to refer to presynaptic and postsynaptic structures and functions.

But even in 1906 Sherrington was typically undogmatic about the reality of the synapse, and was careful to describe it as a hypothetical entity: "*If there exists any surface or separation at the nexus between neurone and neurone, much of what is characteristic of the conduction exhibited by the reflex-arc might be more easily explainable. . . . The characters distinguishing reflex-arc conduction from nerve-trunk conduction may therefore be largely due to intercellular barriers, delicate transverse membranes, in the former. In view, therefore, of the probable importance physiologi-*

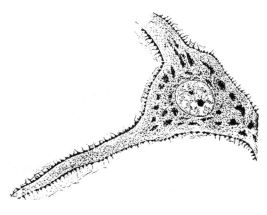

FIGURE 3.16. The first representation of synaptic endings (*Endknöpfchen*) in the central nervous system (facial nucleus; reduced silver preparation; paraffin section, 3 μm thick). From Leopold Auerbach (1898). He concluded that this was evidence in support of the contact theory of nerve connections: "*As I understand that theory, the axon terminals exert their effects on the cell surface of the ganglion cell by means of close contact of the end bulbs, without intervention of any intermediate substance.*"

cally of this mode of nexus between neurone and neurone, it is convenient to have a term for it. The term introduced has been synapse" (Sherrington, 1906).

The hypothesis of the synapse (for it was no more than a hypothesis in 1897) has been seen as the apotheosis of the neuron theory (see Section 4.4.5). It may be so, but Sherrington himself saw it otherwise—as a necessary inference from the observations showing that conduction in reflex arcs is different from conduction in nerve fibers. For example, reflex arc conduction is characterized by longer latency, irreversibility, temporal summation, and greater susceptibility to anoxia and drugs. Sherrington also arrived at the idea of the synapse as a deduction from the cell theory. At the very beginning of his masterpiece, Sherrington wrote that *"nowhere in physiology does the cell-theory reveal its presence more frequently in the very framework of the argument than at the present time in the study of nervous reactions"* (*The Integrative Action of the Nervous System*, 1906, p. 1).

Physiological experiments showed that conduction in nerve fibers is bidirectional whereas conduction in reflex pathways is unidirectional. Sherrington (1900b, p. 798) proposed that one-way conduction in the reflex arc is the result of a valvelike property of the junction between neurons. Thus, the unidirectional conduction in the CNS was conceived by Sherrington to be a function of one-way conduction at the synapse and not the result of the "dynamic polarization" of the entire neuron from dendrites to axon, as conceived by van Gehuchten (1891c) and Ramón y Cajal (1891b, 1895). Additional physiological evidence supporting the concept of the synapse was provided by measurements of the reflex conduction time, which showed a delay of 1 to 2 msec more than could be accounted for by the conduction time in the nerve fibers (Sherrington, 1906, p. 22; Jolly, 1911; P. Hoffmann, 1922; Lorente de Nó, 1935a, 1938a). The historical development of the concept of synaptic action will be discussed in Chapter 4 in relation to Sherrington's work.

The question of whether transmission of excitation from neuron to muscle is electrical or chemical was first raised by du Bois-Reymond (1877, p. 700 *et seq.*), who concluded in favor of chemical transmission, largely on the basis of the effects of curare. A theory of synaptic receptors was first proposed by Langley (1906) from his experiments on the effects of nicotine on neuromuscular transmission in the chicken. Langley (1906) stated that the *"receptive substance. . . . combines with nicotine and curari [sic] and is not identical with the substance which contracts."* This theory was included in a research program that had started with the simultaneous discovery by Claude Bernard and Albert Koelliker, about 1844, that curare blocks transmission at the neuromuscular junction (Bernard, 1878, pp. 237–315). More than a century later the receptor theory culminated in purification of the nicotinic acetylcholine receptor, its molecular cloning, and elucidation of its primary structure (reviewed by Changeux *et al.*, 1984; Schuetze and Role, 1987).

The accumulation of evidence in favor of chemical transmission at the vagus endings in the heart (Loewi, 1921, 1933), at the neuromuscular junction (Dale, 1935, 1937), and in autonomic ganglia (Feldberg and Gaddum, 1934; Dale, 1935) dealt directly only with transmission at synapses in the peripheral nervous system. Otto Loewi (1921, 1933) showed that the inhibitory action of the vagus on the heart was mediated by ACh. However, there was a very long latency of about 100 msec between the vagal nerve impulses and the slowing of the heart, which seemed to indicate that all chemical synapses have such long delays (Brown and Eccles, 1934). This led to the idea that the very short synaptic delay of about 1 msec measured at the neuromuscular junction and about 3 msec for transmission in the CNS and sympathetic ganglia must preclude chemical transmission and must therefore be caused by electrical transmission (Adrian, 1933; Eccles, 1937). It is now known that transmission at the vagal synapses on the cardiac pacemaker is by the muscarinic action of ACh, which involves the binding of ACh molecules to receptors on the postsynaptic membrane followed by a relatively slow chain of enzymatic reactions which alter the metabolic state of the cardiac pacemaker cells. By contrast, the action of ACh at the neuromuscular junction involves rapid opening of ionic channels through the membrane, the so-called nicotinic action of ACh.

The first strong evidence opposed to the theory of electrical neuromuscular transmission was obtained from the observations of Luco and Rosenblueth (1939) that, in the curarized cat, neither the action potential of the nerve nor the electrical excitability of the muscle was impaired. The crucial physiological evidence proving the existence of chemical synaptic transmission was obtained by means of a revolutionary technique, namely intracellular microelectrode recording from single muscle fibers near the neuromuscular junction (Fatt and Katz, 1951; Fatt, 1954), from single motoneurons in the vertebrate spinal cord (Brock *et al.*, 1952a,b), and from single neurons in the abdominal ganglion of *Aplysia* (Tauc, 1955). The theory of electrical transmission at synapses was finally refuted when intracellular microelectrode recording showed that the end-plate potential (EPP) at the neuromuscular synapse and the excitatory postsynaptic potential (EPSP) recorded intracellularly in the motoneuron had characteristics which could not be accounted for by electrotonic transmission from neuron to neuron. For example, the EPP was prolonged by application of anticholinesterases, thus making it certain that it was mediated by chemical synaptic transmission. This evidence settled the long-standing controversy between electrical and chemical theories of synaptic transmission reviewed by Eccles (1959, 1964, 1975, 1990), who reversed his early position in support of electrical transmission to finally accept the general validity of chemical transmission at central synapses.

The biophysical mechanisms of the inhibitory synaptic activity were revealed by recording intracellularly from spinal cord motoneurons. Eccles (1953) first showed that the hyperpolarizing postsynaptic potential, which was the mirror image of the EPSP, is always associated with inhibition, and thus named it the inhibitory postsynaptic potential (IPSP). Previously Brooks and Eccles (1947) had conjectured that spinal cord cells with short axons were responsible for inhibition via electrically coupled synapses. This was intended to account for the finding of increased latency of inhibitory compared with the excitatory reflexes. This was a radical modification of Sherrington's model of reciprocal innervation, discussed in Section 4.4.4, in which afferent axon terminals formed both excitatory and inhibitory synapses directly on motoneurons (Fig. 3.17A). With intracellular recording of synaptic potentials, it was found that the reflex latency of the IPSP was about 0.8 msec longer than the latency of the EPSP at the spinal motoneuron. This could be interpreted in terms of a model in which an additional inhibitory neuron was interposed between the spinal afferent and the motoneuron

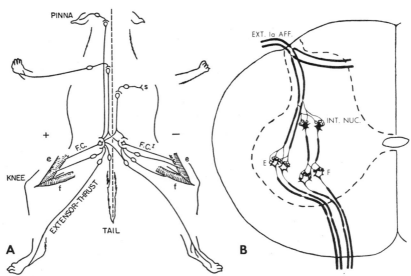

FIGURE 3.17. Two models constructed to explain the neuronal mechanisms of reciprocal innervation of antagonistic muscles. The Sherringtonian model (**A**, from Creed *et al.*, 1932) explains reciprocal reflexes in terms of convergence and integration of excitatory and inhibitory nerve endings on the spinal motoneurons (the final common path, F. C.) innervating flexor and extensor muscles (f, e). Inhibitory and excitatory effects on the motoneuron are produced by branches of the same spinal interneuron or spinal afferent. This possibility was later refuted by evidence that the same synaptic transmitter is released by all presynaptic terminals of the same neuron (Dale, 1934, 1935, 1937). Panel **B** shows the modified model of the function of the interneuron, as a means of converting excitation to inhibition as conceived by Eccles *et al.* (1953, 1956). The inhibitory interneuron was postulated to account for monosynaptic excitation of extensor motoneurons (E) and direct inhibition of flexor motoneurons (F) as a result of impulses entering the spinal cord via group Ia afferent nerve fibers. In A and B the excitatory and inhibitory nerve endings are shown making synaptic connections on the cell bodies, but the dendrites have been eliminated (see Fig. 1.4).

(Fig. 3.17B). The interposition of an interneuron in the inhibitory pathway was consistent with Dale's (1935) hypothesis that each neuron releases only one type of synaptic transmitter, either excitatory or inhibitory, at all its synaptic terminals. The inhibitory effect of antidromic stimulation of spinal motoneurons was shown to be mediated by recurrent collaterals of motoneurons releasing ACh at their synapses on interneurons which then released inhibitory neurotransmitter at their synapses on the motoneurons (Eccles *et al.*, 1954; Eccles, 1964). The axon collaterals of Golgi (1880a,b, 1883c) and Ramón y Cajal (1890d) that had for so long languished as premature discoveries, had at last been included in a mature neuroscience theory.

The preliminary discoveries of inhibitory interneurons in the intermediate nucleus of the spinal cord (Eccles *et al.*, 1954) were followed by discovery of other inhibitory interneurons such as the basket cells of the hippocampus (P. Anderson *et al.*, 1963, 1964, 1969), the basket cells of the cerebellar cortex (Eccles *et al.*, 1966), and inhibitory neurons in the neocortex (Phillips, 1959; Creutzfeldt *et al.*, 1966; Toyama *et al.*, 1974). A new model of neuronal organization of nervous systems was constructed in which each type of neuron has either excitatory or inhibitory effects on other neurons, without any intermediate types (summarized by Eccles, 1964). By contrast with its unique output function, each type of neuron may receive many inputs from both inhibitory and excitatory types of neurons. The model was refined by showing that neurotransmitters released at inhibitory synapses are different from those released at excitatory synapses. In the CNS, for example, glutamate is the most common excitatory neurotransmitter, which binds to at least three different types of glutamate receptors to open some channels with different properties. By contrast, gamma-aminobutyric acid (GABA) is the common inhibitory synaptic transmitter in the brain (Roberts, 1986) while glycine mediates inhibitory synaptic transmission in the spinal cord and brainstem, and these two inhibitory neurotransmitters open ionic channels which have different properties (reviewed by McCormick, 1990). These are of course only a few of the advances that have been made in reduction of synaptic functions to chemistry and physics. I mention these advances because it has become conventional to claim that continued progress along these lines (e.g., Dennett, 1991) will inevitably show the identity of neural events and mental events. The validity of such claims is examined in Chapter 2.

It is ironic that no sooner had the dispute been resolved in favor of chemical synaptic transmission than the first report appeared of an authentic case of electrical transmission in the crayfish giant fiber to motor fiber synapses (Furshpan and Potter, 1957). When potassium permanganate fixation allowed membrane structure to be resolved with the electron microscope, it became evident that the presynaptic and postsynaptic membranes are in close contact at electrically coupled synapses, whereas a synaptic cleft is characteristic of chemical synapses (Robertson, 1963, 1965; Pappas and Bennett, 1966).

The techniques of electron microscopy and intracellular microelectrode recording finally showed that there are entities with some or all the attributes of synapses present in both central and peripheral nervous systems. But those techniques did not spring into existence fully armed, but were gradually improved. Until the invention in 1950 of glass knives for cutting ultrathin sections, the electron microscope revealed little more than the light microscope, namely that there was what appeared to be a single membrane separating the neurons at the synapse (reviewed by Robertson, 1987). Later electron microscopic observations of synapses made with tissues fixed in osmium tetroxide, embedded in methacrylate, and sectioned with glass knives, showed the presynaptic and postsynaptic membranes separated by a synaptic cleft about 20 nm wide (Robertson, 1953; Palade and Palay, 1954). Synaptic vesicles, which characterize presynaptic nerve endings (Palade and Palay, 1954; De Robertis and Bennett, 1955), were immediately recognized as possible storage sites of chemical transmitters (del Castillo and Katz, 1954). After the introduction in 1956 of epoxy embedding and potassium permanganate fixation, it became possible to see the structure of the presynaptic and postsynaptic membranes and to obtain more accurate measurements of the width of the synaptic clefts. Even before glutaraldehyde fixation (Sabatini et al., 1963) provided reliable pictures of cytoplasmic structure, the first attempts were made to find ultrastructural differences between excitatory and inhibitory synapses (Gray, 1959; P. Anderson et al., 1963; reviewed by Eccles, 1964).

The revelation that the surface of nerve cell body and dendrites is virtually covered with focal contacts from other nerve cells, the "terminal buttons" (*Endknöpfchen*) of Leopold Auerbach (1898), opened up three very important problems that led to significant theoretical advances. First, it again raised the problem of how the numbers of axon terminals was regulated—was it regulated by overproduction followed by selective loss of contacts? Was it determined by conditions that exist in and around the terminals? Was it determined by the nerve cells contacted by the axon terminals? Was it regulated by the trophic influence of the cell body to which the axonal terminals belong? Second, it raised the possibility of specificity of spatial distribution of terminals with different functions (Langley, 1895, 1897). Third, the discovery of vast numbers of axon terminal boutons contacting a single nerve cell opened up the question of functional integration of the physiological activities of the terminals. The latter two concepts were soon to be assimilated in Sherrington's concept of functional integration of inhibitory and excitatory synaptic effects—summation, occlusion and the subliminal fringe of synaptic activities as shown in Fig. 4.8. We shall return to this again in Chapter 4 when considering Sherrington's contributions to neuroscience.

We should now consider the theory that synapses initially develop in excess and are later eliminated selectively. This theory has recently matured and is now driving a large research program (Landmesser and Pilar, 1972; Ronnevi and Conradi, 1974; Cragg, 1975; Winfield, 1981; Rakic *et al.*, 1986; Huttenlocher and de Courten, 1987). Underpinning the theory of synapse elimination is the concept of competition and selection on the basis of "fitness," "adaptiveness," and "competitiveness" which derives from Charles Darwin. Once this selectionist idea was grasped it could be extrapolated to deal with populations of molecules, cells, nerve fibers, synapses, or any other parts of the organism. The first to do so was Wilhelm Roux in 1881 in his book *Der Kampf der Theile im Organismus* (*Struggle between the Parts of the Organism*). Charles Darwin considered this "*the most important book on Evolution which has appeared for some time*" and noted that its theme is "*that there is a struggle going on within every organism between the organic molecules, the cells and the organs. I think that his basis is, that every cell which best performs its functions is, in consequence, at the same time best nourished and best propagates its kind*" (Darwin, 1888, Vol. 3, p. 244). As Roux recognized, competition is keenest between individuals that are similar, and will finally result in one type completely displacing the other. In his 1881 book, Roux introduced two other principles of biological modifiability and plasticity: *trophische Reizung* ("trophic stimulation") and *funktionelle Anpassung* ("functional adaptation"). In his autobiography Roux (1923) noted that he had shown that these are "*also applicable as a partial elucidation of adaptation during learning in the spinal cord and brain*" (1881, p. 196; 1883, p. 156; 1895, Vol. 1, pp. 357, 567).

In a single theoretical construct, Roux included competition, trophic interactions, and functional adaptation, as causes of plasticity. This was a premature theory in the sense that it was too far in advance of the facts to be of immediate use in constructing research programs. The significance of Roux's theoretical construct was well known to his contemporaries, but it was not possible to test the theory with techniques available during the 19th century. Starting in the 1960s, the technical methods were devised that could be used to test this theory experimentally and include it in a research program. The theory of competition and selection was then reinvented in more modern terms.

Theories of competitive elimination of synapses are based on the assumption that presynaptic terminals compete with one another for necessary molecules in limited supply such as trophic factors (Ramón y Cajal, 1910, 1919, 1928; Changeux and Danchin, 1976; M. R. Bennett, 1983); or that synapse elimination occurs as a result of secretion of inhibitory or toxic factors by the postsynaptic neuron (Marinesco, 1919; Aguilar et al., 1973; O'Brien et al., 1984; Connold et al., 1986). In 1919 Cajal noted that these factors could be produced by and act upon presynaptic or postsynaptic elements, or both, and that neurotrophic factors could also be secreted by glial cells.

Cajal was aware of Roux's theory of cellular competition and selection, and showed that overproduction of axonal and dendritic branches represents a normal phase of development in which excessive components are eliminated. He tells us that *"we must therefore acknowledge that during neurogenesis there is a kind of competitive struggle among the outgrowths (and perhaps even among nerve cells) for space and nutrition. . . . However, it is important not to exaggerate, as do certain embryologists, the extent and importance of the cellular competition to the point of likening it to the Darwinian struggle. . . ."* (Ramón y Cajal, 1929a). The last sentence indicates the influence of Roux's theoretical position, which is the origin of so-called neural Darwinism (Edelman, 1987). Ramón y Cajal (1910, 1919, 1928) also adopted Roux's idea of trophic agents in the mechanism of competitive interaction, survival of the fittest, and elimination of the unfit nerve terminals, synapses, and even entire nerve cells. Since then selectionist mechanisms have been proposed for development of functionally validated synaptic connections (Jacobson, 1974b; Hirsch and Jacobson, 1975; Changeux and Danchin, 1976), for development of connections between sets of neurons by various forms of competitive interaction between nerve terminals, and for development of behavior and learning (Changeux, 1986; Edelman, 1987).

The first evidence of specificity of formation of synaptic connections was obtained by J. N. Langley (1895, 1897) who showed that, after cutting the preganglionic fibers of the superior cervical ganglion, selective regeneration of presynaptic fibers occurs from different spinal cord levels to the correct postganglionic neurons. Thus, stimulation of spinal nerves T1 dilates the pupil but does not affect blood vessels of the ear whereas the opposite effect is produced by stimulation of T4. Spinal nerves T2 and T3 have both effects, but to different degrees. Langley (1895) proposed the theory that preganglionic fibers recognized postganglionic cells by a chemotactic mechanism. Guth and Bernstein (1961) concluded that this selection was made on the basis of competition between the presynaptic terminals.

Experimental tests of competition between cells or cellular elements are very difficult to do. When one structure supplants another during development the deduction is often made that one has been eliminated as a result of competition. However, there are cases in which one structure is replaced by another without any competition, for example, the pronephros by the mesonephros and the latter by the metanephros. In that case there is not even a causal relationship between the three kidneys that develop in succession. In general, mere succession is not evidence of causal relationship and is thus not evidence of mechanism, competitive or otherwise (see Section 2.13).

An experimental test of neuronal competition was first done by Steindler (1916) by implanting the cut ends of the normal and foreign motor nerves into a denervated muscle. Steindler found no selective advantage of the normal nerve. When two different nerves supply a muscle, the resulting pattern is a mosaic in which individual muscle fibers are innervated at random by one nerve or the other. Steindler's observations have been repeatedly corroborated (Weiss and Hoag, 1946; Bernstein and Guth, 1961; Miledi and Stefani, 1969). Similarly, when two optic nerves are forced to connect with one optic tectum their terminals segregate to form stripes and patches in the optic tectum of the goldfish (Levine and Jacobson, 1975) and the frog (Constantine-Paton, 1981).

Several theories of the possible mechanisms of competitive exclusion or elimination of synapses and of synaptic strengthening by function have been proposed. The oldest of these is the theory of formation of selective connections between neurons that have correlated activities (Cook, 1991). This is an extension of the psychological theory of association of ideas. That theory, deriving from the epistemology of John Locke and David Hume, was first given a neurological explanation by David Hartley. The rule of association was the first learning rule to be conceived by John Locke in 1690, and formulated in strictly neurobiological terms by David Hartley in 1749. The rule has been formalized by others, notably by Donald Hebb in 1949, but it should be called Locke's rule or the Locke-Hartley rule, not Hebb's. In his *Observations on Man* (first published in 1749), Hartley proposed that mental associations form as a result of corresponding vibrations in nerves (an idea that Newton had thrown out in the last paragraph of his *Principia*). The step from a psychological to a neurophysiological theory of association appears to have been made before the mid-19th century, as evidenced by Herbert Spencer's statement: "*As every student of the nervous system knows, the combination of any set of impressions, or motions, or both, implies a ganglion in which the various nerve-fibres concerned are put into connection*" (*Principles of Psychology*, 1855). Tanzi (1893), the principal defender of the neuron theory in Italy at the time, was the first to propose that modifications of nerve connections resulting from use are the basis of learning and memory. This concept was proposed again by Cajal in his theoretical paper of 1895 on the anatomical basis of ideation, association, and attention. The hypothesis that synapses form or become altered between neurons whose electrical activities coincide has become widely accepted in approximately the way in which it was formulated by Ariëns Kappers *et al.* (1936): "*The relationships which determine connections are synchronic or immediately successive functional activities.*"

The general idea that learning is predicated by selective strengthening of synapses (Ramón y Cajal, 1895) has been accepted and elaborated in various forms (Hebb, 1949, 1966; J. Z. Young, 1951; Eccles, 1964; Anokhin, 1968; Beritoff, 1969). Neurophysiological theories of strengthening of synapses between neurons that have synchronous functional activities imply that linkages initially are extensive but become more restricted, functionally and anatomically, as a result of functional activity. In this view, the final arrangement is the result of cooperative interactions between neurons. This view has been extended to include competitive functional interactions between neurons—neurons with equal activities being able to maintain connections with a shared postsynaptic target, while functional imbalance results in the more active neuron excluding the less active neuron from a share of the postsynaptic space (Guillery, 1972; Sherman *et al.*, 1974; Sherman and Wilson, 1975; C. Blakemore *et al.*, 1975; Edelman, 1987).

The theory of adaptive neuronal modification by selective depletion (Jacobson,

1970a,b, 1974a,b) was proposed at a time when the *Zeitgeist* nourished theories of selective neural accretion. For example, the consensus was that learning results in formation of more synapses, a theory that followed closely on the notion of synaptic strengthening as a result of correlated activity (Hebb, 1949). Regressive events such as cell death and elimination of redundant dendritic and axonal branches were commonly misconceived entirely in terms of elimination of errors. By contrast, the theory of "*neuronal modification by selective depletion*" (Jacobson, 1974b) states that overproduction of synapses is genetically programmed, and the final configuration occurs as the result of elimination of synapses on the basis of epigenetic cell interactions and functional criteria of fitness: "*The effect of these interactions is to reduce the initial redundancy. . . . The reduction of structures, including synapses, which this theory predicts, results in increased matching of the functions of the nervous system with the conditions of the world in which the creature lives. As Herbert Spencer put it, it is a process which brings the inner and outer relations of the organism into closer correspondence*" (Jacobson, 1974b).

3.13. CONCLUSIONS

My understanding of the history of neuronal connections and of the neuron theory is different from the generally prevalent view. I see the neuron theory reaching its canonical form, as we finally understand it, only gradually as a result of construction of a more-inclusive theoretical model of nervous organization by a process of convergence and coalescence of many less-inclusive theoretical models rather than as a revolutionary overthrow of one theory by another.

The contact theory of neuronal connections replaced the reticular theory because it explained more observations and was a more consistent predictor of discoveries and ultimately because electron microscopic evidence falsified the reticular theory. The theory of scientific revolutions expounded by Kuhn (1962, 1968, 1970, 1974) does not provide a satisfactory explanation of how the contact theory of nerve connections ultimately replaced the reticular theory. It is said by Kuhn (1970) that scientific paradigms are created under the influence of the climate of opinion (*Zeitgeist*, a term invented by Goethe to account for events that occur "*neither by agreement nor by fiat, but self-determined under the multiplicity of climates of opinion*"). Yet the history of the contact and reticular theories shows that both existed in the same *Zeitgeist*. Some, like Nansen, Koelliker, and Waldeyer, held to parts of both the contact and the reticular theories, at least until the latter became untenable. Many people continued to believe in the reticular theory long after the passing of the *Zeitgeist* in which the theory had grown up. There were also some central figures who remained indifferent to either theory, which shows that the climate of opinion did not affect them crucially. As an example of this indifference it is instructive to quote Wilhelm Wundt (1904), the founder of physiological psychology: "*Whether the definition of the neurone in general, and whether in particular the views of the interconnexion of the neurones promulgated especially by Ramón y Cajal will be tenable in all cases, cannot now be decided. Even at the present day, the theory does not want for opponents. Fortunately, the settlement of these controversies among the morphologists is not of decisive importance for a physiological understanding of nervous functions."* This declaration could be made only because Wundt's model of nervous functions was not a cellular model but was a hydraulic or electrical model. Wundt represented those who conceived of brain functions in terms of waves and gradients, not in terms of discrete cellular units. This idea had enormous influence and it survived especially in the work of Karl Lashley (1929, 1937, 1958; see Titchener, 1921; Boring, 1950). Wundt's reactionary view of the neuron theory is consistent with his archaic exposition of neuroscience which occupies the first volume of his very influential textbook *Principles of Physiological Psychology* (5th ed., English transl., 1904) in which the important advances in neurophysiology, neuroanatomy, and neurocytology of the previous decades are completely ignored.

The synapse was conceived as a theoretical necessity by Sherrington in 1897 more than 50 years before it was conclusively shown to exist and function as a physical entity. The concepts of the synapse as a one-way valvelike contact between neuron and neuron, and as the basis for spatial and temporal integration of excitatory and inhibitory signals, were the most important advances toward understanding the functional organization of nervous systems. The implications of the synapse concept are examined more fully in Chapter 4. Here it may be noted that the synapse concept was a logical deduction from the neurophysiological observations that conduction in peripheral nerve trunks is different from conduction in reflex arcs. Sherrington deduced that there are barriers in reflex arcs that are absent from nerve trunks. He showed that those barriers function as one-way valves, fatigue easily, and are sensitive to toxins and anoxia. He surmised that those effects occur at the points of contact between one neuron and another. The synapse concept owed little to the neuroanatomical data and Sherrington did not cite the work of Cajal or any other neurocytologist when he constructed his explanatory model of the synapse (see Section 4.4.5). Although Sherrington was familiar with neuroanatomy he could have arrived at the concept of the synapse without consulting the anatomical data. Sherrington conceived of function before structure, and he arrived at understanding of structure in terms of function. By contrast, Cajal and the other neurohistologists put structure before function and so failed to understand the functional significance of their own data. The key to that understanding was Sherrington's concept of the synapse and its implications for integrative action, as we shall see in Chapter 4.

None of the 19th century proponents of the different theories of connections between neurons lived to witness the final solution of their problem. Ironically, neither the application of electronic instruments to neurophysiology nor electron microscopy owed anything to the rival theories of nerve connections—the stimuli for their invention came from other sources and those techniques were first applied to other problems before being put to use in solving the problem of neuronal connections. The problems were resolved by opportunistic application of any techniques that seemed likely to work, and finally by means of microelectrode recording and electron microscopy, not merely as a result of wrangling about different theories.

We have examined different theoretical models as parts of several research programs, in which different techniques and different values were favored by their supporters. Techniques in neuroscience are rarely neutral with respect to theories, but are often used because they are known, consciously or not, to provide evidence favorable to one of two rival theories. The techniques that were favored by the neuronists, especially the Golgi methods, corroborated their position by showing free nerve endings. The reticularists favored techniques, for example the gold technique of Apáthy, and neurofibrillar stains, which showed fine fibrils directly connecting nerve cells. The neuronists conceived of strict localization of function in the nervous system, whereas the reticularists conceived of diffusely distributed functions. The neuron theory can be regarded as a stage in the mechanization of the brain picture and as the continuation of the tradition of Cartesian certainty which was a bulwark against relativism. For several decades following the definition of the neuron theory in 1891, neuroscientists could either support the neuron theory and bask in the comforting glow of Cartesian certainty, or they could support the reticular theory and accept the discomforts of uncertainty. These differences were based on different conceptions of relations between structure and function, especially on specificity of functional localization in the brain. These differences were, in general, also related to different values, for example, as reflected in the debates about mechanism versus vitalism, nature versus nurture, and neuronal rigidity versus plasticity.

The adversarial theoretical positions occupied by the supporters of rival theories had a powerful heuristic effect, driving them on to seek evidence to corroborate their own position and to refute that of their adversaries. The research program progressed by means of a dialectical interaction between empirical observations and theoretical models. Because empirical evidence lags behind theory, and because each side held tenaciously to its own theoretical position for as long as possible in the face of counter evidence, both theories continued to be contested long after the reticular theory became impossible to defend. The reticular research program, with its untenable theory, selective techniques, and special worldview, became a degenerating program in the sense that it accumulated artifacts and could not accommodate much of the evidence. The reticular theory was considered to be false when it had to be modified repeatedly and became increasingly complex in order to square with the facts. By contrast, the neuron theory was considered to be true because it explained, unified, and simplified knowledge from many different sources, and because it successfully predicted new observations such as the cell membrane and the synapse, which were eventually demonstrated by the powerful techniques of electron microscopy and electrophysiology.

CHAPTER 4

Hero Worship and the Heroic in Neuroscience

> *Universal history, the history of what man has accomplished in this world, is at bottom The History of the Great Men who have worked here. They were the leaders of men, these great ones; the modelers, patterns, and in a wide sense creators, of whatsoever the general mass of men contrived to do or to attain; all things that we see standing accomplished in the world are properly the outer material result, the practical realization and embodiment of thoughts that dwelt in the Great Men sent into the world. . . .*
> Thomas Carlyle (1795–1881), *On Heroes, Hero-Worship, and The Heroic in History,* Lecture One, 1840.

4.1. THE ARGUMENT

In the past century the research institute has become the temple in which scientific truth is revealed, and the idols are great scientists. The scientific world-view has become a replacement for religion. Many scientists who reject religion bow down in a state of awe before the mysteries of the universe and kneel in adoration before the idols of science. In place of the saints of religion they have substituted the savants of science.

Hero worship is a sort of idolatry that is associated with deference to authority and is opposed to habits of free thought and critical inquiry. Therefore, one deplores the conditions that create the need for symbolic figures, and pities the science that needs heroes. Hero worship is often a symptom of a deeper underlying disorder of its adherents. It is often symptomatic of intolerance of the work of less successful scientists, and portrays them as less worthy of respect or as if they were scientists without a history, or worse, as the history of losers. Hero worship is often associated with a dangerous tendency to simplify the world—to see everything in terms of truth and falsehood, of thesis and antithesis. The cult of heroes is often a symptom of ideological commitment that requires its adherents to support the beliefs of its heroes and of their own community of believers. It may then become a serious and dirty business of suppressing the beliefs of rivals. Arrogance and intolerance are among the toxic products of the mindset that also fosters hero worship. Perhaps we should look at hero worship as an oncologist looks at cancer: the dangers are clear, the remedies ineffectual. If it cannot be eradicated it should be contained and managed.

The need to look up to idols has deep psychological and evolutionary roots. The dog looking up with adoration at its master, in Fig. 4.1, shows the first signs of idolatry and the first stage of learning the habit of deference to authority. As Darwin portrays it, expressing the emotion of adoration of a superior being, the dog also relinquishes its freedom. Scientists lose freedom of thought whenever they defer to the authority of great men. The antidote to hero worship is unrelenting

FIGURE 4.1. The dog expressing the emotion of adoration of its master shows the first signs of hero worship. (From Darwin, 1872.)

critique of authority. We must also subject the personal idiosyncrasies, personal styles, ideological prejudices, and values of great scientists to constructive critique. It is easiest to form an estimate of those factors by considering their effects on the research of indisputably important neuroscientists, like Santiago Ramón y Cajal (1852–1934; Fig. 4.2) and Charles Scott Sherrington (1857–1952; Fig. 4.3). It is probably not worth stripping such heroes below the neck, but to reduce them to nothing but their thoughts surely omits their motivations, emotions, personal styles, and values.

The greatness of such men lies in the ample scope of their originality and the bulk of their achievements rather than in the outstanding merit of any single discovery. Others would have made their discoveries if they had never lived. It is the whole body of their work, and the intelligence and willpower necessary to accomplish it, that show their superiority. They also belong to the small band who have the scientific attitude necessary for them to see their work as an end in itself. The majority of scientists, to varying degrees, consciously or unconsciously, regard their work as a means to public and personal advantages. That attitude is approved by small achievements, whereas great scientific discoveries encourage a view of science as an end in itself. I believe that all scientists are stimulated by the desire to gratify the ego. However, ego gratification may be linked with a thirst for knowledge as an end in itself, and it may also be linked with the desire for knowledge as a means to extrascientific ends. How the need to gratify the ego can be reconciled with the ends of science is one of the problems of the ethics of science that I shall discuss in the final chapter.

That psychological factors come into play in science is now rarely disputed. The debating points are whether such factors can profitably be excluded and, if not, whether personal style, values, and other psychological factors have significant effects on the outcome of a scientist's work. I'll not go so far as to claim that values are the person, but I can claim to show that a neuroscientist's values determine such significant matters as choice of research problems, and the conception of the problem in terms of theories, methods, goals, and interpretation of the results. For this purpose it is not sufficient to give a psychological biography of a particularly significant neuroscientist, which conventionally is presented apologetically to show the "human" side of the otherwise fully rational scientist. My intentions are different. To relate the science to unique personalities, it is necessary to link scientists' autobiographical disclosures and confessions to research programs, and to disclose the significant historical events and value systems which are related to scientific discoveries. My aim is also to show how scientific concepts are formed and also distorted by the actions of powerful intellects and imaginations.

If these assessments of mine are in disagreement with others, they will have fulfilled one of my aims. I believe that the study of history is in a perpetual state of revision. There is no "correct" ideological position from which the historian can gain a privileged view. The job of the historian, as I understand it, is to promote the process of revision of our views of history. A history is useless if it merely restates what everybody already believes. The historian's work is justified when it results in discovery and reinterpretation of facts, when it increases our understanding of our positions in historical traditions, and compells us to examine our positions, and to question our own motives and intentions as well as we question those of others. There cannot be an exact science of history that allows us to see a pattern in the past and to predict one in the future. The notion that history is determined by the actions of great heroes is no better than the notion that history is a succession of fortuitous events that occur at random or are determined by unknown forces.

Questions such as whether scientific discoveries really did occur, when and where they happened, and who were involved are necessary points of departure but are not sufficient for the historical critique of scientists and their scientific research programs. To obtain information required for critique, we have to ask many more questions about the structure of the scientific research program itself, its underlying theoretical concepts, values, and its research techniques. In addition, questions have to be asked about the prevailing beliefs, assumptions, patterns of thought, conventions, and social conditions in which the scientific research program was embedded. Science is a social activity in which participants communicate, share experiences, and engage in dialogue, out of which they construe meanings. New empirical data are not merely brought into purely objective logical relationships within a theoretical construct. Instead, they are arranged, selected, evaluated, interpreted subjectively, and fitted into an explanatory network which is shaped by personal, social, political, and other extrascientific conditions. Those conditions permit, inform, and instruct the meanings of science.

4.2. THE CULT OF HEROIC SCIENTISTS

We may try to reject idolatry, but it is impossible to discard all our illusions: The elimination of illusion is itself an illusion. We should also guard against the danger of making an idol out of the rejection of idolatry. Anti-idolatry can be made into the opposite half of the larger idolatry—of the excessive reverence for pure science devoid of human values.

There are good reasons to respect heroic figures from the past: to exemplify virtues that we want to perpetuate, to counteract the provincialism of the present, and to add to a fair and just critique. Other reasons are not so good: to revise history in their favor, to identify ourselves with heroic figures, and to form cults and mythologies around them. It is deplorable when people sentimentalize about the past because they are too prissy to affirm its nastiness. We need a corrective to the "*O, altissimo!*" sentimentalities of people like Carlyle and Emerson who had the notion that an institution is the shadow of a great man. To assert that there are individuals whose energy, imagination and inspiration have moved others to believe them and follow them is not the answer, it is the question. Fatuous adoration of great men will not enable us to answer the question: Is it necessary to have hypertrophy of the ego to rise to greatness? Cajal's case makes it seem to be so, but Sherrington's case proves otherwise.

It is possible to present a scientist in a way so revisionary as to turn him into someone else: Cajal or Sherrington into a version of the *fons et origo* of neurobiology.

Such games relinquish respect for others who preceded those men. Such extreme points of view are also notoriously resistant to criticism. For example, Sherrington has been said to have *"almost singlehandedly crystallized the special field of neurophysiology"* (Fulton, 1952). Coming from a disciple who tried to adopt his master's methods and values, this statement can be construed as the rhetoric of a sincere admirer, as a desire to be identified with a symbolic figure, or as a bit of both.

Scratch a hero to find a villain, scratch a villain to find a hero, it is said with some truth, but that is not the way in which I have approached the so-called heroes and villains of neuroscience. I have approached them with sympathetic understanding, trying to separate them from the layers of learned papers in which they have been wrapped both by admirers and by critics. The essence of a great scientist such as Cajal is that after all the fancy wrappings have been removed he emerges as indispensable for our understanding of neuroscience: You can be for Cajal or against him but not without him!

The greater the public acclaim of the man, the more necessary is the critique. The alternatives to critique are various degrees of "fusion" of the reader with great figures who are understood merely as part of the reader's own experience, as objects of the reader's unfulfilled wishes, and as symbolic figures with whom the reader wishes to be identified. By contrast, critique aims at reaching an objective and compassionate understanding of the processes of differentiation of a person's beliefs and values.

Unrelenting critique ought to be part of any scientific research program. Critique is part of the recurring cycle that also includes reflection and communication, as I explain in Section 5.4. Imaginitive, sympathetic, and rational reflection give critique emotional as well as cognitive dimensions. Science is not only the discovery of new and useful information. It is also the experience of participation in a historical process and of forming sympathetic bonds with other participants in that process. It is natural to form rich bonds with those closest to oneself, including those regarded as one's close spiritual and intellectual ancestors. It is better to feel sympathy with those and ignore the rest, than to love humanity and not care for those within one's personal circle and close intellectual tradition. Unqualified sympathy naturally extends to one's close personal circle, not to one's intellectual ancestors. Yet critique of them also includes sympathy, even when it may lead to disapproval and disagreement. Engaging in recurring cycles of reflection, communication, and critique can raise one's level of consciousness of the relations of individuals, including oneself, to a historical tradition. One becomes more fully conscious of the efforts and conduct of individuals who have advanced or retarded the progress of science.

Critique of acclaimed neuroscientists shows that neuroscience is particularly vulnerable to the personal idiosyncrasies of individuals. This follows from the peculiarities of theory construction in neuroscience, especially the easy admission of qualitative and analogical thinking, to which we have alluded earlier. Cajal's thought was entirely in the analogical-correlative mode; Sherrington's was more in the causal-analytical mode. Like all essential distinctions this one is not absolute. Nevertheless, there are important differences between Sherrington's method of solving problems, which was driven by deductions made from observations, and Cajal's more inductive creative process, which was strongly analogical (for data-driven methods see P. Langley, 1981; for analogical problem solving see Section 1.2.6 and Arber, 1944; Gick and Holyoak, 1980; Holyoak, 1984; Holyoak *et al.*, 1984).

They differed in other significant ways. They represented two typical types of 19th century intellectuals—Sherrington the classically trained scholar, Cajal the autodidact, a literary stylist and philosopher with no formal literary or philosophical training. They offer an opportunity for us to learn how personal style may either illuminate or obscure scientific content. We should not forget that the differences in style of the two men arise as much from differences in their personal characters as from their national characteristics. They show that great scientists are rarely great inventors or technical innovators, but have a knack of knowing how to adopt and modify techniques to suit their own purposes. Both completed their best scientific research before 1900, during the age of optimism, before serious doubts had arisen about the moral worth of science. They also show that scientists do their most original work when they are young. Apparent exceptions to that rule are theoreticians and older scientists who direct research teams in which the most original work is done by the younger scientists whose originality is too often submerged in multiple authorship.

They both changed with age from the athletic to the aesthetic, but they were quite different in their hearty and their arty styles: Sherrington formed deeply fulfilling human relationships that enhanced both his and their scientific work; Cajal was centered on himself, unable to form creative scientific relationships. Those different character traits were expressed early. For example, as a youth, Sherrington excelled in team sports, while Cajal preferred exercising alone at what he describes as a "gymnastic mania." Cajal expressed the defenses and feelings of the narcissist, as they are described by Stephen Johnson (1987): His feelings of isolation and depression were associated with the defenses of self-involvement, perfectionism, pride, and grandiosity of style. These expressions of loneliness and Cajal's fear of intimacy, and lack of empathy, are in contrast with Sherrington's friendly involvement with others. Sherrington preferred writing poetry anonymously, and he delayed the publication of his book of poems, *The Assaying of Brabantius*, until he reached the age of 68 in 1925. Out of genuine modesty he asked his admirers to refrain from writing his biography.

Sherrington's most important work was completed in the decade before 1900, between the ages of 35 and 45. Of his 53 publications from 1890 through 1900, only 4 were collaborative. It was unusual for a research report to have more than two authors in those times. Sherrington published fewer than one-fifth of his papers in collaboration with another person, and only three times with more than one collaborator.* The first paper that Sherrington had not written himself was written in collaboration with Eccles in 1929 (Eccles, p. 54, in Eccles and Gibson, 1979). The fact that he worked without immediate scientific collaborators and his natural modesty account for the dearth of biographical accounts of the most creative period of Sherrington's life. After he moved to Oxford in 1912, at the age of 55, his laboratory attracted numerous students and

---

*1885, with C. S. Roy and J. Graham Brown; 1911, with F. W. Mott and E. Schuster; 1932, book with R. S. Creed *et al.*

visitors, although it never became a research institute in the old Germanic and new American style, with the famous scientist surrounded by epigones. That pattern had been developed in Germany in the mid-19th century (McClelland, 1980). The model was Carl Ludwig at the Institute of Physiology at the University of Leipzig, surrounded by laboratory assistants and young research students from many countries (Lombard, 1916; Schröer, 1964). Ludwig obtained the funds and defined the scope of the research program. Students and assistants did the experiments and shared the credit with Ludwig. During Sherrington's visit to Germany in 1886–1887 he had some rather unpleasant experiences of the operation of the research groups of Rudolph Virchow and Robert Koch at Berlin (Swazey, 1969).

In any event, Sherrington seems to have developed a dislike for working with large groups and after his return to England he continued working on his own with his devoted laboratory assistant George Cox, and with an occasional scientific collaborator. Even after moving to Oxford he continued doing experiments alone or with a succession of individual young collaborators: A. S. F. Leyton, formerly called Grünbaum, had worked with Sherrington at Liverpool; T. Graham Brown, and E. G. T. Liddell were his other collaborators during the first decade at Oxford. Thereafter, in his sixties and seventies, Sherrington continued to attract very good physiologists to work in his laboratory at Oxford, of whom J. C. Eccles and R. Granit would later be awarded Nobel prizes.

Some of these younger men have given us accounts of their experiences (Graham Brown, 1947; Denny-Brown, 1952, 1957; Granit, 1952, 1966; Liddell, 1952; Viets, 1952; Eccles, 1957; Penfield, 1957, 1962; Eccles and Gibson, 1979). They show the septuagenarian in the mellow Oxford setting, but they hesitate to admit that the work of Sherrington's final period is little more than an extended footnote to *The Integrative Action of the Nervous System*, published in 1906. They no more succeed in depicting the whole character in a historical framework than a picture made by joining the dots can ever succeed as a masterpiece of portraiture. By contrast, Cajal's autobiography is indeed a masterpiece of imaginative reconstruction, loaded with special pleading to woo unwary readers, and mined with misinformation with which to entrap them. I, too, was once briefly entrapped into making a foolish statement about Cajal's prophetic vision (Jacobson, 1970a, p. v).

The 19th century was an adversarial age for which Cajal was temperamentally suited. Cajal was a natural loner who worked best without collaborators. He thrived in opposition. Even after his ideas were generally accepted, he used the opportunity of his autobiography to settle some old scores, not only with his scientific opponents such as Golgi and Nissl, but also with allies such as Forel, His, and Koellicker. Wherever he looked he saw potential competitors for priorities of discoveries which he regarded as his own. Cajal viewed science as a competitive more than a cooperative enterprise—himself against almost all the others; the embattled Spanish against the powerful Germans and "Anglo Saxons," as he liked to refer to the British and Americans. In his autobiography (p. 414) Cajal portrays himself grandiosely as an inspired interpreter of *"the work of a divine artist."* Such a character invites criticism while Sherrington's amiability and modesty disarm the critics.

With two notable exceptions, his polemic in 1892 about vivisection, and his 1894 polemic with Victor Horsley about priority, Sherrington avoided public arguments.* He also stood aside from the intellectual conflicts—science versus religion, materialism versus idealism, utilitarianism versus Kantianism, and so forth - which fitfully illuminated the intellectual landscape of late 19th-century Europe. Undogmatic to a fault, Sherrington qualified every statement that might be construed as a final verdict. As a result his literary style is tentative and gently persuasive in contrast to the grandiose, assertive style of Cajal. But Sherrington and Cajal shared one very important virtue—their ability to get the best of both the possible worlds of scientific rigor and artistic imagination. Both approached the nervous system as they would have approached a work of art to be interpreted with sympathy and imagination, not only to be investigated rationally. They show that the best scientists never grow up—it requires a certain childlike capacity for an adult to continue a vigorous life of the imagination. Scientists who grow up become administrators.

It is important to point out that neither Cajal nor Sherrington tried to justify their work in terms of clinical relevance or other utilitarian values. They were aware of the possible practical applications of their discoveries—both were trained in clinical medicine and both began their research careers in pathology—yet neither felt the need to give extrascientific justifications of their research. They regarded scientific research as an end in itself. This was one reason why Sherrington rarely referred to the work of Hughlings Jackson or of any other clinical neurologist or psychiatrist. Likewise, Cajal's voluminous work on nervous degeneration and regeneration contains very few references to its possible clinical relevance. Cajal saw nerve degeneration and regeneration, not as problems of clinical relevance, but as purely scientific problems to be solved in terms of the neuron theory.

It should by now have become obvious that I am not only tracing the sources of knowledge and the conceptual innovations of Sherrington and Cajal, but I am also attempting to form hypotheses about their characters. Were their scientific conceptual innovations triumphs of ideas over matter—the taming of recalcitrant matter by ideas? Or were they the adaptation of mind to the given materials—perception of patterns

---

*The surgeon Lawson Tait asserted that experimental animals experience pain on *"the slightest touch. . . . applied to almost any part of the peritoneum"* (*Br. Med. J.*,15 Nov. 1892). Sherrington (1892b) contradicted him, pointing out that Tait was ignorant of the fact that the peritoneum is *"quite insensitive to mechanical and other modes of excitation."* This little polemic should be seen in relation to the threat to experimental physiology posed by the antivivisection movement in Britain (see Section 4.4).

The other polemic was with the neurosurgeon Victor Horsley about priority of discovery. The polemic was sharpened by the different moral and other values of the protagonists: Horsley was a muscular Christian, a crusader for abstinence from alcohol and tobacco, and also strongly opposed to the painting of the female nude (Davis, 1964; Tepperman, 1970). In 1894 Sherrington published a note in the *Lancet* describing degeneration of the pyramidal tract after cerebral ablations performed by him on monkeys at the Brown Institute. The following issue of the *Lancet* carried a letter from Horsley claiming that he had suggested the experiments to Sherrington, and that they had subsequently been carried out by one of Horsley's assistants at University College. Horsley demanded an investigation. Sherrington responded by refusing to cooperate, saying that he would not *"admit for one moment that Professor Horsley has any exclusive right to the research."* Nevertheless, Sherrington did not continue that research until after Horsley's death. In similar circumstances Cajal would have continued, not shrinking from confrontation.

already there in the materials? Or can scientific concepts, like those of philosophy, be merely the disguised representations of powerful experiences? In one of his naturalistic frames of mind, that is what Nietzsche thought: *"every great philosophy has so far been the self-confession of its originator, and a kind of involuntary, unconscious autobiography (Beyond Good and Evil, I, 6).* The history of neuroscience shows that all three are parts of the process of conceptual innovation: imaginative as well as logical reconstruction of experiences are episodes in the minds of unique personalities.

4.3. CAJAL'S CASE: "THE BONFIRE OF THE VANITIES"

When a cult grows around a symbolic figure such as Cajal, its language becomes charged with a combination of awe and hyperbole which apparently gives satisfaction to members of the cult but which others find distasteful. In either event, the resulting loss of objectivity distorts the view of the symbolic figure. It may be invidious to select one example when others are more flagrant, but the following recent statement by a distinguished neuroanatomist is typical: "... *during the last decade of the century neuroanatomy advanced more rapidly than any of the other disciplines. The phenomenal growth was due almost entirely to the work of one man, Santiago Ramón y Cajal, who from 1888 to 1911 almost single-handedly created modern neuroanatomy"* (S. L. Palay, Nature 341:493, 1989). But that cannot be correct if, as Edinger (1891, pp. 4–5) claims, it was Benedikt Stilling (1842, 1843, 1846, 1856–1859, 1864–1878) who really "*laid the foundations of our knowledge of the anatomy of the pons, the cerebellum, the medulla oblongata, and the spinal cord, in a series of masterly works, showing an industry never surpassed. . . . which will surely remain forever a monumentum aere perennius. . . .*" Remember that 2000 years ago Horace was accused of vanity and condemned for his ode "*Exegi monumentum aere perennius,*" and that the achievements of individuals are even more perishable now than ever before.

To say that Cajal constructed modern neuroanatomy singlehandedly is to ignore that Golgi provided Cajal with his most useful technique; Wilhelm His showed Cajal what the embryonic nervous system could reveal; Forel and His gave him the idea of the individuality of the nerve cell; Meynert and Koelliker taught Cajal that there was much more to be discovered about the architectonics of the cerebral cortex, and so on. Cajal was in a position to remake the past, but only in continuity with his predecessors. He did so by working in fields already plowed by others. He took up many problems of development, regeneration, and structural organization of the nervous system known before his time and made use of many of the techniques practiced by his contemporaries. In attacking on such a broad front, it was inevitable that Cajal would suffer some retreats, but they are outnumbered by his advances. That is why we hail Cajal a progressive, but we should not make a household god of him. The misconception that Cajal invented the neuron theory and established neuroscience single-handedly is partly the result of Cajal's aggressive claims to priority, as I have pointed out in Section 1.5, and to his seductive romantic style. Koelliker and His laid down the evidence for the neuron theory in the calm and measured words of the classicists. Cajal wrote in the impassioned style of a romantic and a prophet and gave the idea wings.

FIGURE 4.2. Santiago Ramón y Cajal (1852–1934) at about 70 years of age, in a characteristic pose, photographing himself. Some of his many self-portraits, taken when he was younger, are shown by courtesy of the Museo Cajal, Instituto Cajal, Madrid. These self-portraits were intended for posterity which Cajal believed was waiting for him.

The foundations of modern neuroscience had been laid before Cajal's entrance to the field in 1888. Accordingly, I have devoted Chapter 3 entirely to a review of the construction of the neuron theory, the discovery of the modes of development of axons and dendrites, and the discovery of neuronal connections by synaptic contact. All those discoveries were made by others, although Cajal supplied evidence that had significant corroborative value. Cajal deserves credit for having shown, short of electron microscopy, that neurons connect by contacting one another, and for gaining a consensus for including that lower-level theory as one of the components of the higher-level neuron theory. He also improved the existing descriptions of regional microanatomy of the CNS, and he helped to secure the concept of each neuron as an input-output unit. He was much less successful in advancing knowledge about how each region is connected with other parts of the system. One of his principal contributions, shared with van Gehuchten, was to generalize the concept of individual neurons as input-output systems, but to develop that concept was beyond his abilities because he never understood the valvelike function of the synapse. Indeed the word "synapse" does not appear in his Nobel lecture of 1906 or in the third edition of his scientific autobiography which appeared near the end of his career. Almost all of Cajal's discoveries were also made nearly simultaneously by others. The originality and power of Cajal's style almost always surpass those of the content. Perhaps it is closer to the truth to say that the power of his style enhanced the content in terms of both simplification of the relationships between cells in the ensemble and selective distortion and exaggeration of certain details. The result was to create an image that was both realistic and symbolic—a work of art as much as empirical fact.

4.3.1. Fast Track to the Neuron Theory

Cajal's scientific career had four main periods. The first began in 1880 at the University of Zaragoza and continued until 1887 at the University of Valencia. In this period he studied the pathology of inflammation, the microbiology of cholera, and the structure of epithelial tissues and muscle, using conventional histological techniques. The second period began with his adoption of the Golgi technique for studies of the cytoarchitecture of the nervous system. This coincided with his professorship at the University of Barcelona (1887–1892) where he made his most important discoveries on the structure of the neural tube, spinal cord, cerebellum, brainstem, diencephalon, retina, hippocampal formation, and cerebral cortex. This work was done at the peak of his creative powers, between the ages of 35 and 40. He summarized these discoveries in his Croonian Lecture, published in 1894, in the short monograph *Les nouvelles idées sur la structure du système nerveux* (1894b), and crowned this period with his masterpiece, the three-volume *Textura del sistema nervioso del hombre y de los vertebrados* (1897b, 1899–1904; French translation, 1909–1911).

The third period started with his move to the University of Madrid in 1892 at the age of 40, and lasted until about 1905. This period came to a fitting conclusion in 1906 when he shared with Golgi the Nobel prize for physiology and medicine. During the third period he was mainly concerned with the histological changes that occur during degeneration and regeneration of the nervous system, using methylene blue vital staining and the reduced silver method which he had perfected in 1903. At Madrid, Cajal was assisted increasingly by younger collaborators, especially Tello, who did much of the work on degeneration, regeneration, and transplantation of peripheral nerve ganglia to the CNS, and Sanchez, who assisted with the studies of the insect and cephalopod nervous systems which were started after the outbreak of the Great War in 1914. These and other young assistants assumed major responsibility for research in Cajal's laboratory after 1906. During that fourth period he rounded off his studies on the histology of the cerebral cortex and some subcortical centers. He began his studies of neuroglia which were hampered by technical limitations that he overcame partially by developing his gold sublimate technique in 1916 and a modification of the Bielschowsky silver method in 1920. During that period he was involved in a dispute with del Río Hortega about the functions of oligodendroglial cells in myelination in the CNS (see Section 1.4.4). Much of his effort was aimed at defining the significance of the Golgi apparatus and the neurofibrils. Those studies had started during the earlier years in Madrid. They were aimed at combating the theories of Apáthy, Bethe, Held, Bielschowsky, and Nissl, regarding the function of neurofibrils as elements conducting nervous activity from neuron to neuron. This period was crowned, as usual, by a monumental work, the two-volume *Estudios sobre la degeneración y regeneración del sistema nervioso* (1913–1914; English translation, 1928).

The end of Cajal's creative career, like Sherrington's, coincided with the Great War of 1914 to 1918. That cataclysm ended the careers and lives of many of the great European neuroscientists: van Gehuchten, Dejerine, Alzheimer, Brodmann, Edinger, Nissl, and Retzius among others. A list of names cannot portray the intellectual disarray and moral despair into which neuroscientists were thrown by the Great War. Progress appeared to be an illusion. The age of certainty and optimism had ended, the era of doubt and pessimism had begun, even for those, like Cajal and Retzius, whose countries were not directly involved in the war.

Gustaf Retzius was a model of the completely self-confident 19th century scientist—confident of his superiority of intellect, culture, and ethics. In ethics he was a utilitarian and optimist, in religion a deist and anticlerical. Politically he was a liberal. Living in neutral Sweden he was not directly involved in the Great War, but he had close friends on both sides of the conflict which resulted in the disintegration of his intellectual, cultural, and ethical worlds. In his last letter to Cajal in 1919 Retzius wrote "*I die despairing, for I have lost my faith in the destinies of humanity*" (cited in Ramón y Cajal, 1923, Chap. VII, fn. 20). As Cajal thought of that time: "*For six years I remained cut off from communication with foreign laboratories and reduced to a monologue in which disgust and dejection were the fundamental key-notes. . . . in my will, shocked by the catastrophe, there rose for the first time that terrible 'what is the use?'* " (Ramón y Cajal, *Recollections*, p. 583).

There was no question in the mind of Cajal or of anyone else about the importance of the vast increase of knowledge gained by science. But the certainty was eroded by doubts about the applications of that knowledge to do good and to do no harm. Within science the certainty had been undermined by Darwin, who showed that change and progress are accidental, not inevitable. In physics, the position that objectivity is the same as measurability was shaken after 1927 when Werner Heisenberg published the uncertainty principle. That followed from Einstein's interpretation that uncertainty is a problem of measurement, not fundamental. Only later did the majority of physicists agree that Heisenbergian uncertainty is the result of fundamental fuzziness of the wave function, which quantum theory takes to be objective reality. That is close to saying that objective reality is the same as measurability.

Neuroscientists of the generation of Cajal and Sherrington were temperamentally unsuited to the age of pessimism and uncertainty that followed the Great War. After the war the Golgi technique went into an eclipse from which it emerged briefly 40 years later. Experimental methods of tracing connections largely replaced the descriptive neurohistological methods practiced by Cajal. Electronic instruments replaced the neurophysiological apparatus of the prewar period. The generation of Cajal and Sherrington had prepared the way for those changes, but few of the neuroscientists of their generation continued to implement the postwar program. Those who survived the war, like Sherrington, Cajal, and Pavlov, were little more than emblematic figures, to inspire a new generation of neuroscientists.

American scientists escaped the sense of tragedy that affected the Europeans after the First World War. Only after atom bombs were dropped on Japan, and only after America faced a similar threat from Russia, did American scientists share the doubt that had infected European science 30 years earlier. Scientists now recognize that history has taken a turning point in which they share a terrible responsibility. As Robert Oppenheimer confessed in 1948, *"The physicists have known sin."*

The gloom cast over the world from which Cajal departed in 1934, was in poignant contrast with the brilliant future for the world that was predicted when he entered. The rapid progress of neuroscience in the first half of the 19th century has been reviewed in earlier chapters. The state of knowledge of the structure of the nervous system just before Cajal's first original contributions to neuroscience in 1888 can be very well determined from Obersteiner's textbook completed in October 1887 and published in 1888. On pages 124–125 he discusses Golgi's nerve network theory and contrasts that with the recent work of Forel. Obersteiner explicitly recognized that Forel's observations meant that nerve cells connect with one another by contact only: *"Really different is the relationship [between nerve cells] after Forel's concept. He believes that the finest branches of different nervous elements touch one another like the branches of two neighboring trees, thus not passing directly into one another. Nevertheless, he leaves it unclear how he conceives of these finest free terminals. From a physiological standpoint there is no necessity for the processes to be in direct continuity; we can very well think. . . . that also the mere contiguity that is afforded by multiple contacts between them, is sufficient to enable transmission of stimuli approximately like Ehrlich's finding of the attachment of the spiral fiber on the sympathetic cell."* Obersteiner continued by affirming that *"excellent sublimate preparations [i.e., Golgi preparations] also support the latter concept: They always stain only individual nerve cells with their rich fiber systems, and certainly never show a fine anastomosis between two neighboring cells."* These histological findings were fully understood to be evidence for the contact theory and against the opposing theory of anastomosis. The functional meanings of those observations were thus common knowledge, although not universally favored, before

Cajal's work started. They throw doubt on Cajal's claims to priority for the discovery of the crucial evidence corroborating the theory of nerve contact.

Before Cajal began to work on the histology of the nervous system in 1887 it was already known that nerve cells originate by mitosis (His, 1887); that they can migrate as individual cells as well as coherent cell groups (His, 1886, 1887); that the nerve fiber is part of the nerve cell (Bidder and Volkmann, 1842; Bidder and Kupffer, 1857); that the axon is an outgrowth of the nerve cell (Bidder and Kupffer, 1857; His, 1886, 1887); and that nerve cells may degenerate after injury without affecting neighboring cells (Forel, 1887). All this evidence supported the theory that nerve cells come into contact but not protoplasmic continuity with each other. Both His and Forel had published that conclusion in 1887. The alternative theory that nerve cells are linked directly through fiber networks was supported by much histological evidence discussed in Chapter 3. That evidence was proved to be artifactual only when the synaptic cleft was revealed by electron microscopy in the 1950s. During Cajal's lifetime the two theories were caught in a deadlock, and in Cajal's final paper *"Neuron theory or reticular theory?"* published in 1933, he could argue only that the weight of evidence corroborated the contact theory, not that it conclusively refuted the nerve network theory.

What were Cajal's contributions to resolving the conflict? In a nutshell, his great contribution was to characterize the histological structure of the nervous system uncompromisingly in terms of a theory of anatomical contact between nerve cells. That theory as well as other components of what came to be termed the neuron theory were already in place before Cajal entered the field, as we have seen in Chapter 3.

I believe that Cajal started with the assumption that nerve cells, like all other cells, are absolutely independent units and are only in contact. His earlier training in pathology had exposed him to Virchow's successful extension of the cell theory into the domain of cellular pathology. From the beginning of his work on the nervous system, Cajal showed a single-minded intention to bring the nervous system completely into the domain of the cell theory. From the beginning of his career in neurohistology Cajal took an uncompromising stand in support of the hypothesis of nerve cell autonomy and connection by contact. He discounted or ignored any evidence of anastomoses between nerve cells. Thus, Cajal placed himself in a stronger position than the older generation of neurocytologists who had grown up with the theory of protoplasmic connections between nerve cells. They had to adapt the established theory to new evidence as it trickled in. Usually it is an advantage to support only one of two opposing scientific theories—at least one has a chance of being on the winning side. But to subscribe in part to each theory or to change from one to another, is to be exposed to the censure of both. To subscribe in part to the contact theory and the network theory, as some did such as Nansen, was to be exposed to the criticism of the supporters of both systems.

Analysis of scientific careers provides another means of arriving at answers to questions about the nature of the links between theories, techniques, observations, and values. What are the links that unite those components in the mind of the scientist and that enable those components to fullfil their functions in a scientific research program? Is a theory, or a technique, or an observation, or a value capable of independent existence? What are the effects of each of those on the others? Analysis of the career and personality of a scientist such as Cajal shows that those components are transformed by the scientist's personality, style, and values. For example, the Golgi techniques, especially the rapid Golgi technique which Cajal favored, were one of the keys that unlocked the answers to questions about the relations between nerve cell forms and functions, but they delivered quite different answers to Golgi than to Cajal. Techniques can only be fully understood in the context of the entire research program of which they are a part, which also includes theories and the values of relevant participants.

The rapid Golgi technique was ideal for Cajal's purposes: It is now known that precipitation of silver chromate fills the cell completely but less than 5% of neurons are impregnated, apparently at random without selection of specific types of neurons. Now we know that the Golgi techniques are not neutral with respect to the rival theories of the mode of nerve connections: they produce data favoring a theory of separation of neurons at points of surface contact. Cajal assumed that the Golgi technique was nonselective and that it revealed the nerve cell entirely, but his methods were not quantitative and so he could only assume what he set out to prove—that all nerve cells are entirely separate from one another. He dismissed the critics who claimed that there were structures like intercellular protoplasmic bridges and networks that were not impregnated by the Golgi techniques, under any conditions, but could be shown by means of other histological techniques (Apáthy, 1889, 1897; Bergh, 1900; Nissl, 1903). Thick (100 μm) sections of brain and spinal cord stained by the Golgi method showed entire stained neurons as if they were isolated from the others, or in contact with one another. Cajal rapturously described the appearance of the Golgi preparation: *"On a perfectly translucent yellow field appear thin, smooth, black filaments, neatly arranged, or else thick and spiny, arising from triangular, stellate or fusiform black bodies! One might say they are like a Chinese ink drawing on transparent Japanese paper. The eye is disconcerted, so accustomed is it to the inextricable network stained with carmine and haemotoxylin which always forces the mind to perform feats of critical interpretation. Here everything is simple, clear, without confusion. . . . The metallic impregnation has made such a fine dissection. . . ."* (Ramón y Cajal, 1909, Vol. 1, p. 29). In other words, the histological picture revealed by the Golgi technique was accepted by Cajal as *prima facia* evidence, whereas the evidence obtained with other techniques required "feats of critical interpretation." This belief in the authenticity of his data enabled him to reconstruct a synthetic representation of the cytoarchitecture from neurons seen in many different sections and to generalize the principle of neuronal contact to the entire nervous system.*

Cajal considered 1888 *"my greatest year, my year of fortune,"* and said that *"the years 1890 and 1891 were my Palm Sunday"* (*Recollections*, pp. 321, 373). His ascent was

---

*Cajal could draw nerve cells freehand at high magnifications with considerable accuracy, although he probably used the camera lucida for making drawings of microscopic preparations at low magnifications (see later in this section and Section 1.4.1). As we saw in discussing his analogical mode of thinking (Section 1.2), his method of constructing models of nervous organization required the selection of those neurons he deemed significant. As he admits, individual neurons were not always copied from the same microscopic field but were selected from different sections to represent the essential characteristics of each type of neuron. Cajal's entire work is predicated on his belief in the validity of distinct neuronal types. The general problem of cellular typology and especially its relation to reductionism is discussed in Section 3.6.

rapid—his work on the histology of the nervous system began in 1887 and the first reports of his results date from 1888. His discoveries made in 1888–1892 were extremely important because his conception of how nerve cells are connected together in the vertebrate retina, mammalian cerebellar cortex, and spinal cord in birds and mammals, were in advance of previous observations. They provided additional evidence for the theory that neurons connect by contact, not by continuity.

The historical context of Cajal's discoveries is considered in detail in the earlier chapters. Cajal's epoch-making papers of 1889–1890 contain only the barest minimum of information sufficient to support his claims to priority of discovery. Cajal's habit was to publish rapidly in a local journal to establish priority, and to send the same results to a foreign journal to be published about a year later. But even those papers, intended as communication of his results to the world at large, offer paltry information about methods, numbers of specimens examined, and variability of the results. Our standards of objectivity and of the burden of proof are simply not applicable to Cajal and to the standards of his times.

Cajal and his contemporaries were coy when it came to disclosing quantitative information like the number of specimens they examined and the variability of the observations on which they based their conclusions. Experimental evaluations of alternative theories in neuroscience during the 19th century were often merely a matter of claims made on the basis of a small, often undisclosed, number of observations. Nineteenth century neuroscientists were not yet overwhelmed by the statistical revolution. But their neglect of statistics was not an issue in their disputes about the authenticity of their histological observations. This seems anomalous in relation to the great progress made in biological statistics in the 19th century.

Probability and statistics developed in the 19th century, first as a mathematical theory, beginning with Laplace; second, as population and social statistics, beginning with Adolphe Quetelet, and third, and much later, as applied to experimental science. The application of statistics to experimental biology and medicine was widely resisted, for example by Claude Bernard (1865, pp. 131–139; Olmsted and Olmsted, 1952, pp. 140, 220, 232), and was not generally accepted for most of the second half of the 19th century. This may help to explain the disregard of statistics by neuroscientists at that time.

It is difficult to estimate the effects that disregard of statistical and quantitative analysis had on the progress of neurohistology. Disregard of quantitative analysis must have enabled the early neurocytologist to press ahead opportunistically with qualitative observations without expending the much greater labor of gathering data sufficient for quantitative analysis. On the debit side was the habit of making sweeping theoretical generalizations on the basis of few observations, usually obtained by means of a single method. Quantitative analysis of neurohistological observations was not done routinely until after the pioneering work of Bok (1936, 1959) and Sholl (1956a,b).

To communicate his findings widely and rapidly, Cajal founded his own journal, the *Revista trimestral de Histología normal y patológica*. In 1888, the first year of publication of the new journal, Cajal wrote all six papers and also made the drawings for the lithographic plates. Sixty copies were printed, to be distributed mostly to foreign scientists. The two papers on the cerebellar cortex, published in May and August 1888, reported the discovery of the mode of termination of the climbing fibers, described the mossy fiber endings, and correctly described the termination of the basket cell axons as nests surrounding the Purkinje cell bodies (Golgi had discovered them but he believed that the baskets also contributed to a nerve network in the granule layer as shown in Fig. 3.18B). In 1890 Cajal discovered that the parallel fibers are the axons of granule cells.

Cajal's first papers on the structure of the bird's retina, published in 1888 and 1889, reported the discovery that the centrifugal fibers end in the inner plexiform layer, in relation to amacrine cells (called spongioblasts by Cajal), and not on blood vessels as Duval had claimed. In 1891 he showed that the centrifugal fibers end as nests surrounding the amacrine cells. In 1888 he showed the rods and cones ending by contacting the dendrites of bipolar cells. He found that ganglion cell dendrites come into contact with bipolar cell axons in the internal plexiform layer, challenging the report by Tartuferi (1887a,b) that the internal plexiform layer is a diffuse nerve network.

In his autobiography Cajal claims that these observations, particularly the discovery of climbing fibers, "*formed the final proof of the transmission of the nerve impulse by contact*," but his observations could not have supported that claim. That idea may have started germinating in Cajal's mind in 1888 but he never saw the contact zone, only the nerve endings, and he conceived of the contact as very intimate. He did not conceive of a cleft at the site of contact. As we have seen in Section 1.2 he depicted a wide gap there, only to emphasize his opposition to the rival theory of anastomosis between nerve cells. Moreover, Cajal's concept failed to take into account the functional requirements of the contact zone, and that was one of the reasons why his ideas were ignored by neurophysiologists, as we shall see later in this section.

Why was Cajal able to make those discoveries? He attributes his success to his choice of embryonic or immature animals, and to his double-impregnation modification of the rapid Golgi technique. These technical advances were undoubtedly important, yet I question whether they are sufficient to account for Cajal's phenomenal record of original discoveries in the first 5 years after beginning his neurohistological investigations. Others had used the comparative-developmental method before Cajal, as I have shown in Section 1.15, and others had used the Golgi technique (Golgi, 1882–1885; Nansen, 1886–1887; Tartuferi, 1887a,b) but they had not succeeded in accomplishing what Cajal achieved in 1887 to 1890. We have to look deeper into Cajal's scientific background, his values and especially his personality, for the real causes of his originality. We shall find them in his historical position in the construction of the neuron theory, his egotism, his restless intellectual and physical energy, and his romantic imagination.

4.3.2. The Making of a Neuroscientific Mind

It is unjustifiable to claim that further analysis of the psychological processes of Cajal or any other great figure would have little relevance to his scientific work and would perhaps only pander to prurient interest. Some may prefer to admire his monumental achievements as purely intellectual constructs formed in a social and emotional vacuum. I understand Cajal's autobiography as an attempted apologia for the disadvantageous position of the Spanish scientist in European science, and as an expression of his narcissistic personality, his romantic temperament, and his genius as a scientist. Therefore, in addition to analysis of scientific content, one must subject to critical analysis the psychological and societal dynamics that are revealed in its language, daily practices, institutional roles, political and economic conditions. Efforts have hardly begun to mine that rich lode.

Economic power was virtually nonexistent in 19th century Spain. The keys to power were either aristocratic lineage or association with established institutions of church and state. The brain was a domain that could be claimed equally by the temporal and spiritual authorities, by both state and church. Cajal rose to power by claiming authority in that domain. In modern parlance, he occupied a "power vacuum" that nobody else in Spain was qualified to enter.

Cajal's Spain (unlike the present consumer society) was an austere country in which people of Cajal's social class sought refuge from poverty and economic stasis in patriotism, idealism, romanticism, and mysticism. These ideological gestures were not meant to be expressions of a need for money—that would have betrayed a narrow spirit. They were aimed either at pleasure or at less frivolous or even serious aspirations in art, politics, the church, or science. Cajal's aspirations were always serious. His autobiography does not betray the slightest hint of frivolity.

Cajal was not a bourgeois gentleman. He was born in 1852 of stock that thrived in the harsh country of Aragon, which endowed him with tremendous physical and intellectual vitality. Catapulted by his energy and egotism out of the rocky soil of his birthplace in Aragon, he moved rapidly to a succession of provincial posts until he obtained the coveted Professorship in Madrid. The cost was great—he had spent the main forces of his creative energies before leaving Barcelona for Madrid in 1892, at the age of 40.

Cajal was primarily interested in the anatomical relations between nerve fiber terminals and nerve cells within selected regions. By exploiting the Golgi technique and by studying the changing anatomical relations between fiber terminals, dendrites, and nerve cell bodies at different stages of development, he rapidly surpassed all others in the accuracy and detail of his observations. He showed little interest in the connections between centers and in the fiber pathways in the CNS. That is why he paid scant attention to the thalamus and to the extrapyramidal system, and why he neglected to use the Weigert and Marchi methods for tracing myelinated nerve fibers. He invented a reduced-silver method that stained unmyelinated nerve fibers excellently, yet he used his method only for cytological studies and for studying nerve regeneration and degeneration, not for tracing pathways.

From 1887 to 1892 he concentrated all his formidable intelligence and energy on the single problem of how nerve fibers end on nerve cells. The great neurohistologist proves the aphorism of Nietzsche, who said that purity of mind is to be able to concentrate for a long time on one great thing. After 1892 Cajal's attentions started turning to many other things connected with his professorship at Madrid, and his energy was dissipated by writing and illustrating his monumental textbook of neurohistology, published in three volumes in 1897, 1899, and 1904. After that effort his powers of observation diminished, flashes of originality illuminated his works infrequently, and only his fluency with words, his vanity, and his lack of humor remained to the end in 1934. Those characteristics are enshrined in his autobiography *Recuerdos de mi vida* (*Recollections of My Life*), a product of his final period as a national institution.

Scientific memoirs (e.g., Koelliker, 1899; Ramón y Cajal, 1901–1917; His, 1904b; Roux, 1923; Spemann, 1924; Watson, 1968; Crick, 1988) can reveal the roles of the worldviews of the writers and the determinative effects of extrascientific factors. Cajal offers a particularly good example because of his pivotal role in the history of neuroscience and because of the ample records. Biographical texts may be compared and contrasted with contemporary records of events and opinions. The findings may be interpreted to reveal meanings not made explicit in the original text. Influences and prejudices that may have been hidden or denied may be discovered and interpreted. To see how far historians of neuroscience have still to go, we have only to look at the defective English translation by Craigie of Ramón y Cajal's autobiography *Recollections of My Life* (1st ed. 1901–1917, 3rd ed. 1923).* This has been republished in 1989 as an exercise in hero worship rather than as an opportunity for critical analysis of influences, prejudices, and meanings, and as a means of analysis of the workings of a great man's mind. I regard the widespread uncritical admiration of Cajal's biography as evidence that many neuroscientists wish to be identified with Cajal and to use him as an object of their unfulfilled ambitions and desires. I see that as part of the contemporary "*Culture of Narcissism*" (Lasch, 1979). Hero worship is only one of the many defenses of the narcissist (S. M. Johnson, 1987), and worship of another narcissist may be the strongest defense against feelings of inability to relate empathetically with others.

4.3.3. Science as Self-confession

Interpretation of the logic of Cajal's scientific discoveries cannot be made from a literal reading of his autobiography because in it Cajal created a legendary life, as much an artistic creation as a scientific chronicle. In so doing he invites the kind of critical response that is appropriate for all works of literary art. This invitation has not yet been accepted because of the delicacy of the task and because of a justifiable reluctance to intrude into the workings of a passionate temperament. Cajal's autobiographical account of his creative processes is more concerned with persuasion than with confession.

His *Reglas y consejos para investigación científica* (1st ed., 1897; 6th ed. 1920; English translation: *Precepts and Counsels on Scientific Investigation*, 1951) shows how difficult it must have been for him to discipline his unrepentent vanity. This is a book about how to succeed professionally—a self-help manual in the manner of the 17th century Spanish Jesuit priest Baltasar Gracián who wrote his *Art of Wordly Wisdom* at Zaragoza. Gracián's book was very popular during Cajal's time, and as Cajal obtained his medical training at Zaragoza it is likely that he modeled his *Reglas* on Gracián's little

---

*At the beginning of Chapter 13 of his autobiography Cajal says that "*I am fully aware that I am writing for two different publics.*" Unfortunately, the English translation by Craigie chose to cater mainly for the lay public, leaving out most of the technical details that are of interest mainly for the neuroscientist.

book. Both convey the same message: *"Strive for perfection: No one is born that way."* Both teach how to succeed by one's wits without seeming to try.

Cajal was too much engaged in justifying himself and in discrediting his opponents to write a reliable history of his times. Therefore, his autobiography deserves to be treated with as much skepticism as respect. In it (*Recollections of My Life*, English transl., 1937, p. 355–357) Cajal describes his debut at the Berlin Medical Congress in 1889 and the enthusiasm of Koelliker, His, and others for his microscopic preparations of cerebellum, retina, and spinal cord. Liddell (1960, p. 27) remarks that *"this particular demonstration of results which meant so much to the sensitive Spaniard and is described by him in so much detail, is not even mentioned by Kölliker in his own full autobiography."* In fact Koelliker describes the social event in his autobiography but not the significance of the scientific demonstration (1899, p. 233): *"Don Santiago Ramón y Cajal took part in the Berlin International Medical Congress in 1889, where he showed a series of such excellent preparations, especially those demonstrating the spinal cord, that it appeared to me to be an important duty to introduce the Spanish savant, who was not conversant with German, to our anatomists, namely His, Flechsig, Waldeyer and Schwalbe."* Koelliker does not even mention Cajal's histological preparations of the retina and cerebellum which were regarded by Cajal as the best evidence for the contact theory of nerve connections. Others, like Bonin (1970) and Shepherd (1991), uncritically following Cajal's account of the Congress, have overestimated the importance of Cajal's demonstration, and have implied that Koelliker and others hastened to corroborate Cajal's findings, using the Golgi technique he had revealed to them. But if you have the necessary degree of skepticism, you will gather from Cajal's accounts of this and other events in his autobiography, that as his confidence in the correctness of his views increased, his claims for priority became more extravagant, and his generosity toward other claims diminished.

A Cajal cult originated in North America and flourished there until transplanted to Europe in the 1960s. As students in the 1950s we were taught how to recognize the authentic giants, Cajal among many. We learned from Edward Young (1742) that *"Pygmies are still pygmies, though perched on Alps; And pyramids are pyramids in vales."* (*Night Thoughts*, vi, line 309). Not long after moving to the United States in 1965, I was astonished to come across the fervent Cajal cultists with their Cajal Club, translations of some of Cajal's scientific papers into English, as an act of pious adulation more than anything else, and praise of the Giant by many who had taken no more than a perfunctory peek into his scientific works. The chorus of attestations of Cajal's virtues by his admirers is contrived and, given his success, is unnecessary. Many authors cite Cajal only for his talismanic value, as a sign that they pay respect to the historical origin of their own ideas.

Paths have been deeply eroded by processions of pious pilgrims to Cajal's monuments. Americans who had not thought of erecting memorials to their own giants of neuroscience made pilgrimages to Spain where they tended to dominate conferences in honor of Cajal. Perhaps those people have felt more comfortable celebrating a foreigner because they feared the accusation that they would be chauvinists to celebrate a white American male. Ironically, that is exactly what Cajal was, if we are to judge from his statements about the proper place of women in society and in the life of the man of science—views that were shared by most men in his culture (Ramón y Cajal, *Recollections*, p. 491–492; *Precepts and Counsels*, p. 118–131). We may well ask why we now regard these as ridiculous attitudes and why Cajal regarded them as normal, and what the effects of such attitudes were on his life and work.

It was not only on his scientific colleagues that Cajal's egotism was imposed, but also on his wife and children. Numerous photographs (Albarracin, 1982) show them grimly facing the paternal camera, testimony to the wife's business of procreation and of unstinting devotion to the great man's needs. In his autobiography, Cajal gives the following account of his wife's role, revealing a total lack of insight into his egotism and her predicament: "*It is an essential condition for peace and harmony in married life that the wife should accept willingly the ideal of life pursued by the husband [p. 271]. I found in my helpmate only assistance to pay for and satisfy my pursuits and to continue my career. . . . There was no money for fine clothes. . . . but there was enough for books, periodicals and laboratory equipment. My wife condemned herself joyfully to obscurity. . . . and without other aspirations. . . . than the happiness of her husband and children [p. 272]. Needless to say the vortex of publications entirely swallowed up my income. . . . Before that desolating cyclone of expenditure, my poor wife, taken up with caring for and watching five little demons (during the first year of my residence in Barcelona, another son was born to me), determined to get along without a servant. She divined no doubt that there was something unusual gestating in my brain. . . . and, discreetly and self-sacrificingly, avoided any suggestion of rivalry or competition between the children of the flesh and the creatures of the mind [p. 326].*"* In Cajal's defense I can say that he could not have made all his discoveries if he had spent much of his time and salary on his family rather than on his research. But I believe that others would have made those discoveries had Cajal never lived. The content of Cajal's discoveries could have been replaced, but his style was irreplaceable.

The fact that many of Cajal's discoveries were made independently by others, often within the same year, shows that even the best scientist, even the genius, can be replaced by another good scientist. The individual scientific genius is not the functional equivalent of a team of scientists, as Merton (1961) has proposed.

Like many geniuses, Cajal was formed of discordant elements: He could be vehemently passionate and coldly dispassionate, wildly romantic and calmly rational. The question is whether there was a dichotomy between Cajal the man and Cajal the scientist, or whether both were formed of the same mixture of discordant elements. I believe that the person who wrote scientific papers under the name Cajal was a large part of Cajal the man. His scientific work dominated his personal existence. Even the death of his three-year-old daughter in 1890 provided him with a creative stimulus, as he recounts the event in his recollections (pp. 380–381): "*Poor Enriquetta—her pale and suffering image lives in my memory, associated, by a singular and bitter contrast, with*

---

*Page references to Cajal's autobiography are given to the English translation by E. H. Craigie and J. Cano (*Recollections of My Life*), which is easily accessible, although not entirely reliable, rather than to the original Spanish edition, but where necessary, I have corrected the citations and quotations from the original.

one of my most beautiful discoveries: The axis-cylinders of the cerebellar granule cells and their continuity with the parallel fibers of the molecular layer. . . . one ill-starred night, as the shadows began to fall on an innocent being, the splendor of a new truth suddenly illuminated my mind."

This is the appropriate place to attempt an extended analysis of his personality as it effected his scientific work. The time has come to point out, even at the cost of offending the idolaters of Cajal, some of the effects that his frailties, eccentricities, thwarted ambitions, and failures had on his research program. You will understand that I do not utter these critical sentiments from the lofty sanctuary of the professional historian, but from the vulnerable position of the working scientist, and so I can easily find more to praise than to blame. I know that those who are on the outside looking in, or on the inside looking out, are in much safer positions than I am, on the inside looking in.

We have no right to expect a genius to be amiable. The majority of them are singularly unamiable characters, and Cajal was no exception. It is not a matter of forgiving his flaws but of understanding them and asking whether they extended from his character into his science. We have to consider Cajal's vanity and acceptance of himself as an oracle. To what extent did these character traits deprive him of friends and students, although apparently not of uncritical posthumous admirers? What roles did Cajal's nationalistic values, or other value goals declared in his autobiography, play in his research program? What is the validity of his claim to have invented the phonographic recording disk before Edison (see Ramón y Cajal, *Recollections*, pp. 501–502)? What are we to make of the belated and unverifiable claim, in his autobiography, of priority of discovery of the Golgi apparatus? Cajal says that he made the discovery in 1891 but delayed publication to 1904 because of the uncertainty of his results: *"Had it not been for such considerations, the so-called reticular apparatus of Golgi, which the neurologist of Pavia discovered in 1898 (by means of a formula, indeed, which is notably uncertain), it would figure today among my assets and under my name"* (Ramón y Cajal, *Recollections*, p. 572). Rarely did Cajal refer to Golgi without damning him with faint praise, which may be interpreted as a sign of a lack of intellectual generosity often associated with a deep sense of insecurity. He also claims that he invented the idea of a Darwinian struggle between the cells of the organism (he does not give a date but it seems to have been between 1885 and 1888, well after Roux had first put forward the idea in 1881 in his book *Der Kampf der Theile im Organismus*; see Ramón y Cajal, *Recollections*, pp. 296–297). Another claim for priority made by Cajal in his autobiography is that he discovered passive immunity—*"The experimental proof of the formation of antibodies, that is to say, of the possibility of protecting animals from the toxic effects of the most virulent bacillus by previously injecting hypodermically a certain quantity of a culture which has been killed by heat."* Cajal claims to have demonstrated this in September, 1885, before the publication by Salmon and Smith in February, 1886 (Ramón y Cajal, *Recollections*, pp. 288–289).*

---

*Compare this claim with Sherrington's accomplishment: on October 15, 1894 Sherrington was the first person in England to treat a child (his nephew) mortally ill with diphtheria, by successfully injecting the boy with serum he had prepared from the blood of a horse that had been immunized with increasing doses of diphtheria bacteria (Sherrington, 1953, pp. 550–551).

Is it fair to regard the egotism, even narcissism, displayed in Cajal's autobiography as a form of compensation for a hidden sense of inferiority?* Does Cajal exhibit the hypertrophy of the will that goes with a profound sense of insecurity?† Or are these the usual expression of a romantic style? Cajal's style is romantic in terms of his aspiration to express himself as a unique personality, in terms of the author's awareness of his relationship with his reader, and his often self-conscious striking of attitudes. The dominant Cajalian stylistic qualities are intense zeal, the tendency to turn the drama of neural development into a melodrama, associated with a devastating lack of humor. It is precisely for those qualities that we are fascinated by Cajal's style. It is a strong antidote to the dulling effects of contemporary scientific style. Nowadays the editorial pencil skillfully removes the highlights from the scientist's writings, to give different papers the same uniformity, like pale shadows of their authors cast on a dull surface. In Cajal the genius of the artist and scientist were combined to a unique degree. He had the gift, usually granted only to the artist, of incorporating vague and chaotic elements of experiences into an orderly synthesis. The artist is more or less free to adopt, modify, or invent a language to represent and express his experience. The scientist is not usually so free and often lacks the originality and courage that are necessary to liberate himself from the assumptions of his times.

4.3.4. Morphology Cast in a Romantic Mold

The feverish fit of intense creative activity is characteristic of the romantic artist as it is of the romantic scientist. Cajal admits his "*romantic eagerness—a thing of long standing in me. . . .*" (*Recollections*, p. 204). From early 1888 to late 1890 Cajal worked, as he exclaims in his autobiography, "*no longer merely with earnestness, but with fury. . . . I attacked my work with positive fury. . . . It was a delicious rapture, an irresistible enchantment.*" He says that "*a fever for publication devoured me*" (*Recollections*, p. 325). During that time Cajal published some 40 monographs. Exactly at the same time, from February 1888 to June 1890, the romantic composer Hugo Wolf composed with furious intensity 174 songs, including some of his greatest masterpieces. I could give several other parallels between Cajal's style and that of leading romantic artists of his time and compare his romantic style with the classical style of some of his great contemporaries, suggesting how the differences in styles affected their scientific research.‡

---

*Cajal's autobiography received a penetrating review by Sherrington, under the title "*Scientific Endeavour and the Inferiority Complex,*" in *Nature*, Suppl. 140:617–619 (1937).

†I was told the following story by a person who had been a student of Cajal: "*Never,*" Cajal told his student, "*write a book of less than a thousand pages. A large book like my* Histology of the Nervous System, *is certain to impress, but a short book, however good, is certain to be underestimated.*" This story was spread around, with inevitable changes, to make it appear as if all his life Cajal struggled with a fear of not impressing. His multi-thousand-page books may be seen as a heroic struggle to overcome that fear.

‡On p. 180 of his *Recollections* Cajal says that in 1871–1873 "*Among novelists, our idol was Victor Hugo; in the lyric genre, Espronceda or Zorrilla, and in oratory Castelar.*" On p. 293 of *Recollections* Cajal says that his style in 1884 to 1885 "*was inspired by the exuberant and verbose manner of the great Castelar.*" The effect of

Cajal's exuberant style included a tendency to make sweeping generalizations, sometimes at the expense of details of fact, as recollected by Sherrington (1949). In 1894 Sherrington was helping Cajal to choose some microscopic preparations with which to illustrate his Croonian Lecture at The Royal Society. In one of his preparations of the chick embryo, Cajal pointed to some fibers descending into and ending in the spinal cord, calling them pyramidal tract. When Sherrington objected that birds lack a pyramidal tract, Cajal's response was *"Bien, c'est la même chose."* Sherrington concluded: *"My remark, though correct, touched a detail too trivial for him to regard."*

Romanticism regarded as important the beautiful and sublime, not the useful. As a result, the romantics took more interest in form than in function and attached more importance to morphology than to physiology. I think that is the main reason why Cajal failed to recognize the significance of inhibition and of its functional role in the integrative action of the nervous system. Cajal's conception of neuroscience was essentially structural. His mind worked best on problems of structural organization but his grasp of function was not above average. His dogmatic proclivity was expressed most freely when he thought that structure comes before function, for example when he advanced his theory that impulses travel in one direction in the nerve cell because of the intrinsic "dynamic polarization" of the cell. Cajal tended to be dogmatic when asserting the primacy of structure over function but was at his most skeptical when faced with evidence of the converse. An example is his portrayal of the growth cone as a battering ram, structurally adapted to perform certain functions. He could not have thought in terms of extracellular biochemical factors determining the structure of the growth cone, and he was skeptical when faced with evidence showing the primacy of function. In Section 1.4.2 we noted his reluctance to promote the theory that trophic factors can influence the direction of growth of axons and can determine the forms of neurons.*

Cajal's romanticism was at the heart of his research program. His work belongs in the tradition of idealistic morphology, which was closely related to the romantic movement, in which form is the primary concern and structures are seen as the causes of function. Cajal was apparently unaware that he was not moving with the times against idealistic morphology. Centuries of confusion and misconception are associated with the meaning of structure and function and how they are related to one another. In terms of molecular biology and molecular genetics the fallacy of the structure-function dichot-

---

Castelar was more than to inspire—it was no less than to determine Cajal's fate. Emilio Castelar y Ripoll (1832–1899) was a Spanish republican politician, historian, and writer of historical romances. Cajal was drafted into the Army Medical Service at the age of 21 following the accession of Castelar to the presidency, as a result of the compulsory military service instituted by Castelar in 1873. Cajal was infected with malaria while on military service in Cuba. After his discharge from the army, because of his illness, he went into a teaching post in anatomy at the University of Zaragoza rather than into medical practice. Cajal's first scientific paper in 1880 was about inflammation, and for the following 8 years his work ranged across the fields of general pathology and histology. The turning point in his career occurred after he learned the Golgi technique in 1887, resulting in his first papers on the histological structure of the retina and cerebellum in 1888.

*Compare this with Koelliker's understanding of the effects of formative stimulation on the morphology of neurons: *"So I am finally forced to the conclusion that all nerve cells at first possess the same function, and that their differentiation depends solely and entirely on the various influences or excitations which affect them, and originates from the various possibilities that are available for them to respond to those contingencies"* (Koelliker, 1896, p. 810).

omy is quite obvious. Structure and function cannot be separated in molecules. To pose the dichotomy in terms of molecular structures and functions has no validity nowadays (just as the "problem" of the priority of the chicken or egg is now irrelevant in terms of molecular genetics), although it has historical relevance. People used to say that "form follows function" or the reverse. Now they say that "product follows process." Neither form nor function are ends in themselves. Both are ends of evolutionary, genetic, and developmental processes.

The debate at the Paris Académie des Sciences in 1830 between Georges Cuvier and Etienne Geoffroy Saint-Hilaire was about whether structure or function is primary. Cuvier took the position that function is primary, and that teleology is the guiding principle, and comparative anatomy the method of choice. Functional relations between the parts determine the basic structural plan and not vice versa. Geoffroy defended the priority of structure. He aimed at creating a science of pure morphology in which the unity of plan is the guiding principle, and relationships can be determined on the relations of the parts rather than on the similarity of functions. Both those positions could be defended from a knowledge of gross structure and function. However, observations of histological sections placed the functionalist position at a disadvantage, whereas the structuralist position was more easily supported by the evidence of microscopic anatomy.

As microscopic anatomy advanced rapidly after about 1850 the evidence of structural plan was obtained far more easily than evidence of cellular functions. Neurohistologists could see evidence of structural plan directly but they had to infer functions from the evidence of structure. Inferences of function from structure were risky, and prudence dictated that structure, which could be observed directly, was given priority over function, which could only be inferred. That was the approach taken by Ranvier in his great two-volume *Leçons sur l'histologie du système nerveux*, published in 1878, and used by the young Cajal as his guide. Also the priority of function over structure was associated with the reaction of vitalists and holists against reductive materialists.

Curiously, problems of the relations between structure and function in the nervous system, and problems of functional localization in the nervous system, were regarded as marginal to the larger debate about evolution of form in relation to function (e.g., "Brain," "Mind," or "Nervous System" are not listed in the index of any of the three books dealing with the subject of structure-function relations by E. S. Russell, 1916, 1930, 1946). In that debate, which raged through the 19th century, the followers of Lamarck argued that function determines structure, while the converse was argued by the followers of Darwin. However, both sides agreed that nervous function can determine nervous structure, and both taught the doctrine that the inheritance of acquired characters played an important role in the evolution of the nervous system (see Section 5.6 for references). Neither Cajal nor Sherrington ever openly questioned that doctrine. In the only publication in which Sherrington dealt directly and at length with the structure-function problem, he emphatically rejected teleology, and insisted that *"the question **why?** is not answered by positive science, but only the question **how?** and sometimes the question **how much?** The physiologist cannot say **why** a muscle contracts. . ."* (Sherrington, 1899b). Sherrington referred to the relation of structure and function in many of his papers, but always with exasperating brevity. Perhaps his most explicit statement on that subject was made in his Hughlings Jackson Lecture published in 1931: *"The nervous system is indeed both a form and a series of events. These have to be confronted together even for inquiry which concerns itself with function as its chief aim."*

Historians have seen the structure-function dichotomy in terms of a dichotomy between materialism and vitalism (E. S. Russell, 1916), or between Darwinism and Lamarckism (Mayr, 1982), or between empiricism and rationalism (Appel, 1987). On all counts, as a materialist, Darwinist, empiricist, and a romantic, Cajal tended to give priority to structure before function. His disinterest in clinical neurology or experimental work on the CNS virtually precluded a functionalist view of nervous structure. Cajal approached closest to that view in his work on axonal growth and on axonal degeneration and regeneration. His neurotropic theory is a theory of function inferred from observations of structure. Another such theory was the theory of dynamic polarization of the neuron, developed by Cajal and van Gehuchten. As we shall see, that theory was not a radical departure from idealistic morphology which did not deny the necessary relationship of structure and function, but rather insisted that structure determines function and not vice versa.

Cajal's contributions were anatomical, and his ventures into functional concepts were not successful. His main functional concept was that nerve cells are intrinsically polarized to enable nerve impulses to pass in one direction only, from the dendrite to the axon. He put forward this "law of dynamic polarization of the neuron" in 1891. It met with immediate criticism and he had to modify the "law" (Ramón y Cajal, 1895). The most serious objection came from the observations of Retzius showing that in the dorsal root ganglion cells of vertebrates the peripheral (afferent) nerve fiber is an axon (see Fig. 3.1). Proof that neurons can conduct in the direction opposite to that required by Cajal's "law of dynamic polarization" was soon forthcoming when physiologists showed that antidromic conduction can occur in the CNS (e.g., Sherrington, 1897c).

At the time it was well known that axons can conduct in either direction. F. H. Bidder first showed in 1842 that both sensory and motor nerves are capable of conducting in either direction. He concluded that their functional specificity must therefore reside in the properties of their peripheral and central endings. Bidder's principal experiments consisted of showing that nerve conduction was restored, but functional specificity was perturbed, after surgical cross union of the cut central end of a sensory nerve (lingual nerve) to the cut peripheral end of a motor nerve (hypoglossal nerve). The ability of nerves to conduct action potentials in either direction was proved by recording electrical activity in peripheral nerves, first reported by du Bois-Reymond in 1849 (Vol. 2, p. 587–591). From that evidence he extrapolated to the effects of crossing of the optic and auditory nerves: *"Two sensory nerves should be able to replace each other completely. In the case of the optic and auditory nerves healed together crosswise we would, if the experiment were possible, hear lightning with the eye as a bang and see thunder with the ear as a series of visual impressions"* (du Bois-Reymond, 1872, in du Bois-Reymond, *Reden*, Vol. 1, p. 445, 1912). This *Gedankenexperiment* was related, of course, to the "law of specific Nerve Energies" that had been proposed in 1826 by Johannes Müller, du Bois-Reymond's teacher in Berlin. Müller's "law" holds that the specificity of sensations are functions of the central connections of sensory nerves, not of the quality or quantity of nervous activities conducted by them from the sense organs.

A great deal of misunderstanding exists about the development of the concept of unidirectional conduction in the nervous system. One of the fictions that has been fostered by some historians is that the proponents of the reticular theory of nerve connections believed that nerve impulses traveled at random in the so-called "diffuse nerve net." Actually supporters of the reticular theory did not think at all like that, because they were well aware of physiological observations on reflex arc conduction which showed that conduction in the CNS must occur preferentially in one direction, from sensory inputs to motor outputs. The concept of one-way preferential conduction along pathways in the CNS was generally held from the mid-19th century. Even those who conceived of nerve cells joined by networks of protoplasmic processes (e.g., Koelliker, 1852, 1867; Schroeder van der Kolk, 1859) thought of preferential conduction of nervous excitation through pathways leading from the input to the output, from afferent to efferent nerves. This is well illustrated in Fig. 3.9 taken from the fifth edition of Koelliker's *Handbuch der Gewebelehre* published in 1867. At that time Koelliker subscribed to the network theory of connections between nerve cells, but he clearly conceived of preferential conduction in one direction in the nerve net. Physiologists had shown that nerve fibers can conduct in either direction but there was abundant neurophysiological evidence that conduction in the CNS is normally unidirectional. The concept of the reflex arc introduced by Marshall Hall (1850) was predicated on the belief in unidirectional and preferential conduction in the CNS. Sherrington formulated the concept of a one-way valve action of the synapse as an explanation of the difference between conduction of nerve impulses in peripheral nerves and conduction in reflex arcs, as we shall see in the following section.

Cajal conceived of nervous transmission at the contact zone as a purely excitatory process. He appears to have been unaware of the work of the Webers on inhibition of the heart by the vagus nerve and of inhibition of spinal reflexes by higher centers, reported by Sechenov in 1863. Cajal never seems to have grasped either the concept of synaptic delay or of inhibitory synaptic function because he at first thought of neurons in direct physical contact and he conceived of transfer of nervous energy like electrical conduction between two metallic conductors in contact with one another. Later, after Sherrington clarified the problem of the synapse as a barrier to conduction of nerve impulses, Cajal added the possibility of electrical induction as a model of synaptic function. Those two modes of electrical synaptic transmission were the only ones he considered when he gave his Nobel lecture in 1906, and when he wrote his autobiography in 1917, and he did not revise his opinion in the third edition published in 1923.

The nature of transmission at the contact area was correctly grasped by E. A. Schäfer in 1893: "*The same nerve-impulses do not necessarily pass from one element of a nerve-chain to the next, but. . . . more probably new impulses. . . . are generated in the successive elements of the chain.*" This statement was made in an article with the title "*The nerve cell considered as the basis of neurology*" in which Schäfer did not cite Cajal. Schäfer's concept of the contact zone between nerve cells was functional as shown by his understanding of its role in adding to the latency of conduction in the reflex arc: "*Physiologically it may be shown that there is always a partial block to the passage of a nervous impulse at the conjunction of one cell with another. . . . This lost time which occurs at the junctions of the cell chain represents the period of latent excitation of the nerve cell."* This statement was published 4 years before Sherrington first used the term *"synapsis."* It is very significant that Sherrington derived that term from the Greek verb meaning the action of making contact. This shows that he conceived of the synaptic connection as a functional entity, although some structural basis was conjectured in the form of *"delicate transverse membranes"* (Sherrington, 1906). As we shall see in the following section, Sherrington thought of synaptic junctions as inductive inferences from observations of the differences between conduction in peripheral nerves and conduction in reflex arcs. Such entities were entirely hypothetical, but they provided explanations for one-way valve action of CNS conduction, reflex delay, summation of subliminal excitations, integration of excitation and inhibition, and convergence of different inputs at the final common path.

Sherrington did not give Cajal any credit for the concept of synaptic connections between neurons. In *The Integrative Action of the Nervous System* (1906), Sherrington refers to Cajal's work at least ten times in relation to such topics as ameboidism of dendrites, dynamic polarization of the nerve cell, convergence in the retina and olfactory bulb, architectonics of the cerebral cortex, and binocular vision. But Sherrington is completely silent on Cajal's contributions to understanding of the nature of connections between neurons. We can now understand why the neurophysiologists ignored Cajal's ideas about the nature of the contacts between nerve cells. Cajal's ideas were entirely morphological, they were entirely conjectural (he had seen the nerve endings but not the contact zone), and his explanations were inadequate to deal with neurophysiological observations such as reflex delay, summation, and inhibition. Cajal's achievements can be seen as pointed illustrations of the limitations of morphology uninformed by physiology.

Cajal's debt to idealistic morphology is shown by his efforts to discover the modular structures composing the retina, cerebellar cortex and cerebral cortex, and by his use of the concept of neuronal types. Indeed, one of the advantages of regarding structure as prior to function is that it leads to easy acceptance of the validity of reduction of complex structures to simpler structural modules, and of those to various types of neurons. The morphologist sees no loss in the reduction, whereas everything is lost in the view of those who place function before structure.

The turning point away from idealistic morphology and vitalism toward more explicitly mechanistic models is marked by a work published in 1852 which aimed to find causal relationship between physiology and morphology—*An Anatomical and Physiological Survey of the Animal World* (*Übersicht des Thierreiches*) by Carl Bergmann and Rudolph Leuckart (see Lenoir, 1982). Their program was adopted and greatly extended by Wilhelm His who set out to discover the mechanical processes resulting in the form of the nervous system and of the embryo as a whole. The conclusions were published in 1874 as a brief book titled *Unsere Körperform und das physiologische Problem ihrer Entstehung* (Our Body Form and the Physiological Problem of Its Origin). Initially, His thought only in terms of the causes of change of form such as differential growth and folding, but later he included cell division, cell migration (which he discovered), cell growth, and cell differentiation. His work on the morphogenesis of the neural tube and neural crest and the outgrowth of axons and dendrites was part of that program of cell mechanics. The cell mechanics of His were uncompromisingly devoid of teleological explanations. Not so the *Entwicklungsmechanik* (developmental mechanics) of Wilhelm Roux who, as a student of Haeckel, equated cause (*Ursache*) with purpose (*Grund*). Roux was a mechanist, but he set out to solve the causes of morphogenesis in terms of general categories of formative factors such as cell interactions, and functional adaptation. Roux believed chemistry and physics were only indirectly related to the mechanisms of morphogenesis (Churchill, 1966). I find it difficult to comprehend why Cajal, whose research program contains the last vestiges of idealistic morphology, is so often mistaken for the main forerunner of modern neuroscience. Equally incomprehensible is the neglect of His, who was indeed a founder of mechanistic-reductionistic neuroscience, leading directly to the present day.

The assumption of a purpose in the organization of the nervous system is an intellectual habit that is hard to break. It is a vestige of the doctrine of final causes that has been denounced by most philosophers from Bacon on. It is difficult to understand its persistence, unless it is because of intellectual laziness. The concept that all biological structures and functions have evolved with a purpose easily leads people astray. For example, what then is the survival value of the nerves that subserve cardiac ischemic pain? How could this function have evolved only to be expressed once in a lifetime or never at all? The answer seems to me to be that because of neutral mutations there are certain structures that have latent functions. Such latent functions only come into play under contingencies that may never have arisen before in the evolutionary history of the species. It is easy to start with the belief that every component of an organized system must have some use and is adapted for that purpose and that nothing happens by chance. The problem is not solved by saying that "*an organized product of nature is that in which all the parts are mutually ends and means*" (Kant, *Critique of Judgement*). That notion does not allow for persistent errors, for vestigial organs, for the element of chance, and for spontaneous activity. Darwin gradually rid himself of teleological thinking (Ospovat, 1978, 1981). But since E. S. Russell's brilliant but misguided defense of that mode of thought, some philosophers and historians of science have succumbed to its blandishments and seductions (e.g., Lenoir, 1982; Appel, 1987).

Comparison between the depictions of the histological organization of parts of the CNS made by different neuroscientists also shows that Cajal's depictions were idealized. Wilhelm His drew precisely what he saw, artifacts and all. By contrast, the drawings of Camillo Golgi are notable for what they leave out as much as for what they reveal. Golgi's omission of dendritic spines and the depiction of perfectly smooth dendrites reminds me of the witticism about Jane Austen's novels being remarkable for not having mentioned the most significant event of her time, the French Revolution. Golgi's drawings of neurons of any single type, say the hippocampal pyramidal cells, are drawn almost identical to one another and they are smooth and stiff—representations of an ideal type in Golgi's mind. The neurons drawn by Cajal are synthetic reconstructions, as he admits, drawn to combine his experiences of many neurons of the same type. They are intended to be realistic, but they frequently have an element of caricature—

exaggeration of the feature which Cajal wanted to emphasize—large growth cones, for example, and his self-conscious depiction of a wide gap between neuron and neuron at the sites of contact between them. Cajal believed that neurons connected tightly, without any cleft between them, but he showed a gap in order to emphasize his theory of contact and to deemphasize the opposing theory of continuity, as shown in Fig. 1.4. Cajal's histological drawings were models of his theory of the organization of the nervous system. They were also models designed to show the significant differences between his preferred theory and the opposing theories. Constructing such models always gives rise to some degree of conflict between the representation of raw empirical data and their transformation in the mold of a preferred theory. Cajal was not exempt from that conflict.

Cajal's unique gift was his ability to grasp in a novel synthesis the relationships between neurons that were seldom if ever seen in a single view through the microscope. Justifying this method, he wrote that *"in order to decrease the number of figures artists are sometimes forced to combine objects which are scattered in two or three successive sections"* (Ramón y Cajal, 1929a). No doubt there were neurons in reality that were close to the ideal type as he represented it. That is to be expected if he selected a single neuron or combined parts of several to represent the ideal form of what is really a distribution of forms in a population. We need not join Ernst Mayr (1982) in his blanket condemnation of "typological thinking," nor others in their uncritical praise of Cajal's methodology. Our intentions should be to understand the merits and demerits of his methods of approaching the problem of representation of complex neuronal organizations. He could not transcend the limitations of methods of modeling and of symbolic representation that were available at that time. Cajal represented the ideal form of each neuronal type magnificently: his achievement deserves applause but not imitation.

4.3.5. Metaphors and Models of the Microcosm

I was struck many years ago on first reading the works of Cajal and Sherrington by the predominant analogical-correlative mode of Cajal's thought, which is so different from the causal-analytical mode of Sherrington's thinking. Cajal thought of neurons in morphological terms of bushes and trees, and in functional terms he conceived of them anthropomorphically as willing or struggling beings. By so doing he greatly reduced his ability to understand the true functions of the structures he described so picturesquely. He simply got the wrong analogies: by conceiving of cellular functions in terms of inner drives and tropisms analogous with human motivation, he failed to understand them in terms of physical and chemical mechanisms. When Cajal attributed change to the action of voluntary agents he was moving away from science to a fantastic domain in which occult forces operate normally. Cajal's view of the behavior of the growing nerve cell is like the view taken by H. S. Jennings (1906) of the amoeba as a conscious, willing being. In that view, if the neuron or the amoeba could be enlarged to the size of a dog it would behave like a dog. Even if the physical problems of scaling could be overcome (Calder, 1984; Schmidt-Nielsen, 1984), the fallacy of that view would remain—the purposes of the single cell are not the same as the purposes of a complex organism like the dog, and the means for achieving the purposes of individual cells are different from those used by dogs to achieve their purposes. Therefore, resemblances between the behavior of the cell and the behavior of the dog are fortuitous and at most are formal resemblances, not evidence of homologies of structures and functions.

Analogical-correlative thinking allows abundant place for sympathetic and imaginative reflection, and Cajal takes full advantage, going too far in his sympathy for the life struggles of the nerve cells. Indeed, I know of no other example in modern neuroscience of sympathetic reflection taken to such extremes. The projection of his own personality into microscopic objects of his research may be seen as a compensation for his failure to identify with other people's emotions, struggles, and ideas. Cajal's form of analogical thinking involves personification—the interpretation of the behavior of cells in terms of concepts derived from human society and human behavior. Such projection of human feelings and strivings into nature is poetic and even mystical but it is not scientific. In humanizing nature Cajal merely committed the so-called pathetic fallacy. He did not proceed to commit the more serious fallacy of panpsychism by implying that the activities of cells, appearing to be like those of people, reveal a more profound underlying reality of the universality of mind and of will, from electrons to man.

Cajal tried to solve the problem of neuronal form in terms of formative forces rather than in terms of reduction to physical mechanisms. That is one of the reasons for his frequent use of anthropomorphic metaphors as when he speaks in his autobiography of *"protoplasmic kisses, the intercellular articulations, which seem to constitute the final ecstasy of an epic love story,"* or when, in the same work, describing the outgrowth of axons from a spinal ganglion he says, *"a bundle of precocious bipolar cells strikes with its cones, like battering rams, on the posterior basal membrane and opens a breach in it. Other sensory fibers, differentiating later, make use of this opening, and assault the interior of the spinal cord along its dorsal portion."* This trait was also noted by Sherrington (1949, pp. xxiii–xxiv): ;*"He treated the microscopic scene as though it were alive and were inhabited by beings which felt and did and hoped and tried even as we do."* In Sections 1.1 and 1.2.6 I discuss the advantages and disadvantages of using allegorization, namely the representation of one thing under the image of another.

Cajal's allegorization shows that for him the human brain is the microcosm in which the larger reality of the universe (the macrocosm) may be discerned. For this concept Cajal was also directly indebted to Schopenhauer to whom the human will, the human body, and the entire universe are the same thing viewed under different aspects. In Cajal's writings the nerve cells have wills of their own—striving, struggling, and even loving. It is, therefore, highly significant that Cajal's library contained several of Schopenhauer's works including *Die Welt als Wille und Vorstellung* (*El mundo como voluntad y representacion*, listed in a part of Cajal's library catalogue, now in the Cajal Museum, Madrid).

Those who admire Cajal's allegorization have failed to notice that it represents a kind of biology akin to that of the idealistic morphologists in three important characteristics. The first was his use of analogy between the behavior of a microcosm populated by cells and the behavior of a macrocosm populated by large organized systems, not only people and other living organisms but also machines and physical systems.* Second, Cajal regarded form as the cause of function. Finally, he regarded form teleologically, as if designed for its purpose. A single example can illustrate all those characteristics. Cajal drew arrows in his drawings of neuronal cytoarchitecture, and these have been noted as evidence of his amazing prophetic insight—at that time there was no experimental evidence showing the direction of conduction in nervous networks. The arrows were, of course, Cajal's way of illustrating the theory of dynamic polarization of the neuron (van Gehuchten, 1891c; Ramón y Cajal, 1891a,b, 1892, 1895) according to which the neuron has an intrinsic polarity which enables conduction to occur in only one direction, from dendrites to cell body and then to the axon. The linkage of neurons in circuits, with arrows showing the direction of conduction, is the microcosm represented by the macrocosm of the railway network or the telegraph system, which were the models to which the nervous system was most often compared at that time. The arrows are intended to show that function of the system is determined by the form of the neurons, especially the features that make axons differ from dendrites. The entire system is depicted as if designed for a definite purpose, allowing nervous activity to flow in an orderly pattern through the regular sequence of conductors from axons to dendrites. When this regularity was not evident, Cajal resorted to the romantic image of the jungle, in which the profusion of forms and their complex interlacing made it impossible to see individual trees.

Cajal, the romantic, saw "*that apparent disorder of the cerebral jungle, so different from the regularity and symmetry of the spinal cord and of the cerebellum. . . .*" (*Recollections*, p. 395). In his Nobel prize lecture, delivered on December 12, 1906, Cajal resorted to the jungle metaphor: "*the tangled forests of the brain, of which we imagine we have discovered the last leaves and branches, may still possess some bewildering system of filaments binding together the neuronal mass as lianas bind the trees of tropical forests.*" As I have previously noted in my essay "Through the jungle of the brain: Neuronal specificity and typology re-explored" (Jacobson, 1974a), the jungle represents many things in Cajal's conceptual processes. If the tropical jungle is the macrocosm, the jungle of the brain is the microcosm. The tropical jungle is thus a simulacrum, or a model of the types of nervous structure and of their relationships, branching, intertwining, and forming a variety of contacts with each other. The wild and luxurious undergrowth could be pruned to more orderly patterns, and so forth. Then too, the jungle trope must have evoked memories of his Cuban adventure. And, not least,

---

*The image of the body as a point-to-point representation of the cosmos, or of the microcosm of the body as a representation of the macrocosm of the state, has very ancient origins, probably from Babylon (Berthelot, 1949), and from before the 4th century B.C. in China and Greece (A. Meyer, 1900; Conger, 1992; Needham, 1959, p. 294; Lévi, 1989). This sort of imagery was characteristic of the analogical and correlative thinking, out of which grew the causal-analytical thinking of experimental science (Arber, 1944; Temkin, 1949). Analogical-correlative thinking made it natural to relate ethics to protophysiology. Granet (1934, pp. 342, 361) shows the importance of the ancient Chinese belief in the symbolic correlations between the ethics of human actions and the behavior of heavenly bodies. This naturalistic ethics, related to astrology, continued to play a constructive part in motivating physiological research as late as the 17th century (Pagel, 1935). It is represented in the 19th century by *Naturphilosophie*, and in the 20th by theosophy and related cults.

contemplation of the primeval jungle, awesome, sublime, and mysterious, must have set up sympathetic reverberations in the romantic parts of Cajal's mind. To employ the same trope we can say that Cajal's approach to the complexity of the nervous system was to see a way through it by hacking away the luxuriant growth with the energy and ruthlessness of a pioneer in a tropical jungle.

Cajal's methods of drawing from the microscope were consistent with his romantic style. Sherrington (1949, pp. xiii–xiv), describing Cajal's methods, pointed out that "*his preparations. . . . were to appearance roughly made and rudely treated—no cover glass and as many as half a dozen tiny scraps of tissue set in one large blob of balsam and left to dry, the curved and sometimes slightly wrinkled surface of the balsam creating a difficulty for microphotography. He was an accomplished photographer, but, so far as I know, he never practiced microphotography. Such scanty illustrations as he vouchsafed for the preparations he demonstrated were a few slight, rapid sketches of points taken here and there—depicted, however by a master's hand.*" In other words, Cajal's method and style were those of the artist, not the technical draftsman. Cajal refers briefly to his use of a camera lucida in his paper on the dendritic spines (Ramón y Cajal, 1891a, legend to figure 1). In Cajal's paper on the structure of the retina published in *La Cellule* in 1892, he says that "most" of the figures were drawn with the aid of a camera lucida but others were composed of drawings of "*cells which were taken from different sections.*" I have discussed this in Section 1.4.1. To suggest that he used such a technical drawing aid habitually (DeFelipe and Jones, 1988, 1992) is like saying that a life preserver is needed by a powerful swimmer within reach of the shore. Cajal describes different models of camera lucida in his *Manual de Histologia Normal*, but he also describes other instruments such as the microspectroscope and the polarizing microscope, which he probably never used. Further evidence against his habitual use of the camera lucida is that the latter is neither mentioned in Cajal's autobiography nor visible in the numerous photographs that he took of himself at his worktable (Albarracin, 1982).

Cajal's line drawings of Golgi or silver preparations were evidently made with a metal pen or a goose quill, with which the width of the line can be shaped by varying pressure on the point. Penfield (1954) saw Cajal writing with a goose quill (but not drawing with one or wiping it on his bedsheets, as stated by DeFelipe and Jones, 1988). For halftone figures, he used pencils, crayon, and fine paintbrushes (Penfield, 1977, p. 104). Cajal was well aware also of the special artistic effects obtainable with paper of different grades and textures. I believe that he could have used the camera lucida for laying out the picture at low magnification, but after that he drew the details freehand, keeping one eye and hand on the microscope while using the other hand and eye for drawing. This was the way in which students were trained to use the monocular microscope for making histological drawings, and it was also the principal method recommended by Cajal. He notes in his *Manual de Histologia Normal* (p. 36, 4th ed., 1905) that this method "*requires a facility for copying from nature as well as artistic taste which, alas, does not always coexist in the dedicatees of the natural sciences.*" Cajal would undoubtedly have found this direct method no less accurate and much less cumbersome than using a camera lucida attached to his microscope, especially when a very strong source of light is required for viewing the image of a Golgi preparation 100 μm thick. From his own evidence and that of Sherrington quoted earlier it is certain that his preferred method of freehand drawing would have been inhibited and frustrated by the use of a camera lucida.

In discussing Cajal's style it is important to recognize that his work exemplifies the distinction between artistic expression and scientific meaning which is separate from the distinction between truth and error. Style is not the same as personal identity; it does not replace identity but completes it. In neurohistology, originality of artistic style should not be mistaken for originality of scientific content. Style is particularly important here because it enters into the graphic depiction of microscopic preparations. Slight changes of thickness and intensity of the lines, little hooks and tremors at the ends of his strokes, betray his nervous energy and his occasional doubts and hesitations, and above all reveal Cajal's marvelous spontaneity of graphic expression. Expert neurohistologists such as Koelliker, Cajal, Retzius, Lenhossék, and van Gehuchten, who were contemporaries and used similar techniques, had different styles of depicting the same subject, say a Golgi preparation of retina or cerebellar cortex.

Anyone who knows the subject can now distinguish the drawings of Cajal from those of the others, just as an art expert can distinguish drawings of Rembrandt from those of Rubens or anyone else, even when the content is similar, say a landscape. Even when the same artist makes several depictions of the same landscape—Cézanne's depictions of Mont Sainte-Victoire come quickly to mind—they each uniquely emphasize or omit certain elements. Precisely the same occurs when neurohistologists depict a complex microscopic preparation—similar criteria of inclusion or exclusion operate in depicting both the macrocosm and the microcosm. Without foreknowledge of the meaning of the different elements, and without total inclusivity, but with some theory in mind, the scientific artist has to practice selective inclusion, accentuation, and distortion. These forms of selection as well as the fluency with which they are transferred to paper, are the main constituents of personal style that enable us to distinguish a drawing by Cajal from a drawing of the same subject by one of his great contemporaries. Such drawings are representations of theories, namely conceptual models.

4.4. SHERRINGTON'S CASE: "THE CORRECTIONS AND RESTRAINTS OF ART"

In one of his essays in the *Spectator* (Sept. 3, 1711), Joseph Addison made the distinction between two kinds of geniuses, which apply to Cajal and Sherrington rather well. The first, or Cajalian kind, *"draw the admiration of all the world upon them, and stand up as the prodigies of mankind, who by the mere strength of natural parts. . . . have produced works that were the delight of our own times and the wonder of posterity. There appears something nobly wild and extravagant in these great natural geniuses, that is infinitely more beautiful than all the turn and polishing of what the French call a bel esprit. . . ."* Addison compares this with another kind, which I shall call the Sherringtonian kind of genius, *"who have submitted the greatness of their natural talents to the corrections and restraints of art. . . . The genius in both these classes of authors may be equally great, but shows itself after a different manner. In the first it is like. . . . a*

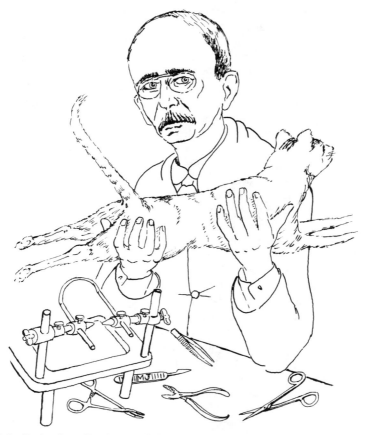

FIGURE 4.3. Charles Scott Sherrington (1857–1952), at about 50 years of age. He is demonstrating a decerebrate cat to a class of medical students at Liverpool University where he was Professor of Physiology from 1895 to 1913.

whole wilderness of noble plants rising in a thousand beautiful landscapes. . . . In the other it. . . . has been laid out in walks and parterres, and cut into shape and beauty by the skill of the gardener." Sherrington's whole concern is about organization and order of nervous systems, and the relations that enable the individual neurons to function as an integrated whole.

To appreciate Sherrington's original contributions we must understand what was known before him, the controversies surrounding that knowledge, the state of research apparatus and other methods available to him, and the world in which he worked. Social conditions in Britain before and during Sherrington's time were different from conditions in the other European countries in which neuroscience developed during the 19th century. Science stagnated in Britain during the first half of the century, as a result of the delayed development of science as a profession in England, and the economic depression following the Napoleonic wars. The years known as "the hungry forties" were bleak for British neuroscience, especially when compared with the resurgence of neuroscience that started in the 1840s in France, Italy, and Germany (Rothschuh, 1952, 1973; Ben-David, 1970, 1971). In the judgment of Edward Sharpey-Shafer (1919), *"Physiology had ceased to exist as an active science"* in England during the 1860s.

Some of the factors in the stagnation of English physiology, discussed by Geison (1972, 1978), were lack of professionalism, lack of institutions with suitable laboratory facilities, and the effective antivivisection movement in England (see also Bretschneider, 1962). Resurgence of neuroscience in Britain began only after about 1880, and Sherrington was fortunate in being in the right place at the right time. An important factor was increased circulation of books and scientific periodicals which became relatively cheap and widely distributed for the first time during the last quarter of the 19th century. No doubt economic prosperity also helped: Britain was the world's richest country for the half-century after 1860 (*The Economist*, June 20, 1992, p. 107). The average purchasing power per person in Britain was twice as large as in Germany in 1870, yet the progress of science in the two countries was the inverse of their economic prosperity. Social rather than economic factors must be found to account for the decline of science in Britain, especially in England at the time of Sherrington's birth in 1857.

Charles Babbage pointed out in his *Reflections on the Decline of Science in England* (1830) that professionalism was one of the missing factors in England. Amateurism and dilettantism persisted in English science long after scientific research had been established as a professional career in France and Germany (Mendelsohn, 1964). The 18th century notion that experiments are more elegant when done by gifted amateurs using improvised apparatus persisted in England throughout the 19th century.

The antivivisection movement was far stronger in England than anywhere else during Sherrington's time. Early in his career as a neurophysiologist, Sherrington (1892b) became involved in a polemic with the surgeon Lawson Tait about how much pain was caused to experimental animals (see footnote 4.2). Disagreement in public was not in Sherrington's style. He was drawn into this polemic unwillingly, but he must have regarded the charge of inflicting pain on experimental animals of sufficient importance to justify a public rebuttal.

The opposition to cruelty to domestic animals can be traced in England to the 17th century, although that opposition hardly extended to blood sports. The antivivisectionists were adamantly opposed to the idea that animals are machines, mere automata, lacking the soul possessed only by humans. This idea had been formulated definitively by Descartes in his *Discourse*, and in the replies to objections appended to his *Meditations*, and in his correspondence with Henry More, one of the principal adherents of Cartesianism in England. The English Cartesians objected to the idea of the animal-as-machine because they clung to a belief in the souls of animals (L. D. Cohen, 1936). The sentimental attachment of the English to their domestic animals was later to be one of the mainsprings of the antivivisection movement, which retarded the progress of experimental physiology in England, especially in the 19th century (Stevenson, 1956; Bretschneider, 1962; Geison, 1972). I doubt whether the opposition would have been as strong if physiologists had vivisected sheep or cattle instead of cats and dogs.

Like his mentor Michael Foster (Geison, 1978), Sherrington was a Cartesian dualist, and both adhered to the doctrine of the animal-as-machine essentially in the original Cartesian form, namely that the brain is a physical mechanism in liaison with the nonphysical soul. Reflex action was at the center of the mind-body problem. Descartes recognized that many purposive movements are carried out without intervention of the soul. He regarded actions such as the involuntary withdrawal of the hand from the fire, and the involuntary movements of the eyes, as purely mechanical. Just as mechanisms can be analyzed in terms of their component parts so too can reflex behavior be subdivided, but mind is indivisible, as Descartes argued at length in his sixth Meditation. Sherrington's entire research program was predicated on the assumption that mind is indivisible but reflexes are divisible. Sherrington was able to believe that it would ultimately be possible to give a complete account of reflex action. If we had the knowledge of a reflex, like the knee jerk, in terms of all the neural events, we should be able to specify the behavior entirely from observations of neural events without observing the behavior. However, dualists, like Sherrington, deny that the same can be done for mental events and brain events: even if we could specify the mental events entirely in terms of neural events, they believe that something is left out, namely "mind."

When Cajal visited England in 1894 he observed astutely that *"in England the man is more important than the organization,"* which is related to the *"method of educating much and instructing abstemiously."* He thought that *"in the land of the Teutons the educational organization is more important than the man,"* and this is consistent with their *"method of instructing much and educating little"* (Ramón y Cajal, 1937, *Recollections*, p. 427). Sherrington illustrates an important difference between the British, mainly empiricist, and the continental, mainly rationalist, schools of neuroscience. Empiricism often goes with liberalism in politics and relativism in ethics, whereas rationalism more often goes with authoritarianism in politics and absolutism in ethics. In the empiricist style, Sherrington deduced conclusions from a large number of experimental observations. The style of some of his continental counterparts, as Sherrington observed of Virchow, for example, was to erect a vast edifice of induction on the basis of few facts. One should not carry this distinction too far: actually there were strong empiricist elements in French neuroscience, for instance Magendie and his pupil Claude Bernard, who opposed speculative theories and system building.* Nevertheless, there is nothing in English quite like the systematizations of the French and Germans unless it is the works of Herbert Spencer and William Whewell.†

---

*Bernard is an interesting case of a scientist who did not seem to understand that he was being philosophical when he denied the influence of philosophy on science: *"men of science achieve their discoveries, their theories and their science apart from philosophers . . . for scientific methods are learned, in fact, only in laboratories"* (*Introduction to the Study of Experimental Medicine*, 1865). Bernard's laboratory notebooks have been carefully studied by Grmek (1973) and Holmes (1974) who find that he did not usually follow the "scientific method" that he advocated. Bernard showed great tenacity in continuing to work on theories in spite of strong evidence against them. When he published he sometimes represented uncertain results as decisive results and slanted the data in support of his theory against rival theories. To portray his experiments in a logical progression in his publications he sometimes changed the order in which they were actually done.

†Whewell, the polymath and coiner of the word "scientist," was a supporter of natural theology and redoubtable opponent of utilitarianism and positivism. He was also that rarity in Britain, a Kantian, opposed to the British empiricist tradition.

A characteristic of German and French science before the First World War was a penchant for erecting rigorous systems. That was noted by the French physicist Pierre Duhem (1915), who also claimed that while the French are peculiarly imaginative and intuitive, the Germans are weak in imagination but skilled at painstaking deduction. Duhem also thought that the trademarks of Anglo-American science are the accumulation of facts and building of models. No doubt Duhem was prejudiced by the fact that his country was at war with Germany, but he observed accurately that Anglo-American scientists tend to eschew grand systems. Sherrington is at one with Hume that experience is the sole basis of his reasoning. However, Hume rejects teleology, which Sherrington accepts with some reservations, as we saw when considering his interpretation of the reflex as a fraction of a goal-directed function (Section 1.3.8).

4.4.1. Ascent from Mechanism to Mind

A scientist's career can be seen either as a continuous movement to its end or as a succession of episodes. Both points of view are valid: What one person sees as a singular event or a period, another sees as part of a continuous process. In retrospect, Sherrington's career can be viewed in four successive periods.

The first, from 1884 to 1891, was an apprenticeship, initially with J. N. Langley at Cambridge and then with Friedrich Goltz at Strasbourg. During this early stage, his mind had already lighted on the problem of coordination of movements and the role of the cerebral cortex in control of spinal cord reflexes. When he became the superintendent of the Brown Institution of the University of London in 1891, he chose the knee jerk as his problem: a spinal reflex that was known to be inhibited by higher levels as well as by appropriately timed sensory inputs to the spinal cord.

His work on the knee jerk marks the beginning of an intensive research project of spinal reflexes. By starting his research on the knee jerk, a relatively simple reflex, Sherrington placed himself in an advantageous position. In terms of a reductionist program of research, the spinal reflex is at a level of organization about midway between complex behavior that stands above it, and the cellular mechanisms that underlie it. Investigation of spinal reflexes can be approached with equal advantages from top-down as from bottom-up. This project occupied the second period, beginning in 1891 and continuing after he became professor of physiology at Liverpool in 1895.

During this period from 1891 to 1901, Sherrington accomplished most of his best work: (1) anatomical and physiological studies on the peripheral and central distribution of the fibers in spinal roots supplying the hind limb; (2) discovery of the sensory nerves to muscle, their role in the reflex control of posture and movement, leading to his theory of proprioception; (3) his study of reciprocal innervation of antagonistic muscles, which led to the concept of integration of excitatory and inhibitory states at the motoneuron, and his theory of the final common path; and (4) his studies of reflex inhibition in the decerebrate or spinal cat, leading to the conclusion that inhibition is an active state. During this period he also began a research project on visual psychophysics, especially on the problem of binocular fusion, within the same conceptual framework as his work on spinal reflexes, namely the concepts of coordination and integration. The harvest of that most fruitful period he gathered in his masterpiece, *The Integrative Action of the Nervous System*, published in 1906 but based on lectures given at Yale University in 1904.

After 1900, his interests shifted to the role of the cerebral cortex in control of voluntary and reflex movements. This problem was to occupy much of the third period of Sherrington's career, lasting from 1901 to the outbreak of the Great War of 1914–1918. The emphasis was not so much on localization of function as on the role of the cortex in coordinating and varying the action of groups of muscle in purposeful movements. More than any of the cortical mappers who preceded him, Sherrington was aware of the variability of localization of the cortical control of purposeful movements. During the later years at Liverpool he continued working on the projects that had been started so brilliantly during the earlier period. There were some notable advances, for example, his analysis of the scratch reflex, but the law of diminishing returns had set in.

The fourth period corresponded with his professorship of physiology at Oxford (1913 to 1935). The number of his review articles and public lectures far exceeded the original scientific research papers published during this period, which lasted until his retirement in 1935. It was distinguished by a series of collaborations with younger assistants, using improved methods of myography. Even when he was close to retirement, and had very able assistants, he was reluctant to delegate experimental work to his assistants, and he took a hand in virtually all the experimental procedures in his laboratory. "*One of the striking things about Sherrington's scientific career has been the intensely personal nature of all his investigations; if he is going to count the fibres in a nerve, he dissects out the nerve itself, and prepares it, stains it, and sections it by his own hands, and counts the fibres himself thru his own microscope. He is rather scornful of the people who have a dozen assistants running around making sections and doing counts, etc.,*" Fulton wrote in his diary, and he noted that Sherrington "*emphasized several times today [April 2, 1930] that the only people who really understand the nervous system were those who made their own preparation*" (cited by Swazey, 1969). But as we shall see shortly, Sherrington was not technically innovative, and he was slow to take advantage of the improved electrophysiological instruments that were available at the beginning of the 20th century. He valued simplicity and elegance of experimental design and detailed analysis of his observations more than technical virtuosity. He was suspicious of the Germanic claim that "*die technische Vollendung ist Alles.*" He showed a disregard for novel apparatus that is most unusual in an empiricist. One of the signs of the empiricist is a tendency to tinker with and perpetually redesign laboratory apparatus, but that was not Sherrington's way. Mistrust of techniques was expressed in Sherrington's obituary of Langley: "*He was always on his guard against deception by technique*" (Sherrington, 1925b). No doubt his mistrust of technique was sharpened by his own experience with the so-called "angle," which was part of the curve produced by the mirror-myograph used in his laboratory in the 1920s. For years the "angle" was regarded by Sherrington as a sign of the accuracy of the technique until Eccles discovered that it was an artifact (Eccles and Sherrington, 1929; Eccles, p. 59, in Eccles and Gibson, 1979).

It is important to recognize the limits of Sherrington's interests: he was not interested primarily in nervous disease. That is one reason why Sherrington and Hughlings Jackson hardly interacted and rarely mentioned each other's work. Second, he was not concerned with nerve excitation and conduction, and his interest in the synapse was mainly in its role in one-way transmission and integration of reflex action, and only incidentally in the mechanisms of synaptic function and structure. Third, he was not primarily interested in functional localization in the CNS. The problem of cerebral localization of function remained a German and French monopoly until taken up by Ferrier in England in the 1870s. Sherrington's investigations on motor functions of the cerebral cortex, in collaboration with Grünbaum (later called Leyton), were primarily concerned with cortical functions in voluntary movement, not with localization *per se* (Grünbaum and Sherrington, 1901, 1902, 1903; Leyton and Sherrington, 1917). Their emphasis was on the role of the cortex in the coordination of voluntary movements, on the variability of the physiological effects of cortical stimulation, and on cortical inhibition of spinal reflexes, not on the stability of the motor representation in the cerebral cortex (see also Graham Brown and Sherrington, 1912). Their work led to the conclusion that neurons at any one point in the cortex are dynamically connected with neurons at other positions in various combinations, enabling them to participate in many different complex movements. The same muscle in the lips or tongue can participate in vocalization, eating, and expression of the emotions, and can change rapidly from one type of movement to its opposite. These abilities are more evident in primates than in carnivores, and stronger in the gorilla than in the chimpanzee, in baboon than in macaque, and in the latter than in the gibbon.

Sherrington was never averse to inventing a new term for a new concept: *perikaryon*, *proprioception*, *synapse*, *motor unit*, *final common path*, *subliminal fringe*, *recruitment*, *occlusion*, and *reciprocal innervation* are some of his terms. Continuous use of these terms to the present is one measure of the influence of Sherrington's concepts. These are all theory-laden terms, as we have defined that notion in Section 1.2. They are terms for designating items of Sherrington's model of the functional organization of nervous systems. Each term is an abstraction from a mass of empirical data, and we require a dictionary to define each term in relation to observed phenomena, and to predictions from Sherrington's models of the nervous system. To understand those models adequately in their present states requires some knowledge of the historical and philosophical context in which they have been constructed over the past century.

Sherrington, like Cajal, started out in pathology and bacteriology which were then the frontiers of medical research, with such practitioners as Koch, Pasteur, Behring and Virchow. They show the advantages of not having overspecialized at too early an age—a burden that is increasingly placed on students nowadays. Cajal was 35, Sherrington 24, at the start of their careers in neuroscience. Sherrington was diverted into neuroanatomy by J. N. Langley (then lecturer, later professor of physiology at Cambridge), who asked him to examine the brain and spinal cord of a dog that had been operated on by Goltz in 1881. That was a critical event in Sherrington's career as it directed him toward the problem of the role of the cerebral cortex in the control of voluntary movement. At the Seventh International Medical Congress at London in 1881, David Ferrier and Friedrich L. Goltz disagreed about the localization of motor function in the cerebral cortex (Ferrier and Goltz, 1881). Goltz exhibited a dog which appeared to have no functional impairment after removal of regions of its cerebral cortex. Goltz argued against strict localization of function in the cortex. Ferrier argued the opposite, and showed a monkey that was apparently quite deaf after bilateral removal of the superior temporal gyri, and another monkey that was hemiplegic after a lesion of the contralateral cortex.

A committee was appointed to examine one of Ferrier's monkeys and Goltz's dog in an attempt to resolve the dispute. Some of their findings, which were not conclusive, were reported in 1881–1882, *J. Physiol. (London)* 4:231–236. One of the jurors was John Langley, who asked his student, the 24-year-old Charles Sherrington, to assist him with the examination of the right half of the medulla and spinal cord of Goltz's dog. Their results, published 3 years later, were Sherrington's first scientific publications (Langley and Sherrington, 1884a,b). Their findings did little to resolve the dispute between Goltz and Ferrier, but earned Sherrington a long visit to Goltz's institute in Strasbourg, which resulted in his first independent paper (Sherrington, 1885).

Sherrington spent almost 9 months in Goltz's laboratory in Strasbourg in 1884–1885 and he made several more visits between 1885 and 1895. That experience led to a series of publications dealing with degeneration in the spinal cord following cerebral cortical lesions (Sherrington, 1885, 1889, 1890a–c), and to the polemic with Victor Horsley in 1894 over priority for showing that fibers in the pyramidal tract degenerated after lesions in certain regions of the motor cortex (see footnote 4.2). Those papers show Sherrington's ability to find the histological technique that was necessary for dealing with his problems which were the functional roles of the long fiber tracts and their clinical significance. Cajal showed remarkably little interest in either of those problems. More importantly, Sherrington's relationship with Goltz led him to study spinal reflexes in monkey, dog, and cat, and to his analysis of spinal shock and the role of inhibition in spinal reflexes. Goltz had previously used dogs to study those problems, but he did not use monkeys, in which spinal shock is much more severe than in dogs. Typically, Sherrington gave Goltz a pair of monkeys after he completed his first visit to Strasbourg in 1885.

4.4.2. Reflexes Are Fractions of Behavior

When Sherrington began his research in 1890 on the knee jerk, a great deal was known about reflex action of the spinal cord (Hoff and Kellaway, 1952; Canguilhem, 1955; Liddell, 1960; Clarke and Jacyna, 1987). The concept of reflex action was already more than a century old: The terms *afferent* and *efferent* (*aufleitend, ableitend*) had been introduced by Johann Unzer (1771). The sensory and motor functions of the posterior and anterior spinal nerve roots had been correctly demonstrated by François Magendie in 1822. The term *reflex arc* had been coined in 1850 by Marshall Hall. The underlying anatomical organization of spinal reflex was conceived in terms of the prevailing notion that nerve cells in the gray matter were extensively interconnected to form a nervous network into which the afferent nerves fed their impulses and which drained its activity via efferent nerves. It was well recognized that the gray matter was not a random nerve network but rather that its *organization* was responsible for the *time delay* between stimulus and response, the *coordination, summation,* and *inhibition* of reflexes, and the *localization* of reflex response in relation to the stimulus locus. These were the basic phenomena that were recognized and conceptualized within the framework of the theory of nerve networks, as we have discussed it in Chapter 3. Experimental analysis of those phenomena and efforts to give causal explanations for them were well advanced before Sherrington began his research. In other words, Sherrington joined an ongoing research program (or programs), he did not have to start from scratch (Gault, 1904; Fearing, 1930; Liddell, 1960; Clarke and Jacyna, 1987). But he also had to contend with the prevailing misconceptions which were obstacles to further understanding of the structures and functions of the CNS.

There were methodological and conceptual obstacles to progress. Most significant among the limitations of technique were inadequate methods of quantitative recording of reflex activity by means of myography, and inadequacies of methods for electrical recording directly from the relevant nervous structures. Rapid advances in electrophysiological techniques in the second half of the 19th century put Sherrington at an advantage which he did not hasten to exploit, as we shall see shortly. The principal conceptual obstacles were the misconceptions about the organization of nerve cells in the spinal cord; the mistaken notion that sensory afferents to the spinal cord project directly to the brain; and the related misconception that "muscle sense," that is, the sense of position and tonus of muscles, is an exclusive function of the cerebral cortex. Finally, the misconception that skeletal muscles are supplied with inhibitory efferents from the spinal cord, was an obstacle to understanding how coordination of reflex inhibition and excitation was effected.

One of the most serious obstacles faced by him was the notion that the spinal cord nerve cells are connected in the form of a nerve network. The reticular organization of the spinal cord was portrayed in authoritative textbooks of physiology, for example, Ludimar Hermann's *Lehrbuch der Physiologie* (1892), and the 1896 edition of *The Brain as an Organ of the Mind* by H. C. Bastian, with whom Sherrington was well acquainted.

It was fortunate for Sherrington to have started his work on spinal reflexes after 1887 when knowledge of the cellular organization of the spinal cord advanced rapidly as a result of the use of the Golgi technique by several very able neurohistologists. By 1894 the modern conception of the neuronal organization of the spinal cord had been built on the earlier studies of Golgi (1880a, 1882–1885) by Ramón y Cajal (1889a,b), Koelliker (1890a,b), Lenhossék (1890a,b, 1893), and van Gehuchten (1891a,b). Their findings can be summarized briefly in the following points illustrated in Fig. 4.4. (1) The dorsal root fibers are direct continuations of the central processes of spinal ganglion cells. (2) Soon after entering the cord the dorsal root fibers divide into a descending and an ascending branch. (3) The descending branch ends within a few segments whereas the ascending branch runs for a long distance in the white matter before terminating in the gray matter at higher levels. Each fiber gives off numerous collaterals (discovered by Golgi in 1881) that enter the gray matter at many different levels of the cord. (4) The terminals and collaterals of the dorsal root fibers end on the dendrites or cell bodies of spinal cord neurons. This model was immediately adopted by Sherrington as the basis for his thinking about spinal reflexes. The fourth of these points was to be the focus of Sherrington's thinking about coordination of reflexes as a result of convergence of many afferent nerve fibers on each spinal cord neuron. The intersegmental distribution of

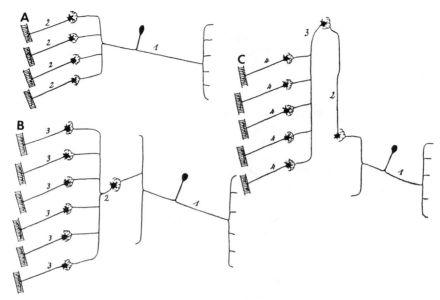

FIGURE 4.4. Organization of spinal cord reflex pathways according to the neuron theory. (**A**) Simple reflex arc; 1, afferent neuron, 2, motor neuron. (**B**) Complex reflex arc; 1, afferent neuron, 2, interneuron, 3, motor neuron. (**C**) Very complex reflex arc; 1, afferent neuron, 2, ascending interneuron, 3, descending interneuron, 4, motor neuron. (From van Gehuchten, 1894.)

afferents was soon to be the subject of a study of the long spinal pathways involved in coordination of limb and body movements in postural reflexes involved in standing and stepping (Laslett and Sherrington, 1903).

Understanding how spinal reflexes function as parts of complex movements required further knowledge in at least four domains: (1) more understanding of the peripheral origins of sensory nerves capable of evoking reflexes experimentally and evidence that they performed the same functions during normal movements; (2) greater knowledge of the organization of central connections within a single spinal segment, between spinal segments, and between spinal cord and higher centers; (3) it was necessary to map the distribution of nerves from motoneurons to muscles, individually and in groups; and (4) the pathways for inhibition were unknown and were misconceived in terms of inhibitory nerves running from spinal cord to muscles.

Understanding of the central origins of nerve fibers innervating individual skeletal muscles had advanced rapidly in the decades before Sherrington began his research on the problem of the nervous control of contraction of skeletal muscles during voluntary and reflex actions. During the first decades of the 19th century it was shown that nerve fibers do not anastomose in the brachial and lumbar plexuses (Kronenberg, 1839). This advance was stimulated by the doctrine of specific nerve energies (Müller, 1826), which required private nervous paths for specific sensations. It was also shown that each muscle receives fibers from several anterior spinal roots forming the plexuses, and that considerable variability of limb muscle movements occurs when a single anterior spinal root is stimulated electrically (Eckhard, 1849, in frog; Peyer, 1853, in rabbit).

Those results were extended by Ferrier and Yeo (1881) in rat, cat, dog, and monkey. They stimulated ventral roots electrically and reported that highly coordinated movements of a definite group of muscles occurred. Since each muscle is innervated, in most cases, from several ventral roots, and since section of a single root does not cause paralysis of individual muscles, they concluded that the distribution of fibers in the brachial and lumbar plexuses is arranged for the purpose of innervating muscles that cooperate in movements. That conclusion was shown to be false by Sherrington (1892c,d) when he found that in the rhesus monkey the complex movement of the limb produced by stimulation of a ventral root was not a purposeful action but rather a purposeless movement produced by co-contraction of muscles belonging to the same myotome. In other words, it was a movement based on anatomical metamerism rather than coordinated function (Sherrington, 1894c, 1898c).

Sherrington showed that the true functional unit was a longitudinal column of motoneurons in the spinal cord supplying a muscle. For this purpose he used the method of residual function: Three motor roots were cut above and three below the root to be investigated. After sufficient time for the nerves to degenerate, usually 28 days, the function of the residual root was investigated by electrical stimulation. He used the method of remaining aesthesia for mapping out the peripheral distribution of one residual dorsal root in the monkey (Sherrington, 1894c, 1898c). He found that each dorsal root serves a segmental area of skin with considerable overlap of the area supplied by neighboring sensory nerve roots (Fig. 4.5). Later he showed that the discrepancy between the arrangement of distribution of sensory and motor roots was caused by the fact that part of the sensory distribution was to subcutaneous tissues and to muscles (Sherrington, 1894b,c, 1898c).

When he started the work summarized above, Sherrington knew that spinal motor neurons involved in coordinated movements are clustered together in the anterior horn of the spinal cord (Waldeyer, 1888; Kaiser, 1891). It was also known that, in general, the nerve fibers supplying antagonistic muscles exit through the same ventral roots (J. S. R. Russell, 1894). The accuracy of the localization of the motoneurons connected with

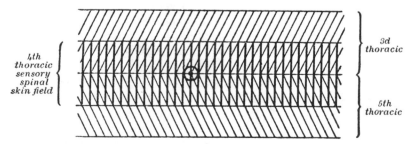

FIGURE 4.5. Segmental areas of skin supplied by neighboring spinal dorsal roots overlap. Diagram of the position of the nipple in the skin sensory fields supplied by the 3rd, 4th, and 5th thoracic roots, showing overlap of cutaneous fields of T3 and T4, and of T4 and T5, but no overlap of T3 and T5. (From Sherrington, 1894c.)

various muscles was immensely improved by the method introduced by Nissl in 1894. He discovered that basophilic granules in the neuronal cytoplasm disintegrated after axotomy, and the nucleus was dislocated to the side of the nerve cell body (Nissl, 1894a,b). This method of mapping the location of motor nerve cell bodies after cutting their axons in nerves to specific muscles confirmed that motoneurons innervating each muscle are grouped together in the anterior horn of the spinal cord (Fig. 4.6). It also showed that muscles involved in coordinated movements have their motoneurons grouped together. These were the anatomical foundations upon which Sherrington was able to build the functional concepts of coordination of movements and of interaction between motoneurons supplying antagonistic muscles.

4.4.3. Behavior Is the Integration of Neural Events

The concept with which Sherrington is most frequently associated, of integrative action of the nervous system, was crystallized by him from previously vague and confused ideas. It is most significant that he defined integration as a function of the whole organism, not only in terms of neuronal circuitry, as the *"action in virtue of which the nervous system unifies from separate organs an animal possessing solidarity, an individual. . . . the due activity of the interconnexion resolves itself into the co-ordination of the parts of the animal mechanism by reflex action. . . ."* (*The Integrative Action of the Nervous System*, 1906, pp. 2, 5). There were many sources of this idea of the nervous system as the coordinating center for the entire organism, and the origins and

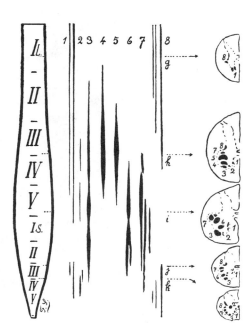

FIGURE 4.6. Columnar groupings of motoneurons in the lumbosacral segments of the spinal cord of the cat determined by the Nissl method after cutting nerves to individual muscles. Each muscle is supplied by a column of motoneurons and each segment of the spinal cord contains columns supplying a number of different muscles. Electrical stimulation of a single ventral root results in a complex movement. Some examples of the innervation of individual muscles: Column 4, quadriceps femoris muscle; column 5, pectineus and adductors of thigh; column 6i, knee flexors, popliteus and triceps surae muscles; 7i, extensors of toes, peronei, tibialis posticus; 7j,k, intrinsic muscles of the foot. (From Sano, 1898; in Barker, 1899, Fig. 586.)

development of the concept of nervous coordination could quite easily be expanded to occupy an entire chapter. For example, one of Sherrington's predecessors, the great British neurologist Thomas Laycock (1860, pp. 339, 342), stated that: "*The functions of the nervous system aim to fulfil an idea; viz., the co-ordination of the parts or organs of an organism into one. . . . The cerebro-spinal axis is the great co-ordinating center of the whole.*"

The notion of coordination can be traced to Cuvier, who at the very beginning of *Le règne animal* (1817) says: "*As nothing can exist if it does not combine all the conditions which render its existence possible, the different parts of each being must be co-ordinated in such a manner as to render the total being possible . . . this is the principle of the **conditions of existence**, vulgarly called the principle of **final causes**.*" In other words, the principle of coordination or integration is a teleological principle: the nervous system does not just have *some plan* but it has an *intelligible purpose*. The attempt to reconcile morphology and teleology, for example, as undertaken by Richard Owen, proved to be misguided when Darwin introduced the element of chance as the guiding principle of evolution. Pierre Flourens (1824), the successor to Cuvier's position, developed the concept of nervous coordination. Sherrington (1900, p. 909), in his contribution to Schäfer's *Text-Book of Physiology*, gives Flourens credit for originating the concept of nervous coordination: "*He translated the disturbances ensuing on destruction of the cerebellum to mean loss of a part possessing ability to coordinate the innervations which guide and execute complex movements. By his doing so, the idea of nervous co-ordination was, it seems to me, formally introduced into physiology.*"

I think it is worth pointing out, if only because it has not been done before, that Sherrington vacillated all his life between the teleological notion that nervous organization is fitted for the purposes of function and action, which is a theory of final causes, which he derived from Cuvier and Flourens, and the opposing notion, derived from Darwin and T. H. Huxley, that the fitness of nervous organization has evolved by chance, by natural selection and survival of the fittest. At no time did Sherrington adhere to the theory that structure and function of animals can be understood by their resemblance to archetypal organizations. That theory was proposed by Étienne Geoffroy Saint-Hilaire as the *Theory of Analogies*, namely that there is a universal system of structural analogies in the morphological organization of animals. In his theoretical work, *Philosophie Anatomique* (1818), Geoffroy claimed that he could connect every part of one species of animal with a part of another, even an invertebrate with a vertebrate, by application of his "*principle of connections*": "*The sole general principle one can apply is given by the position, the relations, and the dependencies of the parts.*" This concept was adopted by Cajal, with minor modifications. Darwin had replaced the idea of morphology in relation to systems of structural analogies with the idea of homologies based on descent from a common ancestor. Yet the old idea persisted in Cajal's attempts to establish a neuronal typology based on analogies (see Sections 2.3 and 3.6).

Flourens deduced that "*three great laws rule nervous action: The first is the specificity of action; the second is the subordination of nervous functions; the third is the unity of the nervous system*" (1824, p. 235). Defining the first, Flourens stated that each part has a distinct function or a special mode of action which is different from the others: "*the medulla, the main center which determines movements of respiration; the cerebellum, the main center coordinating movements or locomotion; and the cerebral lobes, center, and exclusive center, of the intelligence*" (p. 243). By subordination he meant that the spinal cord and peripheral nerves, "*which act spontaneously*," are subordinated to the higher parts of the CNS which are "*regulatory*." Thus, "*the nerves and spinal cord are subordinate to the brain; the nerves, the spinal cord and the brain are subordinate to the medulla, or, more exactly, a vital and central point of the nervous system, located in the medulla*" (p. 243). With regard to the "*unity of the nervous system*," Flourens (p. 243) stated: "*Not only are all parts of the nervous system subordinate one to another. They are all subordinate to the whole.*" The laws of subordination and unity would later be adopted by Thomas Laycock and his student John Hughlings Jackson.

Michael Foster presented the concept of coordination with the emphasis on reflex coordination of sensation and movement in his *Text-Book of Physiology* (3rd ed., 1879, p. 792, the edition used by Sherrington as Foster's student at Cambridge): "*The co-ordinating mechanism. . . . is constructed out of diverse afferent impulses of various kinds arriving at the co-ordinating centre from various parts of the body, ... in fact the co-ordination taking place at the centre is the adjustment of efferent to afferent impulses. . . . It cannot be too much insisted upon that for every bodily movement of any complexity afferent impulses are as essential as the executive efferent impulses. . . . and when we say 'they are guided', we mean that without the sensations the movements become impossible*". Much later, in 1894, Mott and Sherrington were to reach the same conclusion when they observed that a monkey did not move its arm after all the dorsal spinal roots or all the sensory nerves from the arm had been cut, yet the limb moved in response to stimulation of the motor cortex. They concluded that voluntary movements, especially fine movements, cannot be initiated normally by the cortex in the absence of sensory information. This conclusion was based on accurate, but limited observations: either they never restrained the normal arm or they damaged the blood supply to the spinal cord and thus produced more extensive lesions than they planned. Later work has showed that a monkey will make essentially normal use of a deafferented arm if the deafferentation is done at birth, or if it is done in adults in which the use of the normal arm is prevented (Taub *et al.*, 1973; Polit and Bizzi, 1979; Glendinnen *et al.*, 1992).

This project was motivated by Sherrington's curiosity about the coordinating mechanisms between sensation, perception, and movement. That was also one of the motives for his work on visual fusion which led him to support the theory of the "superimposed eyes" or "cyclopian eye" of Ewald Hering (1868) and Helmholtz (1909). He disagreed with Cajal's scheme of the reconstitution of the two halves of the visual image as separate representations in the left and right cerebral hemispheres. He concluded that animals with binocular vision have common efferent paths from corresponding points of both eyes to the eye muscles of both eyes (Sherrington, 1906, pp. 385–386). "*We have therefore to alter such a scheme as that furnished by Cajal by attaching his convergent paths to efferent paths, and by divesting their supposed nodal cortical point of its hypothetical powers as a sensual Deus ex machina. And we thus meet another instance of convergence of afferent paths leading to motor synthesis, but not, or* *only remotely, to sensual. Seen in this light the gulf between sensation and movement looms even wider. . . ."**

How unified visual perception is created by convergence of nerve impulses from corresponding points of both homonymous retinal halves, even during rapid eye movements, was conceived by him as a problem of sensory-motor integration. He was opposed to the idea that the perceptual process ends in the visual cortex, or in some "pontifical neuron," or in any sensory representation encoded in a single neuron or a localized neuronal cluster, as we have considered that notion in Section 2.10. Columns of neurons that specifically respond to complex visual forms have been found in the inferotemporal cortex of the monkey (I. Fujita *et al.*, 1992). Such findings raise once more the questions about localized versus distributed representation in the brain, and about the nature of the end point or culmination of neural processes which relate to cognitive processes. As a believer in the essential duality of mind and matter, Sherrington believed that it was fallacious to reduce a mental process to a physical end point. In Chapter 2 we ask the question: what is the end point or the final neural event that *is* a mental event? If n neural events are necessary, why not $n - 1$ or $n + 1$?

We should consider the creative tension between Sherrington's belief in the separate identities of mind and matter in the brain and his belief in the integration of neural function and neural structure. Sherrington argued that reflex actions were fractions of behavior, and that in the end all fractions have to be summed in the complete pattern of behavior. However, he also argued that some complex patterns of behavior, such as the performance of a piano sonata, are more than the sum of the activities in the neurons that move the fingers over the piano keys. Like Descartes, Sherrington believed that the soul (mind) programmed the whole behavior pattern. But perhaps there is no central programmer, only a sequence of events entrained only by local coupling. But then there have to be control mechanisms to ensure spatial and temporal integration.

The mechanism of inhibition remained the biggest missing part of this puzzle. Sherrington viewed this as the problem of the role of inhibition in the sequences of contractions and relaxations of antagonistic muscles which result in coordinated movements. He also viewed inhibition as a necessary factor in the tonus of muscles involved in maintenance of posture and execution of smooth movements. He conceived of those activities in the context of the large problem of integration of nervous activities. However, he recognized that a successful analysis could not be done on the entire system but only on some relevant part—the individual reflex arc, conceived as a fraction of the machinery of integrated action.

From the beginning of his study of the knee jerk Sherrington's attention was drawn to its functional variability. That was a well-recognized feature of the knee jerk that was under investigation at the time. Observations of inhibition of the knee jerk by voluntary movement and by psychological conditions were made by Lombard (1888, 1889),

---

*The concept of sensory-motor coordination can be traced to Sherrington's predecessors in British physiology. For example, William Carpenter discussed *"symmetry and harmony of muscular movements"* in his *Principles of Human Physiology* (4th ed., 1853) in terms of sensory-motor coordination. In his discussion of the course of the fibers at the optic chiasma, Carpenter (1853, p. 717) argued that *"the purpose of this decussation may be, to bring the visual impressions, which are so important in directing the movements of the body, into proper harmony with the motor apparatus."*

Bowditch and Warren (1890), and A. D. Waller (1890). Sherrington continued that tradition by conceiving of variable function in terms of variable mixtures of excitation and inhibition. He held firmly to that concept and expressed it late in his life in the following terms: *"In reflexes, even under simple spinal or decerebrate conditions, interplay between excitation and inhibition is commonly induced even by the simplest stimulus. It need not surprise us therefore that variability of reflex result is met with by the experimenter. . . . This variability seems under-estimated by those who regard reflex action as too rigid to provide a prototype for cerebral behaviour"* (Sherrington, 1934).

4.4.4. Reciprocal Innervation and the Concept of Active Inhibition

The concept of reciprocal innervation of antagonistic muscles was developed by Sherrington from observations of the inhibition of reflex movements of a limb as the result of sensory stimulation of the skin of the limb on the opposite side (Fig. 3.17). As a result the limb previously flexed was relaxed concurrently with contraction of the extensors. This coordination of the action of antagonistic muscles, Sherrington named reciprocal innervation (Sherrington, 1892c, 1893b, 1897a,b, 1898a,b). Sherrington saw reciprocal innervation as part of the general coordinating functions of the nervous system, not as unique to movements of the limbs.

Reciprocal innervation was well known to Descartes and his contemporaries. Descartes pointed out that when a muscle contracts, e.g., the lateral rectus muscle of the eye, its antagonist relaxes, e.g., the medial rectus muscle (Fig. 4.7). It was also known that agonists and antagonists have different innervation. Descartes also showed different nerves to the extensors and flexors of the forearm. The history of reciprocal muscle innervation and inhibition of voluntary muscle action is reviewed by Tilney and Pike (1925).

There can be no doubt that Goltz stimulated Sherrington's interest in inhibition. Goltz and Freusberg (1874) had studied reflexes in dogs after lumbar spinal cord transection. After spinal shock wore off, they observed inhibition of spinal reflexes by appropriately timed strong stimulation of the skin. Goltz conjectured that inhibition might be the result of competition between the afferent excitation that normally arrived at the reflex center and other afferent excitation that was not normally involved in the reflex. Several other theories were formulated to account for inhibition of spinal reflexes (reviewed by Gault, 1904; Dodge, 1926). They all conceived of inhibition as an absence of excitation as shown in Fig. 1.8. It was Sherrington's great achievement to arrive at the correct idea of inhibition as an active process.

Until Sherrington's 14 papers on reciprocal innervation of antagonistic muscles published between 1893 and 1909, it was believed that vertebrate skeletal muscles were innervated by both excitatory and inhibitory nerves. Sherrington demonstrated that the inhibition is not produced by efferent inhibitory peripheral nerves but via afferents from muscles to CNS. In a letter written in 1918 to Henry Head regarding W. H. Gaskell, Sherrington writes: *"To both him and me it always seemed that the taxis of vol. muscles was impossible without inhibition. But we both expected **efferent** inhibitory nerves to them. . . . Gradually, in view of one's reflex experiments, it burst upon me that for the*

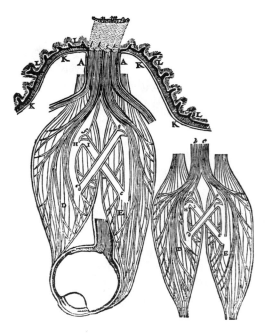

FIGURE 4.7. Descartes's model of the nervous mechanism of reciprocal inhibition of voluntary movement was based on the notion of flow of animal spirits in hollow nerve fibers in which valves could differentially alter the flow through anastomoses between nerves to antagonistic muscles, e.g., lateral and medial rectus muscles of the eye. (From Descartes, *L'homme*, 1664.) "*If we judge Descartes from the severe standpoint of exact anatomical knowledge, we are bound to confess that he, to a large extent, introduced a fantastic and unreal anatomy in order to give clearness and point to his exposition. . . . If we substitute in place of the subtle fluid of the animal spirits, the molecular changes which we call a nervous impulse, if we replace his system of tubes with their valvular arrangements by the present system of concatenated neurons . . . Descartes' exposition will not appear so wholly different from the one which we give today*" (Michael Foster, 1901).

vol. muscles, the inhibitories play not on the muscle direct but on the spinal motor cells driving them, and in that sense are all afferent or central" (Eccles and Gibson, 1979, appendix 6).

Sherrington's mind had been prepared by his mentor W. H. Gaskell (1847–1914) for the concept of inhibition as an active state that has the opposite sign to excitation. In 1887 Gaskell had shown that when the heart of the tortoise was inhibited to a standstill by vagal stimulation, the electrical potential recorded at the sinoatrial node reversed its sign (i.e., to positive from the negative galvanometer deflection recorded from the same region of the excited heart).

His success in studying inhibition against a background of excitation was mainly the result of his use of the decerebrate cat. It was known from earlier work that decerebration produced increased tonus of the muscles, the extensors of the limbs and neck, that keep the cat standing, and thus resulted in marked extensor rigidity.*

---

*Sherrington was scrupulous in giving credit to his predecessors, but when he first used the term decerebrate rigidity in 1898, he lapsed when he wrote that "*the condition is one possessing considerable physiological interest, but I have not succeeded in finding any description of it prior to the above mentioned*" (his own). It had been described by Vulpian (1866) in his *Leçons sur la physiologie du système nerveux*, which Sherrington used when he was an undergraduate, and Liddell (1960) cites earlier work, including a description of decerebrate rigidity in the rabbit by Magendie in 1823. Sherrington first described decerebrate rigidity, without using the term, in his 1897 paper "On reciprocal innervation of antagonistic muscles. Third note" (*Proc. R. Soc. London* 60:414–417). The initial use of the term occurs in his paper published in 1898, "Decerebrate rigidity, and reflex co-ordination of movements" (*J. Physiol. (London)* 22:319–332). That paper had been submitted in 1896 and presented in part as a Croonian Lecture in 1897.

Sherrington recognized the advantage of this preparation for studying inhibition of the extensor reflexes. The extensor rigidity relaxed when a flexion reflex was evoked. This was an example of reciprocal innervation of antagonistic muscles (Sherrington, 1897a, 1898b). Use of the decerebrate cat had great advantages for Sherrington's research. First, the CNS was not depressed after the effects of the anesthetic had worn off; yet the animal was unconscious and could not feel pain. This was a very important ethical consideration and it also dealt with the principal objection of the antivivisectionists. Second, if the surgery was skillfully performed, the animal lost little blood and had normal vital activities. Third, the section of brain stem above the level of the pons (the midcollicular section) and removal of the cerebral hemispheres resulted in excitation of the spinal motoneurons against which inhibition could be measured and analyzed. Sherrington did this by attaching antagonistic muscles to a myograph and recording the tensions they exerted when an afferent nerve was stimulated.

The only index of inhibition available to Sherrington was its effect on excitation. As he well understood, inhibition is an activity that can neutralize excitation. Its effect could be graded and could be produced in several different ways—either by total inhibition of activity in a few units, or by lesser inhibition of activity of a larger number of units, or by various combinations of subliminal excitation and inhibition (Sherrington, 1913, 1934). He reduced these uncertainties without eliminating them, by using a standard excitatory twitch reflex followed at varying intervals by a second excitatory twitch reflex. The diminution of the tension achieved by the second reflex gave a measure of the inhibition latent in the first reflex.

With the decerebrate preparation Sherrington showed that inhibition is an active process, not merely the absence of excitation. From those observations Sherrington deduced that there must be separate inhibitory and excitatory synaptic connections in the CNS. In 1906 in *The Integrative Action of the Nervous System* he wrote: "*In denoting one set of central terminations of an afferent arc **specifically inhibitory**, it is here meant that by no mere change in intensity or mode of stimulation can they be brought to yield any other effect than inhibition.*" He also understood that "*it appears unlikely that in their essential nature all forms of inhibition can be anything but one and the same process*" (Sherrington, 1906).

4.4.5. The Synapse: The Consummating Concept

Neurophysiology has been in a century-long age of the *synapse* that started with Sherrington inventing that term for a concept that had emerged in the final decade of the 19th century. The concept began to be formed when the reflex delay was first recognized as a function of the gray matter. At first the delay was conceived in terms of a network theory of nervous connections, as a slowing of the nervous impulse as it traversed the fine nervous network. But that conception changed with the knowledge of spinal cord organization that was gained by means of the Golgi technique, and by 1894 it was clear that the site of the reflex delay was at the contacts between neurons.

Sherrington coined the term *synapse* in 1897 after the free axonal endings had been seen but before the contact zone had been demonstrated anatomically. In the 1897 edition of Foster's *Text-Book of Physiology*, Sherrington wrote: "*So far as our present knowledge goes, we are led to think that the tip of a twig of the arborescence is not continuous with but merely in contact with the substance of the dendrite or cell-body on which it impinges. Such a special connexion of one nerve cell with another might be called a synapsis.*" This was written in the same year, but before Held had demonstrated the calyciform axonal endings surrounding nerve cell bodies in the trapezoid nucleus—the best anatomical evidence for the contact theory of nerve connections at that time (Fig. 3.15). The synapse was to remain a hypothetical entity until the 1950's when electron microscopy proved its existence (Palay and Palade, 1955; Palay, 1958; reviewed by De Robertis, 1959).

The synapse was inferred from the differences in conduction in reflex arcs and in nerve trunks. As defined by Sherrington (1906) these differences were: (1) Slower speed in reflex arcs, as measured by the latent period between application of the stimulus and muscle contraction. This difference was greater for weak stimuli than for strong. (2) An after-discharge in reflex arcs, as indicated by continuation of the reflex movement after cessation of the stimulus. (3) In reflex arcs there is less close correspondence between the rhythm of the stimulus and the rhythm of the effect. (4) Reflexes exhibit less close correspondence between the grading of the intensity of the stimulus and the grading of the intensity of the effect. (5) Reflex arcs show temporal summation, namely resistance to passage of a single nerve impulse but not to a train of impulses. (6) Irreversibility of direction of conduction in reflex arcs but reversibility of conduction in nerve trunks. (7) Reflex arcs fatigue quickly compared with the greater resistance to fatigue of nerve trunk conduction. (8) Much greater variability of the threshold for stimulation of reflex arcs than of nerve trunks. (9) Reflex arcs show a refractory period and inhibition not found in nerve trunks. (10) Reflex arcs have much greater dependence on blood supply, and much greater susceptibility to anesthetics. These physiological characteristics could be explained anatomically if there were transverse surfaces of separation in the reflex arc not found in nerve trunks. But the concept of a synapse was primarily functional, namely a partial block of transmission of nerve impulses with a one-way valvelike action, resulting in a slight delay in transmission from neuron to neuron. Sherrington had arrived at that concept before there was any convincing anatomical evidence of structures that could subserve the ten functional conditions stated above. They were unobservable entities in the sense that they could not be observed with methods then available. In the previous chapter I have argued that unobservable entities have played important roles in the construction of theoretical models of nervous organization.

In 1925 Sherrington published a largely speculative paper in the *Proceedings of the Royal Society*, entitled "Remarks on some aspects of reflex inhibition." There he introduced the concept of central excitatory and inhibitory states, which were unobservable entities. He conjectured that these states are complementary, can summate, and can persist for longer than the duration of the presynaptic discharge. At that time there was no evidence of any differences between inhibitory and excitatory synapses, so Sherrington conjectured that differences between excitation and inhibition occur exclusively in the postsynaptic neuron. He believed that inhibition exists as a function different from excitation and is not merely an interference with excitation as proposed by Lucas, Forbes, and Adrian in their interference theory of nervous inhibition (see Forbes, 1922).

Sherrington's theory was supported by experimental evidence showing that graded reflex activity occurs as the stimulus is increased, and it was also consistent with evidence of inhibition of reflex activity, crossed extensor reflexes, and with the prolonged duration of excitatory or inhibitory states after cessation of stimulation. However, the crucial evidence supporting the theory could not be obtained until intracellular microelectrode recording from spinal motoneurons showed that the inhibitory or excitatory states are caused by hyperpolarization or depolarization of the postsynaptic neuronal membrane (Brock et al., 1952a,b).

Sherrington had to obtain evidence for synaptic function indirectly from observations on spinal reflexes, specifically from the reflex contraction of a muscle recorded isometrically. This method of myography had been invented by Adolf Fick (1882; also see Tigerstedt, 1911–1912, for a detailed history of methods of recording muscle and nerve activity). The muscle was prevented from shortening so that the tension developed by the contraction was recorded by a myograph when different afferent nerves were stimulated electrically.

An example will help to show the logic of such experiments, which were performed on decerebrate cats. When the motor nerve was stimulated directly the muscle developed a tension of 2500 grams, for example. But the muscle tension reached only 1000 grams when it contracted reflexly in response to stimulation of afferent nerve A, and only 500 grams in reflex response to stimulation of afferent nerve B. However, when A and B were stimulated simultaneously the muscle tension was 2000 grams. Sherrington concluded from such results that stimulation of either A or B elicited fractional effects on the total pool of motoneurons supplying the muscle, meaning that neither nerve could cause all the motoneurons to fire. Some are fired by one afferent nerve, others by a second afferent nerve, and some are shared. If one set of motoneurons was fully active, additional stimulation by another afferent could fire only those motoneurons that were not shared. The first set was said to be "occluded" (Cooper et al., 1926, 1927). Stimulation of A and B together resulted in a greater muscle tension than the sum of the tension of A and B stimulated separately. Therefore, stimulation of both had recruited motoneurons that could not be discharged by either nerve alone. Sherrington termed those motoneurons the "subliminal fringe" (Fig. 4.8). Stimulation of afferent nerve C alone caused no muscle contraction. Stimulation of A, B, and C together resulted in a smaller tension (about 200 grams) than stimulation of either A or B alone, indicating that nerve C had inhibitory effects that reduced the excitatory effect of stimulation of A and B. Using experiments of this general sort, it was observed that excitation and inhibition sum algebraically. Therefore, inhibition is an active state, not merely the absence of excitation.

From such experiments Sherrington developed the concept of convergence of the conduction paths of the nervous system and of integration of afferent excitatory and inhibitory states in the motoneuron pools supplying antagonistic muscles (Sherrington, 1929; Creed et al., 1932). Sherrington developed his theory of synaptic function in his Ferrier Lecture of 1929, entitled "Some functional problems attaching to convergence." There he advanced the theoretical concept that convergence of two or more afferents on motoneuron pools supplying a muscle results in excitation of many neurons in common. Thus, the contraction tension developed is less than the sum of the contraction tension

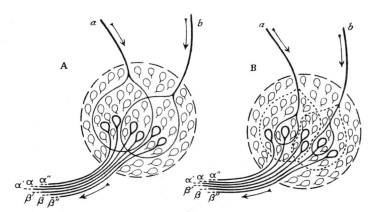

FIGURE 4.8. Interaction of reflexes: **(A)** When two afferent nerves excite many motoneurons in common, their interaction results in occlusion, namely a reflex muscle contraction that is less that the arithmetical sum of the contractions of each afferent. By itself afferent nerve a maximally activates four units, and likewise b by itself, but when stimulated together $a + b$ activate only six units, i.e., a contraction deficit caused by occlusion of units α' and β'. **(B)** Weaker stimulation of a and b, restricting their supraliminal effects in the motoneuron pool of a muscle, results in activation of one unit when a by itself is stimulated, and likewise with b. But stimulation of $a + b$ together activates four units owing to summation of subliminal effects in the overlapping region of the subliminal fringes of their fields. [From Sherrington's (1929) Ferrier Lecture, "Some functional problems attached to convergence."]

resulting from stimulation of each afferent separately. He termed this *occlusion*. However, when there is a large overlap in the subliminally excited neurons, convergence results in facilitation, that is, a larger contraction tension than is produced by stimulation of each afferent alone. The concept of the subliminal fringe and of occlusion were depicted in a diagram that was reproduced for the following 30 years or more (Fig. 4.8).

During the Oxford period the work of Sherrington and his associates (Liddell, Creed, Cooper, Denny-Brown, and Eccles) was guided by the concept of convergence of reflex pathways on the final common paths and their associated motor units. The motor unit was defined as the muscle fibers supplied by the axonal branches of a single spinal motoneuron. The sizes of motor units were measured by counting the number of fibers in nerves to a muscle after section of the dorsal roots and degeneration of the sensory nerves from the muscles. Attempts that were made to correlate the size of the motor units with the muscle contraction tensions were frustrated by inadequate knowledge of the branching of motor axons and especially by failure to recognize the significance of the small-diameter axons that were later shown to innervate the intrafusal muscle fibers of the muscle spindles (Leksell, 1945). This oversight is surprising because Sherrington had recognized the motor innervation of muscle spindles in the 1890s, but had not grasped its function (e.g., Liddell and Sherrington, 1924).

In the concluding remarks in *Reflex Activity of the Spinal Cord* (Creed et al., 1932, pp. 157–159), Sherrington gives a masterful review of the concepts of reciprocal innervation, central inhibitory and excitatory states, and the final common path. The following is typical of his style: "*under intact natural conditions we have to think of each motoneurone as a convergence-point about which summate not only excitatory processes fed by converging impulses of varied provenance arriving by various routes, but also inhibitory influences of varied provenance and path; and that there at that convergence-place these two opposite influences finally interact. The two convergent systems themselves, one excitatory, one inhibitory, make of the entrance to the final common path, which we may accept the motoneurone as constituting, a collision-field for joint algebraically summed effect."*

We should remember that the concepts of occlusion, convergence, subliminal fringe, integration, and the final common path were formed within the neuron theory. It is characteristic of a high-level theory like the neuron theory to explain and unify a wide variety of observations that may be explained by several lower-level, less-inclusive theories. William Whewell, in *The Philosophy of the Inductive Sciences* (1847), invented the term *consilience*, meaning the power of a theory to include and unify a wide range of observations. Consilience is one of the main criteria of a useful theory such as Sherrington's theory of nervous integration.

Sherrington's model explained reciprocal reflexes in terms of inhibitory and excitatory effects on the motoneuron resulting from the actions of different branches of the same axon, either from different branches of a spinal afferent nerve fiber or of a spinal interneuron (Fig. 3.17A). Only in the 1930s did the evidence show that the same synaptic transmitter is released at all branches of any one axon (Dale, 1934, 1935, 1937). Another 20 years elapsed before it was realized that excitation is converted to inhibition by mediation of an inhibitory interneuron (Fig. 3.17B; Eccles *et al.*, 1954, 1956).

4.4.6. Values of a Neuroscientist in a World of Facts

Attempts to form hypotheses about how Sherrington's personality and values related to his scientific research are more difficult than usual because of the dearth of direct evidence about his scientific work habits before 1901. We have Harvey Cushing's testimony about Sherrington's surgical technique and work habits during his Liverpool phase: *"He operates well for a 'physiolog'. . . . It's a great surprise all through physiological work to find that practically all observations are open to dispute or various interpretations. . . . The whole thing referable to experimental neurology much to my surprise is still in a most crude condition. The problems offered are immense. Sherrington goes at them too fast. Few notes are taken during the observations, which is bad. Sherrington says himself he has a bad memory—putters around the laboratory till after 7 in the evening trying to catch up on things and then is used up and doesn't begin till ten or eleven the next day"* (Letter of 17 July, 1901, cited in Fulton, 1946, pp. 196–197). Cushing observed Sherrington's experiments on motor functions of the cerebral cortex of anthropoid apes, and Cushing performed craniotomies on a gorilla, orangutan and several chimpanzees that were the subjects of Sherrington's experiments at Liverpool in 1901.

"Sherrington was of the visionary type, more inspired by new ideas than by technical possibilities," says Ragnar Granit (1966), who was one of Sherrington's students at Oxford. Tinkering with experimental apparatus was not Sherrington's

manner. He was not technically innovative. Although he is one of the moderns in theoretical concepts, and in his causal-analytic mode of thinking about experimental neurophysiology, his techniques were vestiges of the premodern era. Almost all the experimental apparatus used by him as late as 1927 had been invented more than 50 years earlier. In 1927 his laboratory was equipped with a pendulum switch of the type used by Helmholtz in 1850 for measuring the speed of conduction of the nerve impulse. For stimulating muscle and nerve he used induction coils not much different from those developed by du Bois-Reymond in the 1860s. A string galvanometer of the type invented by Einthoven in 1903 was not installed in Sherrington's laboratory until late in 1929 (Eccles and Gibson, 1979, p. 48).

It was not as if Sherrington was working during a period of slow development of experimental techniques. Rather, development of electrophysiological techniques occurred very rapidly during the 19th century (Tigerstedt, 1911–1912; Bernstein, 1912; Hoff and Geddes, 1957; Geddes and Hoff, 1961; Rothschuh, 1971; Geddes, 1984). Important advances in neurophysiological techniques were achieved during the 50 years before Sherrington began his neurophysiological research. Great improvements were made in methods of recording nervous and muscular activity that enormously increased the analytical power of the techniques. Those advances are associated with Hermann Helmholtz (1821–1894), Emil du Bois-Reymond (1818–1896), and Eduard Pflüger (1829–1910) in Germany and with Étienne Jules Marey (1830–1904) in France.*

The moving magnet galvanometer invented in 1821 was not sensitive enough to record nerve action potentials (Humphreys, 1937). In 1830 Nobili was the first to demonstrate electrical activity in muscle using a physical instrument, his moving magnetic needle galvanometer. In spite of improvements in galvanometer design over the next 70 years, galvanometers remained incapable of measuring the shape of action potentials until Einthoven introduced his string galvanometer in 1903. During the interim period it was possible to detect nerve and muscle electrical activity by using the frog sciatic nerve-gastrocnemius muscle preparation as an indicator—the rheoscopic ("current seeing") frog introduced by Carlo Matteucci in 1840. A twitch could be seen in the muscle of the rheoscopic preparation when its nerve was placed in contact with another nerve or muscle in which activity was being observed. This preparation served for recording electrical activity in nerve and muscle until the introduction of the capillary electrometer into electrophysiology in 1876.

When Sherrington first formed an elementary theoretical model of the synapse between 1897 and 1906, he introduced the terms and the concepts of monosynaptic, disynaptic, and polysynaptic reflex pathways. He surmised that the knee jerk is monosynaptic and the scratch reflex is disynaptic (Sherrington, 1906). However, the synaptic delay could not be measured with the best instruments then available, namely the capillary electrometer invented by Lippmann in 1871 and first used in 1876 by Marey for recording the electrical activity of cardiac muscle. In 1888 Gotch and Horsley

---

*The great classical work on graphic recording of biological movements is Marey's *La méthode graphique dans les sciences expérimentales* (1878). Marey invented instruments for recording movements of skeletal and cardiac muscles, pressure changes in the cardiovascular system (Marey, 1863), and he laid the foundations for graphic analysis of biological movements.

had used the capillary electrometer to make the first photographic record of the nerve action potential. They were also the first to record electrical activity in the CNS—in the spinal cord during cortically induced epileptic seizures (see Geddes and Hoff, 1961). The principle of the capillary electrometer was the change in shape of a drop of mercury when an electric current passes through it. This enabled the sign and amplitude but not the shape of the potential to be measured by photographing the shadow of the mercury meniscus (Bernstein, 1912). This technical limitation changed after introduction of electronic valve amplifiers and the cathode-ray oscilloscope by Gasser and Erlanger in 1922. Then it became possible to accurately determine conduction velocities of fibers from the compound action potential, and to show that the conduction velocity is proportional to the diameter of the axon. Cathode-ray oscillographs and electronic amplifiers were truly revolutionary instruments, as I have pointed out near the end of Section 1.3.3. They made it possible to determine that the synaptic delay in the quadriceps stretch reflex (knee jerk) is about 1 msec, and so confirmed Sherrington's conjecture that the knee jerk is a monosynaptic reflex.

Sherrington never took to electronic instruments. After electronic instruments were generally used in neurophysiology during the 1930s and 1940s, the sort of apparatus that Sherrington had used so successfully for stimulating and recording was consigned to the storerooms of numerous physiology departments. The string galvanometer, kymograph, torsion myograph, induction coils, differential rheotome, and Keith Lucas pendulum, similar in design to those in use in Sherrington's laboratory at Oxford, continued to be used for many years in some student classes in mammalian physiology. They were still in use at Edinburgh when I was a student there in the 1950s. In those laboratory classes we also used Sherrington's *Mammalian Physiology: A Course of Practical Exercises* (1919; revised in collaboration with E. G. T. Liddell, 1929), excellent for its time, but not much use for the era of electronic instrumentation.

Literary style can be an accurate reflection of the writer's personality, when allowances are made for the limitations of the genre, the purposes of the work, and the conventions of the time. Sherrington's style of writing is precise although not concise. Occasionally it is awkward, but rarely imprecise. The complexity of his sentences matches the complexity of his ideas, and the flow of thoughts is naturally related to the flow of words. This style is no longer fashionable. Another characteristic is his inoffensive tone, calculated to persuade his opponents rather than score points off them. This style is now perhaps seen as quaint but it was quite typical of the literati of that time, perhaps related to their early schooling in the Latin and Greek classics. It was quite normal for such people to talk as they wrote. The most extreme case is Henry James who "*always spoke with an air which I can only call gracefully groping; that is, not so much groping in the dark in blindness as groping in the light in bewilderment, through seeing too many avenues and obstacles*" (G. K. Chesterton, 1975). Sherrington's lecturing style suffered from his tendency to digress through seeing too many sides to any problem, his distrust of dogmatism, and his fear of speculation. All of us who have thoughts beyond our powers of expression can sympathize with Sherrington's hesitations, circumlocutions, and his efforts to express ideas not yet come to rest, that are, so to say, still in flight.

By all accounts Sherrington's lectures were difficult to follow: "*He eschewed *dogmatism, he fought for. . . . the word or phrase to convey precisely his meaning so that he appeared unduly hesitant; he qualified his statements, he inserted parentheses; indeed he gave the impression that his thoughts were not really in the lecture room but already contemplating and designing another experiment*" (Cohen, 1958, p. 14). "*He was never known as a good lecturer. . . . The subject was approached from many angles, and many reservations precluded pronouncement and simplification. . . . he went on, absent mindedly, as if listening to some inner voice. . . .*" (Granit, 1966, p. 19). When Sherrington gave the Silliman Lectures at Yale in 1904, the large audience at the first lecture had dwindled to four people at the tenth and last lecture (Fulton, cited by Swazey, 1969). In spite of his deficiencies as a lecturer, Sherrington had no doubt of the superiority of the lectures given by research scientists over those delivered by skilled professional teachers: "*Considering the teacher merely in his rôle of lecturer, instructive lectures of a kind have undoubtedly been given by men of ability, the whole of whose knowledge of a lecture subject was secondhand, but it is more than doubtful whether the real life of any science can be felt, still less communicated by one who has not himself learned it by direct enquiry from nature*" (1899a, *Br. Med. J.* 1:878).

Sherrington must have developed his interest in the relationship between mind and body from his boyhood studies of "Divinity" at Ipswich School. He knew Descartes's position, and expounded it in the Silliman Lectures delivered at Yale in 1904. He owned a rare first edition of Descartes's *De homine* (1662), which he gave to Sir William Osler in 1917 (see Fig. 4.7). Unfortunately, he was unwilling to give a detailed account of his understanding of the mind-body relationship, although he referred to that problem often in his later years, for example, the 1922 lecture on "Some Aspects of Animal Mechanism," the 1933 Rede Lecture, and the Gifford Lectures of 1937–1938, published in 1941 as *Man on His Nature*. He never explained his position, but as far as one can tell from brief statements it was that the body and mind exist separately, but in an integrated relationship: "*In all of those types of organisms in which the physical and psychical coexist, each of the two achieves its aim only by reason of a contact between them. And this liaison can rank as the final and supreme integration completing the individual*" (Sherrington, 1941, p. xix). Apparently he accepted the Cartesian distinction between the mind as a thinking substance and matter as extended substance, with two-way causal relations between mind and matter (Descartes, *Meditations*, I and II, *Discourse on the Method*, 5). I discuss this further in Section 2.11.

Mind, in those terms, could be the same kind of entity in all animals that have brains that permit the liaison between brain and mind. Sherrington agreed with the view expressed by Darwin in *The Descent of Man* (1871, Chapters 3 to 5) of the continuity of mental evolution from animals to man, with increase in power but not in kind (see Angell, 1909). The main proponent of mental evolution in his time was G. J. Romanes in his *Mental Evolution in Man* (1893a) and *Mental Evolution in Animals* (1895). Sherrington hesitated to say when mind appeared during evolution but he believed that all vertebrates probably have mind (*Man on His Nature*, pp. 212–213). He did not seem to think that memory is necessary for mind. I agree with him that consciousness is possible without memory, although without memory we could only be conscious of the present. This may be interpreted to mean that the physical basis of consciousness is different from that of memory, or that consciousness does not have a material cause.

It is hard to believe that Sherrington had not developed his own detailed understanding of the mind-body relationship long before he began to make public statements on the subject, only after his fame was already secure. I believe that Sherrington was reluctant to expose his dualist beliefs because of his natural shyness and because the topic had a long and controversial history. To have engaged in that dispute would have placed him in the arena of unprofitable controversy over a matter that is not yet and may never be in the power of science to resolve.

He believed in the existence of the mind as an entity different from, but operating with, nervous mechanisms: "*I have therefore to think of the brain as an organ of liaison between energy and mind, but not as a converter of energy into mind*" (*Man on His Nature*, p. 318). In several places Sherrington uses the word liaison to describe the relation between mind and brain (Foreword to 1947 edition of *The Integrative Action of the Nervous System*, pp. xxi–xxiv; *Man on His Nature*, pp. 318–319, 350–357). I can only surmise that he chose the word *liaison* to imply a causal connection between mind and brain. Such vague and unsystematic pronouncements would not merit serious attention if they had been made by a less illustrious neurophysiologist.

He embraced dualism because he thought that it offered no less compelling an explanation than monism: "*That our being should consist of **two** fundamental entities offers I suppose no greater inherent improbability than that it should rest on one only*" (Foreword to *The Integrative Action of the Nervous System*, 1947, p. xx). This amounts to little more than an apology, not an argument for dualism. His objections to psychophysical monism are weak, for example, his assertion that physiology has merely "*brought us to the brain as a telephone exchange*" (*Man on His Nature*, p. 282). But the alternative, as he expressed it, was close to mysticism: "*But mind seems to come from nothing and to return to nothing*" (*Man on His Nature*, p. 267). True, mind may be analogous to an electron appearing and disappearing in a Wilson cloud chamber, apparently without cause, but Sherrington does not use that analogy.

Unlike most scientists, he did not exclude ethical judgments from the final assessment of the meaning of science. Values were an important part of the motives for the conduct of his science and for guiding it in certain directions. He stated this very clearly: "*Granted the scope of natural science be to distinguish true from false, not right from evil, that simply makes the man of science as such, not the whole man but a fractional man; he is not the whole citizen but a fraction of the citizen. The whole man now that his mind has 'values' must combine his scientific part-man with his human rest. Where his scientific part-man assures him of something and his ethical part-man declares that something to be evil it is for the whole man in his doing not to leave it at that. Otherwise in a world of mishap his scientific knowledge and his ethical judgment become two idle wheels spinning without effect*" (Sherrington, 1941, *Man on His Nature*, p. 375).

In the England of Sherrington's time, scientific rigor was often seen as opposed to emotion, feeling and imagination (e.g., J. S. Mill, *Autobiography*, 1873). That dichotomy did not exist for Sherrington. In his laboratory there were frequent discussions of the poetry of Shakespeare and Keats and the art of the Italian Renaissance (Eccles and Gibson, 1979, pp. 57–58). Among his favorite books were Middleton Murray's *Keats and Shakespeare* (1930) and *The Testament of Beauty* by the English poet Robert

Bridges, himself a good example of a fruitful meeting of the emotional and rational, who practiced as a surgeon until the age of 36. *The Testament of Beauty* was a best-seller when it appeared in 1929, but like Sherrington's poetry it strikes one now as abstracted from authentic emotion, with its unremitting high-mindedness and spirituality. Nevertheless, Sherrington's example goes to show that although science cannot take the place of literature and art, it may transmit the values of both. Those were the values in Sherrington's mind when he wrote *"Only amid human fellowship can 'values' be listened to and shared. In him evolution has shaped a social animal par excellence. . . . Social life is his opportunity as field for his 'values.' To know that he is a life being evolved which carries with it the 'values' can imbue his social system with a purpose. Human fellowship thus emerges as something of unique worth. Fellow human mind is the sole mind to understand and to share to the full with his, and be shared"* (Man on His Nature, 1941). This comes close to the argument of Joseph Butler (1692–1752) that virtue is expressed in human companionship and *"that mankind is a community, that we all stand in relation to each other, that there is a public end and interest of society, which each particular is obliged to promote."* The debt to Butler is evident in the final chapter of Sherrington's *Man on His Nature*, which expounds the idea that ethical values, especially altruism, have evolved uniquely in the human species.

In Victorian England there was a widespread belief in the manifest destiny of mankind ensured by religion with the support of science. As Sherrington put it: *"Natural Religion consults Natural Science,"* and he affirmed that *"Natural religion, along with the great religions holds 'truth' a value"* (Man on His Nature, 1941, p. 402). By natural religion one may mean no more than the idea of ethics apart from established religion, without supernatural revelation, as originally proposed by Shaftesbury (1699, 1711), and Butler (1736). Or one may mean more than that, namely that God is revealed in nature, or that God continues to act in the world, not supernaturally, but through the normal processes of nature. For those who hold that belief, ethical behavior is the same as living in harmony with nature, and ethical understanding is knowing how human behavior is founded on natural laws as they have been discovered by science. But the notion that ethics comes from nature is not the solution, it is the problem, as I argue in Section 5.1.

The notion that science is at the service of religion can be found in William Whewell's *History of the Inductive Sciences* (1857, 3rd ed., Vol. 3, p. 392) which had enormous influence on English intellectuals (Ducasse, 1951). That notion was endorsed by Queen Victoria and by Albert, the Prince Consort. Albert expressed it in his address to the British Association for the Advancement of Science in 1859, when he spoke of scientists *"like pious pilgrims to the Holy Land. . . . in search of truth—God's truth—God's Laws as manifested in His works."* This was quoted with approval by Thomas Laycock in 1860 in his *Mind and Brain*. That work had considerable influence on British neuroscientists, especially Hughlings Jackson, who was Laycock's student.

In view of Sherrington's grounding in Latin, Greek and in theology, it is likely that his notions of truth and beauty were informed by his knowledge of the views of Plato, St. Augustine, St. Thomas Aquinas, and Bishop Joseph Butler as much as by J. S. Mill and Herbert Spencer. Joseph Butler's *Sermons* (1726) and *The Analogy* (1736) may have been sources of inspiration for both Hughlings Jackson and Sherrington. In the *Sermons*, Butler expounded the idea of the tripartite division of human nature which may have inspired Hughlings Jackson's concept of functional metamerism of the nervous system. Butler's three levels are, in ascending order, passions and affections, self-love and benevolence, and conscience, with the former subordinate to the latter and subject to inhibition by it. Butler may have anticipated the concept of integrative action when he wrote: *"Whoever thinks it worth while to consider this matter thoroughly should begin by stating to himself exactly the idea of a system, economy or constitution of any particular nature; and he will, I suppose, find that it is one or a whole, made up of several parts, but yet that the several parts, even considered as a whole, do not complete the idea, unless in the notion of a whole you include the relations and respects which these parts have to each other."*

It is also remarkable that Butler first used the term *"Reflex"* as in the *"Reflex Principle of Approbation"* to mean inhibition of the passions by the conscience—a conceptual intimation of central inhibition! Sherrington excelled at "Divinity" while he was a student at the Ipswich school from 1872 to 1875 (Swazey, 1969, Appendix B). It is quite likely that he read Butler's works at that time. Butler's *Sermons* and *The Analogy* were widely circulated in the 19th century, therefore Sherrington was likely to have read them (there was an 1896 edition of the *Sermons* edited by the former British Prime Minister, W. E. Gladstone).

Sherrington regarded the evolution of the perception of pain and suffering as the necessary basis for protective reflexes ("nociceptive" was his term) and as the necessary basis for distinguishing good from evil: *"Human life has among its privileges that of pre-eminence of pain."* This is not the view of utilitarians such as Jeremy Bentham and J. S. Mill who identified happiness with an absence of pain and an increase of comfort. But happiness is independent of comfort and even pain. The opposite of pain is not pleasure, it is anesthesia. We could not have a notion of vice and of evil without having experienced pain and its integral anguish. Nor could we form a notion of virtue and good without having experienced benevolence, love, and pleasure. Sherrington recognized that Nature is not all nice and cosy, the constructive forces of nature are in balance with destructive forces, pain and death are as much a part of nature as pleasure and procreation: *"To perceive Nature's beauty is perhaps among the loveliest of all human privileges. . . . But, beside the lovable, our world has in it also the hateful. . . . the whole scheme of destruction from the killer-whale to the ichneumon fly. This and much more, not to speak further of human cruelty and hate. . . . in the scheme of Nature to attack and prey upon others is for the majority of animals, even likest ourselves, their charter of existence. . . . Nature sows and reaps vast harvest of pain and fear all over our planet, on earth, in sky and sea."* Sherrington wrote this in 1942, in the third year of the most destructive and bestial of all wars, in his 85th year. Sherrington believed that the ability to feel pleasure and pain are among the surest tests of whether a being has a mind. Pain and pleasure are uniquely personal perceptual experiences without any counterparts in the world apart from the brain. To the mechanical materialists' assertion, "no currents: no mind," I think he would have replied, "no ecstasy, no anguish: no mind." He would have said that machines, unable to experience moods and emotions, have no mind, regardless of any powers of rational intelligence they may possess.

CHAPTER 5

Ethics in Science

He who would do good to another must do it in Minute Particulars.
General Good is the plea of the scoundrel, hypocrite and flatterer;
For Art and Science cannot exist but in minutely organized Particulars.
                William Blake (1757–1827), *Jerusalem, f.55, l.54.*

5.1. THE ARGUMENT

Science is not a vehicle for taking us to heaven but it should include precautions against going to hell. Without self-conscious awareness of moral goals and self-conscious integration of moral and empirical knowledge, we can only solve practical tasks but cannot guide science to goals that we can define as just and humane. Separation of ethics from empirical science has been promoted as a policy aimed at increasing the freedom of science, when it actually diminishes it. A science which professes moral principles but which is really morally unemployed has the worst of both worlds.

The doctrine of value-free science is founded on the false assumption that science holds a privileged position in society and history, transcending the values of any time and place. On that assumption it seems as if science needs no justification in terms of extrascientific values. In fact, science has never been entirely self-justifying. It has always required justification in terms of utility, morality, and other extrascientific values. The problem is not how to exclude those values, because that is impossible, but how to integrate them effectively with science.

In this chapter I shall discuss the causes and consequences of the policy of separation of ethics from science, and consider how ethics and science may be integrated. I shall discuss three questions dealing with the relations between ethics and science: (1) How does ethics function in the conduct of scientific research programs? (2) What are the similarities and differences between ethical understanding and scientific understanding? (3) How are brain processes related to development of ethical understanding and moral behavior?

I arrive at the conclusion that it is possible to have both an ethics of neurobiology and a neurobiology of ethics. With respect to the former I understand that a complete science requires admission of ethical values in addition to its empirical methods and theories. The debate may never be concluded about whether ethics is fundamental and necessary in science, or is only convenient and useful. But it is undeniable that ethics plays important roles in scientific research programs, and should not be ignored. This applies to all sciences, but neuroscience has a special relationship to ethics insofar as it can give neurobiological explanations of moral understanding and conduct.

A scientific research program is composed of observations, theories, techniques, and values. A research program is incomplete when it is deficient in one of those components. The ethical content of science cannot be given exact numerical values. Nevertheless, there are methods for handling inexact and "fuzzy" information that may be used for quantifying value judgments in ways that enable us to deal with them as empirical knowledge.

The old moral theories rest on the assumption that intuition gives us direct insight into the moral aspects of our behavior and gives us the knowledge to select moral goals. In that view, ethical claims are essentially *a priori* and that is what gives them normative character. My personal conviction is that we normally find ethics only in experience, especially early childhood experience, and we have no moral intuition that can lead us to moral goals. We do not seek moral goals because we intuitively know them to be good, but on the contrary, we define goals as moral and seek them because they promote our survival, because we learn to desire them, and because society condones and encourages them. From the time of birth we learn to make moral choices by encounters with others, by emotional and cognitive responses to those encounters, by relating their effects to our states of well-being, and to rules taught by social convention. We construct mental models in which individual experiences and learned moral rules can be connected in such a way as to promote our preservation. Those models may be universal to the extent that the brain processes involved in the moral experiences of all people are determined by the same evolutionary and developmental processes. Conceptual models can be effective in dealing with ethical problems for the same reasons that they can be effective in dealing with scientific problems: because our brains have evolved as part of reality and that is where science and ethics both come from. In both science and ethics the consistency of a conceptual model with new experiences and its predictive reliability are the principal rules for the choice of one model over another, and in both the proof of the model lies in the future; it is not given *a priori*.

I do not assent to the currently fashionable view that "*scientists are salesmen of ideas and gadgets, they are not judges of truth and falsehood. Nor are they high priests of right living*" (Feyerabend, 1981, Vol. 2, p. 31). The dangers of that view are that it denies the intrinsic value content of science and it sees scientists as having lost their functions as generators of values and meanings. In that view scientists are not responsible moral agents with an obligation to know the truth, to serve justice, and to decide for themselves what is good or bad, true or false. Instead, scientists are regarded like any other producers of commodities for profit, or like any other paid laborers whose ideas and discoveries become the "intellectual property" of their employers. In that view, ethical constraints on science are to be imposed entirely from outside the practice and theory of science itself. Then science becomes hedged by legal restrictions that properly apply to manufacturers of products for public consumption. The gadgets and remedies that result from scientific discovery should be regulated by laws to protect consumers, but not science itself because science is not a product: it is a process of understanding and of communicating meanings of observations of phenomena. This is a vital distinction to make if science is to be protected by the principles of freedom of expression.

The scientist in search of knowledge proceeds on the belief that perfecting knowledge is itself good whether or not it is in itself useful or advantageous. Evil comes from lack of knowledge and from outside knowledge itself. No fact is inherently good or bad, but it may become so depending on its context and on the use that is made of it. Values are not properties of things but are given to them by human needs and desires. Products of the human brain—works of art, literature, history, philosophy, politics, religion, technology, and science—are the vehicles for transmission of values within a society and from it to succeeding generations. They are continually reinterpreted, criticized, condemned, or canonized.

Empirical data alone are meaningless—to expect the raw data on their own to yield any meaning without the addition of imaginative exegesis is to expect light to be generated by mirrors facing one another. Values enter at the moment the imagination perceives a definite pattern, and the scientist asks whether the pattern is true or false, beautiful or ugly, useful or useless, good or bad. There is no science that is ideologically neutral. Data are so, but interpretations and practical applications of data are loaded with ideological freight. This is not understood by those who think that science can be brought to a value-free norm, or by those who assert that concern about the objective truth of the facts is the only value that should be admitted in order to do good science, and that all other values should be excluded because they are unnecessary and possibly harmful. The problem is not how to exclude values, since that is impossible, but how to expose them to critique. This applies even to such declared intentions as the promotion of progress and the improvement of nature. As I contend in Section 5.4, one of the purposes of self-reflection, communication and critique is to determine whether motives and intentions have positive, negative, or neutral ethical effects.

Ethical commands or statements expressing moral attitudes have no useful functions in scientific discourse, but are used by extrascientific authorities to control science from outside. This is one of the justifications commonly given for excluding values from science. The alternative is for ethics to be perceived as an intrinsic part of the foundations of science, thus enabling ethics to guide science to appropriate goals from inside its own structure. That can occur because ethical propositions can be formulated as testable hypotheses that can make reliable predictions. The task of ethics in science is to provide testable hypotheses about alternative choices of action that can lead scientific research programs to goals which are defined conventionally as just and humane, not only as useful and otherwise desirable. That procedure should enable us to find empirical solutions to such questions as whether it is right to alter the genome to improve the human brain and behavior, or even to make a human. But neither ethical solutions of scientific problems nor scientific solutions of ethical problems are necessarily consistent with the prevailing sociocultural norms. Conflicts between science and society can occur even when science is cultivated as a program that is intrinsically ethical.

In a pluralistic society, it is inevitable that the goals of any scientific research program will conflict with the world views of some people, including some within the scientific community. Scientific research programs have to contend with the diversity of assumptions, conventions, and patterns of thought prevalent in society. To do so effectively requires a dialogue about the alternative goals of scientific research programs. The dialogue ideally involves full disclosure, deliberation and reflection, modifying and enhancing the scientific research program in ways which create consensus and stimulate progress.

Scientists cannot expect nonscientists to assent to their point of view, however "correct" it may be, merely because of their claims to superior knowledge. Unless scientists are able to justify their claims in generally intelligible terms, the best that they can expect is for nonscientists to tolerate their activities because they recognize their practical benefits, not because they recognize the intrinsic values of scientific ideas. There is considerable validity in the opinion that the masses cannot be expected to share the views of the intellectual elite. But without effective sharing of views there is a great danger of domination of our culture by anti-intellectuals who accept only the gadgets and remedies while distrusting, distorting, and finally denying the values of science. It is, therefore, vital for the future of science that scientists are able to go beyond justifying themselves merely as purveyors of useful gadgets and remedies, and communicate the greater significance of science as a program that is intrinsically ethical.

Belief in science as a vehicle for material progress introduces contradictions in scientific attitudes and intentions. The contradiction arises most often because the attitudes and intentions of science are different from those of technology, yet science and technology are interdependent. Witness the contradictions that have arisen in the attitudes of geneticists with regard to eugenics, of neuroscientists with regard to psychosurgery, and of physicists with regard to development of nuclear weapons.

A scientific attitude is necessary to construct conceptual models of the objects of experiences and to bring the models into increasing synergy and consilience with the realities that they represent. This requires a belief in the perfectability of our models of reality but not a belief in perfecting the real world. The intention to hasten material progress and to change nature in a preconceived way is not part of the scientific attitude: it is a technical program that may come into conflict with the desire to know the truth and to act on that knowledge as far as that is possible and ethical. It needs to be emphasized that for science to progress as an ethical program it is necessary for scientists to think about ethics, but it is not necessary for them to think about ways to change nature or even for them to believe in material progress. Failure to understand this is one of the main causes of the disintegration of ethics in science, both in the behavior of individuals and in the behavior of entire scientific research programs.

Expectations of cumulative benefits gained by the applications of science are baseless in the absence of the will to raise our ethical values to parity with our theories and our techniques. Can we look forward to almost limitless scientific-technological advances without moral evidence for preferring some to others? Science can allow us to select the means to any goal and to calculate the probability of arriving there, but science cannot tell us what moral goals to pursue. The ends are decided by what we want, which depends on the moral values we embrace. But value goals have no finality. We do not possess the intuitive knowledge on which to select ultimate value goals. They can be selected provisionally and approached tentatively and justified by results, that is, by the development of evidence.

5.2. NEUROSCIENTIFIC ETHICS—ETHICAL NEUROSCIENCE

5.2.1. Integrating Science and Ethics

There is a curious misunderstanding about the workings of moral values in science which has led many to deny their importance or to regard them as alien to science. There is the doctrine that ethical values are entirely distinct from truth as conceived by science. Then there is the doctrine that the task of science is only to describe and not to justify. A critique of those doctrines will be given later, but here I state flatly that the belief that science can be brought to a value-free norm is delusory. It can indeed be harmful because it discourages an open declaration of values, and thus helps people to disguise or conceal those values that promote the misuse of science. Moral neutrality in science is also dangerous because authorities like the U.S. National Academy of Sciences communicate the wrong message when they withhold moral judgment. The message that they send is that ethical values cannot be integrated with science as an ongoing process but must be imposed on science by extrascientific authority.

What are the ethical values which affect scientific research programs? Are they our perceptions of universal values inherent in the given situations and materials? Are values imposed on research programs by our need to obtain moral gratification—hostages given by reason to the emotions? Or are they justifications for social, political, and other extrascientific agendas? The practical question is whether science is often or usually value-ridden, the theoretical question is whether it is inevitably and inescapably so. Is it possible to do science without some prior moral values? Do the moral values held by scientists affect the objective validity of their conclusions?

The answers will depend on the theory of moral values which one adopts. It is not possible to identify "the correct" ethical theory, but it is possible to state the objections to theories from a neurobiological position. All ethical theories are open to some objections. In relation to neurobiology, realism, as defined in Section 1.2.4, poses the least serious objections and offers the greatest opportunity for integration of moral philosophy with neurobiology. By that I imply that ethical knowledge is fundamentally the same as scientific knowledge, that they are arrived at in similar ways and can be investigated by similar methods. I believe that the nature of ethical truth or falsity is the same as scientific truth and falsity: a proposition is provisionally true if it is not falsified by experience. Ethical realism is not subject to the critique leveled against ethical naturalism by G. E. Moore (1903, 1912, 1922, 1929). He defined naturalism as the view that ethical statements can be confirmed by scientific methods. That definition is inadequate because realism makes no claims to be able to confirm theories of any kind, but claims only to be able to construct theories that can approach progressively closer to reality. Ethical realism holds that ethical knowledge is like empirical knowledge, but it does not claim that ethical knowledge can be shown to be either absolutely true or absolutely false, because that claim cannot be made for scientific knowledge.

Things are said to have "values" when they are perceived to be in the self-interest of an evaluator, and when, by exercise of the evaluator's sympathy, they are perceived to be in the public interest. Conduct is said to have moral value when its effects benefit the evaluator primarily, and when the effects may benefit others to whom the evaluator's sympathy extends. Perceptions of benefits and interests are largely determined by culture. However, sympathy is founded on universal neurobiological conditions. That is what gives normative character to moral claims.*

What then are the advantages of integration of values and empirical neuroscience? From that integration we may learn not so much how to avoid error as the knowledge to deal correctly with it. We are not empowered to predict or preselect the empirical content of a future science. We have not discovered any laws of development which characterize the growth of scientific thought and we do not have methods which can be used to make predictions about the future of science. There is no theory of history which will enable us to predict whether a much greater understanding of the human brain will ultimately prove to be the greatest success or the greatest folly of mankind. Without increased understanding of ethical values, the advances of science are as likely to be used for bad as for good purposes. Without understanding how ethical values may function in scientific research programs we lack the knowledge to guide science to morally acceptable goals.

5.2.2. Definition of Values Affecting Science

We recognize three classes of values: practical values including commercial and economic values; aesthetic values which include values of the appetites, of pleasure, and of psychological interests; and ethical values which are concerned with good relations between the self and the nonself, and with the ultimate good of mankind. As I shall show shortly, these three classes of values unavoidably come into play in all scientific research programs, unless scientists work without morals, pleasure, profit, or any personal advantage. These values are already in the mind of the scientist

---

*There have been many different definitions of the meaning of "value." The reader who wishes to know what I understand by "value" should also know what I am reacting against. I do not follow the theory, originating with Aristotle, which says that values reside in things with respect to their "essences." This is considered further in Section 3.6 in connection with essentialism and theories of neuronal typology. I also reject both phenomenological and existentialist theories of value because they both, for different reasons, deny that ethical values have any relation to truth as science understands it, and therefore deny the possibility of ever understanding values in terms of evolution and neurobiology. I do not use the word "value" in the sense defined by A. N. Whitehead as "*the word I use for the intrinsic reality of an event*" (*Science and the Modern World*, p. 131, 1925). According to Whitehead, novelty is not a mere reconfiguration of the old but the creation of new intrinsic value.

before starting to do science, and I doubt whether it is possible to do good scientific research without some prior values. An inner drive to know the world and the desire to apply that knowledge to improvement of human life, as well as the desire for fame and material advantages, are the common mixture of motives for the choice of science as a profession.

Most scientists would like to work on a program that gives useful results, other things being equal. There have always been some who would work on a program even if they thought that it might do harm. Few would prefer to work on a favorite problem if it seemed to lead to no practical advantage. Far fewer would say with Lessing that if God offered him the truth he would refuse it, loving the search so much. To gain knowledge for its own sake and in the process to improve one's character is part of a humanist ethos that is now regarded as quaint at best, and at worst as irrelevant and misguided in the world of scientific grantsmanship.

There are two large sets of problems of the relationship of neuroscience to values. The first is to attempt to define and explain values. This includes such problems as the extent to which values are intuitive or are learned; whether values are relative; how they have changed and whether they continue to change and possibly to progress. The idea of progress is a reflection of human hopes and aspirations at various times, not a theory of history (Bury, 1921; Ginsburg, 1953). The idea of moral progress is related to a belief in the perfectibility of human nature. That belief is at variance with the evidence of history (Lecky, 1894; Westermarck, 1906–1908, 1932; Dawson, 1929; Frankena, 1976). Therefore, the idea of progress cannot be a basic assumption of ethics constructed on the empirical model. Indeed, an ethical system built on empirical foundations must regard the belief in progress as one of the illusions that we construct to shield us from what we perceive to be the horrors of the present times.

The second large set of problems is how values enter neuroscience research programs and the modes of operation of values in those programs. Any scientific research can and should be judged and justified with reference to practical, esthetic and moral values. Practical values require little further consideration here, because their importance is well known and scientists usually attend to practical values first and foremost. Science asks the question, "Is it true?"; technology, "Does it work?" It remains for aesthetics to ask, "Is it beautiful?," and for ethics to ask, "Is it good?" The current rule is to justify a research program mainly or only on the strength of its practical applications as perceived by those whose money supports the program. Their first question is: "Can it be done, and what are its uses, advantages, and profits?" Second, they may ask: "What good and what harm will it do, and ought it to be done?" When the answers to the first and second questions are conflicting, the second question is often dropped.

The guiding moral values of science are, or should be, to search for the truth, to seek to do good, and to avoid doing harm. The intention to do good is not a sufficient guide to ethical practice. Because good intentions can result in harm it is necessary to admit the intention not to do harm. These moral values imply a certain ethical attitude, an ability and willingness to make value judgments, namely approval or disapproval of particular cases if the relevant facts are known. These values are not absolute in the sense that they exist without a mind to know them. They do not reside in objects but in minds that perceive them. They are not innate moral ideas but are learned, as we shall see in Section 5.6.

How can we know what is right or wrong if, as I contend, we do not have innate moral sense? If moral judgments are not founded on absolute moral knowledge they can only be calculations of probabilities of results being good or bad, based on impartial observations. However, complete impartiality is unattainable, and therefore personal prejudice, and interest in the outcome must be estimated and taken into the calculation of the predicted effects of actions (see Section 5.4). By right or wrong we mean approval or disapproval of the results of actions. If we could not show relationships between moral judgments and actions, we could approve of what is wrong and condemn what is right without ever discovering the error. Therefore, we corroborate the correctness or error of moral judgments by the predicted or actual results of the actions they judge. In that respect a moral judgment is a hypothesis open to corroboration or refutation as I shall argue in Section 5.3. Only by their actual or perceived effects can we condemn or approve actions. In some cases the effects of actions may be neither entirely good nor entirely bad, and the judgment may follow the doctrine of aiming for the greater good and the lesser evil. Of course, that is quite different from taking an ethically neutral position.

Whatever definition of ethical understanding one accepts, there still remains to inquire what reasons there are for science to embrace ethics. An ethical attitude toward nature forms as part of the development of consciousness as a process of knowing. Conscious experience, which includes scientific and moral theorizing, hinges on the connected processes of perception, memory, motivation and volition. Moral consciousness develops as the result of an ongoing process in which mental models are constructed and modified on the basis of experience, reflection, and communication, to explain certain relations between the self and the external world, especially as they relate to an understanding of virtue. In Chapter 1 we have seen that scientific theories are conceptual models made to explain phenomena in the external world as we personally perceive them, collect them in memory, are motivated to communicate, and collectively attempt to explain and to unify the conceptual models of different observers. This requires development of attitudes and intentions with respect to our perceptions of reality, and our ability to relate to the external world in a certain manner which may be called a scientific attitude and a moral attitude. This includes the ability to sustain a critical interest in phenomena; to use methods to observe their identities, relations, and patterns; to reflect on those observations, and to construe their meanings. The process is facilitated by an ability to overcome the subject-object barrier by entering into sympathetic relationships with the objects of experience. Sympathy is the basis of development of ethical consciousness, and sympathetic reflection enables us to cultivate science as a program which is intrinsically ethical.

Consciousness, which includes ethical consciousness, is based on the principles of synergy and consilience. The minimum conditions for development of consciousness are the synergy of self-awareness with awareness of the external world, and bundling together (consilience) of multiple brain representations of the external world and of the self to form a unified representation. This is a continuously active process which repeatedly recasts our models of reality in the molds of new experiences. An ethical attitude toward nature is part of the scientific attitude. The fundamental reason for ethics to enter science is that we and our minds exist only as part of the wider structure of nature. As a result we have the ability to sympathize with natural phenomena because we are one with them.

Practical needs for ethics to enter science arise because of the conflict between self-interest and public interest and the conflict between freedom and power that arise from possession of knowledge that can be used to gain practical advantages. The sincere desire to find the truth is the highest value in science but it leads to conflicting goals because the truth is both liberating and empowering. One of the functions of ethics is to reconcile the aim of gaining freedom (from ignorance, superstition, disease, etc.) with the aim of gaining power (over nature, over people, etc.). Another is to find ways of resolving conflicts between self-interest and public interest. Ethical understanding arises from reflection on the goodness of motives and intentions, and on the goodness of the consequences of actions. Motives are usually formed from self-interest although the declared intentions may speak of public interest. The private and public interests of scientists may be reconciled either with or without moral understanding. Private or public interests, or both, may be immoral or moral. The problem is whether it is possible to decide between morality and immorality without absolute axioms of ethical conduct.

Motives are the psychological conditions promoting action to a goal; intentions are the desired results of actions. But selfish motives may conflict with good intentions, and good intentions may have unforeseen harmful results. That is one reason why the intention to do no harm should have the same status as the intention to do good. The scientist's avowed intentions may be to advance knowledge and public welfare, but his motives may be to gain funds, promotion, and fame. Like others, scientists seek those because it gives them egoistic pleasure. Selfish motives can function as legitimate means if they do no harm and they stimulate greater efforts to attain legitimate ends. For utilitarians, motives that are aimed at obtaining egoistic gratification are justified if they are means to contributing to public benefit, which is regarded as the ultimate end of science no less than of politics. Of course there are mistaken notions of public benefit which are discriminatory, unjust, and inhumane. But those mistaken notions, justified by one absolute ideology or another, weigh most heavily against absolute ethics. How could they have been tolerated for so long if innate moral sense or the categorical imperative dictate the moral truth?

How the desire for personal security and the need to gratify the ego can be reconciled with public duty is among the main problems of the ethics of science. The ethical dilemmas that arise in connection with certain actions are well known but the effects of inaction are less well recognized. Both action and inaction may have positive, negative, or neutral moral results, meaning that they may do good, or do harm, or do neither. In any particular situation it is necessary to ask oneself whether action, and more especially inaction, has a positive or a negative moral result. In science, orthodoxy, intellectual timidity, and lack of originality are always demerits although they often serve short-range objectives and may thus be mistaken for merits. This raises the question whether an absence of demerits ever adds up to a merit. It may do so in politics but not in science. For example, politicians are elected because they know how to say what their audiences like to hear rather than what they ought to hear. But it is cowardly and hypocritical for scientists not to tell the truth only because that might displease some people. On the other hand the scientist may decide not to reveal the truth if it might harm some people.

5.2.3. Calculating the Values of Science

Classification of values into three types, namely practical, aesthetic, and moral, each with positive, neutral, or negative effects may be further developed to provide a calculus of values that can be applied to actual programs. Such a calculus was originally suggested by several people in the 17th century (Bredvold, 1951). Shaftesbury (1671–1713), who first used the term "moral sense," also conceived of a "Moral Arithmetic" in his *Characteristics* (1711, pp. 78, 93, 173). The concept was given a hedonistic utilitarian form by Francis Hutcheson: *"that action is best which procures the greatest happiness for the greatest numbers, and that worst which in like manner occasion misery"* (*Inquiry Concerning the Original of our Ideas of Virtue*, 1725, Section ii, pp. 177, 178, 189). Bentham and Mill in the 19th century developed the thought that moral judgment is based on a calculation of the probability of actions resulting in happiness or unhappiness. Such a moral calculus of probabilities applied to scientific research would give a measure of a research program's probability of doing good or harm and safeguard us from collapsing into the uncertainties of extreme relativism.

The ethical content of a scientific research program cannot be represented by exact numerical values. Yet there is a need for representation of such uncertain or even vague information, and a need to deal with it scientifically. There are now two well-known methods for handling inexact information, and for quantifying value judgements in such ways that they can be computed. The first is the Bayesian calculus of probability (Howson and Urbach, 1989). The other is Fuzzy Systems Theory (Zimmerman, 1985; Terano et al., 1992). These methods have been useful for making predictions from uncertain information. They may also be useful for diagnosis and prognosis of the ethics of scientific research, and for computing the effects of values on scientific research programs.

I am not going to propose a calculus that can be put into immediate practice in order to make ethical diagnosis and prognosis that would be objective beyond dispute. That is an undertaking that is far beyond my competence. I am only proposing that ethical judgments would reach a higher level of objectivity if they were put on a quantitative footing. The use of a moral calculus, or any other general moral decision-making procedure, would not eliminate the need for sympathy and for caring about others. It would not lead us to act as if we were computers for deciding moral questions because our subjective evaluations could never be excluded.

One way to approach the problem of subjective valuation is from the Bayesian theory of probability calculus propounded by Thomas Bayes in 1763 (Pugh, 1977) which admits the element of subjectivity into the calculation of probability as a criterion of rational action. A calculus of probabilities can be either objective or subjective. Probabilities are objective when they are determined by logic alone, independent of our subjective attitudes. Alternatively, because total objectivity is unattainable, it is necessary to admit the proper elements of subjective beliefs. According to the Bayesian view, the validity of a claim is continually re-evaluated by re-calculating its probability on the basis of changing information. The Bayesian method starts out with a prior distribution of information which will vary from person to person, depending on their subjective estimation. Then, as data accumulate, persons with open minds will tend to agree on a posterior distribution which is determined by the new data and not by the prior distribution (Edwards *et al.*, 1963, p. 527; Howson and Urbach, 1989).

Methods for making judgments under uncertainty cannot avoid disputes about alternative goals and their relative desirability. Such disputes go beyond simple calculations of the ratio of costs to benefits and other measures of efficiency. The diversity of religions and personal ethical predilections has to be respected. That problem would not arise if all people had the same intuitive understanding of fairness, justice and good behavior, and if we were all capable of exercising those ethical intuitions freely. But, in practice, we make different ethical judgments: not only because we are coerced by rewards and punishments and because we differ in our perceptions of desirable goals, but also because we learn different standards of ethical behavior.

5.2.4. Can Science Be Value-free?

The widespread notion that values should be excluded from science arises from the belief that values are emotional and that therefore they are in conflict with reason, which many believe to be the only secure basis of philosophy and science. That is a vestige of the stoic doctrine that all emotions enslave the soul and should thus be brought under the control of reason. The doctrine persisted in Descartes' final work, *The Passions of the Soul* published in 1649, which concludes with the assertion that "*the chief use of wisdom lies in teaching us to be masters of our passions and to control them. . . .*" In ancient China the idea that the sage should not be ensnared by emotions was held by Confucianist and neo-Confucianist schools of philosophy (Fung, 1947, 1948).

That all values are emotive, not cognitive, is one of the tenets of logical positivism that has become widely accepted and has reinforced the present ethos in which scientists suppress the value content of their work. "Meaning" was defined by logical positivism in terms of the "verification principle," first formulated by Wittgenstein and members of the Vienna circle in the late 1920s: "*The meaning of a proposition is the method of its verification*" (Carnap, 1932; Schlick, 1936). A second way of giving meaning to a statement is Popper's falsification criterion of the empirical value of a hypothesis. Arguments for and against that criterion are given in Sections 1.3 and 1.4. Like the verification principle, the falsification principle excludes values from the criteria of validity of a hypothesis.

The verification principle requires that a statement is regarded as meaningful only if it is possible to give evidence for or against it. In logical positivist terms, meaningful statements are either empirically verifiable by direct observation, or analytically verifiable by analysis of the meaning of the words or mathematical symbols. The meaning of a statement is a function of its empirical content either directly in the form of sense data, or indirectly in terms of effects that can be observed if the statement is true. Logical positivists regard as meaningless or metaphysical all statements that do not pass the verifiability test. By that test, statements about values are rendered meaningless (Schlick, *Positivism and Realism, Philosophical Papers*, Vol. II, p. 270, 1925–1936; Ayer, 1946). This argument is now commonly used against admitting values into scientific research.

Now it is worth mentioning that statements can be made that have meaning about feelings, beliefs, demands, and promises but do not pass the verifiability test. "I believe in God," "I like being a scientist," "I promise to tell the truth," and "I am concerned about the lack of values in science" are such statements, called emotive statements by Ogden and Richards (1927) in contradistinction to cognitive statements. The words *emotion* and *emotive* are open to various interpretations, and there are several theories of emotional development (James, 1884; Izard, 1971; Panksepp, 1982; Plutchik, 1980; Sroufe, 1979; Campos and Barrett, 1984; Frijda, 1986; Izard and Malatesta, 1987; Leventhal and Scherer, 1987; Oatley and Johnson-Laire, 1987; Fogel and Reimers, 1989; LeDoux, 1989, 1992; Harris, 1989), and theories of the role of emotions in making value judgments (Gardiner *et al.*, 1937; Black, 1948).

Eventually logical positivism ran afoul of its own narrow definition of meaning in terms of verification of individual statements. Quine (1953) pointed out that the empirical consequences of a theory are consequences of the theoretical matrix in which individual observational statements are embedded, not consequences of any single statement. Therefore, the meaning of a statement will change as the entire theoretical matrix changes. After the decline of logical positivism in the 1960s, its legacy was the continued enthusiasm for logical analysis of empirical data and the continued suspicion of emotive statements, and thus a fear of allowing values to enter pure science.

The distinction made by positivists between emotive and cognitive is parallel with that made by some existentialists between two modes of existence called "non-being" and "being-in-itself" by Heidegger (1957, *Sein und Zeit*, pp. 59–61, 167). In that theory, man has ethical choice that is entirely distinct from truth as conceived by science: valuation belongs to the mode of existence called "non-being," whereas cognition is in the mode of "being-in-itself." There is no place here to do justice to existentialist theories of value, but I think it fair to say that Heidegger conceived of science as fundamentally deficient in the means of apprehending values and therefore believed that the separation of science from values is inescapable. Heidegger certainly never envisaged a world in which ethics could be fully integrated with scientific research.

The unstated assumption behind the doctrine of value-free science is that scientists can remove their values as easily as their overcoats when they enter the laboratory, and, so unencumbered, can proceed to do the right things, guided by pure reason. But the search for truth is not a zero-sum game in which loss of values always means the victory of truth. The paradoxes that follow from those assumptions seem to have escaped scientists who advocate value-free science. How can we know the difference between right and wrong without reference to a system of values? How can we cast off values which are learned by early experience, still less those that some people believe to be given by innate moral sense? Is it not more probable that our values will be concealed and continue to function covertly in scientific research? If so, the efforts to make science value-free in order to allow reason to be the sole guide to the true and good will fail, and may play into the hands of those whose values, carefully concealed, lead to the misuse of science. Because of the risks of concealment of values we should learn to expose and define them, and only then allow them to perform appropriate roles in science.

How far some influential scientists are from understanding these problems is shown by the publication of a report *On Being a Scientist* (1989) by the U.S. National Academy of Sciences, which has the following statement on *"Values in Science"*: *". . . .social and personal values unrelated to epistemological criteria—including philosophical, religious, cultural, political, and economic values—can shape scientific judgments in fundamental ways. . . . The obvious question is whether holding such values can harm a person's science. In many cases the answer has to be yes. The history of science offers many episodes in which social or personal values led to the promulgation of wrongheaded ideas. For example, past investigators produced 'scientific' evidence for overtly racist views. . . . Attitudes regarding sexes can also lead to flaws in scientific judgments. . . . Conflict of interest caused by financial considerations are yet another source of values that can harm science. . . . The above examples are valuable reminders of the danger of letting values intrude into research."* This curious statement is derived from the outmoded notion of a value-free science. The academicians imply that exclusion of bad values from research programs will allow the scientists' benevolent impulses to be channeled toward the common good. The academicians' confusion of positive and negative values is like an accountant referring to a debit as a credit, or a banker referring to a withdrawal as a deposit.

One of the main reasons why the National Academy of Sciences proclaims the advantages of value-free science is that it has undertaken the role of mediating between science and government. The doctrine that the role of science is to seek the truth objectively, independent from extrascientific influences, appears to free science from values and thus to sanitize and neutralize science with respect to religion, commerce, and politics (Proctor, 1991). The hypocrisy of this doctrine is not hard to see: science would like to be well supported by government and industry and would also like to appear to be seeking the truth objectively, without bias resulting from political or commercial advantages. It is curious that the National Academy report avoids any reference to research on weapons while raising the other old moral dilemmas of race, sex, and money. They might have used the old argument about just versus unjust causes, but in doing so they would have exposed the weakness of their argument, not least that, like sex, money, and race, power also has a moral dimension which can only be ignored at some risk. The National Academy's report misses the essential point that since values cannot be excluded they must be clearly defined, exposed, and subjected to critique. A calculus of values would help to determine their effects in each particular case.

Values are inescapable components of scientific research programs, and they cannot be excluded by adoption of technical and logical procedures, as positivists have claimed (Braithwaite, 1953; E. Nagel, 1961; Hempel, 1965a,b). Even if there were some scientific programs in which values could be concealed with negligible effects, there are demonstrably many others from which values ought not to be concealed, even if it were possible to do so. Investigations of human intelligence, of psychiatric illness, and of the genetic control of brain development and function are some neuroscience research programs in which empirical knowledge is insufficient to guide them to just and humane applications—they require ethical guidance. For this purpose it is necessary for neuroscience to acknowledge its ethical content. Undeniably, certain values may obstruct progress of science, just as other values may enhance and justify scientific research. Therefore, while striving for an objectivity that can only be approached but never attained, values must be identified, subjected to analysis, deliberated and reflected upon, and thereby allowed to play appropriate roles in scientific research programs—guiding them to just and humane goals.

5.2.5. Can Scientists Claim Diminished Moral Responsibility?

Dissociation of the technical goals of science from ethical goals is dangerous because it opens the way for scientists to claim diminished responsibility. It is obvious that a scientist cannot perform well without holding special practical values relating to research methodology. A concern about the quality of experimental techniques is necessary in order to obtain accurate data. Some people would claim that scientists have fully discharged their moral responsibilities if they diligently maintain certain technical standards. From that position the moral values of science can be guaranteed by scientists only insofar as they relate to the special practical values of scientific expertise. Scientists who are lazy, careless or otherwise negligent in doing research, or who falsify data, negate the practical values of scientific research expertise. Failure to observe those values calls into question the credibility of science. This is the main reason why scientists are held responsible for upholding their professional values, doing good science, even when their science is morally bad, or when the general ethical values of science are perverted or ignored.

According to that view scientists cannot be held responsible for the general values of science—to search for the truth, to do good, and not to do harm—because those values are unattainable, or are too vague, or are contingent on many social, political, and other conditions, to be of practical significance for the scientist working in the laboratory. From that position, upholding those general values appears to be the responsibility of higher authorities and society in general. In that case, scientists may be free to uphold the general values as members of society, but are not at liberty to admit general values into scientific research. Apart from the absurdity of this requirement for scientists to have two separate moral standards, that argument opens the way for scientists to claim diminished moral responsibility.

The plea that technical experts acting under higher authority have diminished responsibility has been used by scoundrels throughout history. While I doubt that normally intelligent adults can have diminished moral responsibility, I admit that they may be coerced into acting immorally and that coercion may be a mitigating factor. Coercion is most effective on people who have separated their technical from their moral responsibilities, for then they can more easily be coerced into a state of moral amnesia without diminishing their technical competence. The actions of Nazi concentration camp doctors and of psychiatrists in Soviet prisons for political dissidents can be understood partly in terms of coercion of the doctors by higher state authorities. Those doctors have pleaded that they were not accomplices but victims of the system. While their crimes were incomparably greater than scientific fraud of the kinds that have recently been publicized, it is important to recognize the role of coercion in those cases too, and to consider the effects that coercion can have on scientific research.

In "normal" science, coercion takes the familiar forms of promises of various advantages such as promotion and research grants, and the threat of their loss if scientists fail to perform, some may say overperform, competitively. This form of coercion is particularly insidious because extreme pressure placed on scientists to outperform competitors has become a conventional part of scientific research programs. Because the coercion can rarely be traced directly to a single superior authority within the program, it permeates the entire system. Therefore, it protects the superior authority while increasing the vulnerability of the subordinates. This is borne out by the fact that in some recent cases, relatively junior scientists have been punished for scientific fraud but their program directors either have been absolved from direct responsibility or have received no more than mild reprimands. The plea of diminished responsibility has been made by both parties: the subordinates claiming that they were under extreme pressure to produce certain results, and the superiors claiming that the organization of a large research group makes it impossible to monitor the standard of performance of the subordinates. Both miss the point that the real problem is not in the organization or research group but in the moral values of all parties: not so much a failure of communication of empirical knowledge as a failure to share moral knowledge, and to engage in critical dialogue.

5.3. "MORAL RULES NEED A PROOF; ERGO NOT INNATE"

The title of this section is the title of Section 4 of John Locke's *Essay Concerning Human Understanding* (4th Ed., 1700). The whole of moral philosophy from the 17th century to the present lies in the shade of Locke, who tried to refute the idea of innate ideas and specifically of an innate moral faculty which guides us to do good. Locke agreed with Hobbes that our understanding of good and evil is no more than what we learn to associate with pleasure and pain. A consensus about good and evil can be reached because all people are similar in faculties and nature. That was Locke's basis for accepting natural law as expounded by Grotius and his disciple Puffendorf.

The doctrine that all moral knowledge is learned has to account for the uniformity as much as the diversity of moral beliefs. In Section 5.6 I propose that moral beliefs tend to uniformity because they are learned under the relatively invariant conditions of interactions between the mother and young infant, starting *in utero* about three months before birth. Experience of the mother's benevolence gained during fetal development and infancy is expressed as sympathy later in life with such spontaneity that it has been mistaken for the expression of innate moral knowledge.

The main reason for questioning the validity of extreme moral relativism is that the initial human experience that leads most people to agree on what is right, fair, and just is the experience of maternal benevolence during infancy, and the later generalization of that experience. The child's development of sympathy is an important part of its emotional development that is fostered by social interactions. If moral values are learned by most people under relatively uniform conditions during infancy, and if those early experiences result in patterns of moral beliefs that continue to be expressed in later life, the moral beliefs of most adults will tend toward uniformity *regardless of the variability* of their adult experiences. The alternative is that a person's moral beliefs change continually in response to various sociocultural conditions. If that were correct, moral beliefs would always adapt to the prevailing conditions. But the opposite is observed: people's moral beliefs frequently conflict with social, cultural, and scientific programs. Most adults do not adapt their moral principles rapidly to changing circumstances unless they are coerced.

5.3.1. Moral Statements Are Hypotheses

The classical deontological, intuitionist, and utilitarian ethical theories are based on axioms that each hold to be self-evident, from which particular ethical statements can be inferred.* Ethics has been treated as a pure deductive system having a few unalterable axioms from which deductions can be extended infinitely downwards. Although they start with different axioms, they all admit a moral statement only if it can be derived from axioms, using accepted rules of deductive logic by which the statement is made on the stated axioms. Instead of deduction downwards from axioms, natural science stands on a base of empirical observations on which hypotheses are formed by synthetic inferences (inductions) as well as by analytic inferences (deductions). Induction is the inference of a hypothesis from empirical evidence for it, or the inference of a generalization (a higher-order hypothesis) from its premises. The latter may be true but the inductive conclusions false. This is the opposite of deduction in which the conclusion is a logical consequence of the premises, and therefore the conclusions are true only if the premises are true.

Intuitionist moral philosophers all hold as axiomatic that there are a number of intuitively self-evident moral principles such as benevolence, veracity, prudence, fortitude, temperance, justice, and so on. These are accepted primarily because they seem to be intuitively true, not because of their utility, although they may also be useful, whereas some criteria of optimization of social and personal welfare are the basic axioms of utilitarians. Instead of deduction down from those axioms, it is also possible, and I believe preferable, to build an ethical system on an empirical basis. This may start with a hypothesis which is an inductive inference. The initial inference may be false and cannot possibly be ideologically impartial. All that can reasonably be required is to estimate the strength of bias in each case. But bias can be corrected by adequate disclosure and criticism, leading to acceptance or rejection. This process allows revision and development of moral understanding as new information is obtained and new answers, based on evidence, are either rejected or accepted.

Both in science and in ethics the first hypothesis that is formed on the basis of preliminary data is rarely supported by new facts. In both, inductive inferences are

---

*The universality of these preoccupations with the nature of moral understanding can be shown by the fact that the ideas of intuitionist and utilitarian schools of European moral philosophy have counterparts that have developed independently in Chinese philosophy (Fung, 1947, 1948; Chan, 1957, 1963).

made in accordance with the principle that the accumulated evidence is sufficiently good to support a theoretical model that is explanatory and on which reliable predictions may be made, as I have argued in Chapter 1. The criteria for "sufficiently good" includes technical procedures such as adequate controls, correct statistical evaluation of data, and impartial evaluation of data in relation to alternative hypotheses. But it is generally admitted that subjective factors enter into the choice of reputable controls, adequate statistical methods, and the evaluation of data in relation to rival hypotheses. Therefore, some further tests of validity are needed to justify inductive procedures. Rules of inductive inference may be justified because they lead to hypotheses that have simplifying, explanatory, and predictive value—i.e., by what Peirce (1878) called "truth producing virtue," and Braithwaite (1953) called "predictive reliability". Justification of a moral proposition or claim lies in its problem-solving powers, namely its explanatory and predictive value and does not preclude its refutation at a later time. For an inductive procedure to succeed in science as well as ethics it is not necessary that all or even a majority of its hypotheses remain unrefuted. It is sufficient that some unrefuted hypotheses continue to explain some observatons and to make reliable predictions, and that those hypotheses are preferred to others that fail to explain the data and make unreliable predictions. Hypotheses will be retained because they make predictions that help to solve problems.

Science and ethics are both conducted by making hypotheses and then calculating their probability in the light of empirical data. Many historical examples can be given to show that the data accepted into calculations of probabilities of hypotheses in both science and ethics are not determined by purely objective criteria but are biased by the various values held by different people. Because purely objective criteria of truth are unattainable, it is necessary to identify subjective factors which enter into explanations. For the present we shall confine the discussion to ethical values although similar arguments may be applied to aesthetic and practical or utilitarian values. To enable ethical values to function in guiding science toward humane goals, they must be brought into the empirical context of science as hypotheses from which generalizations can be inferred. Ethical values may then be integrated with concepts and techniques into a complete research program in which ethical values may guide the program toward just and humane goals by an internal dialectical process. If conflicts arise in a research program between alternative interpretations of means and ends, and even about the meanings of "just" and "humane." they may be resolved by reiterative self-reflection, open communication, and critique, eventually pointing the way to effective action, allowing the research program to progress.

Moral statements cannot be made under conditions of impartiality. Anyone can understand this who understands the complexity of the economic, social, political, ideological, and cultural domains in which we are embedded. The fundamental problem is to explain the partialities. To do so it is necessary to regard "good," "duty," and other moral words and concepts in terms of those whose interests they serve and whose survival they promote. Those adaptive strategies may be reduced to psychological processes and may be reduced further to physical processes in the brain. The feasibility of such a reductive method and whether reduction entails identity, isomorphism, correspondence, correlation, or causal relationships are discussed in Chapter 2.

If I ask whether ethical theory can be reduced to neurobiological theory, I must specify how the terms of the reduced and reducing theory are related. Ethical-to-neural reduction is difficult because it is a cross-categorical reduction, without synonymous theoretical terms. To avoid that difficulty, the terms of the reduced theory can be altered so that they can be logically deduced from the terms of the reducing theory (Hooker, 1981). That may be done by treating statements about moral obligations and duties in the same terms as statements about empirical observations. For example, we can ask what conditions are required for verification or refutation of moral statements. Such questions are regarded as unanswerable by nonrealists in ethics. They hold that moral knowledge is obtained by a process utterly different from any by which we obtain knowledge of the physical world. It follows from that doctrine that moral statements are fundamentally different from statements about empirical facts because moral statements involve duty, in the use of the word "ought." To take an extreme example: "Thou shalt not murder!" is a categorical commandment which is beyond dispute in terms of verification or refutation. I do not propose to discuss the functions of commands in moral discourse, but I contend that moral commands cannot enter scientific discourse because commands have no function there.

For a moral statement to enter scientific discourse it may either be in the form of a hypothesis, or may function as a question, or may function like a tentative condition-action rule that can replace a command. For example, instead of the imperative "DON'T do such-and-such," I may ask, "IF such-and-such is done, THEN what may follow?" Or I may conjecture, "IF such-and-such is done, THEN so-and-so may follow." Condition-action rules can be made when the initial conditions and the final goals can be defined: "IF the conditions are such-and-such, THEN do so-and-so" to achieve a defined goal. Computers may be able to solve moral problems by using condition-action rules, and can modify their performance under various conditions in relation to achievement of specified goals of a scientific research program. The complexity of both the conditions and the actions can vary greatly in such a condition-action rule (J. R. Anderson, 1983). It is easy to think of examples that are nonsensical. However, nonsensical cases are eliminated because whether the rule is obeyed depends on how it performs in competition with other rules in making predictions about the attainment of goals. Rules that make predictions that are relevant to everyday life situations, that solve ordinary moral problems (not problems posed by philosophers only for the sake of argument), will be facilitated and used again in the future, whereas rules that lead to error will be modified or discarded.

From the foregoing it follows that the imperative "ought" can have no absolute validity—it can only mean that something is desirable and is recommended from evidence of what it has been and from predictions of what it is likely to be. Critics may say that a belief in ethical realism may lead to moral respectability, but not to the absolute moral virtue that is given by the Kantian categorical imperative. My answer is that moral respectability is the best that is possible in a world in which knowledge of absolute virtue is unattainable, and in which the actual and predicted desirable effects of human actions are always uncertain and can be demonstrated only to some degree of approximation. This kind of moral uncertainty about how to reach the moral goal leads to humility and tolerance, which are preferable to the moral arrogance and intolerance of many who are absolutely certain of the correctness of their path and of the universality of their duties. Belief in absolute values and abstract virtues lead more often to immorality than to morality—men kill one another for them. Nothing is calculated to weaken ethical theories more than the adoption of fundamental axioms that are held to be absolute because they appear to be self-evident. In ethics, as in science, self-evidence has proved to be illusory so often that it can be no more than the first of many independent evidences which only gradually converge on and may approach the truth, never arriving at finality.

Ethics is one of the topics that has been considered to be beyond the reach of scientific inquiry and therefore metaphysical, but there are historical examples showing that such topics have been included in empirical science. For example, the inside of the atom was considered to be a metaphysical subject before Rutherford proposed the nuclear model in 1911. Similarly, good and evil will remain metaphysical subjects until they are regarded as states of mind, and are brought into the domain of neuroscience. As it is now widely understood, ethics is said to include a metaphysical element, namely the definition of good and evil as ends in themselves, beyond the requirement for biological adaptation and even beyond the requirement for mere survival. In that view, metaphysical choice of goals supervenes completely on genetic determinism. The great objection to it seems to me to be that ethics is not free of biological necessity because ethical understanding is based on neurobiological processes which predispose people to attribute goodness and badness to things and events in their experiences.

My account of ethics is that our attribution of badness and goodness to all objects and all events is like our attribution of greenness and sourness to unripe tomatoes and of redness and sweetness to ripe tomatoes—it is no more than that the object is disposed to produce the appropriate sort of perception and experience in a relevant class of observers under relevant conditions. This is a dispositional account of experience first proposed in 1690 by John Locke in *An Essay Concerning Human Understanding* (II, viii). Obviously, experience is formed from perception of relations and from the context as well as from distinctive properties of objects. A single property like redness may not be enough because sensory or neurological defects affecting perception of that quality may prevent identification of objects unless the objects have additional relevant properties that are sufficient to dispose observers to identify them and choose between them.

Now let us consider a hypothetical class of observers who have neurological processes that have the effect of making them ascribe goodness to objects and events to which normal observers ascribe badness, and vice versa. All such observers who are inherently constituted so as to share the same experiences of objects (perverted though their experiences may be to normal observers) will arrive at a consensus that is the opposite of the normal consensus about attributions of goodness and badness. This hypothesis is raised here as an objection to the notion that objects have intrinsic values that are different from subjective values, as argued by G. E. Moore (1922). The truth seems to me that values have some fixed relations to the minds of the observers and are not intrinsic to things or events outside the observers.

Moral norms and the definitions of good and bad vary in different societies, yet, for reasons given in Section 5.6, there are general ethical criteria that are accepted in almost all societies. These criteria only appear to be "objective" because many people experience similar conditions of development of morality during infancy and because all people have the same neurological processes which predispose individuals under relevant circumstances to take a moral view of things and events. This moral attitude requires a person to have the intention to benefit and not harm others; it requires a person to perceive the possibility of alternative modes of conduct with respect to alternative moral goals; and may, in addition, require justification of actions in relation to societal benefits. For such a person, a scientific research program is ethical only if it actually or potentially does no harm and also promotes the interests of others in addition to those of its proponents. But some may adopt the attitude that a research program is ethical as long as it does no harm, regardless of who benefits from it. The question then arises whether selfishness is unethical even if it has no significant effects on others. To what extent can such questions be answered by recourse to a moral sense alone, or in conjunction with moral rules, or can they be answered only with reference to experience and empirical evidence?

Some influential philosophers have tried to convince us that morals are innate. Others have argued that morals are learned. Both arguments lead equally to an optimistic or pessimistic prognosis, depending on one's preconceived ideas about the human condition. Anyone who has followed the nativist-empiricist or nature-nurture debate will end up with a bad headache, and a diminished confidence in the ability of philosophers to understand human nature. This is one reason for my conspicuous absence of enthusiasm for purely philosophical solutions to the problem of the biological origins of morality, and the problem of the relation of ethics to science. Therefore, there is a strong urge to disregard the philosophical arguments and to embark immediately on a purely experimental research program. It seems to be simplest to blow up the whole of neurophilosophy and start all over again. However, I doubt whether that is advisable as a first move, although it may finally prove to be the best strategy. To my way of thinking, it is best to begin by examining the philosophical arguments before embarking on a purely experimental research program about the problem of the origins of morality and of the relation of morality to scientific research. Some of the philosophical arguments might well be adopted or modified to form the theoretical components of neuroscience research programs.

The problem of whether science and philosophy can be built independently or whether they are necessarily interdependent has not been solved. I have taken that uncertainty into account in questioning the validity of the claims of philosophy to be able to give an adequate explanation of ethics or of any other product of the human brain, including philosophy itself. Products of the human brain, including philosophy, literature, works of art, and science, are objects of interest to empirical neuroscience. Products of human brain processes cannot be understood adequately without the empirical evidence of neuroscience. In Chapter 2 I have given the main arguments in defense of the assumption that brain processes are related to mental processes which are related to philosophy, art, science, and so forth, but we remain uncertain about the nature of those relationships. That problem cannot be solved by either neurobiologists or philosophers alone. A meeting of minds of philosophers and neuroscientists would be enormously fruitful, but bringing about the union is like trying to mate pandas.

5.3.2. Doubt and Certainty in Ethics

Scientific doubt and ethical doubt both lead to relativism: the status and fate of a hypothesis are contingent on future conditions that are unknowable, and on data that can be known only to a degree of probability. If doubt and uncertainty is the condition of empirical knowledge, how can ethical knowledge be different? This is one reason why I am prepared to meet ethical relativists halfway. In the search for truth one is inevitably forced to change one's ethical as well as scientific concepts from time to time. This is the consequence of any honest effort to search for the truth wherever the search may lead. A change of concepts can never be easy; and it is made with anguish when others regard it as a sign of weakness and lack of conviction. Then it becomes necessary to show one's colleagues that a change of concepts is not a sign of opportunism.

Similar decision rules are used in the choice of one concept over another in science and in ethics: in science and in ethics the consistency of a theoretical concept with new results is the principal rule. The assumptions underlying the decision rules used in the choice of alternative scientific theories are that we can, with some degree of probability, know the difference between true and false and that we are motivated to discover the truth. For ethical theories the fundamental assumptions are that we can learn to understand the difference between morality and immorality, and that we are normally motivated by benevolence, not by malevolence.

For a concept or theory to be ethical it must promote the interest of others besides or in addition to the interest of the people who advance the concept. Thus, sympathy leading to unselfishness is a necessary condition for the advancement of ethical concepts. Sympathetic reflection results in the general recognition of reciprocal obligations and of equal rights. Those are also necessary conditions for advancement of scientific research programs in which open communication and critique have free play. The alternative is for scientific research programs to limit communication and critique and to be excessively encumbered by legal contracts, patents, and other manifestations of self-interest and mutual distrust.

Ethical experience has the status of evidence for testing ethical theories, just as empirical observations have validity as evidence for testing scientific theories. The choice of one of several alternative ethical theories is made on the consistency with which they explain how certain means lead to more or less beneficial ends. It is assumed that beneficial ends are desirable and that some ends are more beneficial and less harmful than others. It cannot be assumed, but must be ascertained, that judgments based on ethical theories actually or potentially result in more benefit or less harm. The success of an ethical theory can be measured by its effectiveness in making judgments that have beneficial results. Of course, this rests on a foundation of assumptions about what are beneficial or harmful effects of human conduct. Those assumptions are unavoidable but not unalterable. They can only be justified, or modified, or invalidated, on the evidence of experience.

Ethical theories are progressively modified by assimilating more practical experiences that conduce to greater good. The test of validity of an ethical theory is its consistency with practices that are beneficial. Ethical theories are invalidated if they are consistent with practices that are harmful. Because scientific theories are not invalidated if they are harmful, science on its own is unable to determine whether scientific progress is harmful or beneficial—that is an ethical judgment.

5.4. THE PROCESS OF ITERATIVE REFLECTIVE JUDGMENT

As I have contended from the beginning of this work, hypotheses form as part of the development of consciousness. Both scientific and ethical hypotheses are mental models made to explain phenomena and to define and to reconcile relations between the self and the outer world. Ethical models are constructed by connecting learned moral rules with personal experiences in ways that are perceived to promote personal preservation, and may also promote the preservation of others. This process requires development of attitudes and intentions that enable a person to respond adaptively to experiences. Among those are attitudes and intentions that function in carrying out actions that are defined as moral. All definitions of morality are relative to different material conditions and historical traditions under which individuals pursue their self-interest and different people band together to promote their common interests.

The conflict between self-interest and the common interests results in hypocrisy and other well-recognized moral absurdities "*to which no living creature is subject, but man himself*" (Hobbes, *Leviathan*, Chap. 5). But our distinctly human privilege is the ability to sympathize and empathize, which are the fundamental attitudes required for moral behavior. Sympathy and empathy enable us to overcome the barrier between self and nonself, and thus to derive the basic moral principle that we would rather do to others what we would have them do to us than the alternative. This attitude is built into our mental model of the relations between self and others, as a result of experiences during infancy and childhood, as the paradigm of benevolent relations between mother and child. I consider this at greater length in Section 5.6.

Some definitions are now in order. Sympathy is used throughout to mean the ability to respond in a mimetic manner to the mental state, feelings, and emotions of another person (the ability of B to share A's feelings). Sympathy is primarily based on sensation and emotion, but may secondarily acquire a cognitive component after reflection on alternatives. Empathy is the ability to enter into another person's mental state (B shares A's mental state through feeling as one with A). Empathy requires inferential knowledge of other people's minds, and such knowledge is severely limited, as we have seen in Section 2.10. There I argue that I do not have intuitive knowledge even of my own mind and that knowledge of my own mind is inferred from my experience, just as I infer that other people have minds. Sympathy and empathy are feelings *with*, but pity is a feeling *about* another person in distress. Sympathy and empathy are also a person's emotional identification with nonliving things and with natural phenomena, which are part of the scientific attitude as well as ethical and aesthetic attitudes.

The process of forming a mental model in both science and ethics is essentially the same, as I have said earlier in this chapter. Here I reformulate that process in terms of mental processes involving self-reflection, motivation, and emotion in addition to reason. Thus, the similarity between scientific and ethical problem-solving will become clearer, and we shall see how integration of ethics with science may proceed.

Forming a mental model, in science or in ethics, is a process that starts with self-reflection, often based on sympathy and imagination, made before all the arguments have been examined rationally. A preliminary judgment is progressively revised in the light of subsequent arguments and additional evidence. The process of reiterative reflection-communication-critique has the general features of problem-solving strategies that use successive modifications of a model to adapt it to perform a task more effectively. The process of reiterated reflection, communication, and critique continues until, ideally, a consensus is reached. Common consent is rarely attained in practice, or is delayed because evidence is inconclusive and arguments are unpersuasive. The delay has heuristic advantages, preventing premature judgment and maintaining interest. A problem is no longer of interest to the scientist when it has been solved or appears to have been solved, although its mode of solution may be of interest to the philosopher and historian of science.

Self-reflection is a process whereby we are enabled to form conscious mental constructs from both the rational and emotional components of our previous experiences. I recognize three components of reflection: rational, imaginative, and sympathetic.* We form all kinds of judgments—scientific, ethical, aesthetic—by similar processes of reflection. Perhaps we exercise sympathetic and imaginative reflection more, and rational reflection less, when making artistic than when making ethical or scientific judgments. Yet, the initial inductive step, the point at which our attention is directed at a problem, and at which we decide to reflect on it, is the same in all three cases.

Reason predominates in the process of rational reflection in which components of experience are brought into logical relationships and are compared. Reflection is a means of comparing ideas and thus of choosing between them. This is in accord with Hume's argument that reason is capable only of comparing ideas. Rational reflection is a dialogue with oneself for negotiation and mediation between alternative positions. Reflection, followed by open communication and critique, is the preferred method of resolving conflicts between people holding different theories or different values. Purely rational reflection may not be possible and may not be sufficient: when alternative positions appear to be rationally irreconcilable, the sympathetic component of reflection may help to prolong the dialogue through many iterations until new information and deeper understanding enable a synthesis to occur, leading to progress.

In advocating self-reflection I am certainly not advising total withdrawal from the rational contemplation of things into flights of fancy or into some sort of mystical detachment. I am suggesting that cognition cannot be separated entirely from emotion, even at great expense of effort, and that a creative state of mind requires the interplay of reason and imagination, of cognition and emotion. A distinction has to be made between mere fancy and creative imagination, possibly in the manner of Coleridge (*Biographica Literaria*, Chapters 13 and 14, 1817). According to him, fancy is a "mode of memory" in which sense data are recombined arbitrarily, as in dreams, but imagination is an active faculty that arranges sense data to form novel constructs. Imaginative reflection draws on

---

*To see human nature as some combination of faculties like imagination, emotion, will, and reason, has the obvious disadvantage of formally separating things that do not exist separately at any time in any person. That view is only a useful simplifying device that is hard to abandon completely.

memories of previous experiences to form new ideas. The creativity of imaginative reflection arises from rearranging components of previous experiences into meaningful new patterns. In passing, I should point out that an artificial intelligence system programmed to operate logically could possibly exhibit rational reflection, but not imaginative or sympathetic reflection (Dreyfus, 1972).

I am not only saying what the methodology of science and ethics would be like if the process of reflection-communication-critique were to be put into practice, I am also claiming that this process is what scientific and ethical problem-solving is really like. According to this theory, ethical knowledge is like empirical knowledge, but it does not claim that ethical knowledge can always be shown to be either true or false, because the theory does not make that claim for scientific knowledge. Progress of science and ethics occur in essentially the same way—by obtaining evidence from experiments and by submitting the evidence to repeated reflection, communication, and critique, using the well-known procedures of induction and deduction. But those procedures are conventionally regarded as completely rational. Here I emphasize that this problem-solving process involves motivation, intention, and sympathy in addition to reason. Motivation implies a goal, intention specifies a goal, and choices of goals are ethical choices when they go beyond the limited technical goals of science.

Conscious differentiation of our values requires critical self-reflection. One's own judgment of one's motives and intentions is notoriously unreliable. Self-reflection is a mirror in which our moral blemishes tend to be smoothed out. We are unable to examine ourselves as other people would examine us. In order to avoid mere validation of our prejudices, self-reflection has to be done with reference to other people and to external conditions. Without such external references, self-reflection tends to degenerate into total subjectivity, becoming self-fulfilling prophecy. In order to avoid mere ratification of one's own prejudices, self-reflection must be followed by communication.

Communication is a link in the circle of repeated reflection, communication, and critique. It includes the oral and written communication of research methods and data, the peer review system, the academic promotion and tenure system, and the other negative feedback mechanisms which are ostensibly designed to reduce error and to ensure the survival of only the fittest forms of scientific life. But for those purposes we require less feedback and more critique. The person who is able to exercise critique has the element of doubt necessary to make a good scientist and the degree of honesty necessary to be genuinely critical.

Critique and feedback are quite different structurally and functionally. Feedback involves the return of some part of the output back to the input, either adding or subtracting it from the input but without addition of anything new. Constructive critique aims at strengthening the system by selectively augmenting its strengths as well as by eliminating its weaknesses. Negative feedback is built into the central structure of the system; critique comes from eccentric positions, from outside the system, or at least from outside its central structure. Feedback mechanisms such as the referee and peer review systems are anonymous; they function to restore the system to status quo; they tend to act as forms of coercion, reducing the degrees of freedom of expression and action. Critique is open: it functions to reorganize the system; it suggests and demands additional degrees of freedom. At the level of action now prevailing in neuroscience there is little critique in which scientists critically examine their concepts and their cognitive and associated intellectual processes and basic assumptions, not only as they are now held but as they were formed by historical forces, and have become value-laden.

5.5. ETHICAL CONDUCT FOR REACHING CONSENSUS

Scientists with access to the same data, but with different world views and historical perspectives, tend to arrive at different conceptual positions and different value goals which, if fully disclosed and adequately communicated, enrich dialogue leading to creative consensus. This is the dialectic by which scientists reach increasingly higher levels of conscious understanding and progressively greater harmony between thought and action, that is to say, between theory and practice. A similar procedure is called the method of reflective equilibrium by John Rawls in *A Theory of Justice* (1971). Rawls accepts the Kantian theory of moral intuition, however, and does not recognize that moral consciousness is learned, not innate as Kant believed. That is one reason why rational people in possession of the same information do not all form the same mental models or select the same moral goals of science or the same "just" system of government (Barry, 1973). Rawls defines justice as an abstraction that is devoid of reference to human history, culture, and psychology. It lacks any justification in biology, anthropology, or neuroscience. But that is precisely the justification needed for any theory of justice that can be integrated with science and that can lead science to just goals.

Rationalists and intuitionists define the conscience as the arbiter of right versus wrong, justice versus injustice, self-interest versus duty. In their view conscience is where moral causes-and-effects are connected by necessity. Realists and empiricists understand that what appear to be moral causes-and-effects are no more than relations of components in mental models: relations between self and nonself, especially relations that require sympathy and empathy; relations between learned moral rules and experiences of pleasure or pain. To a large extent, conscience is the ability, learned from childhood on, to engage in sympathetic reflection about our actions with respect to their causing pleasure or pain in others. Evidence shows that children only begin to show sympathy for others at about one year of age, and that they have to learn such responsibilities and duties as sharing, obeying, and helping others, usually beginning between the ages of two and three (Murphy, 1937; M. L. Hoffman, 1975; Rheingold *et al.*, 1976; Rheingold and Hay, 1978). This evidence goes some way toward refuting intuitionist theories, the essence of which is that moral obligations are known in the first place by intuition, not by learning. Like other cognitive faculties, conscience can be modified by use and disuse, and it could be of practical as well as theoretical interest to discover whether there is a critical period during which some specific kind of experience is required for the subsequent development of a normal conscience.

The moral and scientific attitude requires that scientists understand that, however critical they may be of one another's theoretical positions, they are not implacable opponents but are all working to solve problems of common interest. This presupposes that in each of us there is a desire to share knowledge with each other, and that every rational person has the ability to understand and judge fairly the claims made by others. Judgment of the validity of scientific hypotheses is an inductive method, meaning that it does not start with self-evident premises but with an imaginative concept which has the potential for development and revision, as new information is accepted or rejected, after reflection, communication, and criticism. It is an adaptive process that requires effective communication between the parties who hold conflicting positions if it is to lead to progress.

The process begins and progresses most effectively when all parties agree in advance to some elementary procedural rules. This code of conduct, which I have adapted from Wessel (1976), would facilitate communication leading to resolution of conflicts between scientists whose theories collide, between contestants for priority of scientific discovery, and also between scientists and anonymous reviewers whose opinions differ about the merits of research reports or grant proposals. These rules are not explanatory nor do they pretend to add meaning to the conflicts that are being negotiated. They function like operating instructions that make it safer to enter negotiations, but they are not ethical principles that can enter the negotiations directly.

1. Ethical integrity will be maintained as a first priority.
2. Values will be disclosed and their roles in the research program will be explained.
3. Information will not be withheld. All data will be disclosed fully.
4. Theoretical concepts will be explained, and simplified if necessary, to achieve maximum communication.
5. Uncertainty and inadequate knowledge will be disclosed voluntarily.
6. Unjustified assumptions will be avoided.
7. Personal prejudice and conflict of interest in the outcome will be disclosed, especially by anonymous reviewers.
8. Dogmatism will be avoided.
9. Unfair tricks designed to mislead will be avoided and the motivation of adversaries will not be impugned.
10. If existing information is inadequate to arrive at a consensus, research will be planned to obtain the additional information necessary to advance toward the resolution of the conflict.

Prescriptive morality treats ethics as a set of instructions for keeping society or any part of it, say science, working efficiently. Ethics is reduced to little more than an instruction manual such as we use for operating a machine. Opinions may be divided as to whether this is any better than having no rules or any worse than any other set of moral precepts. Moral rules and commandments like those prescribed above should not be intended to relieve the individual from making difficult moral judgments voluntarily, after reflection on the present conditions and future consequences.

These prescriptions are intended to facilitate reflection, communication, and critique, leading to resolution of conflicts in scientific research programs. They are not supposed to diminish the importance of personal moral choices. They are intended to enhance moral responsibility and to stimulate the exercise of moral judgments aimed at resolving conflicts, especially when self-interest and public interest collide.

In practice it is difficult to arrange conditions and procedures that encourage resolution of conflicts between scientists. One assumes that people expect the conditions and procedures to be fair, and they may expect some reward for good behavior. However, the theory of utility maximization is not a good predictor of actual behavior (Tversky and Kahneman, 1974; Kahneman and Tversky, 1979). Most people prefer a certain positive reward to a much larger uncertain reward. This makes it easier for most people to accept mediation of disputes.

How interactions between opposing scientific research programs actually occur is considered in Chapters 1 and 2. The conditions which are prevalent now, especially the predominant role of techniques in scientific research, do not encourage debate about concepts or full disclosure of values. The result is that dialogue centers on differences of techniques and squabbles over the validity of data rather than on differences between opposing theories and between alternative values. Consensus about technical goals is often reached too quickly to permit effective operation of the procedures for modifying the prevailing theories and values in a content-increasing way.

Another condition favoring a rapid consensus is that nowadays few scientists dare to express eccentric ideas. The danger of uniform obedience to the authority of the entrenched majority has actually increased since John Stuart Mill warned more than a century ago: *"That so few now dare to be eccentric marks the chief danger of the time"* (*On Liberty*, 1859). Those who hold the secure center are prone to mistake the uniformity of opinion which surrounds them for a sign of strength when it may really be symptomatic of intellectual weakness. The shared values of a group of scientists have often consolidated, sustained, and legitimized their research program. Such a group admire one another's work and frequently establish bonds of friendship and of professional obligations. They form a mutual congratulation society with restrictive entry. Those who fail to gain admission tend to go into opposition. Finally the old established group is overtaken by the young established group, which tends to have more advanced research techniques, and the process repeats itself.

5.6. DEVELOPMENT OF ETHICAL CONSCIOUSNESS

One of the main problems of neurobiology is whether it is possible to explain the development of a person's ethical principles and values in terms of neurobiological functions. The following brief outline does scant justice to the importance of this problem. I intend it to serve only as a means of introducing the concept that universal patterns of value judgments do not necessitate a belief in preformed knowledge, but they can be accounted for by the relative uniformity of conditions under which infants learn to make value judgments.

Emotional and cognitive development, which includes development of ethical values, is seen as an interactive process in which the infant actively constructs its understanding by interaction with objects in its environment, principally with its mother (Piaget, 1952; Brazelton *et al.*, 1974; Bandura, 1977, 1989; Kurtines and Gewirtz, 1987). In terms of the mental model theory discussed in Section 1.2 the child actively builds a mental model that is capable of generating new rules and making predictions that relate its behavior to conditions that have ethical relevance, for example, that relate its feelings of pleasure or pain to other people, namely their pleasure or pain, approval or disapproval as shown by facial expression, gesture, and vocal expression. From birth the infant starts learning that its emotional states and its conduct are associated with approval or disapproval of others, principally its mother. Soon after birth the infant starts imitating the behavior of its mother and incorporates it in its own repertoire of behavior (Meltzoff and Moore, 1977; Kaye and Marcus, 1978; Field et al., 1983). The infant's behavior results in changes in the mother's behavior in ways that either reinforce the infant's feelings of self-efficacy and trust, or result in feelings of frustration and possibly of distrust. Infants become attached to people and objects with which they interact frequently and that gratify their needs, and the loss of such attachments results in unhappiness (Schaffer and Emerson, 1964; Bowlby, 1969). By one year of age children approach things with respect for their integrity. They are not neutral but rather put a value on certain things. Children at age one year can point out toys that are broken and are distressed by the perception.

The ancient prejudice against this concept of the neonatal origins of valid empirical knowledge is found in Descartes's *Principles of Philosophy* (Part 1, p. 71–72) where he claims that "*the chief cause of error arises from the preconceived opinions of childhood.*" This he says is because the mind of the child is a slave to the body and refers everything to it and not to anything outside itself. That doctrine is false, as we now know from the evidence that the fetus is aware of its mother's voice and attends to it preferentially starting immediately after birth (DeCasper and Fifer, 1980), and that from the age of two days the infant prefers to look at its mother's face rather than at the face of another woman (Bushnell et al., 1989). At 4 months of age the baby starts to smile at a face and at 6 months it can easily discriminate between different faces.

A useful distinction can be made between moral behavior learned during infancy, which I think is carried out impulsively, without reflection, and moral behavior learned at a later age, which tends to be performed deliberately, after reflection on the possible consequences of the action. Shaftesbury (1711, *Characteristics*, Vol. 1, p. 129) had a related notion that spontaneous, unreflective ethical impulses are the best because they are directly initiated by a moral sense. Butler thought of natural benevolence as an unreflective, spontaneous impulse, proceeding from a moral sense, but virtue he recognized as reflective, as benevolence controlled by reason (Butler, 1736, *Analogy*, II; Sidgwick, 1902, p. 197). I suggest that the spontaneity of moral acts learned during infancy misled Shaftesbury and Butler, and much later Charles Darwin, to the conclusion that such impulsive moral acts are performed on the basis of an innate moral sense. Nevertheless, Darwin admitted that a moral belief "*constantly inculcated during the early years of life, whilst the brain is impressible, appears to acquire almost the nature of an instinct; and the very essence of an instinct is that it is followed independently of reason*" (*The Descent of Man*, 1874, p. 122).

Darwin made the first systematic attempt at reducing moral philosophy to evolutionary biology and physiology. He says: "*I fully subscribe to the judgment of those writers who maintain that of all the differences between man and the lower animals, the moral sense or conscience is by far the most important. . . . The following proposition seems to me in a high degree probable—namely, that any animal whatever, endowed with well-marked social instincts, the parental and filial affection being here included, would inevitably acquire a moral sense or conscience, as soon as the intellectual powers had become as well, or nearly as well developed, as in man. The feeling of pleasure from society is probably an extension of the parental or filial affections, since the social instinct seems to be developed by the young remaining for a long time with their parents.*" Darwin placed great significance on instinctive moral sense because he believed in the inheritance of habits, including moral habits, and therefore he equated moral sense with a social instinct (*Descent of Man*, 1874, p. 116).

Social and moral behavior, Darwin believed, "*were no doubt acquired by the progenitors of man . . . through natural selection, aided by inherited habit*" (*The Descent of Man*, 1874, p. 130). With respect to the inheritance of altruistic behavior, Darwin believed that "*the habit of performing benevolent actions certainly strengthens the feeling of sympathy which gives the first impulse to benevolent actions. Habits, moreover, followed during many generations probably tend to be inherited*" (*The Descent of Man*, 1874, p. 131). After approving of the doctrine of inherited moral sense, Darwin astutely acknowledged that his "*chief source of doubt with respect to such inheritance, is that senseless customs, superstitions, and tastes, such as the horror of a Hindoo for unclean food, ought on the same principle to be transmitted. I have not met with any evidence in support of the transmission of superstitious customs or senseless habits. . . .*" (*The Descent of Man*, 1874, p. 124).

Darwin believed in the inheritance of learned behavior, and although he does not refer to Lamarck, he adheres to his theory of adaptation, e.g., "*Actions, which were first voluntary, soon become habitual and at last hereditary, and may be performed even in opposition to will*" (*The Expression of the Emotions in Man and Animals*, 1872, Chap. XIV, p. 356). In the same work, Darwin says, "*it seems probable that some actions, which were first performed consciously, have become through habit and association converted into reflex actions, and are now so firmly fixed and inherited, that they are performed, even when not of the least use, as often as the same causes arise, which originally excited them in us through volition*" (pp. 39–40). The doctrine of inheritance of acquired nervous functions appeared to receive support from the experiments of Brown-Séquard (1866, 1871–1872) purporting to show inheritance of acquired epilepsy in guinea pigs. The epilepsy was later found to be caused by an infection of the nervous system transmitted from the mother to her progeny. Experiments in Pavlov's laboratory seemed to show inheritance of conditioned reflexes in mice. Those experiments were reported at the International Congress of Physiology in 1923, but Pavlov later admitted that they "*have been found to be very complicated, uncertain, and moreover extremely difficult to control*" (Pavlov, 1927, p. 285). I mention these briefly because they illustrate that throughout the ages, from Plato's conjectures about *anamnesis*, to Pavlov's inheritance of conditioned reflexes, people have continued to believe in the existence of innate knowledge, in spite of the overwhelming weight of counterevidence. This is another example to show that refutation of a theory does not prevent its resurrection in modified versions. Weismann (1892, 1904) refuted the Lamarckian theory of inheritance of acquired characters but it has been resurrected in the politically motivated version of Lysenko in Russia (Medvedev, 1969), and more recently in the molecular biological version of Steele (1979, 1981).

Some philosophers continue to argue that we have innate knowledge and therefore have substantive *a priori* knowledge. For example, Peter Carruthers (1991) argues that folk psychology provides evidence of the prevalence of innate knowledge. But the fact that folk knowledge about matter, force, momentum, acceleration, and the like is mostly erroneous and that it can easily be corrected by scientific demonstration, is argument for cultural, not genetic, transmission of folk knowledge. Neither folklore nor superstition nor modern habits of scientific reasoning are intuitive but rather are all culturally acquired. An even more fundamental misconception held by some philosophers is that because nervous processes of sensation, perception, and cognition are innately determined, the knowledge related to those nervous processes must also be innate. No doubt we require certain kinds of neural organization, determined genetically, as the means to obtain knowledge. The kind of neural organization may set limits to knowledge, but it does not determine the content of that knowledge, which is entirely contingent on experience. Moreover, development of the neuronal mechanisms of emotion, perception, and cognition is also contingent on relevant experience during critical periods (Jacobson, 1991, pp. 535–537).

Sympathy and empathy are defined in Section 5.4. The child's ability to react to another person's feelings with an emotional response that is similar to the other's feelings is called sympathy. The idea that sympathy for others develops in young children, starting during the first year, and that sympathy forms a basis for later development of social behavior, has been reviewed by Murphy (1937). Most infants from birth through the first year react emotionally to the distress of others (Sagi and Hoffman, 1976), but the cognitive ability to recognize the feelings of others normally develops by the end of the second year (Hoffman, 1981, 1982; Eisenberg and Miller, 1987). The ability to sympathize and empathize requires the recognition that other people have emotions, and that perception develops by the third year. By the end of the second year the child has discovered that it is pleasant to direct its sympathy, its generous emotions, to others, and for others to direct theirs in return. It has also learned that lack of affection and of the esteem of others results in unhappiness. The delights of companionship, friendship, conversation, mutual esteem, and the approbation of others are powerful reinforcements of the child's altruistic behavior such as sharing and cooperation. Altruism is here defined broadly as any kind of behavior that decreases one's own interests in the service of others. It is not used here in the limited sense used by evolutionary biologists, namely behavior that decreases one's own reproductive potential in the service of others. Altruism may be a way of passing more copies of one's genes to future generations, therefore a form of kin selection. Altruism may also be adaptive in a social but nongenetic manner.

There is now considerable evidence showing that children between 1 and 3 years of age develop altruistic social behavior such as sharing, helping, taking turns, cooperating, and obeying (Murphy, 1937; Hoffman, 1975; Rheingold *et al.*, 1976; Rheingold and Hay, 1978). Reciprocity, which is the ability to take turns and to cooperate with voluntary sharing, develops initially as a nonverbal form of communication. Many of those social interactions which foster a moral consciousness have been reviewed by Damon (1988) without recognizing the significance of the prenatal and early postnatal period for integration of the perceptions on which moral values are gradually constructed. Development of moral consciousness during early infancy is also shown by the evidence that 2-year-olds begin to talk about happiness and sadness (Bretherton and Beeghly, 1982; Ridgeway et al., 1985), and by 3 years of age children understand that certain situations elicit happiness or sadness and that the outcome depends on prior expectations (Harris, 1989; Wellman and Bannerjee, 1991). By 3 or 4 years of age children are able to name the appropriate emotions depicted by different facial expressions (Bullock and Russell, 1984).

Evidence has accumulated recently to show that children start learning from birth to empathize, namely to recognize that other people have emotional and cognitive states like their own—to "read the minds" of other people (Bretherton and Beeghly, 1982; Eisenberg and Miller, 1987; Wellman, 1990, 1991; Whiten, 1991). The child's ability to give verbal and written representations of other people's emotional and cognitive states develops rapidly after the age of 3 (Estes et al. 1989; Fox, 1991). During that period the child constructs a mental model of other people's emotional and cognitive processes and makes predictions of other people's motivations and behavior on which it plans and executes its own behavior. The model incorporates rules about how the child's emotional and cognitive states affect other people, and vice versa, and general rules describing the child's relation with other people (Craik, 1943; Gentner and Stevens, 1983; Johnson-Laird, 1983).

Following Holland et al. (1986), the most basic rule of the mental model is a condition-action rule which has the form "IF such-and-such, THEN so-and-so." The mental model has a memory in which relevant "facts" about experience are stored, and once those facts are found to match the condition "IF . . .," the "THEN" part of the rule is executed. One of the features of this formulation is that mental models are dynamic. They can be changed by adding a new rule and deleting an old rule on the basis of ongoing interactions with the environment. Whether a rule is added or deleted, and whether it is executed depends on how it fares in competition with other rules. Several rules may be executed in parallel if they all match the facts and thus have their condition satisfied. Rules that are often activated together will eventually become associated to form a "rule cluster" that represents entities in terms of default hierarchies. A default hierarchy is a cluster of rules organized to establish relations between categories of different degrees of generality or specificity, for example, for establishing the identity of an object under conditions of uncertainty. An example of a default hierarchy is: person, female, mother, smiles, . . . The child infers that the mother approves and then performs the appropriate response. The child responds to the invariant pattern of behavior of its mother in the variable flux of behavior of others surrounding the mother. This process starts before birth: the invariance of the mother's heartbeat and her voice transmitted to the fetus in the uterus, may be important data that the fetus builds into a mental model of its mother.

The emotions of happiness and unhappiness are present from birth (Balnton, 1917; Oster, 1978; Izard and Malatesta, 1987). The emotions soon become associated with a particular motivational state: happiness motivates the infant to act in ways that lead to more happiness, and it is accompanied by facial expressions and body movements that signal to others that the infant wants to continue the interaction with them. Anger has opposite effects. As a result of such interactions the infant rapidly learns to modify the expression of its emotions (Izard and Malatesta, 1987). This concept requires that interactions between mother and baby are predicated on stereotyped patterns of behavior, especially facial expressions of emotional states.

Babies have stereotyped facial expressions associated with different emotions (Darwin, 1872; Ekman and Friesen, 1975, 1978; Izard *et al.*, 1980). Those facial expressions are "innate," meaning that an invariant pattern of reflex actions of facial muscles is evoked by an emotional state. Facial expressions are controlled subcortically (Rinn, 1984). As expected of reflexes, the same patterns of facial muscular actions occur in people of diverse societies and cultures (Ekman, 1973; Ekman and Friesen, 1975). Babies rapidly learn to recognize those facial expressions as signs of emotion in other people (Meltzoff and Moore, 1977, 1989; Hiatt *et al.*, 1979; Field *et al.*, 1983).

One of the most astonishing facts about early childhood is that the experiences of those first years are almost totally forgotten and beyond recall. Perhaps the learning of the fetus and young baby is subcortical and perhaps what is learned subcortically cannot be transferred to the immature cerebral cortex. Development of cortical synaptic connections and growth of cortical neurons, especially their dendrites, continues for several years after birth (reviewed in Jacobson, 1991, pp. 237–283 and 408–430). We do not know whether subcortical learning in the human fetus and possibly even in the baby can be transferred at a later time to the cerebral cortex as that develops after birth. Babies born without a cerebral cortex or without cerebral hemispheres rarely survive long enough for us to observe their learning capabilities (Edinger and Fischer, 1913; Gampes, 1926). In some rare instances, infants with hydranencephaly are able to learn in the apparent absence of the cerebral cortex and with vestigial cerebral hemispheres (Lorber, 1965). Of the 25% of such infants who survive more than 2 years, the majority show no obvious signs of learning, but adequate studies of their memory and learning capabilities have not been done (reviewed by Warkany *et al.*, 1981).

Adults don't know what it is like to be a baby, in some respects as we don't know what it is like to be another species (T. Nagel, 1974). However, there is a big difference between knowing what it is like to be a baby and to be another species: we were once babies. Therefore, although we have forgotten most of what it was like to be a baby, we may still retain some memories of babyhood, and some cognitive and emotional processes that are present in babies persist throughout life. Babies sleep for much of the first year, and it is possible that the desynchronization of electrical activity that occurs during the rapid-eye-movement (REM) phase of sleep may abolish some memories. During REM sleep there is a high level of brain activity characterized by desynchronization of the oscillations of electrical potential that are the EEG, and by repetitive, high-amplitude discharges called ponto-geniculo-occipital (PGO) spikes. During the PGO discharges the child is unconscious. This high level of brain activity starts in the fetus and is almost continuous during the first two weeks after birth. The periods of REM activity decrease during childhood (reviewed by Moruzzi, 1972). Theorists have proposed that PGO discharges are necessary for brain development (Roffwarg *et al.*, 1966; Jouvet, 1972, 1973). Development of consciousness may be related to synchronized 40-hertz oscillation of electrical potentials in the cerebral cortex that reflect the synchronized timing of neurons all responding to the same stimulus but scattered throughout the cortex (Gray *et al.*, 1989; Crick and Koch, 1990; Engel *et al.*, 1992). Maturation of those neurons may not be completed for many months after birth. It seems likely that consciousness of objects in the environment develops as the necessary neural machinery develops to correlate events in different neurons, including the formation of connections between neurons that have similar functional parameters but are separated in different parts of the cerebral cortex (Eggermont, 1990).

Development of human behavior starts prenatally in the form of species-specific patterns of reflexes, as Preyer (1885) and Hooker (1944, 1952) were first to recognize, but the prenatal origin of learning has been recognized only recently (Vince, 1979; Gottlieb, 1981; Birnholz and Benacerraf, 1983). For instance, the fetus learns to prefer the sound of its mother's voice (DeCasper and Fifer, 1980), and I surmise that access of the amniotic fluid to the olfactory and gustatory receptors enables the fetus to learn its mother's "flavor." Following birth it is primed to attend preferentially to its mother's voice, and probably to its mother's taste and odor. This behavior is not "innate" or "intuitive" although it appeared to be so to those who were unaware of the facts of prenatal learning.

Auditory perceptual learning starts *in utero*, in the guinea pig (Vince, 1979; Horner *et al.*, 1987), the sheep (Vince *et al.*, 1982) and in the human after the 6th month of gestation, when the fetus can hear the mother's heartbeat and her voice more clearly because it is conducted through her body (DeCasper and Spence, 1986). In the human fetus the cochlea and sensory end organs of the inner ear have completed development at about 24 weeks of gestation (Bast and Anson, 1949; Ormerod, 1960). Thus, the fetus has the peripheral sensory apparatus necessary for hearing during the final 3 months of gestation. It also has the necessary central nervous mechanisms. Measurements of reactions to acoustical stimulation have been made in premature infants and in fetuses *in utero*, showing that hearing in the fetus during the last trimester of gestation is normal over a wide frequency range (500–4000 Hz), which would enable the fetus to hear voices and other sounds, notwithstanding the attenuation caused by the surrounding tissues and the amniotic fluid (Johansson *et al.*, 1963; Querlieu *et al.*, 1981; Birnholz and Benacerraf, 1983).

During the final 3 months of gestation the fetus is able to perceive its mother's voice as a constantly repeated signal against the variable background of other sounds (DeCasper and Spence, 1986). The mother's voice thus becomes more effective than other sounds in gaining attention and arousing behavior in the fetus and newborn infant. There is evidence that the newborn infant attends preferentially to the sound of its mother's voice (DeCasper and Fifer, 1980; Birnholz and Benacerraf, 1983; DeCasper and Spence, 1986). It is very probable that fetal perception of the mother's voice is mediated by subcortical mechanisms, probably the inferior colliculus, because the auditory cortex is in a very immature state before birth. Similarly, visual perception in newborn babies is both subcortical (superior colliculus) and cortical. As the cortex matures during the postnatal period, auditory and visual perception, especially perception of complex patterns, becomes increasingly a function of the cortex. Yet, despite the relative immaturity of the cerebral cortex, babies as young as 2 days of age prefer to look at their mother's face than at the face of a strange woman (Bushnell *et al.*, 1989).

Moral conduct in children develops from reciprocity between the mother and her child. Like all forms of mutual gratification, it is rarely altruistic, but is doubly selfish.

However, reciprocity is moral when it promotes the interests of both parties equally, or nearly equally. The mother-child relationship may foster possessiveness as well as benevolence, and it has been said that the husband-wife relationship may also have both effects. In the mother-child relationship the conflict between egoism and altruism is negotiated over many years on the basis of social conventions and emotional reactions as well as moral reasoning. The mother's benevolence is only the first causality in a chain of interactions which has been identified as necessary for differentiation of moral standards in the 2- to 4-year-old child.

The infant first learns to relate its own states of pleasure or pain with its mother's actions, and later generalizes this relationship to the pleasure or pain of others in relation to its own feelings and emotions. Later in the child's development, this leads to benevolent actions because it feels good to be generous and kind and because benevolence meets with the approval of others. In the first instance the child develops a perception of the associations between its own behavior and another person's pleasure or pain. Further development results in perception of its responsibility to others for its own acts. The experience of pain and suffering is a basis for distinguishing good from evil, and is also the basis of sympathy for pain of others in distress. The ethical meaning of pain comes from our understanding that if pain is not evil, cruelty is no vice.

The result of a child's negotiation of myriads of particular moral situations is that a child's mental model of morality is constructed pragmatically. The model has a certain form in which the components cohere for pragmatic reasons. The model is not in the form of supreme moral principles of right and justice. Rather, it consists of a variety of principles which have been found by experience to work in different cases. Those various principles need not even be logically consistent with each other. It is sufficient that they are constructed on a broad empirical base, like a pyramid standing on its feet (Fig. 1.23). The pragmatic justifications for the model are that it can serve as a guide in the individual's encounters with moral problems, that it can make reliable predictions about the effects of moral decisions, and that it can be modified progressively and justified in relation to new experiences.

This theory of ethical development avoids the difficulties of theories of moral sense which is inherited in a fully differentiated form. It also avoids the problem of development of morality in the face of a conflict between self-interest and public interest—the interests of the mother and infant are normally in harmony, for if they were in serious conflict the life of the child would be in jeopardy. Patterns of interaction between mothers and their newborn infants in different societies vary normally within narrow limits. That is what gives the development of moral consciousness its universality and gives moral judgments their normative character. One normally finds ethics in experience, especially early childhood experience.

References

Ackerknecht, E. (1953). *Rudolf Virchow: Doctor, Statesman, Anthropologist*. Univ. Wisconsin Press, Madison.
Adrian, E. D. (1926). *J. Physiol., (London)* **61**:49–72. The impulses produced by sensory nerve endings. Part I.
Adrian, E. D. (1928). *The Basis of Sensation. The Action of the Sense Organs*. Cambridge Univ. Press, Cambridge. Reprinted by Hafner, New York, 1964.
Adrian, E. D. (1933). *Ergeb. Physiol.* **35**:744–755. The all-or-nothing reaction.
Adrian, E. D., and Y. Zotterman (1926). *J. Physiol., (London)* **61**:viii. Impulses from a single sensory end-organ.
Adrian, E. K., and R. L. Schelper (1981). Microglia, monocytes and macrophages, pp. 113–124. In *Glial and Neuronal Cell Biology*, (N. Fedoroff, ed.), Liss, New York.
Aguayo, A. J. (1985). Axonal regeneration from injured neurons in the adult mammalian central nervous system, pp. 457–484. In *Synaptic Plasticity* (C. W. Cotman, ed.), Guilford, New York.
Aguayo, A. J., M. Benfey, and S. David (1983). A potential for axonal regeneration in neurons of the adult mammalian nervous system, pp. 327–340. In *Nervous System Regeneration* (B. Haber, J. R. Perez-Polo, G. A. Hashim, and A. M. G. Stella, eds.), Liss, New York.
Aguilar, C. E., M. A. Bisby, E. Cooper, and J. Diamond (1973). *J. Physiol. (London)* **234**:449–464. Evidence that axoplasmic transport of trophic factors is involved in the regulation of peripheral nerve fields in salamanders.
Albarracin, A. (1982). *Santiago Ramón y Cajal o la Passion de España*. Editorial Labor, Barcelona.
Albright, T. D., R. Desimone, and C. G. Gross (1984). *J. Neurophysiol.* **51**:16–31. Columnar organization of directionally selective cells in visual area MT of the macaque.
Allen, F. (1912). *J. Comp. Neurol.* **22**:547–568. The cessation of mitosis in the central nervous system of the rat.
Allen, G. E. (1978). *Life Science in the Twentieth Century*. Cambridge Univ. Press, Cambridge.
Allen, G. E. (1981). *J. Hist. Biol.* **14**:159–176. Morphology and twentieth-century biology: a response.
Alpers, B. J., and W. Haymaker (1934). *Brain* **57**:195–205. The participation of the neuroglia in the formation of myelin in prenatal infantile brain.
Alzheimer, A. (1910). *Hist. u. Histopath. Arb. Nissl-Alzheimer* **3**:401. Beiträge zur Kenntniss der pathologischen Neuroglia und ihrer Beziehungen zu den Abbauvorgangen im Nervengewebe.
Anderson, C. (1992). *Nature* **355**:101. Authorship. Writer's cramp.
Anderson, J. R. (1983). *The Architecture of Cognition*. Harvard Univ. Press, Cambridge, Massachusetts.
Anderson, P., J. C. Eccles, and Y. Løyning (1963). *Nature* **198**:540–542. Recurrent inhibition in the hippocampus with identification of the inhibitory cell and its synapses.
Anderson, P., J. C. Eccles, and Y. Løyning (1964). *J. Neurophysiol.* **27**:608–619. Pathway of postsynaptic inhibition in the hippocampus.
Anderson, P., G. N. Gross, T. Lømo, and O. Sveen (1969). Participation of inhibitory and excitatory interneurones in the control of hippocampal cortical output, pp. 415–465. In *The Interneuron* (UCLA Forum in Medical Sciences, No. 11, M. A. B. Brazier, ed.), Univ. California Press, Berkeley.

Andreoli, A. (1961). *Basler Veröff. Gesch. Med. Biol.* **10**:1–86. Zur geschichlichen Entwicklung der Neuronentheorie.
Andreou, A. G. (1991). *Nature* **354**:501. Electronic arts imitate life.
Andrew, W., and C. T. Ashworth (1945). *J. Comp. Neurol.* **82**:101–127. The adendroglia.
Andriezen, W. L. (1893a). *Br. Med. J.* **2**:227–230. The neuroglia elements of the brain.
Andriezen, W. L. (1893b). *Int. Monatschr. Anat. Physiol.* **10**:533. On a system of fibre-cells surrounding the blood-vessels of the brain of man and its physiological significance.
Angell, J. R. (1909). *Psychol. Rev.* **16**:152–169. The influence of Darwin on psychology.
Angevine, J. B., Jr., and R. L. Sidman (1961). *Nature* **192**:766–768. Autoradiographic study of the cell migration during histogenesis of cerebral cortex in the mouse.
Anokhin, P. K. (1968). *The Biology and Neurophysiology of the Conditional Reflex* (in Russian). Meditsina, Moscow.
Apáthy, S. (1889). *Biol. Zbl.* **9**:625–648. Nach welcher Richtung hin soll die Nervenlehre reformiert werden?
Apáthy, S. (1897). *Mittheil. aus der zool. Stat. zur Neapal.* **12**:495–748. Das leitende Element des Nervensystems und seine topographischen Beziehung zu den Zellen.
Appel, T. A. (1987). *The Cuvier-Geoffroy Debate. French Biology in the Decades Before Darwin.* Oxford Univ. Press, New York and Oxford.
Arber, A. (1944). Analogy in the history of science. In *Studies and Essays in the History of Science and Learning.* Schuman, New York.
Ariëns Kappers, C. U. (1929). *The Evolution of the Nervous System in Invertebrates Vertebrates and Man.* De Erven F. Bohn, Haarlem.
Ariëns Kappers, C. U., G. C. Huber, and E. C. Crosby (1936). *The Comparative Anatomy of the Nervous System of Vertebrates Including Man.* 2 vols. Macmillan, New York.
Arnett, L. D. (1904). *Am. J. Psychol.* **15**:121–200, 347–382. The soul—a study of past and present beliefs.
Aschoff, P. (1924). Reticulo-endothelial system. In *Lectures on Pathology.* Hoeber, New York.
Auerbach, L. (1898). *Neurol. Central bl.* **17**:445–454. Nervenendigung in den Centralorganen.
Augustine, Saint (ca. 397). *Confessions.* Translated by J. K. Ryan. Image Books, New York.
Ayer, A. J. (1946). *Language, Truth and Logic.* 2nd ed. Kegan Paul, London.
Ayer, A. J. (1959). *Logical Positivism.* Free Press, New York.
Baars, B. (1988). *A Cognitive Theory of Consciousness.* Cambridge Univ. Press, Cambridge.
Bailey, P. (1959). *Perspect. Biol. Med.* **2**:417–441. The seat of the soul.
Baillarger, J. G. F. (1840). *Mém. Acad. Royale Méd.* (Paris) **8**:149–183. Recherches sur la structure de la couche corticale des circonvolutions du cerveau.
Bain, A. (1868). *The Senses and the Intellect*, 3rd Edition. Parker, London. (First edn. 1855, 2nd edn. 1864, 4th edn. 1894).
Baker, J. R. (1945). *The Discovery of the Uses of Colouring Agents in Biological Micro-technique.* Williams & Norgate, London.
Baker, J. R. (1948–1955). *Quart. J. microscop. Sci.* **89**:103–125; **90**:87–108; **93**:157–190; **94**:407–440; **96**:449–481. The cell-theory: a restatement, history and critique.
Balfour, F. M. (1881). *A Treatise on Comparative Embryology*, 2 vols., Macmillan, London.
Bandura, A. (1977). *Social Learning Theory.* Prentice-Hall, Englewood Cliffs, New Jersey.
Bandura, A. (1989). *Amer. Psychol.* **44**:1175–1184. Human agency in social cognitive theory.
Barker, L. F. (1899). *The Nervous System and Its Constituent Neurones.* D. Appleton and Company, New York.
Barker, L. F. (1898). *Amer. J. Insanity* **55**:31–49. On the validity of the neurone doctrine.
Barlow, H. (1972). *Perception* **1**:371–394. Single units and sensation: a neuron doctrine for perceptual psychology.
Barnard, R. I. (1940). *J. Comp. Neurol.* **73**:235–264. Experimental changes in end-feet of Held-Auerbach changes in the spinal cord of the cat.
Barnett, S. A. (1991). *Nature* **353**:786. Reductionism.
Barr, M. L. (1940). *Anat. Rec.* **77**:367–374. Axon reaction in motoneurones and its effect upon the endbulbs of Held-Auerbach.
Barry, B. M. (1973). *The Liberal Theory of Justice.* Clarendon Press, Oxford.
Bartelmez, G. W., and N. L. Hoerr (1933). *J. Comp. Neurol.* **57**:401–428. The vestibular club endings in Ameiurus. Further evidence on the morphology of the synapse.

Bast, T. H., and B. J. Anson (1949). *The Temporal Bone and the Ear*. Thomas, Springfield, Illinois.
Bastian, H. C. (1888). *Brain* **10**:1–89. The 'muscular sense'; its nature and cortical localization.
Bastian, H. C. (1896). *The Brain as an Organ of the Mind*. D. Appleton and Company, New York (1st edn. Kegan Paul, London, 1880).
Bateson, W. (1894). *Materials for the Study of Variation*. Macmillan, London.
Bauer, K. F. (1953). *Organisation des Nervengewebe und Neurencytiumtheorie*. Urban & Schwarzenberg, Berlin.
Bayes, T. (1763). *Phil. Trans. Roy. Soc. London* **53**:370–418. An essay towards solving a problem in the doctrine of chances. Reprinted with a biographical note by G. A. Barnard in *Biometrika* (1958) **45**: 293–315.
Beale, L. S. (1860). *Phil. Trans. Roy. Soc.* pp. 611–618. On the distribution of nerves to the elementary fibres of striped muscle.
Beale, L. S. (1862). *Proc. Roy. Soc.* June 19, 1862. Further observations on the distribution of nerves to the elementary fibres of striped muscle (Abstract).
Beams, H. W., and R. G. Kessel (1968). *Int. Rev. Cytol.* **23**:209–267. The Golgi apparatus: Structure and function.
Beard, J. (1896). *Zool. Jahrbücher Abt. morphol.* **9**:1–106. The history of a transient nervous apparatus in certain Ichthyopsida. An account of the development and degeneration of ganglion-cells and nerve fibres.
Bechtel, W., and A. Abrahamsen (1991). *Connectionism and The Mind: An Introduction to Parallel Processing in Networks*. Blackwell, Oxford.
Bechterew, W. von (1894). *Die Leitungsbahnen im Gehirn und Rückenmark*. Translated from the Russian by R. Weinberg, (Second edition, 1899). Verlag von Arthur Georgi, Leipzig.
Beevor, C. E. (1901). *Brain* **24**:331–335. Review of *The Nervous System and its Constituent Neurones*. By Lewellys F. Barker.
Bell, C. (1811). *Idea of a new anatomy of the brain; submitted for the observations of his friends*. Strahan and Preston, London.
Belliveau, J. W.; D. N. Kennedy, R. C. McKinstry, B. R. Buchbinder, R. M. Weisskoff, M. S. Cohen, J. M. Vevea, T. J. Brady, and B. R. Rosen (1991). *Science* **254**:716–719. Functional mapping of the human visual cortex by magnetic resonance imaging.
Belloni, L. (1966). *Physics* **8**:253–266. La neuroanatomia di Marcello Malpighi.
Belloni, L. (1968). *Analecta Medico-Historica* **3**:193–206. Die Neuroanatomie von Marcello Malpighi.
Ben-David, J. (1970). *Minerva* **8**:160–179. The rise and decline of France as a scientific center.
Ben-David, J. (1971). *The Scientist's Role in Society. A Comparative Study*. Prentice-Hall, Englewood Cliffs, New Jersey.
Beniger, J. R., and D. L. Robyn (1978). *Amer. Statistician* **32**:1–11. Quantitative graphics in statistics: a brief history.
Benjelloun-Touini, S., C. M. Jacque, P. Dever, F. DeVitry, R. Maunoury, and P. Dupouey (1985). *J. Neuroimmunol.* **9**:7–97. Evidence that mouse astrocytes may be derived from radial glia.
Bennett, M. R. (1983). *Physiolog. Rev.* **63**:915–1048. Development of neuromuscular synapses.
Bennett, M. V. L., E. Aljure, Y. Nakajima, and G. D. Pappas (1963). *Science* **141**:262–264. Electrotonic junctions between teleost spinal neurons: electrophysiology and ultrastructure.
Bergh, R. S. (1900). *Anat. Hefte* **14**:379–407. Beiträge zur vergleichenden Histologie.
Bergmann, C., and R. Leuckart (1852). *Anatomisch-physisch Übersicht des Thierreiches. Vergleichende Anatomie und Physiologie. Ein Lehrbuch für den Unterricht und zum Selbststudium*. Stuttgart.
Beritoff, J. S. (1969). *Structure and Functions of the Cerebral Cortex* (in Russian), Nauka, Moscow.
Berkley, H. J. (1897). *Johns Hopkins Hosp. Rep.* **6**:1–88. Studies on the lesions produced by the action of certain poisons on the cortical nerve cell.
Berlin, R. (1858). *Beiträge zur Strukturlehre der Grosshirnwindungen*, Junge, Erlangen.
Bernard, C. (1865). *Introduction à l'étude de la médecine expérimentale*. J.-B. Baillière, Paris. Translated by H. C. Greene as *An Introduction to the Study of Experimental Medicine*, Macmillan, New York.
Bernard, C. (1878). Le curare, pp. 237–315. In *La Science expérimentale*. Baillière, Paris.
Bernfeld, S. (1950). *Yearbook of Psychoanalysis* **6**:24–50. Freud's scientific beginnings.
Bernstein, J. (1871). *Untersuchungen über den Erregungsvorgang im Nerven—und Muskelsysteme*. Winter, Heidelberg.

Bernstein, J. (1912). *Elektrobiologie*. Friedr. Vieweg and Sohn, Braunschweig.
Bernstein, J. J., and L. Guth (1961). *Exp. Neurol.* **4**:262–275. Nonselectivity in establishment of neuromuscular connections following nerve regeneration in the rat.
Berry, M., and R. Flinn (1984). *Proc. Roy. Soc. London. Ser. B* **221**:321–348. Vertex analysis of Purkinje cell dendritic trees in the cerebellum of the rat.
Berthelot, R. (1949). *La Pensée de l'Asie et l'Astrobiologie*, Payot, Paris.
Berzelius, I. J. (1813). *A View of the Progress and Present State of Animal Chemistry*. Translated from Swedish by G. Brunnmark. John Hatchard; J. Johnson and Co.; and T. Boosey, London.
Bethe, A. (1896). *Anat. Anz.* **12**:438–446. Eine neue Methode der Methylenblaufixation.
Bethe, A. (1898). *Morph. Arb.* **8**:95–116. Uber die Primitivfibrillen in den Ganglienzellen von Menschen.
Bethe, A. (1900). *Arch. mikr. Anat.* **55**:513–558. Uber die Neurofibrillen und die Ganglienzellen von Wirbeltieren.
Bethe, A. (1903). *Allgemeine Anatomie und Physiologie des Nervensystems*. Thieme, Leipzig.
Bethe, A. (1904). *Deutsche Med. Woch.* **30**:1201–1204. Der heutige Stand der Neurontheorie.
Beveridge, W. I. B. (1951). *The Art of Scientific Investigation*. Norton, New York.
Bhaskar, R. (1975). *A Realist Theory of Science*. Humanities Press, London.
Bianchi, L. (1922). *The Mechanism of the Brain and the Function of the Frontal Lobes*. Transl. by J. H. Macdonald. Wm. Wood & Co., New York.
Bichat, X. (1800). *Recherches physiologiques sur la vie et la mort*. Republished 1962 by Alliance Culturelle du Livre, Paris.
Bidder, F. H. (1847). *Zur Lehre von dem Verhältniss der Ganglienkörper zu den Nervenfasern*. Breitkopf & Härtel, Leipzig.
Bidder, F. H., and C. Kupffer (1857). *Untersuchungen über die Textur des Rückenmarks und die Entwickelung seine Formelemente*, Breitkopf & Haertel, Leipzig.
Bidder, F. H., and A. W. Volkmann (1842). *Die Selbständigkeit des sympathischen Nervensystems durch anatomische Untersuchungen nachgewiesen*. Breitkopf & Härtel, Leipzig.
Bielschowsky, M. (1904). *J. Psychol. Neurol.* **3**:169–189. Die Silberimprägnation der Neurofibrillen.
Bielschowsky, M. (1908). *J. Psychol. Neurol.* **10**:274–281. Die fibrillare Struktur der Ganglienzelle.
Bielschowsky, M. (1928). Nervengewebe, Vol. 4, pp. 1–142. In *Handbuch der mikroskopischen Anatomie des Menschen* (W. Möllendorff, ed.). Springer, Berlin.
Birks, R. I., and F. C. MacIntosh (1957). *Br. Med. Bull.* **13**:157–161. Acetylcholine metabolism at nerve endings.
Bisby, M. A. (1982b). *Fed. Proc.* **41**:2307–2311. Functions of retrograde transport.
Black, M. (1948). *Philos. Rev.* **57**:111–126. Some questions about emotive meaning.
Blackstad, T. W. (1965). *Z. Zellforsch.* **67**:819–834. Mapping of experimental axon degeneration by electron microscopy of Golgi preparations.
Blakemore, C., R. C. Van Sluyters, and J. A. Movshon (1975). *Cold Spring Harbor Symp. Quant. Biol.* **40**: 601–609. Synaptic competition in the kitten's visual cortex.
Blakemore, W. F. (1975). *Acta neuropathologica* **6**:Suppl. 273–278. The ultrastructure of normal and reactive microglia.
Blanton, M. G. (1917). *Psychol. Rev.* **24**:456–483. The behavior of the human infant during the first thirty days of life.
Blasius, W. (1964). Die Bestimmung der Leitungsgeschwindigkeit im Nerven durch Hermann v. Helmholtz am Beginn der naturwissenschaftlichen Ära der Neurophysiologie, pp. 71–84. In *Von Boerhaave bis Berger* (K. E. Rothschuh, ed.), Gustav Fischer Verlag, Stuttgart.
Blinzinger, K. H., and G. W. Kreutzberg (1968). *Z. Zellforsch. Mikrosk. Anat.* **85**:145–147. Displacement of synaptic terminals from regenerating motoneurons by microglial cells.
Blum, F. (1893). *Zeitschr. Wissensch. Med.* **10**:314–315. Der Formaldehyd als Härtungsmittel. Vorläufige Mittheilung.
Blunt, M. J., F. Baldwin, and C. P. Wendell-Smith (1972) *Z. Zellforsch. Mikrosk. Anat.* **124**:293–310. Gliogenesis and myelination in kitten optic nerve.
Bodian, D. (1936). *Anat. Rec.* **65**:89–97. A new method for staining nerve fibers and nerve endings in mounted paraffin sections.

Bodian, D. (1937). *J. Comp. Neurol.* **68**:117–159. The structure of the vertebrate synapse. A study of the axon endings on Mauthner's cell and neighboring centers in the goldfish.
Bodian, D. (1948). *Bull. Johns Hopkins Hosp.* **83**:1–108. The virus, the nerve cell and paralysis. A study, of experimental poliomyelitis in the spinal cord.
Bodian, D., and H. A. Howe (1941a). *Bull. Johns Hopkins Hosp.* **68**:248–267. Experimental studies on intraneural spread of poliomyelitis virus.
Bodian, D., and H. A. Howe (1941b). *Bull. Johns Hopkins Hosp.* **69**:79–85. The rate of progression of poliomyelitis virus in nerves.
Boeke, J. (1932). Nerve endings, motor and sensory, pp. 243–315. In *Cytology and Cellular Pathology of the Nervous System*, Vol. 1, (W. Penfield, ed.), Hoeber, New York.
Boeke, J. (1942). *Schweiz Arch. Neurol. Psychiat.* **49**:9–32. Sur les synapses à distance. Les glomerules cérébelleux, leur structure et leur développement.
Bohr, N. (1958). *Atomic Physics and Human Knowledge*. Wiley, New York.
Bok, S. T. (1936). *Proc. Acad. Sci. Amsterdam* **39**:1209–1218. The branching of the dendrites in the cerebral cortex.
Bok, S. T. (1959). *Histonomy of the Cerebral Cortex*. Elsevier, Amsterdam.
Bonin, G. von (1950). *Essays on the Cerebral Cortex*, Charles C. Thomas, Springfield, Illinois.
Bonin, G. von (1970). Rudolf Albert von Kölliker (1817–1905), pp. 51–54. In *The Founders of Neurology*, Second Edition (W. Haymaker and F. Schiller, eds.). Thomas, Springfield, Illinois.
Bonnet, R. (1878). *Morphol. Jahrb. Leipzig* **4**:329–398. Studien ueber die Innervation der Haarbälge der Hausthiere.
Boring, E. G. (1950). *A History of Experimental Psychology*, 2nd edn. Appleton-Century-Crofts, New York.
Bostock, J. (1825–1828). *An Elementary System of Physiology*. 3 vols. Wells and Lilly, Boston.
Bowditch, H. P. (1871). *Bericht. Königl. Sachs. Gesellsch. Wissensch.* **23**:652. Über die Eigenthumlichkeiten der Reizbarkeit welche die Muskelfasern des Herzens zeigen.
Bowditch, H. P. and J. W. Warren (1890). *J. Physiol.* **11**:25–64. The knee-jerk and its physiological modifications.
Bowlby, J. (1969). *Attachment and Loss, Vol. 1: Attachment*. Basic Books, New York.
Boya, J. (1976). *Acta anat.* **95**:598–608. An ultrastructural study of the relationship between pericytes and cerebral macrophages.
Boya, J., J. Calvo, and A. Prado. (1979). *J. Anat. (London)* **129**:177–186. The origin of microglial cells.
Boya, J., J. Calvo, A. L. Carbonell, and E. Garcia-Mauriño (1986). *Acta anat.* **127**:142–145. Nature of macrophages in rat brain. A histochemical study.
Boycott, B. B. and Kolb, H. (1973). *J. Comp. Neurol.* **148**:91–114. The connections between bipolar cells and photoreceptors in the retina of the domestic cat.
Boyd, R. (1972). *Philosophy of Science* **39**:431–450. Determinism, laws, and predictability in principle.
Boyd, R. (1983). *Erkenntnis* **19**:74–105. On the current status of the issue of scientific realism.
Bracegirdle, B. (1978). *A History of Microtechnique*. Cornell Univ. Press, Ithaca, New York.
Bradley, F. H. (1897). *Appearance and Reality. A Metaphysical Essay*. 2nd Edition (revised) with an Appendix. Swan, Sonnenschein and Co., New York: The Macmillan Company.
Braitenberg, V. (1978). Cortical cytoarchitectonics, pp. 443–465. In *Architectonics of the Cerebral Cortex* (M. A. B. Brazier and H. Petsche, eds.), Raven, New York.
Braithwaite, R. B. (1953). *Scientific Explanation. A Study of the Function of Theory, Probability and Law in Science*. Cambridge Univ. Press, Cambridge.
Braus, H. (1905). *Anat. Anz.* **26**:433–479. Experimentelle Beiträge zur Frage nach der Entwicklung peripherer Nerven.
Brazelton, T. B., B. Koslowski, and M. Main (1974). The origins of reciprocity: the early mother-infant interaction. In *The Effect of the Infant on Its Caregiver* (M. Lewis and L. Rosenblum, eds.), Wiley, New York.
Brazier, M. A. B. (1959). The historical development of neurophysiology, pp. 1–58. In *Handbook of Physiology*, Section 1, Vol. 1 (J. Field, ed.). American Physiological Society, Washington, D. C.
Brazier, M. A. B. (1965). *J. Hist. Behav. Sci.* **1**:218–234. The growth of concepts relating to brain mechanisms.
Brazier, M. A. B. (1984). *A history of neurophysiology in the 17th and 18th centruies. From concept to experiment*. Raven Press, New York.

Bredvold, L. (1951). The invention of the ethical calculus, pp. 165–180. In *The Seventeenth Century: Studies in the History of English Thought and Literature from Bacon to Pope*, R. F. Jones, editor. Stanford Univ. Press, Stanford.

Breschet, G. (1836). *Mem. de l'Acad. Royale de Med.* (Paris) **5**:229–523. Recherches anatomiques et physiologiques sur l'organe de l'ouie et sur l'audition dans l'homme et les animaux vertebres.

Bretherton, I., and M. Beeghly (1982). *Develop. Psychol.* **18**:906–921. Talking about internal states: The acquisition of an explicit theory of mind.

Bretschneider, H. (1962). *Der Streit um die Vivisektion im 19. Jahrhundert: Verlauf, Argumente, Ergebnisse*. Fischer Verlag, Stuttgart.

Brittan, G. G., Jr. (1970). *J. Philos.* **67**:446–457. Explanation and reduction.

Broca, P. P. (1861a). *Bull. Soc. d'Anthrop.* **2**:190–204, 309–321. Sur le principe des localisations cérébrales.

Broca, P. P. (1861b). *Bull. Soc. Anat.* **36**:330–357. Remarques sur le siège de la faculté du langage articulé, suivies d'une observation d'aphémie.

Broca, P. P. (1861c). *Bull. Soc. Anat.* **36**:441–446. Nouvelle observation d'aphémie produite par une lésion de la moitié postérieure des deuxiéme et troisiém circonvolutions frontal gauches.

Broca, P. P. (1864). *Bull Soc. d'Anthrop.* **4**:200–208. Localisations des fonctions cérébrales. Siège de la faculté du langage articulé.

Broca, P. P. (1877). *Bull. Soc. Anthrop Paris* **12 (ser. 2)**:646–657. Sur la circonvolution limbique et la scissure limbique.

Broca, P. P. (1878). *Rev. Anthrop. Paris* **1**:385–498. Anatomie comparée des circonvolutions cérébrales. Le grand lobe limbique et la scissure limbique dans la série des mammifères.

Broca, P. P. (1888). *Mémoires sur le Cerveau de l'Homme*. C. Reinwald, Paris.

Brock, L. G., J. S. Coombs, and J. C. Eccles (1952a). *J. Physiol. London* **117**:431–460. The recording of potentials from motorneurones with an intracellular electrode.

Brock, L. G., J. S. Coombs, and J. C. Eccles (1952b). *Proc. R. Soc. London Series B* **140**:170–176. The nature of the monosynaptic excitatory and inhibitory processes in the spinal cord.

Brodmann, K. (1909). *Vergleichende Lokalisationslehre der Grosshirnrinde in ihre Prinzipien dargestellt auf Grund des Zellenbaues*. J. A. Barth, Leipzig.

Brooks, C. McC. and J. C. Eccles (1947). *Nature* **159**:760–764. An electrical hypothesis of central inhibition.

Brown, G. L., and J. C. Eccles (1934). *J. Physiol. (London)* **82**:211–241. The action of a single vagal volley on the rhythm of the heart beat.

Brown-Séquard, C. E. (1860). *Course of Lectures on the Physiology and Pathology of the Central Nervous System*. Collins, Philadelphia.

Brown-Séquard, C. E. (1866). *Arch. Physiol. Norm. et Path.* **2**:211–220, 422–438, 496–503. Nouvelles recherches sur l'épilepsie due à certaines lésions de la moelle épinière et des nerfs rachidiens.

Brown-Séquard, C. E. (1871–1872). *Arch. Physiol. Norm. et Path.* **4**:116–120. Quelques faits nouveaux relatifs à l'épilepsie quon observe à la suite de diverses lésions du système nerveux chez les cobayes.

Brun, R. (1936). *Schweiz. Arch. f. Neurol. u. Psychiat.* **37**:200–207. Sigmund Freud's Leistungen auf dem Gebiete der organischen Neurologie.

Bueker, E. D. (1948). *Anat. Rec.* **102**:369–390. Implantation of tumors in the hindlimb field of the embryonic chick and developmental response of the lumbosacral nervous system.

Buess, H. (1964). Von Beitrag der Schweizer Ärzte zur Geschichte der Neuronentheorie, pp. 186–210. In *Von Boerhaave bis Berger. Die Entwicklung kontinentalen Physiologie im 18. und 19. Jahrhundert* (K. E. Rothschuh, ed.). Gustav Fischer Verlag, Stuttgart.

Bullock, M., and J. A. Russell (1984). *Internat. J. Behav. Develop.* **7**:193–214. Preschool children's interpretation of facial expressions of emotion.

Bullock, T. H. (1959). *Science* **129**:997–1002. Neuron doctrine and electrophysiology.

Bullock, T. H., and G. A. Horridge (1965). *Structure and Function in the Nervous Systems of Invertebrates*. 2 Vols. Freeman, San Francisco.

Bunge, M. B., R. P. Bunge, and G. D. Pappas (1962). *J. Cell Biol.* **12**:448–453. Electron microscopic demonstration of connections between glia and myelin sheaths in the developing mammalian nervous system.

Burdach, K. F. (1819–1826). *Vom Baue und Leben des Gehirns*. 3 vols. Dyk, Leipzig.

Burdach, K. F. (1826–1840). *Die Physiologie als Erfahrungswissenschaft*. 6 vols. L. Voss, Leipzig.

Burrows, M. T. (1911). *J. Exper. Zool.* **10**:63–84. The growth of tissues of the chick embryo outside the animal body, with special reference to the nervous system.
Bury, J. B. (1921). *The Idea of Progress. An Inquiry into its Origin and Growth*. Macmillan, London.
Butler, J. (1726). *Fifteen Sermons upon Human Nature, or Man Considered as a Moral Agent*. London.
Butler, J. (1736). *The Analogy of Religion, Natural and Revealed, to the Constitution and Course of Nature*. M. H. Newman, New York.
Butterfield, H. (1931). *The Whig Interpretation of History*. G. Bell & Sons, London.
Cabanis, P. J. G. (1805). *Rapports du physique et du moral de l'homme*, 2nd ed, 2 vols. Crapart, Caille & Ravier, Paris.
Cajori, F. (1928–1929). *History of Mathematical Notation*, 2 vols. Open Court, Chicago.
Calder, W. A. (1984). *Size, Function, and Life History*. Harvard Univ. Press, Cambridge, Massachusetts.
Campbell, A. W. (1905). *Histological Studies on the Localisation of Cerebral Function*. Cambridge Univ. Press, Cambridge.
Campbell, D. T. (1974). 'Downward causation' in hierarchically organized biological systems, pp. 179–186. In *Studies in the Philosophy of Biology* (F. J. Ayala and T. Dobzhansky, eds.), Macmillan, London.
Campbell, N. R. (1920). *Physics: The Elements*. Cambridge Univ. Press, Cambridge. Republished in 1957 as *Foundations of Science*. Dover Books, New York.
Campos, J. J., and K. C. Barrett (1984). Toward a new understanding of emotions and their development, p. 229–263. In *Emotions, Cognition and Behavior* (C. E. Izard, J. Kagan, and R. B. Zajonc, eds.). Cambridge Univ. Press, New York.
Canguilhem, G. (1955). *La Formation du Concept de Réflex aux XVIIe et XVIIIe Siécles*. Presses Universitaire, Paris.
Carnap, R. (1932). *Erkentniss* **2**:219–241. Überwindung der Metaphysik durch logische Analyse der Sprache. Translated as *The elimination of metaphysics through logical analysis of language*, pp. 60–81. In *Logical Positivism* (A. J. Ayer, ed.), Free Press, New York.
Carpenter, W. B. (1853). *Principles of Human Physiology*. 5th American from the 4th and enlarged London Edition. Edited, with Additions by F. G. Smith. Blanchard and Lea, Philadelphia.
Carruthers, P. (1991). *Human Knowledge and Human Nature. A New Introduction to an Ancient Debate*. Oxford Univ. Press, Oxford.
Carus, C. G. (1814). *Versuch einer Darstellung des Nervensystems und insbesondere des Gehirns, nach ihrer Bedeutung, Entwickelung und Vollendung im Thierischen Organismus*. Breitkopf & Härtel, Leipzig.
Carus, C. G. (1835). *Traité élémentaire d'anatomie comparée, suivi de recherches d'anatomie philosophique ou transcendente sur les parties primaires du système nerveux et du squelette intérieur et extérieur*, 2 vols. Translated by A. J. L. Jourdan from the 2nd edition. Baillière, Paris.
Causey, R. L. (1972). *Synthese* **25**:176–218. Uniform microreductions.
Causey, R. L. (1977). *Unity of Science*. Reidel, Drodrecht.
Chan, W. T. (1957). *Philosophy East and West* **6**:309–332. Neo-Confucianism and Chinese scientific thought.
Chan, W.T. (1963). *A Source Book in Chinese Philosophy*. Princeton Univ. Press, Princeton, New Jersey.
Changeux, J.-P. (1986). *Neuronal Man*. Oxford Univ. Press, Oxford.
Changeux, J.-P., and A. Danchin (1976). *Nature* **264**:705–711. Selective stabilization of developing synapses as a mechanism for the specification of neuronal networks.
Changeux, J.-P., A. Devillers-Thiéry, and P. Chemouilli (1984). *Science* **225**:1335–1345. Acetylcholine receptor: An allosteric protein.
Chappell, W. (1970). *A Short History of the Printed Word*. Alfred A. Knopf, New York.
Chen, D. H., W. W. Chambers, and C. N. Liu (1977). *Exp. Neurol.* **57**:1026–1041. Synaptic displacement in neurons of Clarke's nucleus following axotomy in the cat.
Chesterton, G. K. (1975). In *The Oxford Book of Literary Anecdotes*, p. 286, (J. Sutherland, ed.), Oxford University Press, London.
Choi, B. H. (1981). *Dev. Brain Res.* **1**:249–267. Radial glia of developing human fetal spinal cord: Golgi, immunohistochemical and electron microscopic study.
Churchill, F. B. (1966). *Wilhelm Roux and a Program for Embryology*. Ph.D. thesis, Harvard University, Cambridge, Massachusetts.
Churchland, P. M. (1988). *Matter and Consciousness* (Revised ed.; First ed. 1984). M.I.T. Press, Cambridge, Massachusetts.

Churchland, P.M. (1989). *A Neurocomputational Perspective*. MIT Press, Cambridge, Massachusetts
Churchland, P. S. (1986a). *Mind* **95**:279–309. Some reductive strategies in cognitive neurobiology.
Churchland, P. S. (1986b). *Neurophilosophy*. MIT Press, Cambridge, Massachusetts.
Churchland, P. S., and T. J. Sejnowski (1992). *The Computational Brain*. MIT Press, Cambridge, Mass.
Ciaccio, G. V. (1883). *Quart. J. Microsc. Sci.* **11**:97–105. On the anatomy of nerve-fibres and cells, and the ultimate distribution of nerve-fibres. Three demonstrations delivered by Professor Lionel S. Beale.
Clarke, E. (1968). The doctrine of the hollow nerve in the seventeenth and eigthteenth centuries, pp. 123–141. In *Medicine, Science and Culture* (Stevenson and Multhauf, eds.). Johns Hopkins Press, Baltimore.
Clarke, E., and J. G. Bearn (1968). *J. Hist. Med.* **23**:309–330. The brain 'glands' of Malpighi elucidated by practical history.
Clarke, E., and K. Dewhurst (1972). *An Illustrated History of Brain Function*. Univ. Calif. Press, Berkeley and Los Angeles.
Clarke, E., and L. S. Jacyna (1987). *Nineteenth-Century Origins of Neuroscientific Concepts*. Univ. Calif. Press, Berkeley.
Clarke, E., and C. D. O'Malley (1968). *The Human Brain and Spinal Cord. A Historical Study Illustrated by Writings from Antiquity to the Twentieth Century*. Univ. Calif. Press, Berkeley.
Clarke, J. A. L. (1851). *Phil. Trans. Roy. Soc.* **141**:607–621. Researches into the structure of the spinal cord.
Cohen, I. B. (1985). *Revolutions In Science*. Belknap, Harvard.
Cohen, L. D. (1936). *Ann. Sci.* **1**:48–61. Descartes and Henry More on the beast-machine—A translation of their correspondence pertaining to animal automatism.
Cohen, Lord, of Birkenhead (1958). *Sherrington: Physiologist, Philosopher, Poet*. The University of Liverpool Sherrington Lectures, Vol. IV, Liverpool Univ. Press, Liverpool.
Cohnheim, J. (1867). *Virchow's Archiv.* **38**:343–386. Über die Endigung der sensiblen Nerven in der Hornhaut.
Collingwood, R. G. (1946). *The Idea of History*. Clarendon Press, Oxford.
Conger, G. P. (1922). *Theories of Macrocosm and Microcosm in the History of Philosophy*. Inaugural Dissertation, Columbia Univ., New York.
Connold, A. L., J. V. Evers, and G. Vrbova (1986). *Dev. Brain Res.* **28**:99–107. Effect of low calcium and protease inhibitors on synapse elimination during postnatal development in the rat soleus muscle.
Constantine-Paton, M. (1981). Induced ocular dominance zones in tectal cortex, pp. 47–67. In *Organization of the Cerebral Cortex* (F. O. Schmitt, F. G. Worden, G. Adelman, and S. G. Dennis, eds.). MIT Press, Cambridge, Mass.
Cook, J. E. (1991). *Trends Neurosci.* **14**:397–401. Correlated activity in the CNS: a role on every timescale?
Cooper, S., D. Denny-Brown, and C. S. Sherrington (1926). *Proc. Roy. Soc. Lond. Ser. B,* **100**:448–462. Reflex fractionation of a muscle.
Cooper, S., D. Denny-Brown, and C. S. Sherrington (1927). *Proc. Roy. Soc. Lond. Ser. B,* **100**:262–303. Interaction between ipsilateral spinal reflexes acting on the flexor muscles of the hind limb.
Cozzens, S. E. (1989). *Social Control and Multiple Discovery in Science. The Opiate Receptor Case*. State Univ. New York Press, Albany.
Coss, R. G. (1985). *Behav. Neurol. Biol.* **44**:151–185. The function of dendritic spines: A review of theoretical issues.
Cottingham, J., R. Stoothoff, and D. Murdoch (1984). *Descartes. The Philosophical Writings*, 2 vols. Cambridge Univ. Press, Cambridge.
Cragg, B. G. (1975). *J. Comp. Neurol.* **160**:147–166. The development of synapses in the visual system of the cat.
Craik, K. (1943). *The Nature of Explanation*. Cambridge Univ. Press, Cambridge.
Cranefield, P. F. (1974). *The way in and the way out: François Magendie, Charles Bell and the roots of the spinal nerves*. Futura, Mount Kisco, New York.
Creed, R. S., D. Denny-Brown, J. C. Eccles, E. G. T. Liddell, and C. S. Sherrington (1932). *Reflex Activity of the Spinal Cord*. Clarendon Press, Oxford.
Crick, F. (1988). *What Mad Pursuit: A Personal View of Scientific Discovery*. Basic Books, New York.
Crick, F. H. C., and C. Koch (1990). *Seminars in the Neurosciences* **2**:263–275. Towards a neurobiological theory of consciousness.
Crosby, E. C., and H. N. Schnitzlein, eds. (1982). *Comparative Correlative Neuroanatomy of the Vertebrate Telencephalon*. Macmillan, New York.

Cupples, B. (1977). *Philos. Sci.* **44**:387–408. Three types of explanation.
Cuvier, G. (1800–1805). *Leçons d'anatomie comparée de G. Cuvier.* 5 vols. Vols. 1 and 2 (1800) edited by C. Dumeril, Vols. 3–5 (1805) edited by G. L. Duvernoy. Baudouin, Paris.
Cuvier, G. (1817). *Le règne animal distribué d'apres son organisation.* 4 vols. Deterville, Paris.
Cuvier, G. (1822). *Journal de Physiologie Expérimentale et Pathologique, Paris* **2**:372–384. Rapport fait à l'Académie des Sciences sur des expériences relatives aux fonctions du système nerveux. Also in Flourens (1842), pp. 60–84.
Dale, H. H. (1934). *Brit. Med. J.* **1**:835–841. Chemical transmission of the effects of nerve impulses.
Dale, H. H. (1935). *Proc. Roy. Soc. Med.* **28**:319–332. Pharmacology and nerve endings.
Dale, H. H. (1937). *Harvey Lect.* **32**:229–245. Transmission of nervous effects by acetylcholine.
Dale, H. H. (1937–1938). *J. Physiol. (London)* **91**:4P. Du Bois-Reymond and chemical transmission.
Damon, W. (1988). *The Moral Child: Nurturing Children's Natural Moral Growth.* Free Press, New York.
Darwin, C. (1859). *On the Origin of Species By Means of Natural Selection, Or The Preservation of Favoured Races In The Struggle For Life*, 1st ed. (2nd ed., 1860). John Murray, London.
Darwin, C. (1862). *On the Various Contrivances By Which British and Foreign Orchids are Fertilized By Insects.* John Murray, London.
Darwin, C. (1868). *The Variations Of Animals And Plants Under Domestication*, 2 vols, 1st ed. John Murray, London.
Darwin, C. (1871). *The Descent of Man and Selection in Relation to Sex*, 1st ed. (2nd ed., 1874). John Murray, London.
Darwin, C. (1872). *On the Expression of the Emotions in Man and Animals.* John Murray, London.
Darwin, C. (1874). *The Descent of Man and Selection in Relation to Sex.* (Second edition in one volume, with a 'Note on the resemblances and differeneces in the structure and the development of the brain in man and apes' by T. H. Huxley). John Murray, London.
Darwin, C. (1888). *Life and Letters of Charles Darwin. Including An Autobiographical Chapter*, 3 vols. Edited by His Son, Francis Darwin. John Murray, London.
Davidson, E. H. (1986). *Gene Activity During Early Development*, 3rd Edition. Academic Press, Orlando.
Davidson, E. H. (1990). *Development* **108**:365–389. How embryos work: A comparative view of the diverse modes of cell fate specification.
Davies, A. M. (1988). *Trends Neurosci.* **11**:243–244. The emerging generality of the neurotrophic hypothesis.
Davis, J., and H. Eichenbaum, eds. (1991). *Olfaction as a Model System for Computational Neuroscience.* MIT Press, Cambridge, Massachusettes.
Davis, R. A. (1964). *Surg. Gynecol. Obstet.* **119**:1333–1340. Victor Horsley, Victorian physician-scholar and pioneer physiologist.
Dawson, C. (1929, reprinted 1945). *Progress and Religion: an Historical Enquiry.* Greenwood Press, Westport, Conn.
Debbage, P. L. (1986). *J. Neurol. Sci.* **72**:319–336. The generation and regeneration of oligodendroglia.
DeCasper, A. J., and W. P. Fifer (1980). *Science* **208**:1174–1176. Of human bonding: Newborns prefer their mothers' voices.
DeCasper, A. J., and M. J. Spence (1986). *Infant Behavior and Development* **9**:133–150. Prenatal maternal speech influences newborns' perception of speech sounds.
DeFelipe, J., and E. G. Jones (1988). *Cajal on the Cerebral Cortex.* Oxford University Press, New York.
DeFelipe, J., and E. Jones (1992). *Trends Neurosci.* **15**:237–246. Santiago Ramón y Cajal and methods in neurohistology.
Deiters, O. F. K. (1865). *Untersuchungen über Gehirn und Rückenmark des Menschen und der Säugethiere* (M. Schultze, ed.). F. Vieweg und Sohn, Braunschweig.
del Castillo, J., and B. Katz (1954). *J. Physiol.* **124**:560–573. Quantal components of the end plate potential.
Del Cerro, M., and A. A. Monjan (1979). *Neuroscience* **4**:1399–1404. Unequivocal demonstration of the hematogenous origin of brain macrophages in a stab wound by a double-label technique.
Dennett, D. C. (1978). *Brainstorms: Philosophical Essays on Mind and Psychology.* Bradford, Montgomery, Vermont.
Dennett, D. C. (1987). *The Intentional Stance.* MIT Press, Cambridge, Massachusetts.
Dennett, D. C. (1991). *Consciousness Explained.* Little, Brown, Boston.
Denny-Brown, D. (1952). *Am. J. Psychol.* **65**:474–477. Charles Scott Sherrington.

Denny-Brown, D. (1957). *J. Neurophysiol.* **20**:543–548. The Sherrington school of physiology.
Denny-Brown, D. (1970). Augustus Volney Waller (1816–1870), pp. 88–91. In *The Founders of Neurology*, 2nd edn., (W. Haymaker, and F. Schiller, eds.), Thomas, Springfield, Illinois.
De Robertis, E. (1956). *J. Biophys. Biochem. Cytol.* **2**:503–512. Submicroscopic changes in the synapse after nerve section in the acoustic ganglion of the guinea pig. An electron microscope study.
De Robertis, E. (1959). *Int. Rev. Cytol.* **8**:61–96. Submicroscopic morphology of the synapse.
De Robertis, E., and H. S. Bennett (1954). *Fed. Proc.* **13**:35. Submicroscopic vesicular component in the synapse.
De Robertis, E., and H. S. Bennett (1955). *J. Biophys. Biochem. Cytol.* **1**:47–58. Some features of the submicroscopic morphology of synapses in frog and earthworm.
Descartes, R. (1931). *The Philosophical Works of Descartes Rendered into English*, by E. S. Haldane and G. R. T. Ross. 2 vols. Cambridge Univ. Press, Cambridge.
Descartes, R. (1984). *The Philosophical Writings.* 2 vols. Translated by J. Cottingham, R. Stoothoff, and D. Murdoch. Cambridge Univ. Press, Cambridge.
Devor, M. (1987). *Effects of Injury on Trigeminal and Spinal Somatosensory Systems* (L. M. Pubols and B. J. Sessle, eds.), p. 215–225, Liss, New York.
Diamond, I. T. (1982). The functional significance of architectonic subdivisions of the cortex: Lashley's criticism of the traditional view, p. 101–136. In *Neuropsychology After Lashley* (J. Orbach, ed.), Erlbaum, Hillsdale, New Jersey.
DiStefano, P. S., and E. M. Johnson, Jr. (1988). *J. Neurosci.* **8**:231–241. Nerve growth factor receptors on cultured rat Schwann cells.
Dodds, W. J. (1878). *J. Anat. Physiol. London* **12**:340–363, 454–494, 636–660. On the localization of the functions of the brain: being an historical and critical analysis of the question.
Dodge, R. (1926). *Psychol. Rev.* **33**:106–122, 167–187. Theories of inhibition.
Döllinger, I. (1814). *Beyträge zur Entwicklungsgeschichte des menschlichen Gehirns.* Heinrich Ludwig Brönner, Frankfurt am Main.
Dowling, J., and B. Boycott (1966). *Proc. Roy. Soc. Lond.* **160**:80–111. Organization of the primate retina: Electron microscopy.
Drabkin, D. L. (1958). *Thudichum Chemist of the Brain.* Univ. Pennsylv. Press, Philadelphia.
Dreyfus, H. (1972). *What Computer's Can't Do: A Critique of Artificial Reason.* Harper and Row, New York.
Driesch, H. (1908). *The Science and Philosophy of the Organism*, 2 Vols. Adam and Charles Black, London.
Driesch, H. (1914). *The History and Theory of Vitalism.* Macmillan, London.
du Bois-Reymond, E. (1848–1884). *Untersuchungen über thierische Elektricität.* 2 vols. Reimer, Berlin. Vol. 1 (1848); Vol. 2, part 1 (1849); Vols. 2, part 2, pp. 1–384 (1860); Vol. 2, part 2, pp. 385–579 (1884).
du Bois-Reymond, E. (1875–1877). *Gesammelte Abhandlungen zur allgemeinen Muskel- und Nervenphysik.* 2 vols. Verlag von Veit & Co., Leipzig.
du Bois-Reymond, E. (1912). *Reden*, 2 vols, 2nd edn. Verlag von Veit & Co., Leipzig (1st edn. 1887).
Ducasse, C. J. (1951). *Philos. Rev.* **60**:56–69 and 213–234. Whewell's philosophy of scientific discovery.
Duhem, P. (1906). *La Théorie physique: son objet et sa structure.* Paris. Reprinted 1954 as *The Aim and Structure of Physical Theory* (N. Wiener, transl.), Princeton Univ. Press, Princeton.
Duhem, P. (1915). *La Science allemande.* Librairie Scientifique A. Hermann et Fils, Paris. Translated by J. Lyon as German Science. Open Court, La Salle, Illinois (1991).
Duhem, P. (1954). *The Aim and Structure of Physical Theory.* Translated from the 1906 edition by N. Wiener. Princeton University Press, Princeton, New Jersey.
Dustin, A. P. (1910). *Arch. Biol. Liége* **25**: 269–388. Le rôle des tropismes et de l'odogenés dans la régéneration du systéme nerveus.
Duval, M. (1897). *Précis d'Histologie.* Masson, Paris.
Eccles, J. C. (1937). *Physiol. Rev.* **17**:538–555. Synaptic and neuro-muscular transmission.
Eccles, J. C. (1953). *The Neurophysiological Basis of Mind: The Principles of Neurophysiology.* Clarendon Press, Oxford.
Eccles, J. C. (1957). *Notes and Records of the Royal Society of London* **12**:216–225. Some aspects of Sherrington's contribution to neurophysiology.
Eccles, J. C. (1959). The development of ideas on the synapse, pp. 39–66. In *The Historical Development of Physiological Thought* (C. Mac. Brooks and P. F. Cranefield, eds.), Hafner, New York.

Eccles, J. C. (1964). *The Physiology of Synapses*. Springer Verlag, Berlin.
Eccles, J. C. (1973). *The Understanding of the Brain*. McGraw-Hill, New York.
Eccles, J. C. (1974). Cerebral activity and consciousness, pp. 87–105. In *Studies in the Philosophy of Biology* (F. J. Ayala and T. Dobzhansky, eds.), Macmillan, London.
Eccles, J. C. (1975). Under the spell of the synapse, pp. 159–179. In *The Neurosciences: Paths of Discovery* (F. G. Worden, J. P. Swazey, G. Adelman, eds.), MIT Press, Cambridge, Massachusetts.
Eccles, J. C. (1976). Brain and free will, pp. 101–121. In *Consciousness and the Brain* (G. G. Globus, G. Maxwell, and I Savodnik, eds.), Plenum Press, New York.
Eccles, J. C. (1989). *Evolution of the Brain: Creation of the Self*. Routledge, London.
Eccles, J. C. (1990). *J. Neurosci.* **10**:3769–3781. Developing concepts of synapses.
Eccles, J. C., and W. C. Gibson (1979). *Sherrington His Life and Thought*. Springer Verlag, Berlin, Heidelberg, New York.
Eccles, J. C., and C. S. Sherrington (1929). *J. Physiol. London,* **69**:p.i. Improved bearing for the torsion myograph.
Eccles, J. C., P. Fatt, and S. Landgren (1953). *Aust. J. Sci.* **16**:130–134. The 'direct' inhibitory pathway in the spinal cord.
Eccles, J. C., P. Fatt, and K. Koketsu (1954). *J. Physiol. London* **216**:524–562. Cholinergic and inhibitory synapses in a pathway from motor-axon collaterals to motoneurons.
Eccles, J. C., P. Fatt, and S. Landgren (1956). *J. Neurophysiol.* **19**:75–98. Central pathway for direct inhibitory action of impulses in largest nerve fibres to muscle.
Eccles, J. C., R. Llinas, and K. Sasaki (1966). *Exp. Brain Res.* **1**:1–16. The inhibitory interneurons within the cerebellar cortex.
Ecker, A. (1873). *The Cerebral Convolutions of Man Represented According to Original Observations, Especially Upon Their Development in the Foetus*. Transl. by R. T. Edes from the 1870 German edition. Appleton, New York.
Eckhard, C. (1849). *Zeitschr. f. rat. Med. Heidelberg.* **1**:281–310. Über reflexbewegungen der vier letzten Nervenpaare des Frosches.
Economo, C. (1929). *Zeitschr. ges. Neurol. Psychiat.* **121**:323–409. Wie sollen wir Elitgehirne verarbeiten?
Eddington, A. S. (1935). *New Pathways In Science*. Univ. Michigan Press, Ann Arbor.
Eddington, A. S. (1939). *The Philosophy of Physical Science*. Univ. Michigan Press, Ann Arbor.
Edelman, G. M. (1987). *Neural Darwinism. The Theory of Neuronal Group Selection*. Basic Books, New York.
Edinger, L. (1889). *Vorlesungen über den Bau der nervösen Zentralorgane des Menschen und der Thiere*. 2^e Auflage. Verlag von F. C. W. Vogel, Leipzig.
Edinger, L. (1891). *Twelve Lectures on the Structure of the Central Nervous System*. Second Revised Edition, translated by W. H. Vittum, edited by C. E. Riggs. Davis, Philadelphia.
Edinger, L. (1904). *Vorlesungen über den Bau der nervösen Zentralorgane des Menschen und der Thiere*. 7th edition. Verlag von F. C. W. Vogel, Leipzig.
Edinger, L. (1909). *Einführung in die Lehre vom Bau und Verrichtungen des Nervensystems*. Verlag von F. C. W. Vogel, Leipzig.
Edinger, L., and B. Fischer (1913). *Arch. ges. Physiol.* **152**:1–27. Ein Mensch ohne Grosshirn.
Edwards, W., H. Lindeman, and L. J. Savage (1963). *Psychol. Rev.* **70**:193–242. Bayesian statistical inference for psychological research.
Eekman, F., ed. (1982). *Analysis and Modeling of Neural Systems*. Kluwer, Amsterdam.
Eggermont, J. J. (1990). *The Correlative Brain. Theory and Experiment in Neural Interaction*. Springer-Verlag, Berlin.
Ehrenberg, C. G. (1833). *Annalen der Physik und Chemie* (Poggendorff's) **28**:449–473. Nothwendigkeit einer feineren mechanischen Zerlegung des Gehirns und der Nerven vor der chemischen, dargestellt aus Beobachtungen von C. G. Ehrenberg.
Ehrenberg, C. G. (1836). *Beobachtung einer auffallenden bisher unerkannten Struktur des Seelenorgans bei Menschen und Thieren*. Königlichen Akademie der Wissenschaften, Berlin. English transl. in *Edinburgh Med. Surg. J.* (1837) **48**:258–304.
Ehrlich, P. (1885). *Sauerstoffbedürfnis des Organismus*. Springer, Berlin.
Ehrlich, P. (1886). *Dtsch. med. Wochenschr.* **12**:49–52. Über die Methylenblaureaction der lebenden Nervensubstanz.

Eisenberg, N., and P. A. Miller (1987). *Psychol. Bull.* **101**:91–119. The relation of empathy to prosocial and related behaviors.

Ekman, P. (1973). Cross-cultural studies of facial expression. In *Darwin and Facial Expression* (P. Ekman, ed.), Academic Press, New York.

Ekman, P., and W. V. Friesen (1975). *Unmasking the Face: A Guide to Recognizing Emotions from Facial Clues.* Prentice-Hall, Englewood Cliffs, New Jersey.

Ekman, P., and W. V. Friesen (1978). *Facial Action Coding Systems.* Consulting Psychologists Press, Palo Alto, California.

Emmert, F. (1836). *Ueber die Endigungsweise der Nerven in den Muskeln.* Bern.

Estes, D., H. M. Wellman, and J. D. Woolley (1989). *Adv. Child Dev. Behav.* **22**:41–87. Children's understanding of mental phenomena.

Ewald, A. (1876). *Pflüger's Arch.* **12**:529–549. Ueber die Endigung der motorischen Nerven in den quergestreiften Muskeln.

Farber, P. L. (1976). *J. Hist. Biol.* **9**:93–119. The Type-Concept in Zoology during the First Half of the Nineteenth Century.

Farquhar, M. G., and J. F. Hartmann (1957). *J. Neuropath. Exp. Neurol.* **16**:18–39. Neuroglial structure and relationships as revealed by electron microscopy.

Fatt, P. (1954). *Physiol. Rev.* **34**:674–710. Biophysics of junctional transmission.

Fatt, P. (1959). Skeletal neuromuscular transmission, pp. 199–214. In *Handbook of Physiology*, Section 1. *Neurophysiology*, Vol. 1. Amer. Physiol. Soc. Washington, D.C.

Fatt, P., and B. Katz (1950). *Nature* **166**:597–598. Some observations on biological noise.

Fatt, P., and B. Katz (1951). *J. Physiol. London* **115**:320–370. An analysis of the end-plate potential recorded with an intracellular electrode.

Fatt, P., and B. Katz (1952). *J. Physiol. London* **117**:109–128. Spontaneous subthreshold activity at motor nerve endings.

Fearing, F. (1930). *Reflex Action: A Study in the History of Physiological Psychology.* Williams and Wilkins, Baltimore.

Fechner, G. T. (1860). *Elemente der Psychophysik.* Republished 1964, by E. J. Bonset, Amsterdam.

Federoff, S. (1985). Macroglial cell lineages, pp. 91–117. In *Molecular Bases of Neural Development* (G. M. Edelman, W. E. Gall, and W. M. Cowan, eds.), Neuroscience Res. Found., New York.

Feigl, H. (1958). The 'Mental' and the 'Physical,' pp. 370–497. In *Minnesota Studies in the Philosophy of Science*, Vol. II (H. Feigl, M. Scriven, and G. Maxwell, eds.), Univ. Minnesota Press, Minneapolis.

Feigl, H. (1960). Mind-body not a pseudoproblem. In *Dimensions of Mind* (S. Hook, ed.). New York Univ. Press, New York.

Feldberg, W., and J. H. Gaddum (1934). *J. Physiol. (London)* **81**:305–319. The chemical transmitter at synapses in a sympathetic ganglion.

Ferrier, D. (1873). *West Riding Lunatic Asylum Medical Reports* **3**:30–96. Experimental researches in cerebral physiology and pathology.

Ferrier, D. (1874a). *Proc. Roy. Soc. London* **22**:229–232. The localization of function in the brain.

Ferrier, D. (1874b). *West Riding Lunatic Asylum Medical Reports* **4**:30–62. Pathological illustrations of brain function.

Ferrier, D. (1876). *The Functions of the Brain.* Smith, Elder, London.

Ferrier, D. (1886). *The Functions of the Brain.* 2nd Edition. Smith, Elder, London.

Ferrier, D., and F. L. Goltz (1881). *Transactions of the International Medical Congress* **1**:228–233, 237–240. Discussion on the localization of function in the cortex cerebri. Reprinted, *J. Neurosurg.* **21**:724–733 (1964).

Ferrier, D., and G. F. Yeo (1881). *Proc. Roy. Soc. London* **32**:12–20. The functional relations of the motor roots of the brachial and lumbo-sacral plexuses.

Feyerabend, P. K. (1957). *Against Method.* New Left Books, London.

Feyerabend, P. K. (1962). Explanation, reduction and empiricism. In *Minnesota Studies in the Philosophy of Science*, Vol. 3 (H. Feigl and G. Maxwell, eds.), Univ. Minnesota Press, Minneapolis.

Feyerabend, P. K. (1963a). *Rev. Metaphysics* **17**:49–66. Materialism and the mind-body problem.

Feyerabend, P. K. (1963b). *J. Philos.* **60**:295–296. Mental events and the brain.

Feyerabend, P. (1970). Problems of Empiricism, II. In *The Nature and Function of Scientific Theory*. University of Pittsburgh Press, Pittsburgh.
Feyerabend, P. K. (1975). *Against Method*. Humanities Press, New Jersey.
Feyerabend, P. K. (1981). *Philosophical Papers*. 2 Vols. Cambridge Univ. Press, Cambridge.
Fick, A. (1882). *Mechanische Arbeit und Wärmeentwicklung bei der Muskelthätigkeit*. Brockhaus, Leipzig.
Field, T. M., R. Woodson, D. Cohen, R. Greenberg, R. Garcia, and K. Collines (1983). *Infant Behavior and Development* **6**:485–489. Discrimination and imitation of facial expression by term and preterm neonates.
Fifková, E. (1985). *Cell Mol. Neurobiol.* **5**:47–63. A possible mechanism of morphometric changes in dendritic spines induced by stimulation.
Fink, R. P., and L. Heimer (1967). *Brain Res.* **4**:369–374. Two methods for selective silver impregnation of degenerating axons and their synaptic endings in the central nervous system.
Fiorito, G., and P. Scotto (1992). *Science* **256**:545–547. Observational learning in *Octopus vulgaris*.
Fisher, R. A. (1936). *Ann. Sci.* **1**:115–137. Has Mendel's work been rediscovered?
Fisher, R. A. (1947). *The Design of Experiments*. 4th Edition. Oliver and Boyd, Edinburgh.
Flanagan, O. (1991). *The Science of the Mind*. 2nd edition. MIT Press, Cambridge, Massachusetts.
Flechsig, P. (1876). *Die Leitungsbahnen im Gehirn und Rückenmark des Menschen auf Grund entwicklungsgeschichtlicher Untersuchungen*. Engelmann, Leipzig.
Flechsig, P. (1889). *Arch. Anat. Physiol. (Physiol. Abt.)* **1889**:537–538. Ueber eine neue Färbungsmethode des centralen Nervensystems und deren Ergebnisse bezüglich des Zusammenhanges von Ganglienzellen und Nervenfasern.
Flechsig, P. (1920). *Anatomie des menschlichen Gehirns und Rückenmarks auf myelogenetische Grundlage*. Thieme, Leipzig.
Flew, A. G. N., and A. Vesey (1987). *Agency and Necessity*. Blackwell, Oxford.
Flourens, P. (1824). *Recherches Expérimentales sur les Propriétés et les Fonctions du Système Nerveux, dans les Animaux Vertébrés*. Crevot, Paris.
Flourens, P. (1842). *Recherches Expérimentales sur les Propriétés et les Fonctions du Système Nerveux dans les Animaux Vertébrés*. 2nd edition, Baillière, Paris.
Fodor, J. A. (1985). *Mind* **94**:76–100. Fodor's guide to mental representation: the intelligent auntie's vademecum.
Foerster, O., O. Gagel, and D. Sheehan (1933). *Z. Anat. Entw.-Gesch.* **101**:553–565. Veränderungen an dem Endösen im Rückenmark des Affen nach Hinterwurzeldurchschneidung.
Fogel, A. and M. Reimers (1989). *Monogr. Soc. Res. Child Develop.* **54**:105–113. On the psychobiology of emotions and their development. Commentary on Malatesta *et al.*
Fontana, F. G. F. (1781). *Traité sur le vénin de la vipère*. 2 vols. Florence.
Forbes, A. (1922). *Physiol. Rev.* **2**:361–414. The interpretation of spinal reflexes in terms of present knowledge of nerve conduction.
Forbes, A. (1939). *J. Neurophysiol.* **2**:465–472. Problems of synaptic function.
Forel, A. (1887). *Arch. Psychiat. Berlin* **18**:162–198. Einige hirnanatomische Betrachtungen und Ergebnisse.
Forssman, J. (1898). (*Ziegler's*) *Beitr. Pathol. Anat. Allgem. Pathol.* **24**:56–100. Ueber die Ursachen, welche die Wachstumsrichtung der peripheren Nervenfasern bei der Regeneration bestimmen.
Forssman, J. (1900). (*Ziegler's*) *Beitr. Pathol. Anat. Allgem. Pathol.* **27**:407–430. Zur Kenntnis des Neurotropismus.
Foster, J. (1991). *The Immaterial Self. A Defense of the Cartesian Dualist Conception of the Mind*. Routledge, London.
Foster, M. (1877). *A Text-Book of Physiology*. Macmillan, London. (Later editions: 2nd, 1878; 3rd, 1879; 5th 1888–1890; 7th, 1897).
Foster, M. (1901). *Lectures on the History of Physiology During the Sixteenth, Seventeenth, and Eighteenth Centuries*. Cambridge Univ. Press, Cambridge.
Fox, R. (1991). *Brit. J. Devel. Psychol.* **9**:281–298. Developing awareness of mind reflected in children's narrative writing.
Frankena, W. K. (1976). The ethics of respect for life, pp. 24–62. In *Respect for Life in Medicine, Philosophy and the Law* (O. Temkin, W. F. Frankena, and S. H. Kadish, eds.). Johns Hopkins Press, Baltimore.

Franz, S. I. (1907). *Arch. Psychol. N. Y.* **1**:1–64. On the functions of the cerebrum.
Franz, S. I. (1923). *Psychol. Rev.* **30**:438–446. Conceptions of cerebral functions.
Freeman, W. J. (1975). *Mass Action in the Nervous System*. Academic Press, New York.
Freud, S. (1895). Project for a Scientific Psychology, pp. 347–445. In *The Origins of Psychoanalysis, Letters to Wilhelm Fliess, Drafts and Notes: 1887–1902* (M. Bonaparte, A. Freud, and E. Kris, eds.). Translated by E. Mosbacher and J. Strachey. Basic Books, New York, 1954.
Frijda, N. (1986). *The Emotions*. Cambridge Univ. Press, Cambridge, England.
Fritsch, G., and E. Hitzig (1870). *Arch. für Anat. Physiol. und wissensch. Med.* (Reichert's und du Bois-Reymond's Archiv) **3**:300–332. Ueber die elektrische Erregbarkeit des Grosshirns.
Fujita, I., K. Tanaka, M. Ito, and K. Cheng (1992). *Nature* **360**:343–346. Columns for visual features of objects in monkey inferotemporal cortex.
Fujita, S., and T. Kitamura (1975). *Acta Neuropath. Suppl.* **6**:291–296. Origin of Brain macrophages and the nature of the so-called microglia.
Fujita, Y. (1968). *J. Neurophysiol.* **31**:131–141. Activity of dendrites of single Purkinje cells and its relationship to so-called inactivation response in rabbit cerebellum.
Fulton, J. F. (1943). *Physiology of the Nervous System*. 2nd edition. Oxford University Press, London.
Fulton, J. F. (1946). *Harvey Cushing: A Biography*. Thomas, Springfield, Illinois.
Fulton, J. F. (1952). *J. Neurophysiol.* **15**:167–190. Sir Charles Scott Sherrington, O.M.
Fung, Y. L. (1947). *The Spirit of Chinese Philosophy*. E. R. Hughes, transl. Kegan Paul, Trench, Trubner, London.
Fung, Y. L. (1948). *A Short History of Chinese Philosophy*. Macmillan, New York.
Funkhouser, H. G. (1937). *Osiris* **3**:269–404. Historical development of the graphical representation of statistical data.
Furshpan, E. J., and D. D. Potter (1957). *Nature* **180**:342–343. Mechanism of nerve-impulse transmission at a crayfish synapse.
Furshpan, E. J., and D. D. Potter (1959). *J. Physiol. London* **145**:289–325. Transmission at the giant motor synapses of the crayfish.
Gadamer, H.-G. (1982). Reason in the Age of Science (F. G. Lawrence, transl.). MIT Press, Cambridge, Massachusetts.
Gall, F. J., and J. C. Spurzheim (1810–1819). *Anatomie et physiologie du systéme nerveux en générale et du cerveau en particulier*. F. Schoell, Paris.
Gallistel, C. R. (1989). *Ann. Rev. Psychol.* **40**:155–189. Animal cognition: The representation of space, time, and number.
Galvani, L. (1791). *De viribus electricitatis in motu musculari commentarius*, translated by R. M. Green, E. Licht, Cambridge, Mass. 1953.
Gampes, E. (1926). *Z. Ges. Neurol. Psychiat.* **102**:154–235 and **104**:49–120. Structure and functional capacity of human mesencephalic monster (arhinencephalic with encephalocele); contributions to teratology and fiber system (in German).
Gardiner, H. M., R. G. Metcalf, and J. G. Beebe-Center (1937). *Feeling and Emotion, A History of Theories*. New York.
Gaskell, W. H. (1887). *J. Physiol. London* **8**:404–414. On the action of muscarin upon the heart, and on the electrical changes in the non-beating cardiac muscle brought about by stimulation of the inhibitory and augmentor nerves.
Gaskell, W. H. (1889). *J. Physiol. (London)* **10**:153–211. On the relation between the structure, function and origin of the cranial nerves, together with a theory on the origin of the nervous system of vertebrata.
Gaskell, W. H. (1908). *The Origin of Vertebrates*. Longmans, Green, London.
Gasser, H. S., and J. Erlanger (1922). *Am. J. Physiol.* **62**:496–524. A study of the action currents of nerve with the cathode ray oscillograph.
Gault, R. H. (1904). *Am. J. Psychol.* **15**:526–568. A sketch of the history of reflex action in the latter half of the nineteenth century.
Gaze, R. M. (1970). *The Formation of Nerve Connections*. Academic Press, New York.
Geddes, L. A. (1984). *Physiologist Suppl.* **27**:1–46. A short history of the electrical stimulation of excitable tissue including electrotherapeutic applications.

Geddes, L. A., and H. E. Hoff (1961). *Arch. internat. d'hist. des sciences* **14**:275–290. The capillary electrometer. The first graphic recorder of bioelectric signals.
Geison, G. L. (1972). *Bull. Hist. Med.* **46**:30–58. Social and institutional factors in the stagnancy of English physiology, 1840–1870.
Geison, G. L. (1978). *Michael Foster and the Cambridge School of Physiology. The Scientific Enterprise In Late Victorian Society*. Princeton Univ. Press, Princeton, N.J.
Gentner, D., and A. L. Stevens (eds.) (1983). *Mental Models*. Erlbaum, Hillsdale, New York.
Geoffroy Saint-Hilaire, É. (1818). *Philosophie Anatomique*. Paris.
Georgopoulos, A. P. (1986). *Ann. Rev. Neurosci.* **9**:147–170. On reaching.
Gerber, S. E., and G. T. Mencher, editors (1983). *The Development of Auditory Behavior*. Grune and Stratton, New York.
Geren, B. B. (1954). *Exp. Cell Res.* **7**:558–562. The formation from the Schwann cell surface of myelin in the peripheral nerves of chick embryos.
Geren, B. B., and J. Raskind (1953). *Proc. Natl. Acad. Sci. USA* **39**:880–884. Development of the fine structure of the myelin sheath in sciatic nerves of chick embryos.
Gerlach, J. (1858). *Microscopische Studien aus dem Gebiet der menschlichen Morphologie*. 2nd ed., 1865. F. Enke, Erlangen.
Gerlach, J. von (1872). *Zentralbl. med. Wiss.* **10**:273–275. Über die struktur der grauen Substanz des menschlichen Grosshirns.
Ghez, C., W. Hening, and J. Gordon (1991). *Current Opinion in Neurobiol.* **1**:664–671. Organization of voluntary movement.
Gick, M. L., and K. J. Holyoak (1980). *Cognitive Psychology* **12**:306–355. Analogical problem solving.
Ginsberg, M. (1944). *Moral Progress*. Frazer Lecture at the University of Glasgow. Glasgow Univ. Press, Glasgow.
Ginsburg, M. (1953). *The Idea of Progress: A Re-evaluation*. Methuen, London.
Glass, B. (1959). Maupertuis, pioneer of genetics and evolution, pp. 51–83. In *Forerunners of Darwin, 1745–1859* (B. Glass, O. Temkin, and W. L. Strauss, Jr., eds.), Johns Hopkins, Baltimore.
Glees, P., and J. Cole (1950). *J. Neurophysiol.* **13**:137–148. Recovery of skilled motor functions after small repeated lesions of motor cortex in macaque.
Goldscheider, A. (1898). *Die Bedeutung der Reize für Pathologie und Therapie im Lichte der Neuronlehre*. Barth, Leipzig.
Golgi, C. (1873). *Gazz. med. ital. Lombardia* **33**:244–246. Sulla struttura della sostanza grigia del cervello. English translation: On the structure of the grey matter of the brain, pp. 647–650. In *Golgi Centennial Symposium* (1975, M. Santini, ed. and transl.), Raven, New York.
Golgi, C. (1875). *Riv. sper. Freniatria Med. legal.* **1**:66–78. Sulla fina struttura dei bulbi olfattori.
Golgi, C. (1879). *Archs. Sci. Med.* **3**:1–7. Di una nuova reazione apparentemente nera della cullule nervose cerebrali ottenuta col bicloruro di mercurio.
Golgi, C. (1880a). *Studi istologici sul midollo spinale. Communicatione fatta al terzo congresso freniatrico italiano tenuto in Reggio-Emilia nel Sett. 1880*. Also in *Arch. ital. per le malattie nervose*, anno 18, 1881. Also in *Anat. Anz.* **5**:431.
Golgi, C. (1880b). *Giornale internaz. della Scienze mediche Anno III*. Sulla origine centrale dei nervi. Communicatione fatta alle sezione anatomia del III Congresso medico in Genova nel Sett 1880.
Golgi, C. (1880c). Archivioper le scienze mediche, pp. 149–171. In *Opera Omnia*, Vol. 2 (1903).
Golgi, C. (1880d). *Mem. Roy. Acad. Sci. Tor.* **32**:359–385. Sui nervi nei tendini dell'uomo e di altri vertebrati e di un nuovo organo nervoso terminale musculo-tendineo.
Golgi, C. (1882–1885). *Rivista sperimentale di Freniatria* **8**:165–195, 361–391 (1882); **9**:1–17, 161–192, 385–402 (1883); **11**:72–123, 193–220 (1885). Sulla fina anatomia degli organi centrali del sistema nervoso. Also in *Opera omnia* 295–393, 397–536.
Golgi, C. (1883a). *Alienist and Neurologist* **4**:383–416. Chap. II. Continuation of the study of the minute anatomy of the central nervous system. Chap. III. Morphology and disposition of the nervous cells in the anterior, central, and superior-occipital convolutions.
Golgi, C. (1883b). *Arch. Ital. Biol.* **3**:285–317; **4**:92–123. Recherches sur l'histologie des centres nerveux.

Golgi, C. (1883c). *Atti del IV Congresso freniatrico ital. ten. in Voghera nel Sett. 1883*. La cellula nervosa motrice.
Golgi, C., (1884). *Arch. ital. Biol.* **4**:92–123. Recherches sur l'histolgie des centres nerveux.
Golgi, C. (1885). *Alienist and Neurologist* **6**:307–324. Continuation of the study of the minute anatomy of the central nervous system. Chap. V. On the minute anatomy of the great foot of the hippocampus.
Golgi, C. (1886). *Arch. ital. Biol.* **7**:15–47. Sur l'anatomie microscopique des organes centraux du système nerveux.
Golgi, C. (1890). *Anat. Anz.* **5**:431. Ueber den Feineren Bau des Rückenmarkes. (Republication of Studi istologici sul midollo spinale in Fatta al 3e Congr. freniatr. ital. tenuto in Reggio-Emilia, Set. 1880).
Golgi, C. (1891a). *Trans. in Arch. ital. de biol. Turin* **15**:434–463. La rete nervosa diffusa degli organi centrali del sistema nervoso; suo significata fisiologico.
Golgi, C. (1891b). *Riv. med. Napol.* **7**:193–194. Modificazione del metodo di colorazione deli elementi nervosi col bicloruro di mercurio.
Golgi, C. (1894). *Untersuchungen über den feineren Bau des centralen und peripherischen Nervensystems*. (R. Teuscher, transl.). Fischer, Jena.
Golgi, C. (1898a). *Arch. Ital. Biol.* **30**:60–71. Sur la structure des cellules nerveuses.
Golgi, C. (1898b). *Arch. Ital. Biol.* **30**:278–286. Sur la structure des cellules nerveuses des ganglions spinaux.
Golgi, C. (1898c). *Boll. Soc. Med. Pavia* **1**:3–16. Intorno alla Struttura delle cellule nervose.
Golgi, C. (1898d). *Boll. Soc. Med. Pavia* **2**:5–15. Sulla struttura delle cellule nervose dei gangli spinali.
Golgi, C. (1899). *Arch. Ital. Biol.* **31**:273–280. Di nuovo sulla struttura delle cellule nervose dei gangli spinali.
Golgi, C. (1900). *Verhandl. deutsch. Anat. Gesellsch.* **14**:164–176. Intorno alla struttura delle cellule nervose della corteccia cerebrale.
Golgi, C. (1901). Sulla fina organizzazione del sistema nervoso. In L. Luciani, *Trattato di fisiologia dell' uomo*. Società editrice libraria, Milan. (Also in *Opera Omnia*, 721–733).
Golgi, C. (1903). *Opera Omnia*. Hoepli, Milano.
Golgi, C. (1907). La Doctrine du Neurone. Théorie et Faits. In *Les Prix Nobel en 1906*. Imprimerie Royale, P.-A. Norstedt et Fils, Stockholm. English translation: The neuron doctrine—theory and facts. Nobel Lecture, December 11, 1906, pp. 189–217. In *Nobel Lectures Physiology or Medicine 1901–1921*. Elsevier, New York, 1967
Goltz, F., and A. Freusberg (1874). *Pflüg. Arch. ges. Physiol.* **8**:460–486. Ueber die Functionen des Lendenmarks des Hundes.
Goodman, N. (1979). *Ways of Worldmaking*. Hackett, New York.
Gordon-Taylor, G., and E. W. Walls (1958). *Sir Charles Bell: His Life and Times*. Livingstone, Edinburgh.
Gotch, F. (1902). *J. Physiol. (London)* **28**:395–416. The sub-maximal electrical response of nerve to a single stimulus.
Gotch, F., and V. Horsley (1888). *Proc. Roy. Soc. London* **45**:18–26. Observations upon the mammalian spinal cord following electrical stimulation of the cortex cerebri.
Gottlieb, G. (1981). Roles of early experience in species-specific perceptual development, pp. 5–44. In *Development of Perception* (R. N. Aslin *et al.*, eds.), Academic Press, New York.
Gould, S. J., and N. Eldredge (1977). *Paleobiology* **3**:115–151. Punctuated equilibria: the tempo and mode of evolution reconsidered.
Graham Brown, T. (1947). *Brit. Med. J.* **2**:810–812. Sherrington: The man.
Graham Brown, T., and C. S. Sherrington (1912). *Proc. Roy. Soc. London Ser. B* **85**:250–277. On the instability of a cortical point.
Granet, M. (1934). *La Pensée Chinoise*. Albin Michel, Paris.
Granit, R. (1952). *Nature* **169**:688. Sir Charles Scott Sherrington: An Appreciation.
Granit, R. (1966). *Charles Scott Sherrington. An Appraisal*. Thomas Nelson, London.
Gray, C. M., P. Konig, A. K. Engel, and W. Singer (1989). *Nature* **338**:334. Oscillatory responses in cat visual cortex exhibit inter-columnar synchronization which reflects global stimulus properties.
Gray, E. G. (1959). *Nature* **183**:1592–1593. Electron microscopy of synaptic contacts on dendritic spines of the cerebral cortex: An electron microscope study.
Gray, E. G., and R. W. Guillery (1961). *J. Physiol. (London)* **157**:581–588. The basis for silver staining of synapses of the mammalian spinal cord: A light and electron microscope study.

Gray, E. G., and R. W. Guillery (1966). *Int. Rev. Cytol.* **19**:111–182. Synaptic morphology in the normal and degenerating nervous system.
Greene, J. C. (1971). The Kuhnian paradigm and the Darwinian revolution in natural history, pp. 3–25. In *Perspectives in the History of Science and Technology* (D. Roller, ed.). University of Oklahoma Press, Norman.
Greene, J. C. (1981). *Science, Ideology and World View: Essays in the History of Evolutionary Ideas.* University of California Press, Berkeley.
Gregory, F. (1977). *Scientific Materialism in Nineteenth Century Germany.* Reidel, Dordrecht.
Grmek, M. D. (1973). *Raisonnement expérimental et recherches toxicologiques chez Claude Bernard.* Droz, Geneva.
Grünbaum, A. (1976). *Brit. J. Philos. Science* **27**:105–136. Is the method of bold conjectures and attempted refutations justifiably the method of science.
Grünbaum, A. S. F., and C. S. Sherrington (1901). *Proc. Roy. Soc. London* **69**:206–209. Observations on the physiology of the cerebral cortex of some of the higher apes.
Grünbaum, A. S. F., and C. S. Sherrington (1902a). *Brit. Med. J.* **2**:784–785. A discussion of the motor cortex exemplified in the anthropoid apes.
Grünbaum, A. S. F., and C. S. Sherrington (1903). *Proc. Roy. Soc. London* **72**:152–155. Observations on the physiology of the cerebral cortex of the anthropoid apes.
Grundfest, H. (1959). Synaptic and ephaptic transmission, pp. 147–197. In *Handbook of Physiology. Section 1*: Neurophysiology (J. Field, H. W. Magoun, and V. E. Hall, eds.). American Physiological Society, Washington, D.C.
Gudden, B. von. (1870). *Arch. Psychiat. (Berlin)* **2**:693–723. Experimentaluntersuchungen über das peripherische und centrale Nervensystem.
Gudden, B. von (1889). *Gesammelte und hinterlassende Abhandlungen.* H. Grashey, Wiesbaden.
Guillery, R. W. (1972). *J. Comp. Neurol.* **144**:117–130. Binocular competition in the control of geniculate cell growth.
Guth, L., and J. J. Bernstein (1961). *Exp. Neurol.* **4**:59–69. Selectivity in the reestablishment of synapses in the superior cervical sympathetic ganglion of the cat.
Gutherie, P. B., M. Segal, and S. B. Kater (1991). *Science* **354**:76–79. Independent regulation of calcium revealed by imaging dendritic spines.
Haeckel, E. (1874). *Anthropogenie oder Entwickelungsgeschichte des Menschen.* Verlag von Wilhelm Engelmann, Leipzig.
Haeckel, E. (1883). *The Evolution of Man* (English Translation of the Third German edition of *Anthropogenie*, 1877). Kegan Paul, Trench & Co., London.
Hajós, F., and E. Bascó (1984). *Adv. Anat. Embryol. Cell Biol.* **84**:1–81. The surface contact glia.
Haldane, E. S., and G. R. T. Ross (1931). *The Philosophical Works of Descartes, Rendered into English.* 2 vols, reprinted with corrections. Cambridge Univ. Press, Cambridge.
Hall, M. (1850). *Synopsis of the Diastaltic Nervous System; or the system of the spinal marrow, and its reflex arcs; as the nervous agent in all the functions of ingestion in the animal economy.* J. Mallett, London.
Hall, T. S. (1969). *Ideas of Life and Matter. Studies in the History of General Physiology 600 B.C.–1900 A.D.*, 2 vols. Univ. Chicago Press, Chicago.
Hamberger, A., H. A. Hanson, and J. Sjöstrand (1970). *J. Cell Biol.* **47**:319–331. Surface structure of isolated neurons: Detachment of nerve terminals during axon regeneration.
Hannover, A. (1840). *Arch. für Anat. Physiol. und wissensch. Med.* pp. 549–558. Die Chromsäure, ein vorzügliches Mittel bei mikroskopischen Untersuchungen.
Hannover, A. (1844). *Recherches Microscopiques sur le Système Nerveux.* Paris.
Hardesty, I. (1904). *Am. J. Anat.* **3**:229–268. On the development and nature of the neuroglia.
Harris, P. L. (1989). *Children and Emotion: The Development of Psychological Understanding.* Blackwell, Oxford.
Harrison, R. G. (1904). *Am. J. Anat.* **3**:197–220. An experimental study of the relation of the nervous system to the developing musculature in the embryo of the frog.
Harrison, R. G. (1906). *Am. J. Anat.* **5**:121–131. Further experiments on the development of peripheral nerves.
Harrison, R. G. (1907a). *J. Exp. Zool.* **4**:239–281. Experiments in transplanting limbs and their bearing upon the problem of the development of nerves.

Harrison, R. G. (1907b). *Anat. Rec.* **1**:116–118. Observations on the living developing nerve fiber.
Harrison, R. G. (1910). *J. Exp. Zool.* **9**:787–846. The outgrowth of the nerve fiber as a mode of protoplasmic movement.
Harrison, R. G. (1912). *Anat. Rec.* **6**:181–193. The cultivation of tissues in extraneous media as a method of morphogenetic study.
Harrison, R. G. (1924). *J. Comp. Neurol.* **37**:123–205. Neuroblast versus sheath cell in the development of peripheral nerves.
Hartley, D. (1749). *Observations on Man, His Frame, His Duty and His Expectations*, 2 vols. Leake & Frederick, London.
Hassall, A. H. (1849). *The microscopic anatomy of the human body, in health and disease*. 2 vols. S. Highley, London.
Hatai, S. (1902). *J. Comp. Neurol.* **12**:291–296. On the origin of neuroglia tissue from the mesoblast.
Hebb, D. H. (1949). *The Organization of Behavior: A Neuropsychological Theory*. Wiley, New York.
Hebb, D. O. (1966). *A Textbook of Psychology*, 2nd ed. Saunders, Philadelphia.
Heidegger, M. (1957). *Sein und Zeit*, 8th ed. Niemeyer, Tübingen.
Heisenberg, W. (1958). *Daedalus* **87**:95–108. The representation of nature in contemporary physics.
Held, H. (1897). Zweite Abhandlung. *Arch. Anat. Physiol. (Anat. Abt.)* pp. 204–294. Beiträge zur Structur der Nervenzellen und ihrer Fortsätze.
Held, H. (1905). *Arch. Anat. Physiol.* Leipzig p. 55–78. Zur Kenntnis einer neurofibrillären. Continuität im Centralnervensystem der Wirbelthiere.
Held, H. (1909). *Die Entwickelung der Nervengewebe bei den Wirbelthieren*. Verlag von Johann Ambrosius Barth, Leipzig.
Held, H. (1929). Die Lehre von den Neuronen und vom Neurencytium und ihr heutiger Stand, pp. 1–72. In *Fortschritte der Naturwissenschaftliche Forschung Heft 8 (Neue Folge)* (E. Aberhalden, ed.), Urban und Schwarzenberg, Berlin.
Helmholtz, H. von (1842). *De fabrica systematis nervosi evertebratorum*. Typis Nietackianis, Berlin.
Helmholtz, H. von (1847). *Ueber die Erhaltung der Kraft*. Vortrag in der physikalischen Gesellschaft zu Berlin.
Helmholtz, H. von (1850). *Arch. Anat. Physiol.* (Anat. Abt., Supplement-Bd.) pp. 71–73. Vorläufiger Bericht über die Fortpflanzungsgeschwindigkeit der Nervereizung. Transl. in W. Dennis (1948), *Readings in Psychology*, pp. 197–198, Appleton-Century-Crofts, New York.
Helmholtz, H. von (1909). *Handbuch der physiologische Optik*. L. Voss, Hamburg & Leipzig.
Hempel, C. G. (1965a). *The Philosophy of Natural Science*. Prentice-Hall, Englewood Cliffs, N.J.
Hempel, C. G. (1965b). *Aspects of Scientific Explanation*. Free Press, New York.
Henle, J. (1871). *Handbuch der systematischen Anatomie des Menschen, Bd. 3, Nervenlehre*. F. Vieweg, Braunschweig.
Henle, J. (1879a). *Handbuch der Anatomie der Nervensystems*. Friedrich, Braunschweig.
Henle, J. (1879b). *Handbuch der systematischen Anatomie des Menschen. Heft 3, 2e Teil, Handbuch der Nervenlehre des Menschen*. 2 Edn. Friedrich, Braunschweig.
Hensen, V. (1864). *Arch. für Path. Anat. (Virchow's Archiv)* **31**:51–73. Über die Entwickelung des Gewebes und der Nerven im Schwanze der Froschlarve.
Hering, E. (1868). *Die Lehre vom binocularen Sehen*. Verlag von Wilhelm Engelmann, Leipzig.
Hermann, F. (1884). *Arch. f. mikr. Anat.* **24**:216–229. Beitrag zur Entwicklungsgeschigte des Geschmacksorgans beim Kaninchen.
Hermann, L. (1870). *Arch. ges. Physiol.* **3**:15–34. Weitere Untersuchungen über die Ursache der electromotorischen Erschwindigungen an Muskel und Nerven.
Hermann, L. (1892). *Lehrbuch der Physiologie*. Zehnte, vielfach verbesserte Auflage, August Hirschwald, Berlin.
Hertwig, O. (1893–1898). *Die Zelle und die Gewebe*. 2 vols. Gustav Fischer, Jena.
Hertwig, O. (1911–1912). *Arch. mikr. Anat.* **79**:113–120. Methoden und Versuche zur Erforschung der Vita propria abgetrennter Gewebs- und Organstückchen von Wierbeltieren.
Hertwig, O. and R. Hertwig (1879,1880). *Jena Zeitschr.* **13**:457–460; **14**:39–80. Die *Actinien* anatomisch und histologisch mit besonderer Berücksichtigung des Nervenmuskelsystems untersucht.
Hertz, H. (1899). *The Principles of Mechanics*. English translation by D. E. Jones and J. T. Walley from the German edition of 1894. Republished 1956, Dover, New York.

Hesse, M. (1966). *Models and Analogies in Science*. Univ. Notre Dame Press, Notre Dame, Indiana.
Hesse, M. (1980). *Revolutions and Reconstructions in the Philosophy of Science*. The Harvester Press, Brighton, New Jersey.
Hiatt, S., J. J. Campos, and R. N. Emde (1979). *Child Develop.* **50**:1020–1035. Facial patterning and infant emotional expression: Happiness, surprise and fear.
Hild, W. (1957). *Z. Zellforsch. Mikroskop. Anat.* **46**:71–95. Myelogenesis in cultures of mammalian central nervous system.
Hill, A. (1896). *Brain* **19**:1–42. The chrome-silver method: A study of the conditions under which the reaction occurs and a criticism of its results.
Hill, A. (1900). *Brain* **23**:657–690. Considerations opposed to the 'neuron theory.'
Hines, M. (1929). *Physiol. Rev.* **9**:462–574. On cerebral localization.
Hirano, A. (1968). *J. Cell Biol.* **38**:637–640. A confirmation of oligodendroglial origin of myelin in the adult rat.
Hirano, A., and H. M. Dembitzer (1967). *J. Cell Biol.* **34**:555–567. A structural analysis of the myelin sheath in the central nervous system.
Hirano, A., H. M. Zimmerman, and S. Levine (1966). *J. Cell Biol.* **31**:397–411. Myelin in the central nervous system as observed in experimentally induced edema in the rat.
Hirano, M., and J. E. Goldman (1988). *J. Neurosci. Res.* **21**:155–167. Gliogenesis in rat spinal cord: Evidence of origin of astrocytes and oligodendrocytes from radial precursors.
Hirsch, H. V. B. and M. Jacobson (1975). The perfectible brain, p. 107–137. In *Foundations of Psychobiology* (M. Gazzaninga and C. Blakemore, eds.). Academic Press, New York.
His, W. (1856). *Beiträge zur normalen und pathologischen Histologie der Kornea*. Schweighauser, Basel.
His, W. (1868). *Untersuchungen über die erste Anlage des Wirbeltierleibes: Die erste Entwicklung des Hühnchens im Ei*. Vogel, Leipzig.
His, W. (1874). *Unserer Körperform und das physiologische Problem ihrer Entstehung*. Engelmann, Leipzig.
His, W. (1879). *Arch. Anat. Physiol. Leipzig, Anat. Abth.*, pp. 455–482. Ueber die Anfänge des peripherischen Nervensystemes.
His, W. (1886). *Abh. kgl. sächs. Ges. Wissensch. math. phys. Kl.* **13**:147–209, 477–513. Zur Geschichte des menschlichen Rückenmarks und der Nervenwurzeln.
His, W. (1887). *Arch. Anat. Physiol. Leipzig, Anat. Abth.* **92**:368–378. Die Entwickelung der ersten Nervenbahnen beim menschlichen Embryo. Übersichtliche Darstellung.
His, W. (1888a). *Abh. kgl. sächs. Ges. Wissensch. math. phys. Kl.* **24**:339–392. Zur Geschichte des Gehirns sowie der centralen und peripherischen Nervenbahnen beim menschlichen Embryo.
His, W. (1888b). *Proc. Roy. Soc. Edin.* **15**:287–297. On the principles of animal morphology.
His, W. (1889a). *Abh. kgl. sächs. Ges. Wissensch. math. phys. Kl.* **15**:311–372. Die Neuroblasten und deren Entstehung im embryonalen Mark.
His, W. (1889b). *Abh. kgl. sächs. Ges. Wissensch. math. phys. Kl.* **15**:673–736. Die Formentwickelung des menschlichen Vorderhirns vom Ende des ersten bis zum Beginn des dritten Monats.
His, W. (1890). *Arch. Anat. Physiol. Leipzig, Anat. Abt. Suppl.* **95**:95–119. Histogenese und Zusammenhang der Nervenelemente.
His, W. (1894). *Arch. Anat. Physiol. Leipzig, Anat. Abth.* **1**:1–80. Ueber mechanische Grundvorgänge thierischer Formenbildung.
His, W. (1904a). *Die Entwicklung des menschlichen Gehirns waehrend der ersten Monate*. S. Hirzel, Leipzig.
His, W. (1904b). *Lebenserinnerungen*. Als Manuskript gedruckt, Leipzig.
Hitzig, E. (1874). *Arch. für Anat., Physiol. und wissensch. Med.* (Reichert's und du Bois-Reymond's Archiv) **7**:392–441. Untersuchungen über das Gehirn, neue Folge.
Hitzig, E., and G. T. Fritsch (1870). *Arch. Anat. Physiol.* pp. 300–322. Über die elektrische Erregbarkeit des Grosshirns. English transl. in *J. Neurosurg.* **20**:904–916.
Hobbes, T. (1651). *Leviathan, or the Matter, Form and Power of a Commonwealth*. London.
Hobson, J. A. (1988). *The Dreaming Brain*. Basic Books, New York.
Hodgkin, A. L., and A. F. Huxley (1952). *J. Physiol. London* **117**:500–544. A quantitative description of membrane current and its application to conduction and excitation in nerve.
Hodgson, D. (1991). *The Mind Matters*. Clarendon Press, Oxford.
Hoerr, N. L. (1936). *Anat. Rec.* **66**:81–90. Cytological studies by the Altmann-Gersh freeze-drying method. III. The preexistence of neurofibrillae and their disposition in the nerve fiber.

Hoff, E. C. (1932). *Proc. Roy. Soc. London Ser. B* **111**:175–188. Central nerve terminals in the mammalian spinal cord and their examination by experimental degeneration.

Hoff, H. E., and L. A. Geddes (1957). *Bull. Hist. Med.* **31**:212–234, 327–347. The rheotome and its prehistory: a study in the historical interrelation of electrophysiology and electromechanics.

Hoff, H. E., and P. Kellaway (1952). *J. Hist. Med.* **7**:211–249. The early history of the reflex.

Hoffman, M. (1975). *Dev. Psychol.* **11**:607–622. Developmental synthesis of affect and cognition and its implications for altruistic motivation.

Hoffman, M. (1981). Affective and cognitive processes in moral internalization. In *Social Cognition and Social Behavior: Developmental Perspectives* (E. T. Higgins, D. N. Rubel, and W. W. Hartley, eds.), Cambridge Univ. Press, New York.

Hoffman, M. (1982). Development of prosocial motivation: Empathy and guilt. In *The Development of Prosocial Behavior* (N. Eisenberg, ed.). Academic Press, New York.

Hoffmann, P. (1922). *Untersuchungen über die Eigenreflexe (Sehnenreflexe) menschlicher Muskeln*. Springer, Berlin.

Holland, J. H., K. J. Holyoak, R. E. Nisbett, and P. R. Thagard (1986). *Induction*. MIT Press, Cambridge, Massachusetts.

Holmes, F. L. (1974). *Claude Bernard and Animal Chemistry*. Harvard Univ. Press, Cambridge, Massachusetts.

Holyoak, K. J. (1984). Analogical thinking and human intelligence. In *Advances in the Psychology of Human Intelligence*, Vol. 2 (R. J. Sternberg, ed.). Erlbaum, Hillsdale, New Jersey.

Holyoak, K. J., E. N. Junn, and D. Billman (1984). *Child Development* **55**:2042–2055. Development of analogical problem-solving skill.

Hooker, C. (1981). *Dialogue* **20**:38–60, 201–235, 496–529. Towards a general theory of reduction.

Hooker, D. (1944). *The Origin of Overt Behavior*. University of Michigan Press, Ann Arbor.

Hooker, D. (1952). *The Prenatal Origin of Behavior*. Univ. Kansas Press, Lawrence, Kansas.

Hopfield, J. J., and D. W. Tank (1986). *Science* **233**:625–633. Computing with neural circuits.

Horner, K. C., J. Serviere, and C. Granier-Deferre (1987). *Hearing Res.* **26**:327–333. Deoxyglucose demonstration of in-utero hearing in the guinea pig foetus.

Hornick, K., M. Stinchcombe, and H. White (1989). *Neural Networks* **2**:359–368. Multilayer feedforward networks are universal approximators.

Horsley, V. A. H., and E. A. Schäfer (1888). *Phil. Trans. Roy. Soc. London Ser. B* **179**:1–45. A record of experiments upon the functions of the cerebral cortex.

Howson, C., and P. Urbach (1989). *Scientific Reasoning. The Bayesian Approach*. Open Court, La Salle, Illinois.

Hubel, D. H. and T. N. Wiesel (1962). *J. Physiol. (London)* **160**:106–154. Receptive fields, binocular interaction and functional architecture in the cat's visual cortex.

Hubel, D. H., and T. N. Wiesel (1965). *J. Neurophysiol.* **28**:229–289. Receptive fields and functional architecture in two nonstriate areas (18 and 19) of the cat.

Hubel, D. H., and T. N. Wiesel (1968). *J. Physiol. (London)* **195**:215–243. Receptive fields and functional architecture of monkey striate cortex.

Hubel, D. H., and T. N. Wiesel (1974a). *J. Comp. Neurol.* **158**:267–294. sequence regularity and geometry of orientation columns in the monkey striate cortex.

Hubel, D. H., and T. N. Wiesel (1974b). *J. Comp. Neurol.* **158**:295–306. Uniformity of monkey striate cortex: A parallel relationship between field size, scatter, and magnification factor.

Hubel, D. H., T. N. Wiesel, and M. P. Stryker (1978). *J. Comp. Neurol.* **177**:361–380. Anatomical demonstration of orientation columns in macaque monkey.

Hull, D. (1988). *Science as a Process. An Evolutionary Account of the Social and Conceptual Development of Science*. Univ. Chicago Press, Chicago and London.

Hull, D. L. (1964–1965). *Brit. J. Philos. Sci.* **15**:315–326. The effect of essentialism on taxonomy—two thousand years of stasis (1).

Hume, D. (1739). *A Treatise of Human Nature: Being an Attempt to introduce the experimental Method of Reasoning into Moral Subjects*. John Noon, London.

Humphrey, D. R. (1979). On the cortical control of visually directed reaching: Contributions by nonprecentral motor areas, pp. 51–112. In *Posture and Movement* (R. E. Talbott and D. R. Humphrey, eds.), Raven Press, New York.

Humphrey, D. R. (1986). *Fed. Proc.* **45**:2687–2699. Representation of the movements and muscles within the primate precentral motor cortex: Historical and current perspectives.

Humphrey, G. (1951). *Thinking*. Methuen, London.

Humphreys, A. W. (1937). *Ann. Sci.* **2**:164–178. The development of the conception and measurement of electric current.

Hunter, W. S. (1930). *J. gen. Psychol.* **3**:455–468. A consideration of Lashley's theory of the equipotentiality of cerebral action.

Huttenlocher, P. R., and C. de Courten (1987). *Human Neurobiol.* **6**:1–9. The development of synapses in striate cortex of man.

Huxella, T. (1937–1938). *Anat. Anz.* **85**:168–187. Michael von Lenhossék.

Huxley, T. H. (1863). *Evidence as to Man's Place in Nature*. London. Reprinted, Univ. Michigan Press, Ann Arbor, 1959.

Huxley, T. H. (1874). *Nature* **10**:362–366. On the hypothesis that animals are automata, and its history.

Huxley, T. H. (1894). *Discourses Biological and Geological*. Appleton, New York.

Imamoto, K., and C. P. Leblond (1977). *J. Comp. Neurol.* **174**:255–280. Presence of labelled monocytes, macrophages and microglia in a stab wound of the brain following an injection of bone marrow cells labelled with ^{3}H-uridine into rats.

Izard, C. E. (1971). *The Face of Emotion*. Appleton-Century-Crofts, New York.

Izard, C. E., and C. Z. Maletesta (1987). Perspectives on emotional development. I: Differential emotions theory of early emotional development, pp. 494–554. In *Handbook of Infant Development* (J. O. Osofsky, ed.), 2nd Edition. Wiley, New York.

Izard, C. E., R. Huebner, D. Risser, G. McGinnis, and L. Dougherty (1980). *Develop. Psychol.* **16**:132–140. The young infant's ability to produce discrete emotional expressions.

Jackson, F. (1982). *Philosophical Quarterly* **32**:127–136. Epiphenomenal Qualia.

Jackson, F. (1986). *J. Philos.* **83**:291–295. What Mary didn't know.

Jackson, J. H. (1884). *Brit. Med. J.* **1**:591–593, 660–663, 703–707. Croonian Lecture: Evolution and dissolution of the nervous system.

Jackson, J. H. (1898). *Br. Med. J.* **1**:65–69. Remarks on the relations of different divisions of the central nervous system to one another and to parts orf the body.

Jackson, J. H. (1915). *Brain* **38**:75–79. On the anatomical and physiological localization of movements in the brain.

Jackson, J. H. (1931). *Selected Writings of John Hughlings Jackson*. 2 vols. (J. Taylor, ed.). Hodder and Stoughton, London. Reprinted, Basic Books, New York, 1958.

Jacobsen, C. F. (1932). *Proc. Ass. Res. Nerv. Ment. Dis.* **13**:225–247. Influence of motor and premotor area lesions upon the retention of skilled movement in monkeys and chimpanzees.

Jacobson, M. (1970a). *Developmental Neurobiology*. 1st Edition. Holt, Rinehart and Winston, New York.

Jacobson, M. (1970b). Development, specification and diversification of neuronal connections. In *The Neurosciences: Second Study Program* (F. O. Schmitt, ed.-in-chief). Rockefeller Univ. Press, New York.

Jacobson, M. (1974a). *Ann. New York Acad. Sci.* **228**:63–67. Through the jungle of the brain: Neuronal specificity and typology revisited.

Jacobson, M. (1974b). A plenitude of neurons, pp. 151–166. In *Studies on the Development of Behavior and the Nervous System*, Vol. 2 (G. Gottlieb, ed.), Academic Press, New York.

Jacobson, M. (1991). *Developmental Neurobiology*, 3rd Edition. Plenum Press, New York.

James, W. (1884). *Mind* **9**:188–204. What is emotion?

James, W. (1890). *The Principles of Psychology*. 2 Vols. Holt, New York.

Jefferson, G. (1949). *Irish J. Med. Sci.* 6th series, No. 285, pp. 691–706. René Descartes on the localization of the soul.

Jelliffe, S. E. (1937). *J. Nerv. Ment. Dis.* **85**:696–711. Sigmund Freud as a neurologist. Some notes on his earlier neurobiological and clinical neurological studies.

Jennings, H. S. (1906). *Behavior of the Lower Organisms*. Columbia Univ. Press, New York.

Johansson, B., E. Wedenberg, and B. Westin (1963). *Acta oto-laryng.* **57**:188–192. Measurement of tone response by the human foetus.

John, H. J. (1959). *Jan Evangelista Purkyne, Czech Scientist and Patriot*. American Philosophical Society, Philadelphia.

Johnson, S. M. (1987). *Humanizing the Narcissistic Style*. Norton, New York.

Johnson, W. E. (1924). *Logic*. Cambridge Univ. Press, Cambridge.
Johnson-Laird, P. N. (1983). *Mental Models*. Harvard Univ. Press, Cambridge, Massachusetts.
Johnston, J. B. (1906). *The Nervous System of Vertebrates*. P. Blakiston's Son & Co., Philadelphia.
Jolly, W. A. (1911). *Quart. J. exp. Physiol.* **4**:67–87. The time relations of the knee-jerk and simple reflexes.
Jones, E. (1953). *The Life and Works of Sigmund Freud*, Vol. 1. Basic Books, New York.
Jones, E. G. (1972). *J. Hist. Med.* **27**:298–311. The development of the 'muscular sense' concept during the nineteenth century and the work of H. Charlton Bastian.
Jones, E. G. (1975). *J. Comp. Neurol.* **160**:205–268. Varieties and distribution of non-pyramidal cells in the somatic sensory cortex of the squirrel monkey.
Jouvet, M. (1972). *Ergeb. Physiol.* **64**:166–307. The role of monoamines and acetylcholine-containing neurons in the regulation of the sleep-waking cycle.
Jouvet, M. (1973). *Arch. Ital. Biol.* **111**:564–576. Essai sur le rêve.
Kaas, J. H. (1983). *Physiol. Rev.* **63**:206–231. What, if anything, is S-I?: The organization of the 'first somatosensory area' of cortex.
Kahneman, D., and A. Tversky (1979). *Econometrica* **47**:263–290. Prospect theory: An analysis of decision under risk.
Kaiser, O. (1891). *Die Funktionen der Ganglienzellen des Halsmarkes*. Nijhoff, The Hague.
Kallius, E. (1897). *Ergebn. der Anat. u. Entwick (Merkel-Bonnet)* **6**:26–43. Endigungen motorischer Nerven in der Muskulatur der Wirbelthiere.
Kanigel, R. (1986). *Apprentice to a Genius. The Making of Scientific Discovery*. Macmillan, New York.
Kant, I. (1793). *Critique of Judgement*, 2nd Edition (J. H. Bernard, transl.). Hafner, New York, 1951.
Kant I. (1798). *Critique of Practical Reason and other Works on the Theory of Ethics* (T. K. Abbot, transl.). Longmans, Green, London.
Katz, B. (1962). *Proc. Roy. Soc. London Ser. B* **155**:455–477. The transmission of impulses from nerve to muscle and the subcellular unit of synaptic action (Croonian Lecture).
Kauer, J. S. (1991). *Trends Neurosci.* **14**:79–85. Contributions of topography and parallel processing to odor coding in the vertebrate olfactory pathway.
Kaur, C., E. A. Ling, and W. C. Wong (1987). *J. Anat. (London)* **154**:215–227. Origin and fate of neural macrophages in a stab wound of the brain of the young rat.
Kaye, K. (1982). *The Mental and Social Life of Babies*. Univ. of Chicago Press, Chicago.
Kaye, K., and J. Marcus (1978). *Infant Behav. Develop.* **1**:141–155. Imitation over a series of trials without feedback.
Keele, K. D. (1957). *Anatomies of Pain*. Blackwell, Oxford.
Killackey, H. P., and F. F. Ebner (1972). *Brain Behav. Evol.* **s6**:141–169. Two different types of thalamo-cortical projections to a single cortical area in mammals.
Kim, J. (1978). *Am. Philosoph. Quart.* **15**:149–156. Supervenience and nomological incommensurables.
Kingsbury, B. F. (1922). *J. Comp. Neurol.* **34**:461–491. The fundamental plan of the vertebrate brain.
Kisch, B. (1954). *Forgotten Leaders in Modern Medicine*. American Philosophical Society, Philadelphia.
Knobler, R. L., and J. G. Stempak (1973). *Prog. Brain Res.* **40**:407–423. Serial section analysis of myelin development in the central nervous system of the albino rat: An electron microscopical study of early axonal ensheathment.
Knudsen, E. I., S. duLac, and S. D. Esterly (1987). *Ann. Rev. Neurosci.* **10**:41–65. Computational maps in the brain.
Koch, C. and I. Segev, eds. (1989). *Methods in Neuronal Modeling: From Synapses to Networks*, MIT Press, Cambridge, Massachusetts.
Koch, C. A. Zador, and T. H. Brown (1992). *Science* **256**:973–974. Dendritic spines: Convergence of theory and experiment.
Koelliker, A. (1844). *Die Selbständigkeit und Abhängigkeit des sympathischen Nervensystems, durch anatomische Beobachtungen beweisen. Ein akad. Programm*. Meyer & Keller, Zürich.
Koelliker, A. (1848). *Zeitschr für wissensch. Zool.* **1**:135–163. Neurologische Bemerkungen.
Koelliker, A. (1850–1854). *Mikroskopische Anatomie oder Gewebelehre des Menschen*. Vol. 2, part 1, 1850; Vol. 2, part 2, sect. 1, 1852, sect. 2, 1854. Vol. 1 never published. Engelmann, Leipzig.
Koelliker, A. (1852). *Handbuch der Gewebelehre des Menschen*. Verlag von Wilhelm Engelmann, Leipzig. (The 6th Edition, in 3 volumes, was published in 1896).

Koelliker, A. (1853). *Manual of Human Histology*, transl. and edited by G. Busk and T. Huxley. The New Sydenham Society, London.
Koelliker, A. (1854). *Manual of Human Microscopical Anatomy*. Transl. by G. Busk and T. Huxley, edited, with notes and additions, by J. da Costa. Lippincott, Grambo & Co., Philadelphia.
Koelliker, A. (1867). *Handbuch der Gewebelehre des Menschen*, 5th Edition. Verlag von Wilhelm Engelmann, Leipzig.
Koelliker, A. (1886). *Z. wiss. Zool.* **43**:1–40. Histologische Studien an Batrachierlarven.
Koelliker, A. (1887a). *Anat. Anz.* **2**:480–483. Die Untersuchungen von Golgi über den feineren Bau des centralen Nervensystems.
Koelliker, A. (1887b). *Würzb. Sitz.-Ber.* 21 May, pp. 56–62. Golgi's untersuchungen über den feineren Bau des centralen Nervensystems.
Koelliker, A (1890a). *Z. wiss. Zool.* **49**:663–689. Zur feineren Anatomie des Zentralnervensystems. Erster Beitrag. Das Kleinhirn.
Koelliker, A. (1890b). *Z. wiss. Zool.* **51**:1–54. Zur feineren Anatomie des Zentralnervensystems. Zweiter Beitrag. Das Rückenmark.
Koelliker, A. (1890c). *Sitzungsber. der Würzb. Phys. Med. Gesellsch.* 8 März, 1890. Über den feineren Bau des Rückenmarks.
Koelliker, A. (1890d). *Sitzungsber. der Würzb. Phys. Med. Gesellsch.* 12 Juli, 1890. Über den feineren Bau des Rückenmarks menschlicher Embryonen.
Koelliker, A. (1892). *Sitzungsber. phys.-med. Gesellsch. Würzburg*, No. 1, pp. 1–5. Ueber den feineren Bau des Bulbus olfactorius.
Koelliker, A. (1896). *Handbuch der Gewebelehre des Menschen*. 6th edition, Vol. 2, *Nervensystem des Menschen und der Thiere*. W. Engelmann, Leipzig.
Koelliker, A. (1899). *Erinnerungen Aus Mein Leben*. Verlag von Wilhelm Engelmann, Leipzig.
Koelliker, A. (1905). *Z. wiss. Zool.* **82**:1–38. Die Entwicklung der Elemente des Nervensystems.
Koelliker, A. von (1905). *Lancet* **83**:1514; *Br. Med. J.* Suppl. **2**:1375–1377. Obituary.
Koestler, A., and J.R. Smythies, eds. (1969). *Beyond Reductionism*. Hutchinson, London.
Kohlberg, L. (1984). *Essays on Moral Development, Vol 2: The Psychology of Moral Development*. Harper & Row, New York.
Konigsmark, S. W., and R. L. Sidman (1963). *J. Neuropath. Exp. Neurol.* **22**:643–676. Origin of brain macrophages in the mouse.
Konishi, M. (1986). *Trends in Neurosci.* **9**:163–168. Centrally synthesized maps of sensory space.
Koppisch, E. (1935). *Zeitschr. f. Hygiene und Infektionskrankheiten* **117**:386–398. Zur Wanderungsgeschwindigkeit neurotroper Virusarten in peripheren Nerven.
Korr, H. (1980). *Adv. Anat. Embryol. Cell Biol.* **61**:1–72. Proliferation of different cell types in the brain.
Korsching, S., R. Heumann, A. Davies, and H. Thoenen (1986). *Soc. Neurosci. Abstr.* **12**:1096. Levels of nerve growth factor and its mRNA during development and regeneration of the peripheral nervous system.
Kromer, L. F., and C. J. Cornbrooks (1986). *Proc. Natl. Acad. Sci. USA* **83**:6330–6334. Transplants of Schwann cell cultures promote axonal regeneration in the adult mammalian brain.
Kronenberg, F. (1839). *Arch. für Anat. Physiol. und Wissensch. Med.* pp. 360–362. Versuche über motorische und sensible Nervenwurzeln.
Kruger, L., and D. S. Maxwell (1966). *Am. J. Anat.* **118**:411–435. Electron microscopy of oligodendrocytes in normal rat cerebrum.
Kuhn, T. (1962). *The Structure of Scientific Revolutions*. Chicago University of Chicago Press, Chicago.
Kuhn, T. (1968). History of science, pp. 74–83. In *International Encyclopedia of the Social Sciences*, New York.
Kuhn, T. (1970). Logic of discovery or psychology of research. In *Criticism and the Growth of Knowledge* (I. Lakatos and A. E. Musgrave, eds.), Cambridge University Press, Cambridge.
Kuhn, T. (1974). Second thoughts on paradigms, pp. 459–482. In *The Structure of Scientific Theories* (F. Suppe, ed.), Univ. Illinois Press, Urbana.
Kuhn, T. S. (1977). *The Essential Tension*. Univ. Chicago Press, Chicago.
Kühne, W. (1862). *Über die peripherischen Endorgane der motorischen Nerven*. Engelmann, Leipzig.

Kühne, W. (1869). Nerv und Muskelfaser. In Stricker (1869–1872, **1**:147–169). English trans.: The mode of termination of nerve fibre in muscle. In Stricker (1870–1873, 1:202–234).

Kupffer, K. von (1906). Die Morphogenie des Centralnervensystems, pp. 1–272. In *Handbuch der vergleichende und experimentelle Entwicklungslehre der Wirbeltiere*, Vol. 2, Part 3 (R. Hertwig, ed.). Fisher Verlag, Jena.

Kurtines, W. and Gewirtz, J., eds. (1987). *Moral Development Through Social Interaction*. Wiley, New York.

Lakatos, I. (1978). Falsification and the Methodology of Scientific Research Programmes, and History of Science and Its Rational Reconstructions. In *The Methodology of Scientific Research Programmes: Philosophical Papers of Imre Lakatos*, Vol. 1 (J. Worrall and G. Currie, eds.), Cambridge University Press, Cambridge.

LaMettrie, J. O. de (1747). *Man a Machine*. Including Frederick the Great's 'Eulogy' on LaMettrie and extracts from LaMettrie's *The Natural History of the Soul* (1748). Open Court, La Salle, Illinois, 1961.

Landmesser, L., and G. Pilar (1972). *J. Physiol. (London)* **222**:691–713. The onset and development of transmission in the chick ciliary ganglion.

Lange, F. A. (1877, 1890, 1892). *The History of Materialism and Criticism of its Present Importance*, 3 vols. in one. E. C. Thomas, transl. Introduction by Bertrand Russell. Routledge and Kegan Paul, London (1925).

Langley, J. N. (1895). *J. Physiol. (London)* **18**:280–284. Note on regeneration of pre-ganglionic fibres of the sympathetic.

Langley, J. N. (1897). *J. Physiol. (London)* **22**:215–230. On the regeneration of pre-ganglionic and post-ganglionic visceral nerve fibres.

Langley, J. N. (1906). *Proc. Roy. Soc. London Ser.B* **78**:170–194. On nerve-endings and on special excitable substances in cells: Croonian Lecture.

Langley, J. S., and C. S. Sherrington (1884a). *J. Physiol. (London)* **5**:vi. On sections of the right half of the medulla oblongata and of the spinal cord of the dog which was exhibited by Prof. Goltz at the International Medical Congress of 1881.

Langley, J. S., and C. S. Sherrington (1884b). *J. Physiol. (London)* **5**:49–65. Secondary degeneration of nerve tracts following removal of the cortex of the cerebrum in the dog.

Langley, P. (1981). *Cognitive Science* **5**:31–54. Data-driven discovery of physical laws.

Lanterman, A. J. (1876). *Arch. Mikr. Anat.* **12**:1. Ueber den feineren Bau der markhaltigen Nervenfasern.

Larsell, O. (1920). *Scient. Monthly* **10**:559–569. Gustav Retzius, 1842–1919.

Lasch, C. (1979). *The Culture of Narcissism*. Warner Books, London.

Lashley, K. S. (1924). *Arch. Neurol. Psychiat.* **12**:249–276. Studies of cerebral function in learning. V. The retention of motor habits after destruction of the so-called motor area in primates.

Lashley, K. S. (1929). *Brain Mechanisms and Intelligence: A Quantitative Study of Injuries to the Brain*. Univ. of Chicago Press, Chicago.

Lashley, K. S. (1937). *Arch. Neurol. Psychiatry* **38**:371–387. Functional determinants of cerebral localization.

Lashley, K. S. (1938). *Psychol. Rev.* **45**:445–471. Experimental analysis of instinctive behavior.

Lashley, K. S. (1949). *Quart. Rev. Biol.* **24**:28–42. Persistent problems in the evolution of mind.

Lashley, K. S. (1958). *Res. Publ. Assoc. Nerv. Ment. Dis.* **36**:1–18. Cerebral organization and behavior.

Laslett, E. E., and C. S. Sherrington (1903). *J. Physiol. (London)* **29**:58–96. Observations on some spinal reflexes and the interconnections of spinal segments.

Laudan, L. (1977). *Progress and Its Problems*. University of California Press, Berkeley.

Lawrence, P. A. (1992). *The Making of a Fly. The Genetics of Animal Design*. Blackwell, Oxford.

Laycock, T. (1845). *Brit. & Foreign Med. Rev.* **19**:298–312. On the reflex function of the brain.

Laycock, T. (1851). *Journal of Psychological Medicine*, October. On sleep, dreaming and insanity.

Laycock, T. (1860). *Mind and Brain: The Correlations of Consciousness and Organization*. 2 vols. Sutherland and Knox, Edinburgh.

Lecky, W. E. H. (1894). *History of European Morals from Augustus to Charlemagne*, 11th ed. Longmans, Green, & Co., London.

LeDoux, J. E. (1987). Emotion, pp. 419–460. In *Handbook of Physiology. Section I. The Nervous System, Vol. V, Higher Functions of the Brain*. Amer. Physiol. Soc. Bethesda, Maryland.

LeDoux, J. E. (1989). *Cognition and Emotion* **3(4)**:267–289. Cognitive-emotional interactions in the brain.

LeDoux, J. E. (1992). *Current Opinion In Neurobiol.* **2**:191–197. Brain mechanisms of emotion and emotional learning.

Leeuwenhoek, A. van (1685). *Philos. Trans. Roy. Soc. London* **15**:883–895. An abstract of a letter from Mr. Anthony Leeuwenhoek, Fellow of The Royal Society, concerning the parts of the brain of several animals.

Legendre, R. (1912). *Nature* **40**:359–363. La survie des organes et la 'culture' des tissus vivants.

Lenhossék, M. von (1890a). *Arch. f. Anat. und Physiol. Anat. Abt.* Zur Kenntnis der ersten Entstehung der Nervenzellen und nervenfasern beim Vogelembryo.

Lenhossék, M. von (1890b). *Anat. Anz.* No. 13 & 14. Über Nervenfasern in hinteren Wurzeln welche aus dem Vorderhorn entspringen.

Lenhossék, M. von (1891). *Verh. Anat. Ges.* **5**:193–221. Zur Kenntnis der Neuroglia des menschlichen Rückenmarkes.

Lenhossék, M. von (1893). *Der feinere Bau des Nervensystems im Lichte neuester Forschung.* Fischer's Medicinische Buchhandlung, H. Kornfeld, Berlin (Originally published in *Fortschr. d. Med. Bd.* 10, 1892, revised and enlarged in 1894).

Lenoir, T. (1980). *Isis* **71**:77–108. Kant, Blumenbach and vital materialism in German biology.

Lenoir, T. (1981). *Studies in the History and Philosophy of Science* **12**. Teleology without regrets. The transformation of physiology in Germany: 1790–1847.

Lenoir, T. (1982). *The Strategy of Life: Teleology and Mechanics in Nineteenth Century German Biology.* Reidel, Dordrecht.

Leuret, F., and P. Gratiolet (1839–1857). *Anatomie comparée du système nerveux*, 2 vols. and Atlas. Baillière et Fils, Paris.

Leventhal, H., and K. Scherer (1987). *Cognition and Emotion* **1**:3–28. The relationship of emotion to cognition: A functional approach to a semantic controversy.

Lévi, J. (1989). The body: The daoists' coat of arms, pp. 104–126. In *Zone 3, Part 1. Fragments for a History of the Human Body* (M. Feher, R. Naddaff, and N. Tazi, eds.). Urzone, New York.

Levi-Montalcini, R. (1952). *Ann. N.Y. Acad. Sci.* **55**:330–343. Effects of mouse tumor transplantation on the nervous system.

Levi-Montalcini, R. (1987). *Science* **237**:1154–1162. The nerve growth factor 35 years later.

Levine, R., and M. Jacobson (1975). *Brain Res.* **98**:172–176. Discontinuous mapping of retina onto tectum innervated by both eyes.

Levitt, P., and P. Rakic (1980). *J. Comp. Neurol.* **193**:815–840. Immunoperoxidase localization of glial fibrillary acidic protein in radial glial cells and astrocytes of the developing rhesus monkey brain.

Lewes, G. H. (1859–1860). *The Physiology of Common Life.* 2 vols. Blackwood, Edinburgh.

Lewes, G. H. (1874–1879). *Problems of Life and Mind*, 5 vols. Trübner, London.

Lewis, M. R., and W. H. Lewis (1911). *Anat. Rec.* **5**:277–293. The cultivation of tissues from chick embryos in solutions of NaCl, $CaCl_2$, KCl and $NaHCO_3$.

Lewis, M. R., and W. H. Lewis (1912). *Anat. Rec.* **6**:207–211. The cultivation of chick tissues in media of known chemical constitution.

Leyacker, J. (1927). *Arch. Geschichte Med.* **19**:253–286. Zur Entstehung der Lehre von den Hirnventrikeln.

Leydig, F. (1857). *Lehrbuch der Histologie des Menschen und der Thiere.* Verlag von Meidinger Sohn & Co., Frankfurt.

Leyton, A. S. F., and C. S. Sherrington (1917). *Quart. J. Exp. Physiol.* **11**:135–222. Observations on the excitable cortex of the chimpanzee, orang-utan, and gorilla.

Liddell, E. G. T. (1952). *Obit. Notices Roy. Soc. London* **8**:241–259. Charles Scott Sherrington, 1857–1952.

Liddell, E. G. T. (1960). *The Discovery of Reflexes.* Clarendon Press, Oxford.

Liddell, E. G. T., and C. S. Sherrington (1924). *Proc. Roy. Soc. London Ser. B* **86**:212–242. Reflexes in response to stretch (myotatic reflexes).

Linell, E. A., and M. I. Tom (1931). *Anat. Rec. Suppl.* **48**:27. The postnatal development of the oligodendroglia cell in the brain of the white rat and the possible role of this cell in myelogenesis.

Ling, E. A. (1978). *J. Anat. (London)* **126**:111–121. Electron microscopic studies of macrophages in Wallerian degeneration of rat optic nerve after intravenous injection of colloidal carbon.

Ling, E. A. (1981). The origin and nature of microglia, pp. 33–82. In *Advances in Cellular Neurobiology*, Vol. 2 (S. Fedoroff and L. Hertz, eds.). Academic Press, New York.

Ling, G., and R. W. Gerard (1949). *J. Cell. Comp. Physiol.* **34**:383–396. Normal membrane potential of frog sartorius fibres.

Lister, J. Lord (1870). *Trans. Microsop. Soc. Lond.* **3**:134–143. Of the late Joseph Jackson Lister, F.R.S., F.Z.S., with special reference to his labours in the improvement of the achromatic microscope.

Liuzzi, F. J., and R. H. Miller (1987). *Brain Res.* **403**:385–388. Radially oriented astrocytes in the normal adult spinal cord.

Locke, J. (1690). *An Essay Concerning Human Understanding.* Thomas Basset, London.

Lockwood, M. (1989). *Mind, Brain and the Quantum.* Basil Blackwell, Oxford.

Loeb, L. (1902). *Arch. Entw. Mech. Organ.* **13**:487–506. Ueber das Wachstum des Epithels.

Loewenstein, W. R. (1981). *Physiol. Rev.* **61**:829–913. Junctional intercellular communication: the cell-to-cell membrane channel.

Loewi, O. (1921). *Pflüg. Arch. ges. Physiol.* **189**:239–242. Über humorale Übertragbarkeit der Herznervenwirkung.

Loewi, O. (1933). *Proc. Roy. Soc. London Ser. B.* **118**:299–316. The Ferrier lecture on problems connected with the principle of humoral transmission of nervous impulses.

Lombard, W. P. (1888). *Amer. J. Psychol.* **1**:5–71. The variations of the normal knee-jerk and their relation to the activity of the central nervous system.

Lombard, W. P. (1889). *J. Physiol. (London)* **10**:122–148. On the nature of the knee jerk.

Lombard, W. P. (1916). *Science* **44**:363–375. The life and work of Carl Ludwig (1816–1895).

Longet, F.-A. (1842). *Anatomie et physiologie du système nerveux de l'homme et des vertébrés.* Fortin, Masson et Cie., Paris.

Lorber, J. (1965). *Dev. Med. Child Neurol.* **7**:628–637. Hydrancephaly with normal development.

Lorente de Nó, R. (1935). *Amer. J. Physiol.* **111**:272–281. The synaptic delay of motoneurons.

Lorente de Nó, R. (1938a). *J. Neurophysiol.* **1**:187–194. Limits of variation of the synaptic delay of motoneurons.

Lorente de Nó, R. (1938b). The cerebral cortex: Architecture, intracortical connections and motor projections, p. 291–325. In *Physiology of the Nervous System* (J. F. Fulton, ed.), Oxford Univ. Press, London.

Lovejoy, A. O. (1960). *The Revolt Against Dualism.* Second edition. Open Court, La Salle, Illinois.

Luciani, L. (1891). *Il cervelleto. Nuovi studi di fisiologia normale e patolologica.* Monnier, Florence.

Luciani, L. (1901). *Trattato di fisiologia.* Monnier, Florence.

Luciani, L., and A. Tamburini (1878). *Riv. Sperimentale di Freniatria di Medicina Legale* **4**:69–89, 225–280. Ricerce sperimentali sulle funzioni del cervello.

Luco, J. V., and A. Rosenblueth (1939). *Am. J. Physiol.* **126**:58–65. Neuromuscular 'transmission-fatigue' produced without contraction during curarization.

Lund, J. S., R. G. Boothe, and R. D. Lund (1977). *J. Comp. Neurol.* **176**:149–188. Development of neurons in the visual cortex of the monkey (*Macaca nemestrina*): A Golgi study from fetal day 127 to postnatal maturity.

Luse, S. A. (1956). *J. Biophys. Biochem. Cytol.* **2**:777–784. Formation of myelin in the central nervous system of mice and rats, as studied with the electron microscope.

Lwoff, A. (1957). *J. Gen. Microbiol.* **17**:239–253. The concept of virus.

Mach, E. (1914). *The Analysis of Sensations and the Relation of the Physical to the Psychical.* Republished by Open Court, Chicago.

Mackintosh, N. J. (1990). *Nature* **347**:332. B. F. Skinner (1904–1990).

Magendie, F. (1822). *J. Physiol. exp. path., Paris* **2**:276–279. Experience sur les fonctions des racines des nerfs rachidiens.

Magendie, F. (1839). *Leçons sur les fonctions et les maladies du système nerveux.* Ébrard, Paris.

Magini, G. (1888). *Arch. Ital. Biol.* **9**:59–60. Sur la neuroglie et les cellules nerveuses cérébrales chez les foetus.

Mahowald, M., and R. Douglas (1991). *Nature* **354**:515–518. A silicon neuron.

Mallory, F. B. (1900). *J. Exp. Med.* **5**:15–20. A contribution to staining methods.

Malpighi, M. (1686). *Opera Omnia.* Robert Scott & George Wells, London.

Malpighi, M. (1697). *Opera Posthuma.* A. J. Churchill, London.

Marchi, V. and G. Algeri (1885–1886). *Riv. Sper. Freniatria Med. Legal* **11**:492–494, **12**:208–252. Sulla degenerazioni discendenti consecutive a lesioni sperimentali in diverse zone della corteccia cerebrale.

Marey, E.-J. (1863). *Physiologie médicale de la circulation du sang basée sur l'étude graphique des mouvements du coeur et du pouls artériel avec application aux maladies de l'appareil circulatoire.* A. Delahaye, Paris.

Marey, E.-J. (1878). *La méthode graphique dans les sciences expérimentales et principalement en physiologie et en médicine*. Masson, Paris.
Marinesco, G. (1909). *La cellule nerveuse*. O. Doin, Paris.
Marinesco, G. (1919). *Phil. Trans. Roy. Soc. London Ser. B* **209**:229. Nouvelles contributions à l'étude de la régénération nerveuse et du neurotropisme.
Marinesco, G., and J. Minea (1912a). *Compt. Rend. Soc. Biol.* **73**:668–670. Croissance des fibres nerveuses dans le milieu de culture, *in vitro*, des ganglions spinaux.
Marinesco, G., and J. Minea (1912b). *Acad. nat. méd. Paris. Bull.* ser. 3, **68**:37–40. Culture des ganglions spinaux des mamifères (*in vitro*). Suivant le procédé de M. Carrel.
Marinesco, G., and J. Minea (1912c). *Compt. Rend. Soc. Biol.* **73**:346–348. Culture des ganglions spinaux des mamifères '*in vitro*' suivant la méthode de Harrison et Montrose T. Burrows.
Marinesco, G., and J. Minea (1912d). *Rev. Neurol.* **24**:469–482. La culture des ganglions spinaux de mammifères *in vitro*. Contribution à l'étude de la neurogenèse.
Marinesco, G., and J. Minea (1914). *Anat. Anz.* **46**:529–547. Nouvelles recherches sur la culture 'in vitro' des ganglions spinaux de mammifères.
Marin-Padilla, M. (1969). *Brain Res.* **14**:633–646. Origin of the pericellular baskets of the pyramidal cells of the human motor cortex: A Golgi study.
Marr, D. (1982). *Vision*. Freeman, San Francisco.
Matteucci, C. (1840). *Essai sur les phénomènes electriques des animaux*. Carillian, Goeury & Dalmont, Paris.
Maturana, H. R. (1960). *J. Biophys. Biochem. Cytol.* **7**:107–120. The fine anatomy of the optic nerve of anurans: An electron microscope study.
Maxwell, D. S., and L. Kruger (1965). *Exp. Neurol.* **12**:33–54. Small blood vessels and the origin of phagocytes in the rat cerebral cortex following heavy particle irradiation.
Maxwell, G. (1963). The ontological status of theoretical entities. In *Scientific Explanation. Space and Time* (H. Feigl and G. Maxwell, eds.), Univ. of Minnesota Press, Minneapolis.
May, R. M. (1925). *J. Exp. Zool.* **42**:371–410. The relation of nerves to degenerating and regenerating taste buds.
Mayr, E. (1976). *Evolution and the diversity of life: selected essays*. Harvard University Press, Cambridge, Mass.
Mayr, E. (1982). *The Growth of Biological Thought*. Harvard Univ. Press, Cambridge, Mass.
McCarley, R. W., and J. A. Hobson (1977). *Am. J. Psychiatry* **134**:1211–1221. The neurobiological origins of psychoanalytic dream theory.
McClelland, C. E. (1980). *State, Society, and University in Germany. 1700–1914*. Cambridge Univ. Press, Cambridge.
McClelland, J., D. Rumelhart, and the PDP Research Group (1986). *Parallel Distributed Processing: Explorations in the Microstructure of Cognition*. Vol. 2. MIT Press, Cambridge, Mass.
McCloskey, M. (1983). *Scientific American* **24**:122–130. Intuitive physics.
McCloskey, M., and M. K. Kaiser (1984). *The Sciences* **24**:40–45. Children's intuitive physics.
McCormick, D. A. (1990). Membrane properties and neurotransmitter actions, pp. 32–66. In *The Synaptic Organization of the Brain* (G. M. Shepherd, ed.), Oxford Univ. Press, New York and Oxford.
McDougall, W. (1901). *Brain* **24**:577–630. On the seat of the psycho-physical processes.
McDougall, W. (1903). *Brain* **26**:153–191. The nature of inhibitory processes within the nervous system.
McDougall, W. (1905). *Principles of Physiological Psychology*. London.
McGinn, C. (1990). *Mind* **98**:349–366. Can we solve the mind-body problem?
McKenna, T., J. Davis, and S. F. Zornetzer, eds. (1992). *Single Neuron Computation*. Academic Press, Orlando, Florida.
Mead, C. (1989). *Analog VLSI and Neuronal Systems*. Addison-Wesley, Reading, Massachusetts.
Medvedev, Z. A. (1969). *The Rise and Fall of T. D. Lysenko*. (I. M. Lerner, transl.). Columbia Univ. Press, New York.
Meier, C. (1976). *Brain Res.* **104**:21–32. Some observations on early myelination in the human spinal cord, light and electron microscope study.
Meltzoff, A., and W. Moore (1977). *Science* **188**:75–78. Imitation of facial and manual gestures by human neonates.
Meltzoff, A., and W. Moore (1989). *Develop. Psychol.* **25**:954–962. Imitation in newborn infants: Exploring the range of gestures imitated and the underlying mechanisms.

Mendelsohn, E. (1964). The emergence of science as a profession in nineteenth century Europe, pp. 3–48. In *The Management of Scientists* (Hill, ed.). Harvard Univ. Press, Cambridge, Massachusetts.

Mendelsohn, E. (1965). Physical models and physiological concepts: Explanation in nineteenth century biology, pp. 127–150. In *Boston Studies in the Philosophy of Science* (R. S. Cohen and M. W. Wartofsky, eds.), Humanities Press, New York.

Merkel, F. (1875). *Arch. mikr. Anat. Bonn.* **11**:636–652. Tastzellen und Tastkörperchen bei den Hausthieren und beim Menschen.

Merkel, F. (1880). *Ueber die Endigungen der sensiblen Nerven in der Haut der Wirbeltiere.* Schmidt, Rostock.

Merton, R. K. (1961). *Proc. Amer. Phil. Soc.* **105**:470–486. Singletons and multiples in Scientific discovery: A chapter in the sociology of science.

Merton, R. K. (1965). *On the Shoulders of Giants.* Free Press, New York.

Merz, J. T. (1904–1912). *A History of European Thought in the Nineteenth Century.* 4 vols. William Blackwood & Son, London.

Meyer, A. (1900). *Wesen und Geschichte der Theorie vom Mikro- und Makrokosmos.* Berner Studien zur Philosophie und ihre Geschichte. Band XXV. Verlag von C. Sturzenegger, Bern.

Meyer, A. (1971). *Historical Aspects of Cerebral Anatomy.* Oxford Univ. Press, London.

Meyer, A. (1978). *Brain* **101**:673–685. The concept of sensorimotor cortex. Its early history, with special emphasis on two early experimental contributions by W. Bechterew.

Meyer, K. F.(1970). *J. Neurol. Neurosurg. Psychiat.* **33**:553–561. Karl Friedrich Burdach and his place in the history of neuroanatomy.

Meyer, S. (1895). *Arch. mikrosk. Anat. EntwMech.* **46**:282–290. Die subcutane Methylenblauinjektion ein Mittel zur Darstellung der Elemente des Centralnervensystems von Säugethieren.

Miale, I. L., and R. L. Sidman (1961). *Exp. Neurol.* **4**:277–296. An autoradiographic analysis of histogenesis in the mouse cerebellum.

Miescher, F. J. (1871). *Med. Chem. Untersuch.* **1**:441–460. Ueber die chemische Zusammensetzung der Eiterzellen.

Miledi, R., and E. Stefani (1969). *Nature* **222**:569–571. Nonselective re-innervation of slow and fast muscle fibres in the rat.

Mill, J. S. (1859). *On Liberty.* In *The Philosophy of John Stuart Mill* (M. Cohen, ed.), Modern Library, New York, 1961.

Mill, J. S. (1873). *Autobiography.* Longmans, Green, Reader, and Dyer, London.

Mills, K. R. (1991). *Trends In Neurosci.* **14**:401–405. Magnetic brain stimulation: a tool to explore the action of the motor cortex on single human spinal motoneurones.

Milne-Edwards, H. (1826). *Annales des Science Naturelles, Paris* 9(1e ser.):362–394. Recherches microscopiques sur la structure intime des tissus organiques des animaux.

Milne-Edwards, H. (1857–1881). *Leçons sur la physiologie et l'anatomie comparée de l'homme et des animaux.* 14 vols. and index vol. Victor Masson, Paris.

Milner, P. M. (1974). *Psychol. Rev.* **81**:521–535. A model for visual shape recognition.

Mitchell, T. M. (1982). *Artificial Intelligence* **18**:203–226. Generalization as search.

Monakow, C. von (1885). *Arch. Psychiat Nervenkrank.* **16**:317–352. Experimentelle und pathologisch-anatomische Untersuchungen über die Beziehungen der sogenannten Sehsphäre zu den infracorticalen Opticuscentren und zum N. opticus.

Monod, J. (1971). *Chance and Necessity.* Knopf, New York.

Monod, J. (1974). *On Chance and Necessity*, pp. 357–375. In *Studies in the Philosophy of Biology* (F. J. Ayala and T. Dobzhansky, eds.). Macmillan, London.

Monti, A. (1895). *Arch. Ital. Biol.* **24**:20–33. Sur l'anatomie pathologique des elements nerveux dans les processus provenant d'embolisme cérébral: considérations sur la signification physiologique des prolongements protoplasmiques des cellules nerveuses.

Moore, G. E. (1903). *Principia Ethica.* Cambridge Univ. Press, Cambridge.

Moore, G. E. (1912). *Ethics.* Holt, New York.

Moore, G. E. (1922). The conception of intrinsic value. In *Philosophical Studies*, Routledge and Kegan Paul, London.

Moore, G. E. (1929). *Principia Ethica*, 2nd ed. Cambridge Univ. Press, Cambridge.

Moran, J. (1973). *Printing Presses. History and Development from the Fifteenth Century to Modern Times.* Univ. Calif. Press, Berkeley.

Moravec, H. (1988). *Mind Children.* Harvard Univ. Press, Cambridge, Mass.
Morel, C. (1864). *Traité élementaire d'histologie humaine normal et pathologique precede d'un exposé des moyens d'observer au microscope.* Text vol. and Atlas (by J.-A. Villemin). Baillière et Fils, Paris.
Mori, S., and C. P. Leblond (1969). *J. Comp. Neurol.* **135**:57–80. Identification of microglia in light and electron microscopy.
Morrison, L. R. (1932). *Arch. Neurol. Psychiat.* **28**:204–205. Role of oligodendroglia in myelogenesis.
Moruzzi, G. (1972). *Ergeb. Physiol.* **64**:1–165. The sleep-waking cycle.
Moruzzi, G., and H. W. Magoun (1949). *Electroenceph. Clin. Neurophysiol.* **1**:455–473. Brain stem reticular formation and activation of the EEG.
Mott, F. W. (1894). *J. Physiol. (London)* **15**:464–487. The sensory-motor function of the central convolutions of the cerebral cortex.
Mott, F. W., and C. S. Sherrington (1894–1895). *Proc. Roy. Soc. London* **57**:481–488. Experiments upon the influence of sensory nerves upon movement and nutrition of the limbs. Preliminary communication.
Mountcastle, V. (1979). An organizing principle for cerebral function: The unit module and the distributed system, pp. 21–42. In *The Neurosciences. Fourth Study Program* (F. O. Schmidt and G. Worden, eds.), MIT Press, Cambridge, Massachusetts.
Mugnaini, E., and P. F. Forströnen (1967). *Z. Zellforsch. Mikrosk. Anat. Abt. Histochem.* **77**:115–143. Ultrastructural studies on the cerebellar histogenesis. I. Differentiation of granule cells and development of glomeruli in the chick embryo.
Mugnaini, E., and F. Walberg (1964). *Ergeb. Anato. Entw.-Gesch.* **37**:193–236. Ultrastructure of neuroglia.
Müller, J. (1826). *Zur vergleichenden Physiologie des Gesichtsinnes der Menschen und der Thiere.* Cnobloch, Leipzig.
Müller, J. (1833–1840). *Handbuch der Physiologie des Menschen fur Vorlesungen,* Vol. 1 1833–1834, Vol. 2 1837–1840. J. Holscher, Koblenz.
Müller, J. (1840). *Elements of Physiology,* 2nd edition. 2 vols. Translated from the German [1st edition of 1835 and 3rd edition of 1837] with Notes by W. Baly. Taylor and Walton, London.
Müller, J. (1851). *Manuel de Physiologie.* 2 vols. Trans. by A.-J.-L. Jourdan. 2nd edition, revised and annotated by É. Littré. Baillière, Paris.
Müller, H. (1851). *Z. wiss. Zool.* **3**:234–237. Zur Histologie der Netzhaut.
Müller, W. and J. A. O'Connor (1991). *Science* **354**:73–76. Dendritic spines as individual neuronal compartments for synaptic Ca^{2+} responses.
Munk, H. (1881). *Ueber die Funktionen der Grosshirnrinde. Gesammelte Mitteilungen aus den Jahren 1877–1880.* Hirschwald, Berlin.
Munoz-Garcia, D., and S. K. Ludwin (1986a). *J. Neurocytol.* **15**:273–290. Gliogenesis in organotypic tissue culture of the spinal cord of the embryonic mouse. I. Immunocytochemical and ultrastructural studies.
Munoz-Garcia, D., and S. K. Ludwin (1986b). *J. Neurocytol.* **15**:291–302. Gliogenesis in organotypic tissue culture of the spinal cord of the embryonic mouse. II. Autoradiographic studies.
Murphy, L. B. (1937). *Social Behavior and Child Personality: An Exploratory Study of Some Roots of Sympathy.* Columbia University Press, New York.
Nagel, E. (1961). *The Structure of Science.* Harcourt, Brace and World, New York.
Nagel, T. (1974). *Philosophical Review* **83**:435–450. What is it like to be a bat?
Nageotte, J. (1918). See Ramón y Cajal, S., 1933b, p. 310.
Nansen, F. (1886–1887). *The Structure and Combination of the Histological Elements of the Central Nervous System.* Bergens Mus. Aarsberetning, Bergen.
Nauta, W. J. H. (1950). *Arch. Neurol. Psychiat.* **66**:353–376. Über die sogenannte terminale Degeneration im Zentralnervensystem und ihre Darstellung durch silberimprägnation.
Nauta, W. J. H. (1957). Silver impregnation of degenerating axons, pp. 17–26. In *New Research Techniques of Neuroanatomy* (W. F. Windle, ed.) Thomas, Springfield, Illinois.
Needham, J. (1954). *Science and Civilisation in China, Vol. 1. Introductory Orientations.* Cambridge Univ. Press, Cambridge.
Needham, J. (1959). *Science and Civilisation in China, Vol. 2. History of Scientific Thought.* Cambridge Univ. Press, Cambridge.
Nelson, J. W. (1992). *Science* **258**:948–955. Regulation of cell surface polarity from bacteria to mammals.
Neuburger, M. (1897). *Die historische Entwicklung der experimentellen Gehirn und Rückenmarksphysiologie vor Flourens.* Translated and edited with additional material by Edwin Clarke; *The Historical Develop-*

ment of Experimental Brain and Spinal Cord Physiology Before Flourens. Johns Hopkins Univ. Press, Baltimore.

Nisbett, R. E., and T. D. Wilson (1977). *Psychol. Rev.* **84**:231–259. Telling more than we can know: Verbal reports on mental processes.

Nissl, F. (1894a). *Neurologisches Centralblatt* **13**:507–508. Ueber eine neue Untersuchungsmethode des Centralorgans speciell zur Feststellung der Localisation der Nervenzellen.

Nissl, F. (1894b). *Neurologisches Centralblatt* **13**:676–685, 781–789, 810–814. Ueber die sogennanten Granula der Nervenzellen.

Nissl, F. (1903). *Die Neuronlehre und Ihre Anhänger. Ein Beitrag zur Lösung des Problems der Beziehungen Zwischen Nervenzelle, Faser und Grau.* Verlag von Gustav Fischer, Jena.

Oatley, K., and P. Johnson-Laird (1987). *Cognition and Emotion* **1**:29–50. Towards a cognitive theory of emotion.

Obersteiner, H. (1883). *Biol. Zentralbl.* **3**:145–155. Der feinere Bau der Kleinhirnrinde bei Menschen und Tieren.

Obersteiner, H. (1888). *Anleitung beim Studium des Baues der Nervösen Centralorgane im gesunden und kranken Zustande.* Toeplitz & Deuticke, Leipzig und Wien.

O'Brien, R. A. D., A. J. C. Ostberg, and G. Vrbova (1984). *Neuroscience* **12**:637–646. Protease inhibitors reduce the loss of nerve terminals induced by activity and calcium in developing rat soleus muscles in vitro.

Ochs, S. (1975). *Clio Medica* **10**:253–265. Waller's concept of the trophic dependence of the nerve fiber on the cell body in the light of early neuron theory.

Ochs, S. (1977). *Med. Hist.* **21**:261–274. The early history of nerve regeneration beginning with Cruikshank's observations in 1776.

Ogden, C. K., and I. A. Richards (1927). *The Meaning of Meaning,* 2nd ed. Kegan Paul, Trench, Trubner, London.

O'Keefe, J. and L. Nadel (1978). *The Hippocampus as a Cognitive Map.* Clarendon Press, Oxford.

Olby, R. C. (1966). *Origins of Mendelism* (2nd Ed. 1985), Schocken Books, New York.

Olmsted, J. M. D. (1920a). *J. Comp. Neurol.* **31**:465–468. The nerve as a formative influence in the development of taste-buds.

Olmsted, J. M. D. (1920b). *J. Exp. Zool.* **31**:369–401. The results of cutting the seventh cranial nerve in *Ameiurus nebulosus* (Lesueur).

Olmsted, J. M. D. (1939). *Claude Bernard, physiologist.* Cassell, London.

Olmsted, J. M. D. (1944). *François Magendie. Pioneer in experimental physiology and scientific medicine in XIX century France.* Schuman, New York.

Olmsted, J. M. D., and E. H. Olmsted (1952). *Claude Bernard and the Experimental Method in Medicine.* Cassell, London, Schuman, New York.

Olson, C. R., and S. J. Hanson (1990). Spatial representation of the body, pp. 193–254. In *Connectionist Modeling and Brain Function* (S. J. Hanson and C. R. Olson, eds.). MIT Press, Cambridge, Massachusetts.

Oppel, A. (1913). *Zentr. Zool. allgem. exp. Biol.* **3**:209–232. Explantation (Deckglaskultur, *in vitro*-Kultur).

Oppenheimer, J. (1966). *Bull. Hist. Med.* **40**:525–543. Ross Harrison's contribution to experimental embryology.

Oppenheimer, J. (1971). *Trans. Stud. Coll. Physic. Philadel.,* Ser. 4, **39**:26–33. Historical relationships between tissue culture and transplantation experiments.

Oppenheimer, J. (1972). *Dict. Sci. Biog.* **6**:131–135. Ross Granville Harrison.

Ormerod, F. C. (1960). The pathology of congenital deafness in the child. In *The Modern Educational Treatment of Deafness* (A. Ewing, ed.). Manchester Univ. Press, Manchester.

Ospovat, D. (1978). Perfect adaptation and teleological explanation: approaches to the problem of the history of life in the mid-nineteenth century, pp. 33–56. In *Studies in History of Biology,* vol. 2 (W. Coleman and C. Limoges, eds.), Johns Hopkins Univ. Press, Baltimore.

Ospovat, D. (1981). *The Development of Darwin's Theory: Natural History, Natural Theology, and Natural Selection, 1838–1859.* Cambridge Univ. Press, Cambridge.

Oster, H. (1978). Facial expressions and affect development, pp. 43–75. In *The Development of Affect* (M. Lewis and L. A. Rosenblum, eds.), Plenum, New York.

REFERENCES

Owen, R. (1848). *On the Archetype and Homologies of the Vertebrate Skeleton*. John van Voorst, London.
Owen, R. (1868). *On the Anatomy of Vertebrates*. 3 vols. Longmans, Green & Co., London.
Owsjannikow, P. (1860). *Müller's Arch. f. Anat. Physiol. und wissensch. Med. Leipzig*, pp. 469–477. Ueber die feinere Struktur der Lobi olfactorii der Saugethiere.
Owsjannikow, P. (1864). *Bull. Acad. St. Petersbourg* **7**:58. Cited in Koelliker, 1897, p. 755.
Pagel, W. (1935). *Bull. Int. Hist. Med.* **3**:97. Religious motives in the medical biology of the seventeenth century.
Pagel, W. (1945). *Bull. Hist. Med.* **18**:1–43. The speculative basis of modern pathology.
Pagel, W. (1958). Medieval and Renaissance contributions to knowledge of the brain and its functions, p. 95–114, In *The History and Philosophy of the Brain and its Functions* (F. N. L. Poynter, ed.). Blackwell, Oxford.
Palade, G. E., and S. L. Palay (1954). *Anat. Rec.* **118**:335–336. Electron microscope observations of interneuronal and neuromuscular synapses.
Palay, S. L. (1956). *J. Biophysic. and Biochem. Cytol.* Suppl. **2**:193–201. Synapses in the central nervous system.
Palay, S. L. (1958). *Exp. Cell Res.*, Suppl. **5**:275–293. The morphology of synapses in the central nervous system.
Palay, S. L. (1989). *Nature* **341**:493–494. The neuroanatomist.
Palay, S. L., and G. E. Palade (1955). *J. Biophys. Biochem. Cytol.* **1**:69–88. The fine structure of neurons.
Panem, S. (1989). *Science* **246**:1329–1330. The discovery track.
Panksepp, J. (1982). *Behav. Brain Sci.* **5**:407–467. Toward a general psychobiological theory of emotions.
Pappas, G. D., and M. V. L. Bennett (1966). *Ann. N. Y. Acad. Sci.* **137**:495–508. Specialized junctions involved in electrical transmission between neurons.
Parker, G. H. (1919). *The Elementary Nervous System*. Lippincott, Philadelphia.
Parker, G. H. (1932). *Am. Naturalist* **66**:147–158. On the trophic impulse so-called, its rate and nature.
Passmore, J. A. (1960). *Philosophy* **35**:326–331. Popper's account of scientific method.
Pavlov, I. P. (1927). *Conditioned Reflexes. An Investigation of the Physiological Activity of the Cerebral Cortex* (G. V. Anrep, transl.). Oxford Univ. Press, London.
Peiper, A. (1963). *Cerebral Function in Infancy and Childhood*. Consultants Bureau, New York.
Peirce, C. S. (1868). *J. Speculative Philos.* **2**:103–114, 140–157. Questions concerning certain faculties claimed for man and some consequences of four incapacities. Reprinted in *Charles S. Peirce Selected Writings (Values in a Universe of Chance)*. P. P. Wiener, ed., Dover, New York, 1966.
Peirce, C. S. (1878). *Popular Science Monthly*, January, pp. 286–302. How to make our ideas clear. Reprinted 1958 in *Values in a Universe of Chance* (P. P. Wiener, ed.). Doubleday, New York.
Peirce, C. S. (1905). *The Monist* **15**:161–181. What Pragmatism Is. Reprinted 1958 in *Values in a Universe of Chance* (P. P. Wiener, ed.), Doubleday, New York.
Peirce, C. S. (1931–1935, 1958). *Collected Papers of Charles Sanders Peirce*, 8 vols. Vols. I–VI edited by C. Hartshorne and P. Weiss, Vols. VII and VIII edited by A. W. Burke. Harvard Univ. Press, Cambridge, Mass.
Peirce, C. S. (1958). *Values in a Universe of Chance* (P. P. Wiener, ed.). Doubleday, New York. Reprinted 1966, as *Charles S. Peirce: Selected Writings*. Dover, New York.
Pellionisz, A., and R. Llinas (1986). *Neuroscience* **19**:1–32. Tensor network theory of the metaorganization of functional geometrics in the central nervous system.
Penfield, W. (1924). *Brain* **47**:430–452. Oligodendroglia and its relation to classical neuroglia.
Penfield, W. (1926). *Arch. Neurol. Psychiat.* **26**:212–220. The career of Ramón y Cajal.
Penfield, W. (1954). Ramón y Cajal, an appreciation. In *Neuron Theory or Reticular Theory?* Translated by M. U. Purkiss and C. A. Fox. CSIC Instituto 'Ramón y Cajal,' Madrid.
Penfield, W. (1957). *Brain* **80**:402–410. Sir Charles Sherrington, poet and philosopher.
Penfield, W. (1962). *Notes and Records of the Royal Society of London* **17**:163–168. Sir Charles Sherrington, O.M., F.R.S.: An appreciation.
Penfield, W. (1977). *No Man Alone. A Neurosurgeon's Life*. Little Brown, Boston.
Penfield, W., and L. Roberts (1959). *Speech and Brain Mechanisms*. Princeton Univ. Press, Princeton, New Jersey.
Penrose, R. J. (1989). *The Emperor's New Mind*. Oxford Univ. Press, Oxford.

Percheron, G. (1979a). *Neurosci. Lett.* **14**:287–293. Quantitative analysis of dendritic branching. I. Simple formulae for the quantitative analysis of dendritic branching.
Percheron, G. (1979b). *Neurosci. Lett.* **14**:295–302. Quantitative Analysis of dendritic branching. II. Fundamental dendritic numbers as a tool for the study of neuronal groups.
Perner, J. (1991). *Understanding the Representational Mind*. MIT Press, Cambridge, Massachusetts.
Perry, V. H., and S. Gordon (1988). *Trends Neurosci.* **11**:273–277. Macrophages and microglia in the nervous system.
Peters, A. (1959). *Am. J. Anat.* **93**:177–195. Experimental studies on the staining of nervous tissue with silver proteinate.
Peters, A. (1960). *J. Biophys. Biochem. Cytol.* **8**:431–446. The formation and structure of myelin sheaths in the central nervous system.
Peters, A. (1964a). *J. Anat. (London)* **98**:125–134. Observations on the connections between myelin sheaths and glial cells in the optic nerve of young rats.
Peters, A. (1964b). *J. Cell Biol.* **20**:281–296. Further observations on the structure of myelin sheaths in the central nervous system.
Peters, A. (1966). *Quart. J. Exp. Physiol.* **51**:229–236. The node of Ranvier in the central nervous system.
Peters, A., and A. R. Muir (1959). *Quart. J. Exp. Physiol.* **44**:117–130. The relationship between axons and Schwann cells during development of peripheral nerves in the rat.
Pettigrew, A. G., R. Lindeman, and M. R. Bennett (1979). *J. Embryol. Exp. Morphol.* **49**:115–137. Development of the segmental innervation of the chick forelimb.
Peyer, J. (1853). *Zeitschr. f. rat. Med. Heidelberg.* **4**:52–77. Über die peripherischen Endigungen der motorischen und sensiblen Fasern der in den Plexus brachialis des Kaninchens eintretenden Nervenwurzeln.
Piaget, J. (1952). *The Origins of Intelligence in Children*. International Universities Press, New York.
Piatt, J. (1940). *J. Exp. Zool.* **85**:211–241. Nerve-muscle specificity in *Amblystoma*, studied by means of heterotopic cord grafts.
Piatt, J. (1957a). *J. Exp. Zool.* **134**:103–125. Studies on the problem of nerve pattern. II. Innervation of the intact forelimb by different parts of the central nervous system in *Amblystoma*.
Piatt, J. (1957b). *J. Exp. Zool.* **136**:229–247. studies on the problem of nerve pattern. III. Innervation of the regenerated forelimb in *Amblystoma*.
Piccolino, M. (1988). *Trends Neurosci.* **11**:521–525. Cajal and the retina: a 100-year retrospective.
Picken, L. (1956). *Nature* **178**:1162–1165. The fate of Wilhelm His.
Piéron, H. (1906a). *Bull. Inst. Gen. Physiol. Paris* **6**:40–59. Contribution à la psychologie des Actinies.
Piéron, H. (1906b). *Bull. Inst. Gen. Physiol. Paris* **6**:146–169. Contribution à la psychophysiologie des Actinies. Les réactions de l'Actinia equina.
Plutchik, R. (1980). *Emotion: A Psychoevolutionary Synthesis*. Harper and Row, New York.
Polanyi, M. (1958). *Personal Knowledge*. Routledge and Kegan Paul, London.
Polanyi, M. (1968). *Science* **113**:1308–1312. Life's irreducible structure.
Popper, K. R. (1959). *The Logic of Scientific Discovery*. Hutchinson, London.
Popper, K. R. (1962). *The Open Society and its Enemies*, 4th Edition. 2 volumes. Princeton Univ. Press, Princeton, New Jersey.
Popper, K. R. (1972). *Objective Knowledge*. Clarendon Press, Oxford.
Popper, K. R. (1974). Scientific reduction and the essential incompleteness of all science, pp. 259–284. In *Studies in the Philosophy of Biology* (F. J. Ayala and T. Dobzhansky, eds.), Macmillan, London.
Popper, K. R., and J. C. Eccles (1977). *The Self and its Brain*. Springer International, Berlin.
Prévost, J. L., and J. B. A. Dumas (1823). *J. Phisiol. Expér. Pathol. (Paris)* **3**:301–344. Sur les phénomènes qui accompagnent la contraction de la fibre musculaire.
Preyer, W. (1885). *Specielle Physiologie des Embryo. Untersuchungen ueber die Lebenserscheinungen vor der Geburt*. Th. Grieben's Verlag (L. Fernau), Leipzig.
Prochaska, J. (1784). English translation by T. Laycock, in *The Principles of Physiology, by John Augustus Unzer; and a Dissertation on the Functions of the Nervous System, by George Prochaska*. The New Sydenham Society, London, 1851.
Proctor, R. N. (1991). *Value-Free Science? Purity and Power in Modern Knowledge*. Harvard Univ. Press, Cambridge, Mass.

Pugh, G. E. (1977) *The Biological Origin of Human Values*. Basic Books, New York.
Purkinje, J. E. (1837a). *Opera Omnia* **2**:88. Neueste Beobachtungen über die Struktur des Gehirns.
Purkinje, J. E. (1837b). *Opera Omnia* **3**:45–49. Neueste Untersuchungen aus der Nerven-und Hirnanatomie.
Purkinje, J. E. (1839). *Opera Omnia* **2**:90–91. Ueber die Analogieen in der Struktur-Elementen des thierischen und pflanzlichen Organismus.
Purkinje, J. E. (1918–1973). *Opera Omnia*. 12 Vols. Purkynova Spolecnost, Prague.
Purves, D., D. R. Riddle, and A.-S. LaMantia (1992). *Trends in Neurosci.* **15**:362–367. Iterated patterns of brain circuitry (or how the cortex gets its spots).
Putnam, H. (1975). *Mind, Language and Reality: Philosophical Papers*. Cambridge University Press, Cambridge.
Putnam, H. (1980). Philosophy and our mental life, pp. 134–143. In *Readings in Philosophy of Psychology*, Vol. 1 (N. Block, ed.), Harvard Univ. Press, Cambridge, Massachusetts.
Pylshyn, Z. W. (1984). *Computation and Cognition: Toward a Foundation for Cognitive Science*. M.I.T. Press Cambridge, Massachusetts.
Quastler, H. (1959). *Exp. Cell Res.* **17**:420–438. Cell population kinetics in the intestinal epithelium of the mouse.
Quattrone, G., and E. Jones (1980). *J. Personality and Social Psychol.* **38**:141–152. The perception of variability within in-groups and out-groups: Implications for the law of large numbers.
Querlieu, D., X. Renard, and G. Crepin (1981). *J. Gynecol. Obstet. Biol. Reprod.* **10**:307–314. Perception auditive et réactive foetale aux stimulations sonores.
Quine, W. V. (1953). *From a Logical Point of View*. Harvard University Press, Cambridge, Massachusetts.
Quine, W. V. (1960). *Word and Object*. MIT Press, Cambridge, Massachusetts.
Quine, W. V. (1969). *Ontological Relativity and Other Essays*. Columbia Univ. Press, New York.
Rádl, E. (1930). *The History of Biological Theories*. Oxford Univ. Press, London.
Raichle, M. E. (1991). *Cold Spr. Harb. Symp. Quant. Biol.* **55**:983–986. Anatomical explorations of mind: Studies with modern imaging techniques.
Rakic, P. (1971a). *Brain Res.* **33**:471–476. Guidance of neurons migrating to the fetal monkey neocortex.
Rakic, P. (1971b). *J. Comp. Neurol.* **141**:283–312. Neuron-glia relationship during granule cell migration in developing cerebellar cortex. A Golgi and electronmicroscopic study in *Macacus rhesus*.
Rakic, P. (1972). *J. Comp. Neurol.* **145**:61–84. Mode of cell migration to the superficial layers of fetal monkey neocortex.
Rakic, P. (1981). *Trends Neurosci.* **4**:184–187. Neuronal-glial interaction during brain development.
Rakic, P., J.-P. Bourgeois, M. F. Eckenhoff, N. Zecevic, and P. S. Goldman-Rakic (1986). *Science* **232**:232–235. Concurrent overproduction of synapses in diverse regions of the primate cerebral cortex.
Rall, R. W. (1964). Theoretical significance of dendritic tree for input-output relation, pp. 73–97. In *Neural theory and Modeling* (R.F. Reiss, ed.), Stanford Univ. Press, Palo Alto.
Rall, W., and I. Segev (1988). *Neurol. Neurobiol.* **37**:263–282. Synaptic integration and excitable dendritic spine clusters.
Ramón-Moliner, E. (1962). *J. Comp. Neurol.* **119**:211–227. An attempt at classifying nerve cells on the basis of their dendritic patterns.
Ramón-Moliner, E. (1968). The morphology of dendrites, pp. 205–267. In *The Structure and Function of Nervous Tissue, Vol. 1* (G. H. Bourne, ed.). Academic Press, New York.
Ramón y Cajal, S. (1888a). *Rev. trim. Histol. normal y patol.* **1**:1–10. Estructura de los centros nerviosos de las aves. (Reprinted in *Trabajos Escogidos*, tomo 1, Jimenez y Molina, Madrid, 1924, pp. 305–315).
Ramón y Cajal, S. (1888b). *Rev. trim. Histol. normal y patol.* May 1888. Terminaciones nerviosas los husos musculares de la rana.
Ramón y Cajal, S. (1888c). *Rev. trim. Histol. normal y patol.* **1**:11–32. Estructura de la rétina de las aves.
Ramón y Cajal, S. (1888d). *Rev. trim. Histol. normal y patol.* **1**:33–49. Sobre las fibras nerviosas de la capa molecular del cerebelo.
Ramón y Cajal, S. (1888e). *Gac. méd. Catalana* **11**:449–457. Estructura del cerebelo.
Ramón y Cajal, S. (1889a). *Manual de histologia normal y de técnica micrográfica*. Libreria de Pascuel Aguilar, Valencia.
Ramón y Cajal, S. (1889b). *La medicina práctica, Madrid,* **2**:341–346. Conexión general de los elementos nerviosos. (Reprinted in *Trabajos escogidos*, tomo 1, Jimenez y Molina, Madrid, 1924, pp. 378–487).

Ramón y Cajal, S. (1889c). *Rev. trim. histol. normal y patol.* **1**:79–106. Contribución al estudio de la estructura de la médula espinal.
Ramón y Cajal, S. (1889d). *Anat. Anz.* **4**:111–121. Sur la morphologie et les connexions des éléments de la rétine des oiseaux.
Ramón y Cajal, S. (1890a). *Gac. méd. Catalana* **13**:737–739. Sobre la existencia ce células nerviosas especiales en la primera capa de las circonvoluciones cerebrales. (Republished 1924 in *Trabajos Escogidos* **1**:625–628, Jimenez y Molina, Madrid).
Ramón y Cajal, S. (1890b). *Int. Monatschr. Anat. Physiol.* **7**:12–31. A propos de certains éléments bipolaires du cervelet avec quelques détails nouveaux sur l'évolution des fibres cérébelleuses.
Ramón y Cajal, S. (1890c). *Anat. Anz.* **5**:85–95; 111–119; 609–613; 631–639. Sur l'origine et les ramifications des fibres nerveuses de la moelle embryonnaire.
Ramón y Cajal, S. (1890d). *Anat. Anz.* **5**:579–587. Réponse à Mr. Golgi à propos des fibrilles collaterales de la moelle épinière, et de la structure générale de la substance grise.
Ramón y Cajal, S. (1890e). *Revista trimestral microgràfica* **5**:1–11. Estudios sobre la corteza cerebral humana. III. Cortex motriz.
Ramón y Cajal, S. (1890f). *Gac. san Barcelona*, 11 October. Origen y terminación de las fibras nerviosas olfactorias.
Ramón y Cajal, S. (1890g). *Gaceta Sanitaria de Barcelona*. August, p. 414–419. Sobre la aparición de las expansiones celulares en la médula embrionaria.
Ramón y Cajal, S. (1891a). *Cellule* **7**:125–176. Sur la structure de l'écorce cérébrale de quelques mammifères.
Ramón y Cajal, S. (1891b). *Rev. Cien. méd. Barcelona* **17**:673. Significación fisiológica de las expansiones protoplasmáticas y nervosias de las células de la substancia gris.
Ramón y Cajal, S. (1891c). *Trabajos del Laboratorio Histológico de la Facultad de Medicina, Barcelona* August, pp. 1–56. Pequeñas contribuciones al conocimiento del sistema nervioso. I. Estructura y conexiones de los ganglios simpáticos. II. Estructura fundamental de la corteza cerebral de los batracios, reptiles y aves. III. Estructura de la retina de los reptiles y batracios. IV. Estructura de la médula espinal de los reptiles.
Ramón y Cajal, S. (1892). *Rev. Cien. méd. Barcelona* **18**:numbers 10, 20, 22, 23. Nuevo concepto de la histologia de los centros nerviosos.
Ramón y Cajal, S. (1893). *Cellule* **9**:17–257 La rétine des vertébrés. German translation by R. Greef with additions by Cajal: 'Die Retina der Wirbelthiere,' Wiesbaden (1894). Updated version in French: 'La rétine des vertébrés,' *Trav. Lab. Rech. biol. Univ. Madrid* **28** (1933). English translation by S. A. Thorpe and M. Glickstein: *The Structure of the Retina*, Thomas, Springfield, Illinois (1972).
Ramón y Cajal, S. (1894a). *Die Retina der Wirbelthiere*. Bergmann-Verlag, Wiesbaden.
Ramón y Cajal, S. (1894b). *Les nouvelles idées sur la structure du système nerveux chez l'homme et chez les vertébrés*. (Transl. by A. Azoulay, from Ramón y Cajal, 1892). Reinwald, Paris.
Ramón y Cajal, S. (1894c). *Proc. R. Soc. London Ser. B* **55**:444–467. The Croonian Lecture: La fine structure des centres nerveux.
Ramón y Cajal, S. (1895). *Rev. med. cirurg. práct. Madrid* **36**:497–508. Algunas conjecturas sobre el mecanismo anatómico de la ideación, associación y atención. Translated into German in Arch. Anat. Physiol. Anat. Abt. Leipzig ⅘:367–378 (1895).
Ramón y Cajal, S. (1896a). *Rev. Trim. Microgr.* **1**:123–136. Las espinas colaterales de las células del cerebro teñidas con el azul de metileno.
Ramón y Cajal, S. (1896b). *Beitrag zur Studium der Medulla oblongata, des Kleinhirns und des Ursprung der Gehirnnerven*. Ambrosius Barth, Leipzig.
Ramón y Cajal, S. (1896c). *J. Anat. Physiol., Paris* **33**:481–543. Nouvelles contributions à l'étude histologique de la rétine et à la question des anastomoses des prolongements protoplasmiques.
Ramón y Cajal, S. (1897a). *Rev. Trim. Microgr.* **2**:105–127. Las células de cilindro–eje corto de la capa molecular del cerebro.
Ramón y Cajal, S. (1897b, 1899–1904). *Textura del sistema nervioso del hombre y de los vertebrados*. 2 vols. in 3. N. Moya, Madrid.
Ramón y Cajal, S. (1898). *Rev. Trim. Microgr.* **1**: Marzo, 1898. Estructura del quiasma óptico y teoría general de los entrecruzamientos nerviosos.

Ramón y Cajal, S. (1900). *Rev. Trim. Microgr.* **5**:1–11. Estudios sobre la corteza cerebral humana. III. Corteza motriz.
Ramón y Cajal, S. (1900–1906). *Studien über die Hirnrinde des Menschen.* Aus dem Spanischen übersetzt von Dr. J. Bresler. 5 Hefte: Die Sehrinde (1900); Die Bewegungsrinde (1900); Die Hörrinde (1902); Die Riechrinde beim Menschen und Säugetier (1903); Vergleichende Strukturbeschreibung und Histogenese der Hirnrinde. Anatomisch-physiologische Betrachtungen über das Gehirn. Struktur der Nervenzellen des Gehirns (1906). Verlag von Johann Ambrosius Barth, Leipzig.
Ramón y Cajal, S. (1901–1917). *Recuerdos de mi vida.* Tomo I, *Mi infancia y juventud*; Tomo II, *Historia de mi labor científica.* Third edition 1923 with '*Post scriptum.*' English translation of the third edition by E. H. Craigie and J. Cano, *Recollections of My Life*, American Philosophical Society, Philadelphia, 1937.
Ramón y Cajal, S. (1903). *Trab. Lab. Invest. Biol. Univ. Madrid.* **2**:129–221. Un sencillo método de coloración selectiva del retículo protoplásmatico y sus efectos en los diversos organos nerviosos.
Ramón y Cajal, S. (1905a). *Trab. Lab. Invest. Biol. Univ. Madrid* **4**:119–210. Mecanismo de la regeneración de los nervios.
Ramón y Cajal, S. (1905b). *Trab. Lab. Invest. Biol. Univ. Madrid* **4**:227–294. Génesis de las fibras nerviosas del embrión.
Ramón y Cajal, S. (1906). Les structures et les connexions des cellules nerveux. In *Les Prix Nobel 1904–1906.* Norstedt, Stockholm.
Ramón y Cajal, S. (1907). *Anat. Anz.* **5**:113–144. Die histogenetische Beweise der Neuronentheorie von His und Forel.
Ramón y Cajal, S. (1908). *Anat. Anz.* **23**:1–25; 65–87. Nouvelles observationes sur l'évolution des neuroblasts avec quelques remarques sur l'hypothése neurogénétique de Hensen-Held, p. 71. In *Studies in Vertebrate Neurogenesis* (L. Guth, transl.). Thomas, Springfield, Ill.
Ramón y Cajal, S. (1909–1911). *Histologie du Systeme Nerveux de l'Homme et des Vertébrés*, 2 vols. (L. Azoulay, transl.). Reprinted by Instituto Ramón y Cajal del CSIC, Madrid, 1952–1955.
Ramón y Cajal, S. (1910). *Trab. Lab. Invest. Biol. Univ. Madrid* **8**:63–134. Algunas observaciones favorables á la hipótesis neurotrópica.
Ramón y Cajal, S. (1913a). *Trab. Lab. Invest. biol. Univ. Madrid* **11**:219–237. Sobre un nuevo proceder de impregnacion de la neuroglia y sus resultados en los centros nerviosos del hombre y animales.
Ramón y Cajal, S. (1913b). *Trab. Lab. Invest. Biol. Univ. Madrid* **11**:255–315. Contribución al conocimiento de la neuroglia del cerebro humano.
Ramón y Cajal, S. (1913–1914). *Estudios sobre la degeneración y regeneración del sistema nervioso.* N. Moya, Madrid.
Ramón y Cajal, S. (1916). *Trab. Lab. Invest. Biol. Univ. Madrid* **14**:155–162. El proceder del oro-sublimado para la coloracion de la neuroglia.
Ramón y Cajal, S. (1919). *Trab. Lab. Invest. Biol. Univ. Madrid* **17**:181–228. Acción neurotrópica de los epitelios (Algunas detalles sobre el mecanismo genético de las ramificaciones nerviosas intraepiteliales, sensitivas y sensoriales), pp. 149–200. In *Studies On Vertebrate Neurogenesis* (L. Guth, transl.). Thomas, Springfield, Ill., 1960.
Ramón y Cajal, S. (1920). *Trab. Lab. Invest. Biol. Univ. Madrid* **18**:109–127. Algunas consideraciones sobre la mesoglia de Robertson y Río-Hortega.
Ramón y Cajal, S. (1922). Studies on the fine structure of the regional cortex of rodents. I: Subcortical cortex (retrosplenial cortex of Brodmann). *Trab. Lab. Invest. Biol. Univ. Madrid* **20**:1–30. Translated by J. De Felipe and E. G. Jones in *Cajal on the Cerebral Cortex.* Oxford Univ. Press, New York, 1988.
Ramón y Cajal, S. (1923). *Recuerdos de mi vida: Historia de mi labor científica.* 3.ª edición, Juan Pueyo, Madrid.
Ramón y Cajal, S. (1928). *Degeneration and Regeneration of the Nervous System* (R. M. May, transl.). Hafner, New York, 1959.
Ramón y Cajal, S. (1929a). Étude sur la neurogenése de quelques vertébrés (L. Guth, transl.). In *Studies on Vertebrate Neurogenesis.* Thomas, Springfield, Ill., 1960.
Ramón y Cajal, S. (1929b). *Trab. Lab. Invest. Biol. Univ. Madrid* **26**:107–130. Considérations critiques sur le rôle trophiques des dendrites et leurs prétendues relations vasculaires.

Ramón y Cajal, S. (1933a). *Histology*. Translated by M. Fernán-Núnez from the 10th Spanish edition. Wood, Baltimore.
Ramón y Cajal, S. (1933b). ¿Neuronismo o reticularismo? Las pruebas objectivas de la unidad anatómica, de las cellulas nerviosas. *Arch. Neurobiol. Psicol. Madrid* **13**:217–291, 579–646; French transl. *Trab. Lab. Invest. Biol. Univ. Madrid* **29**:1–137 (1934); German transl. Die Neurononenlehre, *Bumke u. Foersters Handb. Neurol.* **1**:887–994 (1935); English transl. by M. Ubeda Purkiss and C. A. Fox, *Neuron theory or reticular theory? Objective evidence of the anatomical unity of the nerve cells*. Consejo Superior de Investigaciones Cientificas, Madrid, 1954.
Ramón y Cajal, S. (1937). *Recollections of My Life*. English translation by E. H. Craigie and J. Cano from the third (1923) Spanish edition. Amer. Philos. Soc. Philadelphia.
Ramón y Cajal, S. (1951). *Precepts and Counsels on Scientific Investigation* (J. Ma. Sanchez-Perez, transl.). Pacific Press Publ. Assoc., Mountain View, California. English translation of *Reglas y consejos sobre la investigación biológica*, 1st edition, 1897.
Ramón y Cajal, S. (1952). *Trab. Inst. Cajal. Invest. Biol.* **44**:1–8, originally published in *Trab. Lab. Invest. Biol. Univ. Madrid* **1**:1–8. (1901) Significación probable de las células nerviosas de cilindro-eje corto.
Ramón y Cajal, S., and D. Sanchez (1915). *Trab. Lab. Invest. Biol. Univ. Madrid* **13**:1–64. Contribución al conocimiento de los centros nerviosos de los insectos.
Ranson, S. W. (1911). *Am. J. Anat.* **12**:67–87. Non-medullated nerve fibers in the spinal nerves.
Ranson, S. W. (1912). *J. Comp. Neurol.* **22**:159–175. The structure of the spinal ganglia and of the spinal nerves.
Ranson, S. W. (1915). *Brain* **38**:381–389. Unmyelinated nerve-fibers as conductors of protopathic sensation.
Ranvier, L.-A. (1871). *Compt. rend. hebd. Acad. Sci. Paris* **73**:1168–1171. Contributions a l'histologie et a la physiologie des nerfs peripheriques.
Ranvier, L.-A. (1872). *Arch. Physiol. norm. Path.* **4**:129–149. Recherches sur l'histologie du systeme nerveux.
Ranvier, L.-A. (1878–1879). *Leçons sur l'histologie du systeme nerveux*. 2 vols. Savy, Paris.
Rawls, J. (1971). *A Theory of Justice*. Harvard University Press, Cambridge, Massachusetts.
Reichenbach, H. (1951). *The Rise of Scientific Philosophy*. Univ. California Press, Berkeley and Los Angeles.
Reil, J. C. (1809). *Arch. Physiol. (Halle)* **9**:485–524. Das verlängerte Rückenmark, die hinteren, seitlichen und vorderen Schenkel des kleinen Gehirns und die theils strangförmig, theils als Ganglienkette in der Axe des Rückenmarks und des Gehirns fortlaufende graue Substanz.
Remak, R. (1836). *Arch. Anat. Physiol.* pp. 145–161. Verläufige Mittheilung microscopischer Beobachtungen über den innern Bau der Cerebrospinalnerven und über die Entwicklung ihrer Formelemente.
Remak, R. (1837). *Froriep's Neue Notizen* **3**:35–40. Weitere Beobachtungen über die Primitivfasern des Nervensystem der Wirbelthiere.
Remak, R. (1838a). *Observationes anatomicae et microscopicae de systematis nervosi structura*. G. Reimer, Berlin.
Remak, R. (1838b). *Froriep's Notizen* **7**:65–70. Ueber die Verrichtungen des organischen Nervensystems.
Remak, R. (1852). *Müller's Arch. f. Anat. Physiol. wissensch. Medicin*, pp. 47–57. Über extracellulare Entstehung thierische Zellen und über Vermehrung derselben durch Theilung.
Remak, R. (1853). *Ber. Verh. k. preuss. Akad. wiss. Berl.* 293–298. Über gangliöse Nervenfasern beim Menschen und bei den Wirbeltieren.
Remak, R. (1854). *Verhandlungen der Könige Preuss. Akad. der Wissenschaften, Berlin* **19**:26–32. Ueber multipolare Ganglienzellen. English translation in *Monthly J. Medical Sciences, Edinburgh* **18**:362–365. Professor Remak on multipolar ganglion cells.
Remak, R. (1855a). *Untersuchungen über die Entwickelung der Wirbelthiere*. G. Reimer, Berlin.
Remak, R., (1855b). *Dtsch. Klinik* **7**:295. Über den Bau der grauen Säulen im Rückenmark der Säugethiere.
Rescher, N. (1970). *Scientific Explanation*. Free Press, New York.
Rescher, N. (1973). *The Coherence Theory of Truth*. Clarendon Press, Oxford.
Rescher, N. (1987). *Scientific Realism*. Reidel, Dordrecht.
Retzius, G. (1890). *Biol. Untersuch. Neue Folge* **1**:1–50. Zur Kenntniss des Nervensystems der Crustaceen.
Retzius, G. (1891a). *Biol. Untersuch. Neue Folge.* **2**:1–28. Zur Kenntniss des centralen Nervensystems der Würmer.
Retzius, G. (1891b). *Biol. Untersuch. Neue Folge* **2**:29–46. Zur Kenntniss des centralen Nervensystems von Amphioxus lanceolatus.

Retzius, G. (1891c). *Biologiska Foreningens Forhandlingar*, 1891. Ueber den Bau der Oberflächenschichte der Grosshirnrinde beim Menschen und bei den Säugethiere.
Retzius, G. (1892a). *Biol. Untersuch. Neue Folge* **3**:1–16. Das Nervensystem der Lumbricinen.
Retzius, G. (1892b). *Biol. Untersuch. Neue Folge* **3**:17–24. Die nervösen Elemente der Kleinhirnrinde.
Retzius, G. (1892c). *Biol. Untersuch. Neue Folge* **3**:25–28. Die endigungsweise des Riechnerven.
Retzius, G. (1892d). *Biol. Untersuch. Neue Folge* **3**:41–52. Zur Kenntniss der motorischen Nervenendigungen.
Retzius, G. (1892e). *Biol. Untersuch. Neue Folge* **4**:19–32. Die Nervenendigungen in dem Geschmacksorgan der Säugetiere und Amphibien.
Retzius, G. (1892f). *Biol. Untersuch. Neue Folge* **4**:45–48. Ueber die Nervenendigungen an den Haaren.
Retzius, G. (1892g). *Biol. Untersuch. Neue Folge* **4**:49–56. Ueber die neuen Prinzipien in der Lehre von der Einrichtung des sensiblen Nervensystems.
Retzius, G. (1893a). *Biol. Untersuch. Neue Folge* **5**:1–8. Die Cajal'schen Zellen des Grosshirnrinde beim Menschen und bei Säugetieren.
Retzius, G. (1893b). *Biol. Untersuch. Neue Folge* **5**:48–54. Zur Kenntniss der ersten Entwicklung der nervösen Elemente im Rückenmarke des Hühnchens.
Retzius, G. (1894a). *Biol. Untersuch. Neue Folge* **6**:1–28. Die Neuroglia des Gehirns beim Menschen und be Säugethieren.
Retzius, G. (1894b). *Biol. Untersuch. Neue Folge* **6**:29–36. Weitere Beiträge zur Kenntniss der Cajal'schen Zellen der Grosshirnrinde des Menschen.
Retzius, G. (1894c). *Biol. Untersuch. Neue Folge* **6**:61–62. Ueber die Endigungsweise der Nerven an den Haaren des Menschen.
Retzius, G. (1906). *Biol. Untersuch. Neue Folge* **13**:107–112. Zur Kenntniss des Nervensystems der Daphniden.
Révész, B. (1917). *Geschichte des Seelenbegriffes und der Seelenlokalisation*. Verlag von Ferdinand Enke, Stuttgart (republished by E. J. Bonset, Amsterdam, 1966).
Rheingold, H. L., and D. F. Hay (1978). Prosocial behavior of the very young, pp. 105–124. In *Morality as a Biological Phenomenon* (G. S. Stent, ed.). Dahlem Konferenzen, Berlin.
Rheingold, H. L., D. F. Hay, and M. J. West (1976). *Child Dev.* **47**:1148–1158. Sharing in the second year of life.
Ridgeway, D., E. Waters, and S. A. Kuczaj (1985). *Develop. Psychol.* **21**:901–908. Acquisition of emotion-descriptive language: Receptive and productive vocabulary norms for ages 18 months to 6 years.
Rieder, R. (1906). *Carl Weigert und seine Bedeutung für die medizinische Wissenschaft unserer Zeit*. Springer, Berlin.
Riese, W. (1958). Descartes's ideas of brain function, pp. 115–134. In *Symposium on the History and Philosophy of Knowledge of the Brain and its functions*. Blackwell, Oxford.
Riese, W., and E. C. Hoff (1950). *J. Hist. Med.* **1**:50–71. A history of the doctrine of cerebral localization. Sources, anticipations and basic reasoning.
Rinn, W. E. (1984). *Psychol. Bull.* **95**:52–77. The neurophysiology of facial expression: a review of the neurological and psychological mechanisms for producing facial expressions.
Rio-Hortega, P. del (1919). *Bol. Soc. Esp. Biol.* **9**:69–129. El tercer elemento de los centros nerviosos. I. La microglia normal. II. Intervención de la microglia en los procesos patológicos. (Celulas en bastoncito y cuerpos granulo-adiposos). III. Naturaleza probable de la microglia.
Rio-Hortega, P. del (1920). *Trab. Lab. Inv. Biol. Univ. Madrid* **18**:37–82. La microglía y su transformación en células en basoncito y cuerpos gránulo-adiposos.
Rio-Hortega, P. del (1921a). *Mem. de la Real. Soc. Esp. Hist. Nat.* **11**:213–268. Histogenesis y evolución normal exodo y distribución regional de la microglia.
Rio-Hortega, P. del (1921b). *Bol. Soc. Esp. Biol.* **21**:64–92. Estudios sobre la neuroglia. La glia de escasas radiaciones (oligodendroglia).
Rio-Hortega, P. del (1922). *Bol. Soc. Esp. Biol.* **10**. Son homologables la glia de escasas radiacion es y la celula de Schwann?
Rio-Hortega, P. del (1924). *Compt. Rend. Soc. Biol.* **91**:818–820. La glie á radiations peu nombreuses et la cellule de Schwann sont elles homologables?
Rio-Hortega, P. del (1928). *Mem. Real. Soc. esp. Hist. Nat.* **14**:5–122. Tercera aportacion conocimiento morfologico e interpretacion functional de la oligodendroglia.

Rio-Hortega, P. del (1932). Microglia, pp. 483–534. In *Cytology and Cellular Pathology of the Nervous System*, Vol. 2, (W. Penfield, ed.). Hoeber, New York.

Rio-Hortega, P. del, and F. Jimenez de Asua (1921). *Archiv. de Cardiol. y Hematol.* **2**:161. Sobre la fagocitosis en los tumores y en otros procesos patológicos.

Roberts, A., and B. M. H. Bush (1981). *Neurones Without Impulses*. Cambridge Univ. Press, Cambridge.

Roberts, E. (1986). *Exp. Neurol.* **93**:279–290. What do GABA neurons really do? They make possible variability generation in relation to demand?

Robertson, J. D. (1953). *Proc. Soc. Exp. Biol. Med.* **82**:219–223. Ultrastructure of two invertebrate synapses.

Robertson, J. D. (1955). *J. Biophys. Biochem. Cytol.* **1**:271–278. The ultrastructure of adult vertebrate peripheral myelinated fibers in relation to myelinogenesis.

Robertson, J. D. (1963). *J. Cell Biol.* **19**:201–221. The occurrence of a subunit pattern in the unit membranes of club endings in Mauthner cell synapses in goldfish brains.

Robertson, J. D. (1965). *Neurosci. Res. Prog. Bull.* **3**:1–79. The synapse: Morphological and chemical correlates of function: A report of an NRP work session.

Robertson, J. D. (1987). *Int. Rev. Cytol.* **100**:129–201. The early days of electron microscopy of nerve tissue and membranes.

Robertson, W. (1897). *J. Ment. Sci.* **43**:733–752. The normal histology and pathology of neuroglia.

Robertson, W. (1899). *Scottish Med. Surg. J.* **4**:23. On a new method of obtaining a black reaction in certain tissue-elements of the central nervous system (platinum method).

Robertson, W. (1900). *J. Ment. Sci.* **46**:733–752. A microscopic demonstration of the normal and pathological histology of mesoglia cells.

Robinson, A. H. (1967). *Imago Mundi* **21**:95–108. The thematic maps of Charles Joseph Minard.

Roffwarg, H. P., J. N. Muzio, and W. C. Dement (1966). *Science* **152**:604–619. Ontogenetic development of the human sleep-dream cycle. The prime role of 'dreaming sleep' in early life may be in the development of the central nervous system.

Rohde, E. (1923). *Zeitschr. fur Wissensch. Zool.* **120**:325–335. Der plasmodiale Aufbau des Tier- und Pflanzenkorpers.

Rolls, E. T. (1991). *Current Opinion in Neurobiol.* **1**:274–278. Neural organization of higher visual functions.

Romanes, G. J. (1877). *Phil. Trans. Roy. Soc. Lond.* **166**:269–313. Preliminary observations on the locomotor system of *Medusae*.

Romanes, G. J. (1878). *Phil. Trans. Roy. Soc. Lond.* **167**:659–752. Further observations on the locomotor system of *Meduase*.

Romanes, G. J. (1893a). *Mental Evolution in Man*. Appleton, New York.

Romanes, G. J. (1893b). Jellyfish, Starfish, and Sea-urchins. Being a Research on Primitive Nervous Systems. International Science Series. *xii*, 323 pp. London.

Romanes, G. J. (1895). *Mental Evolution in Animals*. Appleton, New York.

Ronnevi, L.-O., and S. Conradi (1974). *Brain Res.* **80**:335–339. Ultrastructural evidence for spontaneous elimination of synaptic terminals on spinal motoneurons in the kitten.

Rosenberg, A. (1985). *The Structure of Biological Science*. Cambridge Univ. Press, Cambridge.

Rosenberg, A. and Hardin, C. (1982). *Philosophy of Science* **49**:604–615. In defense of convergent realism.

Rosenblatt, F. (1962). Strategic approaches to the study of brain models. In *Principles of Self-Organization* (H. von Foerster and G. W. Zopf, eds.). Pergamon, New York.

Rosenblueth, A. (1970). *Mind and Brain*. MIT Press, Cambridge, Massachusetts.

Rosenblueth, A., and N. Wiener (1945). *Philos. Sci.* **12**:316–321. The role of models in science.

Ross, W. N., N. Lasser-Ross, and R. Werman (1990). *Proc. Roy. Soc. London* **240**:173–185. Spacial and temporal analysis of calcium-dependent electrical activity in guinea pig Purkinje cell dendrites.

Rothschuh, K. E. (1971). Die Bedeutung apparativer Hilfsmittel für die Entwicklung der biologischen Wissenschaften im 19. Jahrhundert, pp. 137–174. In *Geschichte der Naturwissenschaften und der Technik im 19. Jahrhundert*. VDI-Verlag, Dusseldorf.

Rothschuh, K. E. (1973). *History of Physiology* (G. B. Risse, transl.). Krieger, Huntington, New York.

Roux, W. (1881). *Der Kampf der Theile im Organismus*. Verlag von Wilhelm Engelmann, Leipzig.

Roux, W. (1883). *Arch. Anat. Physiol. Anat. Abtlg.* pp. 76–160. Beiträge zur Morphologie der funktionellen Anpassung: Nr 1. Über die Struktur eines hochdifferenzierten bindegewebigen Organ (der Schwanzflosse des Delphins).

Roux, W. (1895). *Gesammelte Abhandlungen über Entwickelungsmechanik der Organismen.* Vol. 1: Abhandlung I–XII vorwiegend über funktionelle Anpassung, p. 1–816; Vol. 2: Abhandlungen XIII–XXXIII über Entwickelungsmechanik des Embryo, pp. 1–105. Engelmann, Leipzig.

Roux, W. (1923). Autobiographie, pp. 141–206. In *Die Medizin der Gegenwart in Selbstdarstellungen* (L. R. Grote, ed.), Verlag von Felix Meiner, Leipzig.

Rumelhart, D., J. R. McClelland, and The PDP Research Group (1986). *Parallel Distributed Processing: Explorations in the Microstructure of Cognition,* Vol. 1. MIT Press, Cambridge, Massachusetts.

Russell, B. (1912). *The Problems of Philosophy.* Allen & Unwin, London.

Russell, B. (1921). *Analysis of Mind.* Allen & Unwin, London.

Russell, B. (1927). *Analysis of Matter.* Kegan Paul, London.

Russell, B. (1948). *Human Knowledge, Its Scope and Limits.* Allen & Unwin, London.

Russell, E. S. (1916). *Form and Function: A Contribution to the History of Animal Morphology.* Murray, London.

Russell, E. S. (1930). *The Interpretation of Development and Heredity.* Clarendon Press, Oxford.

Russell, E. S. (1946). *The Directiveness of Organic Activities.* Cambridge Univ. Press, Cambridge.

Russell, J. S. R. (1894). *Phil. Trans. Roy. Soc. London Ser. B* **184**:39–63. An experimental investigation of the nerve roots which enter the formation of the brachial plexus of the dog.

Ryle, G. (1949). *The Concept of Mind.* Hutchinson, London.

Sabatini, D. D., K. Bensch, and R. J. Barrnett (1963). *J. Cell Biol.* **17**:19–58. Cytochemistry and electron microscopy: The preservation of cellular ultrastructure and enzymatic activity by aldehyde fixation.

Sagi, A. and Hoffman, M. L. (1976). *Dev. Psychol.* **12**:175–176. Empathic distress in the newborn.

Sakai, K., and Y. Miyashita (1991). *Nature* **354**:152–155. Neural organization for the long-term memory of paired associates.

Sano, F. (1898). *Les Localisations des foncions motrices de la moelle épinière.* Anvers, Brussels.

Sarton, G. (1937). *The History of Science and the New Humanism.* Harvard University Press, Cambridge, Mass.

Schadewald, M. (1941). *J. Comp. Neurol.* **74**:239–246. Effects of cutting the trochlear and abducens nerves on the end-bulbs about the cells of the corresponding nuclei.

Schadewald, M. (1942). *J. Comp. Neurol.* **77**:739–746. Transynaptic effect of neonatal axon section on bouton appearance about somatic motor cells.

Schäfer, E. A. (1893). *Brain* **16**:134–169. The nerve cell considered as the basis of neurology.

Schäfer, E. A. (1898–1900). *A Text-book of Physiology,* 2 vols. Y. J. Pentland, Edinburgh.

Schäfer, E. A. (1900). The cerebral cortex, Vol. II, pp. 697–782. In *A Text-Book of Physiology* (E. A. Schäfer, ed.). Y. J. Pentland, Edinburgh.

Schaffer, H. R., and P. Emerson (1964). *J. Child Psychiat. Psychol.* **5**:1–13. Patterns of response to physical contact in early human development.

Schaffner, K. (1969). *Brit. J. Philos. Sci.* **20**:325–348. The Watson-Crick model and reductionism.

Schaper, A. (1894a). *Anat. Anz.* **9**:489–501. Die morphologische und histologische Entwicklung des Kleinhirns der Teleostier.

Schaper, A. (1894b). *Morphol. Jahrb.* **21**:625–708. Die morphologische und histologische Entwickung des Kleinhirns der Teleostier.

Schaper, A. (1895). *Anat. Anz.* **10**:422–426. Einige kritische Bemerkungen zu Lugaro's Aufsatz: Ueber die Histogenese der Körner der Kleinhirnrinde.

Schaper, A. (1897a). *Arch. Entw.-Mech. Organ.* **5**:81–132. Die frühesten Differenzierungsvorgänge im Centralnervensystem.

Schaper, A. (1897b). *Science* **5**:430–431. The earliest differentiation in the central nervous system of vertebrates.

Scharf, H. J. (1958). Zur Geschichte der Entdeckung und Erforschung der sensiblen Ganglien, Vol. 3, part IV, p. 4–18. In *Handbuch der mikroskopischen Anatomie des Menschen* (W. Möllendorf, ed.). Verlag von Julius Springer, Berlin.

Scharf, J.-H. (1951). *Morphol. Jahrb.* **91**:187–252. Die markhaltigen Ganglienzellen und ihre Beziehung zu den myelogenetischen Theorien.

Scheich, H. (1991). *Current Opinion in Neurobiol.* **1**:236–247. Auditory cortex: comparative aspects of maps and plasticity.

Schelper, R. L., and E. J. Adrian Jr. (1986). *J. Neuropathol. Exp. Neurol.* **45**:1–19. Monoctyes become macrophages: They do not become microglia: A light and electronic microscopic autoradiographic study using 125-Iododeoxyuridine.

Schiff, M. (1858–1859). *Lehrbuch der Physiologie des Menschen.* 2 vols., M. Schauenburg, Lahr.

Schiff, M. (1875). *Arch. exp. Path. Pharmacol.* **3**:171–179. Untersuchungen über die motorischen Functionen des Grosshirns.

Schiller, F. (1965). *Bull. Hist. Med.* **39**:326–338. The rise of the 'enteroid processes' in the 19th century. Some landmarks in cerebral nomenclature.

Schiller, F. (1969). *Bull. Hist. Med.* **53**:67–84. Stilling's nuclei—turning point in basic neurology.

Schiller, F. (1974). *Bull. Hist. Med.* **48**:276–286. The intriguing nucleus of Deiters. Notes on an eponym.

Schiller, F. (1979). *Paul Broca. Founder of French Anthropology, Explorer of the Brain.* Univ. California Press, Berkeley.

Schlick, M. (1936). Philosophical Review 46:339–369. Meaning and verification.

Schmechel, D. E., and P. Rakic (1979). *Anat. Embryol.* **156**:115–152. A Golgi study of radial glial cells in developing monkey telencephalon: Morphogenesis and transformation into astrocytes.

Schmidt, H. D. (1874). *Monthly Microscopical J.*, 1 March. On the construction of the dark or double-bordered nerve-fibre.

Schmidt-Nielsen, K. (1984). *Scaling. Why Is Animal Size So Important?* Cambridge Univ. Press, Cambridge.

Schrödinger, E. (1951). *Science and Humanism: Physics in Our Time.* Cambridge Univ. Press, Cambridge.

Schroeder van der Kolk, J. L. C. (1859). *On the Minute Structure and Functions of the Medulla Oblongata, and on the Proximate Cause and Rational Treatment of Epilepsy,* (W. D. Moore, transl.). The New Sydenham Society, London.

Schroeder van der Kolk, J. L. C. (1863). *Die Pathologie und Therapie der Geisteskrankheiten auf Anatomisch-Physiologischer Grundlage.* Friedrich Vieweg und Sohn, Braunschweig.

Schröer, H. (1964). Carl Ludwig und die Neurophysiologie, pp. 168–185. In *Von Boerhaave bis Berger* (K. E. Rothschuh, ed.), Gustav Fischer Verlag, Stuttgart.

Schuetze, E., and L. Role (1987). *Ann. Rev. Neurosci.* **10**:403–457. Developmental regulation of nicotinic acetylcholine receptors.

Schultz, R. L. (1964). *J. Comp. Neurol.* **122**:281–296. Macroglial identification in electron microscopy.

Schultz, R. L., E. A. Maynard, and D. C. Pease (1957). *Am. J. Anat.* **100**:369–407. Electron microscopy of neurons and neuroglia of cerebral cortex and corpus callosum.

Schultze, M. (1858a). *Abh. Naturforsch. Gesellsch. Halle* **4**:30–33. Zur Kenntniss der elektrischen Organe der Fische.

Schultze, M. (1858b). *Müller's Archiv* **1858**:343–381. Über die Endigungweise des Hörnerven im Labyrinth.

Schultze, M. (1860). *Centralbl. f. d. med. Wissensch. Berlin* **2**:385–390. Das Epithelium der Riechschleimhaut des Menschen.

Schultze, M. (1862). *Abh. naturforsch. Ges. Halle* **7**:1–100. Untersuchungen über der Bau der Nasenschleimhaut, namentlich die Struktur und Endigungsweise der Geruchsnerven bei dem Menschen und den Wirbelthieren.

Schultze, M. (1866). *Arch. mikr. Anat.* **2**:165–286. Zur Anatomie und Physiologie der Retina.

Schultze, M. (1870). General character of nervous tissue, Vol. 1, pp. 147–186. In *Manual of Human and Comparative Histology* (S. Stricker, ed.), The New Sydenham Society, London.

Schultze, M. (1873). The Eye. The Retina, Vol. 3, pp. 218–298. In *Manual of Human and Comparative Anatomy* (H. Power, transl.), New Sydenham Society, London.

Schwalbe, G. A. (1887). *Lehrbuch der Anatomie der Sinnesorgane.* Besold, Erlangen.

Schwann, T. (1839). *Mikroskopische Untersuchungen über die Uebereinstimmung in der Struktur und dem Wachsthum der Thiere und Pflanzen.* G. E. Reimer, Berlin. English transl. by H. Smith, Kraus Reprint Co., New York, 1845, reprinted 1969.

Schwann, T. (1847). *Microscopical Researches into the Accordance in the Structure and Growth of Animals and Plants.* (H. Smith, transl.). The New Sydenham Society, London.

Searle, J. R. (1980). *Behav. Brain Sci.* **3**:417–457. Minds, brains and programs.

Searle, J. R. (1983). *Intentionality: An Essay on the Philosophy of Mind.* Cambridge Univ. Press, Cambridge.

Searle, J. R. (1990). *Scientific American* **262**:26–31. Is the brain's mind a computer program?

Sedgwick, A. (1895). *Quart. J. microsc. Sci.* **37**:87–101. On the inadequacy of the cellular theory of development, and on the early development of nerves, etc.
Segev, I. (1992). *Trends Neurosci.* **15**:414–421. Single neurone models: oversimple, complex and reduced.
Selander, R.K., S.Y. Yang, R.C. Lewontin, and W.E. Johnson (1970). *Evolution* **24**:402–414.
Sellars, W. (1965). The identity approach to the mind-body problem. In *Boston Studies in the Philosophy of Science*, Vol. II (R. S. Cohen and M. W. Wartofsky, eds.), Humanities Press, New York.
Sellars, W. (1975). The structure of knowledge, in *Action, Knowledge, and Reality: Essays in Honor of Wilfrid Sellars*, (H. Castaneda, ed.). Bobbs-Merrill, Indianapolis.
Senden, M. von (1960). *Space and Sight. The Perception of Space and Shape in the Congenitally Blind Before and After Operation*. Methuen, London.
Setschenow, J. (1863). *Physiologische Studien ueber die Hemmungsmechanismen fuer die Reflextaetigkeit im Gehirn des Frosches*. Berlin.
Shaftesbury, Seventh Earl of (1699). An Inquiry Concerning Virtue or Merit.
Shaftesbury, Seventh Earl of (1711). *Characteristics of Men, Manners, Opinions, Times*. 3 vols. London.
Sharpey, W., A. Thomson, and E. A. Schäfer (1878). *Quains Elements of Anatomy*. 2 vols. 8th edition. William Wood, New York.
Sharpey-Shafer, E. A. (1919). *Nature* **104**:207–208. Developments in physiology.
Shepherd, G. M. (1991). *Foundations of the Neuron Doctrine*. Oxford Univ. Press, New York, Oxford.
Shepherd, G. M. (1992). *Nature* **358**:457–458. Modules for molecules.
Sherman, S. M., and J. R. Wilson (1975). *J. Comp. Neurol.* **161**:183–196. Behavioral and morphological evidence for binocular competition in the postnatal development of the dog's visual system.
Sherman, S. M., R. W. Guillery, J. H. Kaas, and R. J. Sanderson (1974). *J. Comp. Neurol.* **158**:1–18. Behavioral, electrophysiological and morphological studies of binocular competition in the development of the geniculo-cortical pathways of cats.
Sherrington, C. S. (1885). *J. Physiol. (London)* **6**:177–191. On secondary and tertiary degenerations in the spinal cord of the dog.
Sherrington, C. S. (1886). *Brain* **9**:342–351. Note on two newly described tracts in the human spinal cord.
Sherrington, C. S. (1889). *J. Physiol. (London)* **10**:429–432. On nerve-tracts degenerating secondarily to lesions of the cortex cerebri.
Sherrington, C. S. (1890a). *Br. Med. J.* **1**:14. note on bilateral degeneration in the pyramidal tracts resulting from unilateral cortical lesion.
Sherrington, C. S. (1890b). *J. Physiol. (London)* **11**:121–122. Addendum to note on tracts degererating secondarily to lesions of the cortex cerebri.
Sherrington, C. S. (1890c). *J. Physiol. (London)* **11**:399–400. Further note on degenerations following lesions of the cerebral cortex.
Sherrington, C. S. (1891). *St. Thomas's Hosp. Rept.* **21**:145–147. Note on the knee-jerk.
Sherrington, C. S. (1892a). *Brit. Med. J.* **1**:545, 654. Note towards the localization of the knee-jerk; Addendum to note on the knee-jerk.
Sherrington, C. S. (1892b). *Lancet* **2**:1416–1417, 1533, (1893) Lancet 1:221. Experiments on living animals.
Sherrington, C. S. (1892c). *J. Physiol. (London)* **13**:621–772. Notes on the arrangement of some motor fibres in the lumbo-sacral plexus.
Sherrington, C. S. (1892d). *Proc. Roy. Soc. (London)* **52**:333–337. Experiments in examination of the peripheral distribution of the fibres of the posterior roots of some spinal nerves (Preliminary note).
Sherrington, C. S. (1893a). *Brit. Med. J.* **2**:685. Experimental note on the knee-jerk.
Sherrington, C. S. (1893b). *Proc. Roy. Soc. London* **53**:407–420. Note on the knee-jerk and the correlation of action of antagonistic muscles.
Sherrington, C. S. (1894a). *Lancet* **1**:265, 439, 571. Notes on experimental degeneration of the pyramidal tract.
Sherrington, C. S. (1894b). *J. Physiol. (London)* **17**:211–258. On the anatomical constitution of nerves of skeletal muscles; with remarks on recurrent fibres in the ventral spinal nerve roots.
Sherrington, C. S. (1894c). *Philos. Trans. Roy. Soc.* **184B**:641–763. Experiments on examination of the peripheral distribution of the fibres of the posterior roots of some spinal nerves (I).
Sherrington, C. S. (1897a). *Proc. Roy. Soc.* **60**:414–417. On reciprocal innervation of antagonistic muscles. Third note.

Sherrington, C. S. (1897b). The central nervous system, Vol. III, in *A Text-Book of Physiology* (M. Foster, ed.). 7th edition, Macmillan, London.
Sherrington, C. S. (1897c). *Proc. Roy. Soc. London* **61**:243–246. Double (antidromic) conduction in the central nervous system.
Sherrington, C. S. (1898a). *J. Physiol. (London)* **22**:319–332. Decerebrate rigidity, and reflex co-ordination of movements.
Sherrington, C. S. (1898b). *Proc. Roy. Soc. London* **64**:179–181. On the reciprocal innervation of antagonistic muscles.
Sherrington, C. S. (1898c). *Phil. Trans. Roy. Soc. London Ser. B* **190**:45–186. Experiments in examination of the peripheral distribution of the fibres of the posterior roots of some spinal nerves (II).
Sherrington, C. S. (1899a). *Brit. Med. J.* **1**:878. The teaching of physiology and histology.
Sherrington, C. S. (1899b). *Trans. Liverpool Biol. Soc.* **13**:1–20. On the relation between structure and function as examined in the arm.
Sherrington, C. S. (1900a). The muscular sense, Vol. II, p. 1002–1025. In *A Textbook of Physiology*, (E. A. Schäfer, ed.). Pentland, Edinburgh.
Sherrington, C. S. (1900b). The spinal cord, Vol II, p. 783–883. In *A Text-book of Physiology* (E. A. Schafer, ed.). Pentland, Edinburgh.
Sherrington, C. S. (1900c). The parts of the brain below the cerebral cortex, Vol. II, p. 884–919. In *A Text-book of Physiology* (E. A. Schäfer, ed.). Pentland, Edinburgh.
Sherrington, C. S. (1900d). The cerebellum, Vol. II, p. 893–910, In *A Text-book of Physiology*, (E. A. Schafer, ed.). Pentland, Edinburgh.
Sherrington, C. S. (1904). *Brit. Assoc. Rep.* **74**:1–14. The correlation of reflexes and the principle of the common path.
Sherrington, C. S. (1906). *The Integrative Action of the Nervous System*. Yale Univ. Press, New Haven.
Sherrington, C. S. (1913). *Quart. J. Exptl. Physiol.* **6**:251–310. Reflex inhibition as a factor in the co-ordination of movements and postures.
Sherrington, C. S. (1919). *Mammalian Physiology. A Course of Practical Exercises*. Clarendon Press, Oxford (2nd Edition, 1929).
Sherrington, C. S. (1925a). *Proc. Roy. Soc.* **97B**:519–545. Remarks on some aspects of reflex inhibition.
Sherrington, C. S. (1925b). *Brit. Med. J.* **2**:925. J. N. Langley—Obituary.
Sherrington, C. S. (1933). *Inhibition as a co-ordinative factor*. Nobel Lecture, December 12, 1932. Norstedt, Stockholm.
Sherrington, C. S. (1934). Inhibition as a coordinative factor. Nobel lecture delivered at Stockholm, 12th December 1932, pp. 1–12. In *Les Prix Nobel En 1932*, Norstedt, Stockholm.
Sherrington, C. S. (1935). *Obit. Not. R. Soc.* **4**:425–441. Santiago Ramón y Cajal 1852–1934.
Sherrington, C. S. (1941). *Man on his Nature*. Cambridge Univ. Press, Cambridge.
Sherrington, C. S. (1949). A Memoir of Dr. Cajal. In *Explorer of the Human Brain: The Life of Santiago Ramón y Cajal* (D. F. Cannon), Henry Schuman, New York.
Sherrington, C. S. (1953). Marginalia, Vol. II, p. 545–553. In *Science, Medicine and History*. Essays on the Evolution of Scientific Thought and Medical Practice, Written in Honour of Charles Singer, (E. A. Underwood, ed.). Oxford Univ. Press, Oxford.
Sherrington, C. S., and H. E. Hering (1899c). *J. Physiol. (London)* **23**:31. Inhibition of the contraction of voluntary muscles by electrical stimulation of the cortex cerebri.
Shoemaker, S. (1984). *Identity, Cause and Mind*. Cambridge University Press, Cambridge.
Sholl, D. A. (1956a). *The Organization of the Cerebral Cortex*. Methuen, London.
Sholl, D. A. (1956b). *Prog. Neurobiol.* **2**:324–333. The measurable parameters of the cerebral cortex and their significance in its organization.
Sidgwick, H. (1902). *Outlines of the History of Ethics*, 5th edition, MacMillan, London. Reprinted 1988 by Hackett Publishing Company, Indianapolis, Indiana.
Sidgwick, H. (1907). *The Methods of Ethics*. 7th Edition. Macmillan, London.
Sidman, R. L., I. L. Miale, and N. Feder (1959). *Exp. Neurol.* **1**:322–333. Cell proliferation and migration in the primitive ependymal zone: An autoradiographic study of histogenesis in the nervous system.
Siegel, C. (1913). *Geschichte der deutschen Naturphilosophie*. Akademische Verlagsgesellschaft, Leipzig.
Simpson, G. G. (1961). *Principles of Animal Taxonomy*. Harvard Univ. Press, Cambridge, Massachusetts.

Skinner, B. F. (1938). *The Behavior of Organisms: An Experimental Analysis*. Appleton-Century, New York.
Smart, I., and C. P. Leblond (1961). *J. Comp. Neurol.* **116**:349–367. Evidence for division and transformation of neuroglia cells in the mouse brain as derived from radioautography after injection of thymidine-H3.
Smart, J. J. C. (1963). *Philosophy and Scientific Realism*. Routledge and Kegan Paul, London.
Smart, J. J. C. (1984). A case for scientific realism, pp. 8–40, in *Scientific Realism* (J. Leplin, ed.). Univ. of California Press, Berkeley.
Smart, J. J. C. and Williams, B. (1973). *Utilitarianism: For and Against*. Cambridge University Press, Cambridge.
Smith, A. (1759). *The Theory of Moral Sentiments*. A. Millar, in The Strand, London; and A. Kincaid and J. Bell, in Edinburgh. Reprinted in *Adam Smith's Moral and Political Philosophy* (H. W. Schneider, ed.). Hafner, New York, 1948.
Smith, G. Elliot (1902). *J. Anat. Physiol.* London **36**:309–319. On the homologies of the cerebral sulci.
Smith, G. Elliot (1907). *J. Anat. Physiol.* **41**:237–254. A new topographical survey of the human cerebral cortex, being an account of the anatomical distinct cortical areas and their relationship to the cerebral sulci.
Smith, G. Elliot (1910). *Lancet* **1**:1–6, 147–153, 221–227. Some problems relating to the evolution of the brain.
Snyder, S. H. (1989). *Brainstorming. The Science and Politics of Opiate Research*. Harvard Univ. Press, Cambridge, Mass.
Soemmerring, S. T. (1788). *Vom Hirn und Rückenmark*. P. A. Winkopp, Mainz.
Soemmerring, S. T. (1791–1792). *Vom Baue des menschlichen Köpers*. 5 Vols. Varrentrapp and Wenner, Frankfurt am Main.
Soury, J. (1899). *Le Système Nerveux Central. Structure et Fonctions. Histoire Critique des Théories et des Doctrines*, 2 vols. Carré et Naud, Paris.
Spacek, J. (1985). *Anat. Embryol.* **171**:235–243. Three-dimensional analysis of dendritic spines. II. Spine apparatus and other cytoplasmic components.
Spacek, J. (1989). *J. Neurocytol.* **18**:27–38. Dynamics of the Golgi method: A time-lapse study of the early stages of impregnation in single sections.
Spatz, H. (1929). *Arch. Psychiat. Nervenkr.* **87**:100–125. Nissl und die theoretische Hirnanatomie.
Spatz, H. (1952). *Münch. med. Wochenschr.* **94**:1154–1164, 1209–1218, 1255–1262. Neuronlehre und Zellenlehre.
Spector, M. (1970). *Concepts of Reduction in Physical Science*. Temple Univ. Press, Philadelphia.
Speidel, C. C. (1932). *J. Exp. Zool.* **61**:279–331. Studies of living nerves. I. The movements of individual sheath cells and nerve sprouts correlated with the process of myelin sheath formation in amphibian larvae.
Spemann, H. (1924). *Naturwissensch.* **12**:1092–1094. Über Organisatoren in der tierischen Entwicklung.
Spencer, H. (1855). *The Principles of Psychology*. Longmans, London. 2nd edition, 2 vols., 1870–1872; 3rd edition, 1890, Williams and Norgate, London.
Spencer, H. (1864–67). *Principles of Biology*. 2 vols., 2nd edition 1898–99. Williams and Norgate, London.
Spencer, H. (1904). *An Autobiography*, 2 vols. Williams and Norgate, London.
Sperry, R. W. (1966). Brain bisection and the mechanism of consciousness. pp. 298–313 in *Brain and Conscious Experience* (J. C. Eccles, ed.). Springer Verlag, Berlin.
Sperry, R. W. (1968). *Amer. Psychol.* **23**:723–733. Hemisphere deconnection and unity of conscious awareness.
Sperry, R. W. (1969). *Psychol. Rev.* **76**:532–536. A modified concept of consciousness.
Sperry, R. W. (1970a). Cerebral dominance in perception. In *Early Experience in Visual Information Processing in Perceptual and Reading Disorders* (F. A. Young and D. B. Lindsley, eds.). Natl. Acad. Sci., Washington, D.C.
Sperry, R. W. (1970b). *Res. Publ. Assoc. Res. Nerv. Ment. Dis.* **48**. Perception in the absence of neocortical commissures.
Sperry, R. W. (1973). Lateral specialization in the surgically separated hemispheres, pp. 5–19. In *The Neurosciences: Third Study Program* (F. O. Schmitt and F. G. Worden, eds.), M.I.T. Press, Cambridge, Massachusetts.
Sperry, R. W. (1976). Mental phenomena as causal determinants in brain function, pp. 163–177. In *Consciousness and the Brain* (G. G. Globus, G. Maxwell, and I. Savodnick, eds.). Plenum Press, New York.

Sperry, R. W. (1980). *Neurosci.* **5**:195–206. Mind-brain interaction: mentalism, yes; dualism, no.
Sperry, R. W. (1982). Forebrain commissurotomy and conscious awareness, pp. 497–522. In *Neuropsychology After Lashley* (J. Orbach, ed.). Erlbaum, Hillsdale, New Jersey.
Spinoza, B. (1677). *Ethics and On the Improvement of the Understanding* (J. Gutmann, ed.). Hafner, New York (1955).
Spitzka, E. A. (1907). *A Study of the Brains of Six Eminent Scientists and Scholars.* Amer. Philosoph. Soc. Philadelphia.
Sroufe, L. A. (1979). Socioemotional development, pp. 462–516. In *Handbook of Infant Development* (J. Osofsky, ed.), Wiley, New York.
Steele, E. J. (1979). *Somatic Selection and Adaptive Evolution. On the Inheritance of Acquired Characters.* Williams & Wallace, Toronto; and Croom Helm, London (1980).
Steele, E. J. (1981). *New Scientist* **90**:360–361. Lamarck and Immunity: a conflict resolved.
Steinberg, S. H. (1961). *Five Hundred Years of Printing*, 2nd edition. Penguin Books, Harmondsworth.
Steindler, A. (1916). *Am. J. Orthopedic Surg.* **14**:707–719. Direct neurotization of paralyzed muscles, further study of the question of direct nerve implantation.
Stell, W. K. (1965). *Anat. Rec.* **153**:389–397. Correlation of retinal cytoarchitecture and ultrastructure in Golgi preparations.
Sterling, P., M. Freed, and R. G. Smith (1986). *Trends Neurosci.* **9**:186–192. Microcircuitry and functional architecture of the retina.
Sterzi, G. (1914–1915). *Anatomia del Sistema Nervoso Centrale dell'Uomo.* 2 vols. Angelo Draghi, Editore, Padova.
Stevenson, L. G. (1956). *Yale J. Biol. Med.* **29**:125–157. Religious elements in the background of the British anti-vivisection movement.
Steward, O. (1983). *Cold Spring Harbor Symp. Quant. Biol.* **48**:745–759. Polyribosomes at the base of dendritic spines of central nervous system neurons: Their possible role in synapse construction and modification.
Stich, S. P. (1983). *From Folk Psychology to Cognitive Science.* MIT Press, Cambridge, Massachusetts.
Stieda, L. (1899). *Geschichte der Entwicklung der Lehre von den Nervenzellen und Nervenfasern Während des XIX Jahrhunderts. I. Teil: Von Sömmering bis Deiters.* Verlag von Gustav Fischer, Jena.
Stilling, B. (1842). *Untersuchungen über die Functionen des Rückenmarks und der Nerven.* O. Wigand, Leipzig.
Stilling, B. (1843). *Ueber die Medulla oblongata.* F. Enke, Erlangen.
Stilling, B. (1846). *Über den Bau des Hirnknotens oder Varolischen Brücke.* F. Mauke, Jena.
Stilling, B. (1856). *Anatomische und mikroskopische Untersuchungen über den feineren Bau der Nerven-Primitivfaser und der Nervenzelle.* Verlag von literarischer Anstalt, Frankfurt.
Stilling, B. (1856–1859). *Neue Untersuchungen über den Bau des Rückenmarks.* Hotop, Kassel.
Stilling, B. (1864–1878). *Untersuchungen über den Bau des kleinen Gehirns des Menschen.* 3 vols. T. Kay (vols. 1 & 2), T. Fischer (vol. 3), Cassel.
Stöhre, P. (1935). *Z. Anat. Entw Gesch.* **104**:133–158. Beobachtungen und Bemerkungen über die Endausbreitung des vegetativen Nervensystems.
Streeter, G. L. (1912). The development of the nervous system, Vol. 2, p. 1–156. In *Manual of Human Embryology*, (F. Keibel, and F. P. Mall, eds.). Lippincott, Philadelphia.
Stubbe, H. (1963). *Kurze Geschichte der Genetik bis zur Wiederentdeckung der Vererbungsregeln Gregor Mendels.* Gustav Fischer Verlag, Jena.
Stubbe, H. (1972). *History of Genetics from Prehistoric Times to the Rediscovery of Mendel's Laws.* MIT Press, Cambridge, Massachusetts.
Sturmbauer, C., and A. Meyer (1992). *Nature* **358**:578–581. Genetic divergence, speciation and morphological stasis in a lineage of African cichlid fishes.
Sulston, J. E., E. Schierenberg, J. G. White, and J. N. Thomson (1983). *Dev. Biol.* **100**:64–119. The embryonic cell lineage of the nematode Caenorhabditis elegans.
Swazey, J. P. (1969). *Reflexes and Motor Integration: Sherrington's Concept of Integrative Action.* Harvard Univ. Press, Cambridge, Mass.
Swinburne, R. (1986). *The Evolution of the Soul.* Oxford University Press, Oxford.
Swindale, N. V. (1990). *Trends Neurosci.* **13**:487–492. Is the cerebral cortex modular?

Szentagothai, J. (1975). *Brain Res.* **95**:475–496. The 'module' concept in cerebral cortex architecture.
Szentagothai, J. (1976). *Brain Res. (Suppl.)* 282–287. The basic circuitry of the neocortex.
Szentagothai, J. (1978). *Proc. Roy. Soc. London Ser. B* **201**:219–248. The neuron network of the cerebral cortex: a functional interpretation.
Szentagothai, J. (1979). Local neuron circuits of the neocortex, pp. 399–418. In the *Neurosciences Fourth Study Program* (F. O. Schmidt and F. G. Worden, eds.). MIT Press, Cambridge, Massachusetts.
Szentagothai, J., and M. Arbib (1974). *Neurosci. Res. Prog. Bull.* **12**:307–510. Conceptual models of neural organization.
Taniuchi, M, H. B. Clark, and E. M. Johnson, Jr. (1986a). *Proc. Natl. Acad. Sci. USA* **83**:4094–4098. Induction of nerve growth factor receptor in Schwann cells after axotomy.
Taniuchi, M., J. B. Schweitzer, and E. M. Johnson, Jr. (1986b). *Proc. Natl. Acad. Sci. USA* **83**:1950–1954. Nerve growth factor receptor molecules in the rat brain.
Taniuchi, M., H. B. Clark, J. B. Schweitzer, and E. M. Johnson, Jr. (1988). *J. Neurosci.* **8**:664–681. Expression of nerve growth factor receptors by Schwann cells of axotomized peripheral nerves: Ultrastructural location, suppression by axonal contact, and binding properties.
Tank, D. W. (1989). *Seminars Neurosci.* **1**:67–79. What details of neural circuits matter?
Tanzi, E. (1893). *Rivista Sperimentale di Freniatria e di Medicina Legale* **19**:1–54. I fatti e le induzioni nell' odierna istologia del sistema nervoso.
Tartuferi, F. (1887a). *Internat. Monatschr. Anat. Physiol.* **4**:421–441. Sull'anatomia della retina.
Tartuferi, F. (1887b). *Arch. per le scien med.* **11**:335–366. Sulla anatomia della retina.
Taub, E., P. Perrella, and G. Barro (1973). *Science* **181**:959–960. Behavioral development after forelimb deafferentation on day of birth in monkeys with and without blinding.
Tauc, L. (1955). *J. Physiol. Path. Gen.* **47**:769–792. Etude de l'activité élémentaire des cellules du ganglion abdominal de l'Aplysie.
Taylor, A. C. (1943). *Anat. Rec.* **87**:379–413. Development of the innervation pattern in the limb bud of the frog.
Taylor, A. C. (1944). *J. Exp. Zool.* **96**:159–185. Selectivity of nerve fibers from the dorsal and ventral roots in the development of the frog limb.
Taylor, C. P., and F. E. Dudek (1984). *J. Neurophysiol.* **52**:126–142. Excitation of hippocampal pyramidal cells by an electric field effect.
Tello, J. F. (1935). *Cajal y su labor histológica.* Universidad Central, Madrid.
Temkin, O. (1946b). *Bull. Hist. Med.* **20**:322–327. Materialism in French and German physiology of the early nineteenth century.
Temkin, O. (1949). Metaphors of human biology, p. 167. In *Science and Civilization* (R. C. Stauffer, ed.), Univ. Wisconsin Press, Madison, Wisconsin.
Tepperman, J. (1970). *Perspect. Biol. Med.* **13**:295–308. Horsley and Clarke: A biographical medallion.
Terano, T., K. Asai, and M. Sugeno (1992). *Fuzzy Systems Theory and Its Applications.* Academic Press, Hialeah, Florida.
Thagard, P. (1978). *J. Philos.* **75**:76–92. The best explanation: Criteria for theory choice.
Thagard, P., and R. E. Nisbett (1982). *Philosophical Studies* **42**:379–394. Variability and confirmation.
Thorpe, W. H. (1966). Ethology and consciousness, pp. 470–505. In *Brain and Conscious Experience* (J. C. Eccles, ed.), Springer-Verlag, Berlin.
Thorpe, W. H. (1974). *Animal Nature and Human Nature.* Methuen, London.
Threlfall, R. (1930). *Biol. Rev.* **5**:357–361. The origin of the automatic microtome.
Tiedemann, F. (1816). *Anatomie und Bildungsgeschichte des Gehirns im Foetus des Menschen, nebst einer vergleichenden Darstellung des Hirnbaues in den Thieren.* Stein, Nuremberg. English translation by W. Bennett (1826), *The Anatomy of the Foetal Brain.* J. Carfrae, Edinburgh.
Tiegs, O. W. (1927). *Aust. J. Exp. Biol. Med. Sci.* **4**:193–212. A critical review of the evidence on which is based the theory of discontinuous synapses in the spinal cord.
Tigerstedt, R. (1911–1912). *Handbuch der physiologischen Methodik.* 3 volumes in 5. Verlag von S. Hirzel, Leipzig.
Tilling, L. (1975). *Brit. J. Hist. Sci.* **8**:204–205. Early experimental graphs.
Tilney, J. F., and F. H. Pike (1925). *Arch. Neurol. Psychiat. Chicago* **13**:289–334. Muscular coordination experimentally studied in relation to the cerebellum.

Tilney, L. G., E. M. Bonder, and D. J. DeRosier (1981). *J. Cell Biol.* **90**:485–494. Actin filaments elongate from their membrane-associated ends.
Titchener, E. B. (1921). *Amer. J. Psychol.* **32**:161–178, 575–580. Wilhelm Wundt.
Toellner, R. (1971). Naturphilosophische Elemente im Denken Purkyne's, pp. 35–41. In *Jan Evangelista Purkyne 1787–1869* (V. Kruta, ed.), Universita J. E. Purkyne, Brno.
Toldt, C. (1877). *Lehrbuch der Gewebelehre.* Verlag von Ferdinand Enke, Stuttgart.
Tolman, E. C. (1948). *Psychol. Rev.* **55**:189–208. Cognitive maps in rats and men.
Torrey, T. W. (1934). *J. Comp. Neurol.* **59**:203–220. The relation of taste buds to their nerve fibers.
Torvik, A. (1975). *Acta Neuropathol.* **6**:297–300. The relationship between microglia and brain macrophages. Experimental investigations.
Tower, D. B. (1970). Johann Ludwig Wilhelm Thudichum (1829–1901), pp. 297–302. In *Founders of Neurology*, 2nd ed., (W. Haymaker, and F. Schiller, eds.). Thomas, Springfield, Illinois.
Treviranus, G. R. (1835). *Neue Untersuchungen uber die organischen Elemente der thierischen Korper und deren Zusammensetzungen.* J. G. Heyse, Bremen.
Trinchese, S. (1867). *Robin J. Anat.* **4**:485–504. Mémoire sur la terminaison péripherique des nerfs moteurs dans la série animale.
Tuckwell, H. C. (1988). *Introduction to Theoretical Neurobiology.* 2 vols. Cambridge University Press, Cambridge.
Tufte, E. R. (1983). *The Visual Display of Quantitative Information.* Graphics Press, Cheshire, Connecticut.
Turing, A. (1950). *Mind* **59**:433–460. Computing machinery and intelligence.
Turner, W. (1890). *J. Anat. Physiol. London* **25**:105–153. The convolutions of the brain: a study in comparative anatomy.
Tversky, A., and D. Kahneman (1974). *Science* **185**:1124–1131. Judgement under uncertainty: Heuristics and biases.
Ulfhake, B., and S. Cullheim (1986). *Neurosci. Lett.* **26**:492. Direct contacts between dendrites and blood vessels in the cat spinal cord.
Ulfhake, B., and S. Cullheim (1988). *J. Comp. Neurol.* **278**:88–102. Postnatal development of cat hind limb motoneurons. II: In vivo morphology of dendritic growth cones and the maturation of dendrite morphology.
Unna, P. G. (1890). *Monatsschr. Prak. Dermatol.* **11**:366–367. Über die Taenzersche (Orcein-) Färbung des elastischen Gewebes.
Unzer, J. (1771). *The Principles of Physiology, by John Augustus Unzer; and a Dissertation on the Functions of the Nervous System, by George Prochaska*, (T. Laycock, transl.). The Sydenham Society, London, 1851.
Uylings, H. B. M., R. W. H. Verwer, J. Van Pelt, and J. G. Parnavelas (1983). *ACTA Stereol.* **2**:55–62. Topological analysis of dendritic growth at various stages of cerebral development.
Uzman, L. L. (1960). *J. Comp. Neurol.* **114**:137–160. The histogenesis of the mouse cerebellum as studied by its tritiated thymidine uptake.
Vaihinger, H. (1911). *Die Philosophie des 'Als Ob.'* Berlin.
Valenstein, E. (1986). *Great and Desperate Cures: The Rise and Decline of Psychosurgery and Other Radical Treatments for Mental Illness.* Basic Books, New York.
Valentin, G. G. (1835). *Handbuch der Entwickelungsgeschichte des Menschen.* A. Rücker, Berlin.
Valentin, G. G. (1836a). *Nova Acta Acad. Caes. Leopoldina* **18**:51–240, 541–543. Ueber die Verlauf und die letzten Enden der Nerven.
Valentin, G. G. (1836b). *Repertorium für Anatomie und Physiologie, Berlin* **1**:300–316. Feinere Anatomie der Sinnesorgane des Menschen und der Wirbelthiere.
Valentin, G. G. (1838). *Repert. Anat. Physiol.* **3**:72–80. Kritische Darstellung fremder und Ergebnisse eigener Forschung.
Valentin, G. G. (1839). *Arch. f. Anat. Physiol. wissensch. Med.* pp. 139–164. Ueber die Scheiden der Ganglienkugeln und deren Fortsetzungen.
van der Loos, H. (1967). The history of the neuron, pp. 1–47. In *The Neuron* (H. Hydén, ed.). Elsevier, Amsterdam.
van Gehuchten, A. (1890). *Cellule* **6**:395. Contribution à l'étude de la muqueuse olfactive chez les mammifères.

van Gehuchten, A. (1891a). *Cellule* **7**:81–122. La structure des centres nerveux. La moëlle épinière et le cervelet.
van Gehuchten, A. (1891b). *Cellule* **8**:1–43. La structure des centres nerveux la moelle épinière et le cervelet.
van Gehuchten, A. (1891c). *Cellule* **18**:315–390. L'anatomie fine de la cellule nerveuse.
van Gehuchten, A. (1892). *Anat. Anz.* **7**:341–348. Contributions à l'étude de l'innervation des poils.
van Gehuchten, A. (1894). *Anatomie du système nerveux de l'homme*. First edition. A. Uystpruyst-Dieudonne, Louvain.
van Gehuchten, A. (1903). *Névraxe* **5**:3. La dégénérescence dite rétrograde ou dégénérescence Wallérienne indirecte.
van Gehuchten, A (1904). *Nevraxe* **6**:81–116, 219–234. Boutons terminaux et reseau pericellulaire.
van Gehuchten, A. (1906). *Anatomie du système nerveux de l'homme*, 4th Edn. A. Uystpruyst-Dieudonne, Louvain.
van Gehuchten, A., and J. Martin (1891). *Cellule* **7**:205–237. Le bulbe Olfactif chez quelques mammifères.
van Gieson, J. (1889). *N.Y. Med. J.* **50**:57–60. Laboratory notes of technical methods for the nervous system.
van Pelt, J., and R. W. H. Verwer (1982). *Bull. Math. Biol.* **45**:269–285. The exact probabilities of branching patterns under terminal and segmental growth hypothesis.
van Pelt, J., and R. W. H. Verwer (1984). *J. Microsc.* **136**:23–34. New classification methods of branching patterns.
Vartanian, A. (1960). *La Mettrie, 'L'Homme Machine': A Study in the Origins of an Idea*. Princeton Univ. Press, Princeton, New Jersey.
Vaughn, J. E., and D. C. Pease (1970). *J. Comp. Neurol.* **140**:207–226. Electron microscopic studies of Wallerian degeneration in rat optic nerve. II. Astrocytes, oligodendrocytes and adventitial cells.
Veith, I. (1950). *Some Philosophical Concepts of Early Chinese Medicine*. Transaction No. 4, The Indian Institute of Culture, Basavangudi, Bangalore.
Veith, I. (1957). Non-western concepts of psychic function, pp. 29–42. In *Symposium on the History and Philosophy of the Brain and its Functions*. Blackwell, Oxford.
Veith, I. (1966). *Huang Ti Nei Ching Su Wên* (The Yellow Emperor's Classic of Internal Medicine), New Edition. Univ. Calif. Press, Berkeley.
Verwer, R. W. H., and J. Van Pelt (1983). *J. Neurosci. Meth.* **8**:335–351. A new method for the topological analysis of neuronal tree structures.
Verworn, M. (1900). *Arch. f. Anat. u. Physiol.* Zur Physiologie der nervösen Hemmungserscheinungen.
Viets, H. R. (1952). *New England J. Med.* **246**:981. Charles Scott Sherrington, 1857–1952.
Vignal, W. (1889). *Développement des éléments du système nerveux cérébro-spinal*. Paris.
Vince, M. A. (1979). *Anim. Behav.* **27**:908–918. Postnatal effects of prenatal sound stimulation in the guinea pig.
Vince, M. A., S. E. Armitage, B. A. Baldwin, and J. Toner (1982). *Behaviour* **81**:296–315. The sound environment of the fetal sheep.
Vintschgau, M. von (1880). *Arch. Ges. Physiol.* **23**:1–13. Beobachtungen uber die Veränderungen der Schmeckbecher nach Durchschneidung des N. glossopharyngeus.
Vintschgau, M. von, and J. Honigschmied (1876). *Arch. Ges. Physiol.* **14**:443–448. Nervus glossopharyngeus und Schmeckbecher.
Virchow, R. (1846). *Allgem. Zeitschr. Psychiat.* **3**:242–250. Über das granulirte Ansehen der Wandungen der Gehirnventrikel.
Virchow, R. (1858). *Die Cellularpathologie in ihre Begründung auf physiologische und pathologische Gewebelehre*. A. Hirschwald, Berlin. English translation by F. Chance (1863) based on 2nd German edition, Lippincott, Philadelphia. Reprinted 1971, Dover, New York.
Virchow, R. (1885). *Arch. path. Anat. klin. Med.* **8**:1–15. Cellular pathology.
Volkmann, A. W. (1844). Nervenphysiologie, Vol. 2, pp. 476–627. In *Handwörterbuch der Physiologie mit Rücksicht auf physiologische Pathologie* (R. Wagner, ed.). F. Vieweg, Braunschweig.
Vulpian, E. F. A. (1866). *Leçons sur la physiologie générale et comparé du système nerveux*. Germer Baillière, Paris.
Vygotsky, L. S. (1962). *Thought and Language*. MIT Press, Cambridge, Massachusetts.
Wagner, R. (1842–1853). *Handwörterbuch der Physiologie mit Rücksicht auf physiologische Pathologie*. 4 vols. F. Vieweg, Braunschweig.

Wagner, R. (1847a). *Nachr. von d. G. A. Univers. u. d. Kgl. Gesellsch. d. Wissensch. zu Göttingen*, No. 6, April 1847. Fortgesetzte Untersuchungen über die Verbreitung des Nerven im elektrischen Organ des Zitterochens.

Wagner, R. (1847b). *Icones Physiologicae Suppl. Leipzig*. Neue Untersuchungen über den Bau und die Endigung der Nerven und die Struktur der Ganglienzellen.

Waldeyer(-Hartz), H. W. G. von (1891). *Deutsche med. Wochenschrift* **17**:1213–1218, 1244–1246, 1267–1270, 1287–1289, 1331–1332, 1352–1356. Über einige neuere Forschungen im Gebiete der Anatomie des Centralnervensystems.

Wall, J. T. (1988). *Trends Neurosci.* **11**:549–557. Variable organization in cortical maps of the skin as an indication of the lifelong adaptive capacities of circuits in the mammalian brain.

Waller, A. (1850). *Phil. Trans. Roy. Soc. London Ser. B* **140**:423–429. Experiments on the section of the glossopharyngeal and hypoglossal nerves of the frog, and observations of the alterations produced thereby in the structure of their primitive fibres.

Waller, A. (1851a). *Edinburgh Med. Surg. J.* **76**:369–376. Experiments on the section of the glossopharyngeal and hypoglossal nerves of the frog, and observations of the alterations produced thereby in the structure of their primitive fibres.

Waller, A. (1851b). *Compt. rend. Acad. Sci. Paris* **33**:606–611. Nouvelle méthode pour l'étude du système nerveux applicable à l'investigation de la distribution anatomique des cordons nerveux.

Waller, A. (1852a). *Comp. rend. hebd. Acad. Sci. Paris* **34**:842–847. Examen des altérations qui ont lieu dans les filets d'origine du nerf pneumogastrique et des nerfs rachidiens, par suite de la section de ces nerfs au-dessus de leurs ganglions.

Waller, A. (1852b). *Arch. Anat. Physiol. (Liepzig)* **11**:392–401. Sur la reproduction des nerfs et sur la structure et les fonctions des ganglions spinaux.

Waller, A. (1852c). *Compt. Rend. hebd. Acad. Sci. (Paris)* **34**:675–679. Nouvelles recherches sur la regeneration des fibres nerveuses.

Waller, A. D. (1890). *J. Physiol. (London)* **11**:384–395. On the physiological mechanism of the phenomenon termed the 'tendon reflex.'

Warkany, J., R. J. Lemire, and M. M. Cohen (1981). *Mental Retardation and Congenital Malformations of the Central Nervous System*. Year Book Medical Publ., Chicago and London.

Wässle, H., M. Yamashita, U. Greferath, U. Grunert, and F. Muller (1991). *Visual Neurosci.* **7**:99–112. The rod bipolar cell of the mammalian retina.

Watermann, R. (1964). *Vom Leben der Gewebe. Der Weg von der antiken Atomistik über die Zellenlehre bis zur modernen Molekularbiologie*. Koln.

Watkins, J. W. N. (1970). Against normal science, pp. 25–37. In *Falsification and the Methodology of Scientific Research Programmes* (I. Lakatos and A. Musgrave, eds.). Cambridge Univ. Press, Cambridge.

Watson, J. D. (1968). *The Double Helix. A Personal Account of the Discovery of the Structure of DNA*. Atheneum, New York.

Wehner, R. and R. Menzel. (1990). *Ann. Rev. Neurosci.* **13**:403–414. Do insects have cognitive maps?

Weigert, C. (1882). *Z. med. Wissench.* **20**:753–757, 772–774. Ueber eine neue Untersuchungsmethode des Centralnervensystems.

Weigert, C. (1895). *Beiträge zur Kenntniss der normalen menschlichen Neuroglia*. Festschrift zum 50-jährigen Jubil. d. ärztl. Vereins zu Frankfurt am Main.

Weigert, C. (1898). *Zentralbl. Allg. Pathol.* **9**:289–292. Über eine Methode zur Färbung elastischer Fasern.

Weismann, A. (1892). *Das Keimplasma: Eine Theorie der Vererbung*. Verlag von Gustav Fischer, Jena.

Weismann, A. (1904). *The Evolution Theory* (Translation of *Vorträge über Descendenztheorie*, Jena, 1902), 2 vols. Arnold, London.

Weiss, P. (1941). *Third Growth Symp.* **5**:163–203. Nerve patterns. The mechanics of nerve growth.

Weiss, P. (1963). Self-renewal and proximo-distal convection in nerve fibers, pp. 171–183. In *The Effect of Use and Disuse on Neuromuscular Functions* (E. Gutman and P. Hnik, eds.). Czechoslovak Acad. Sci., Prague.

Weiss, P., and H. B. Hiscoe (1948). *J. Exp. Zool.* **107**:315–396. Experiments on the mechanism of nerve growth.

Weiss, P., and A. Hoag (1946). *J. Neurophysiol.* **9**:413–418. Competitive reinnervation of rat muscles by their own and foreign nerves.

Weiss, P., and A. C. Taylor (1944). *J. Exp. Zool.* **95**:233–257. Further experimental evidence against 'neurotropism' in nerve regeneration.

Weiss, P., H. Wang, A. C. Taylor, and M. V. Edds, Jr. (1945). *Am. J. Physiol.* **143**:521–540. Proximo-distal fluid convection in the endoneurial spaces of peripheral nerves, demonstrated by colored and radioactive (isotope) tracers.

Weiss, P., A. C. Taylor, and P. A. Pillai (1962). *Science* **136**:330. The nerve fiber as a system in continuous flow: Microcinematographic and electronmicroscopic demonstrations.

Wellman, H. M. (1990). *The Child's Theory of Mind*. MIT Press, Cambridge, Massachusetts.

Wellman, H. M. (1991). From desires to beliefs: Acquisition of a theory of mind, pp. 19–38. In *Natural Theories of Mind* (A. Whiten, ed.). Blackwell, Oxford.

Wellman, H. M., and M. Banerjee (1991). *Brit. J. Develop. Psychol.* **9**:191–214. Mind and emotion: Children's understanding of the emotional consequences of beliefs and desires.

Wessel, M. R. (1976). *The Rule of Reason: A New Approach to Corporate Litigation*. Addison-Wesley, Reading, Massachusetts.

Westermarck, E. A. (1906–1908). *The Origin and Development of the Moral Ideas*, 2 vols. Macmillan, London.

Westermarck, E. A. (1932). *Ethical Relativity*. Harcourt, Brace, New York.

Wheeler, J. A. (1980). Beyond the black hole. In *Some Strangeness in the Proportion* (H. Woolf, ed.). Addison-Wesley, Reading, Massachusetts.

Whewell, W. (1837). *History of the Inductive Sciences*, 3 vols. J. W. Parker, London (revised edition, 1847).

Whewell, W. (1847). *The Philosophy of the Inductive Sciences*. J. W. Parker, London.

Whewell, W. (1858). *Novum Organon Renovatum*. Part II of the 3rd edition of *The Philosophy of the Inductive Sciences*. J. W. Parker, London.

Whewell, W. (1968). *Theory of Scientific Method* (R. E. Butts, ed.). Univ. Pittsburgh Press, Pittsburgh.

White, J. G., E. Southgate, J. N. Thomson, and S. Brenner (1974). *Phil. Trans. Roy. Soc. London Ser. B.* **275**:327–348. The structure of the ventral nerve cord of Caenorhabditis elegans.

Whitehead, A. N. (1925). *Science and the Modern World*, Macmillan, New York.

Whitehead, A. N. (1929). *Process and Reality*. *Macmillan*, New York.

Whitman, C. O. (1893). *J. Morphol.* **8**:639. The inadequacy of the cell theory of development.

Whytt, R. (1751). *An essay on the vital and other involuntary motions of animals*. Hamilton, Balfour, & Neill, Edinburgh.

Whytt, R. (1768). *The Works of Robert Whytt, M. D. Published by His Son*. Beckett, Hondt, & Balfour, Edinburgh.

Wigner, E. P. (1967). Remarks on the mind-body question, pp. 171–184. In *Symmetrics and Reflections* (E. Wigner, ed.). Indiana Univ. Press, Bloomington. Reprinted in *Quantum Theory and Measurement* (J. A. Wheeler and W. H. Zurek, eds.), Princeton Univ. Press, Princeton, N. J., 1983.

Wilson, E. B. (1896). *The Cell in Development and Inheritance*. Macmillan, New York.

Wilson, E. B. (1925). *The Cell in Development and Heredity*. 3rd Edition. Macmillan, New York.

Wimsatt, W. C. (1976a). Reductive explanation: A functional account, pp. 671–710. In *Proceedings of the Philosophy of Science Association*, 1974 (R. S. Cohen, C. A. Hooker, A. C. Michalos, and J. W. van Evra, eds.). Reidel, Dordrecht, Holland.

Wimsatt, W. C. (1976b). Reductionism, levels of organisation, and the mind-body problem, pp. 205–267. In *Consciousness and the Brain* (G. G. Globus, G. Maxwell, and I. Savodnik, eds.). Plenum Press, New York.

Windle, W. F. (1958). *Biology of Neuroglia*. Thomas, Springfield, Illinois.

Winfield, D. A. (1981). *Brain Res.* **206**:166–171. The postnatal development of synapses in the visual cortex of the cat and the effects of eyelid suture.

Wittgenstein, L. (1953). *Philosophical Investigations*. Oxford Univ. Press, London.

Wittgenstein, L. (1977). *Vermischte Bemerkungen*. Suhrkamp Verlag, Frankfurt am Main.

Wlassak, R. (1898). *Arch. Entw. Mech. Organ.* **6**:453–493. Die Herkunft des Myelins.

Wolpert, L. (1992). *The Unnatural Nature of Science*. Faber, London.

Wong, K. C., and L.-T. Wu (1936). *History of Chinese Medicine*. 2nd ed. National Quarantine Service, Shanghai. Reprinted by AMS Press, New York, 1973.

Wood, P., and R. P. Bunge (1984). The biology of the oligodendrocyte, pp. 1–46. In *Oligodendroglia, Advances in Neurochemistry, Vol. 5* (W. T. Norton, ed.). Plenum, New York.

Wundt, W. (1904). *Principles of Physiological Psychology, Vol. 1*, Translated from the Fifth German Edition (1902) by E. B. Titchener. Swan Sonnenschein, Macmillan, New York.

Wurz, R. H. and J. E. Albano (1981). *Ann. Rev. Neurosci.* **3**:189–226. Visual motor function of the primate superior colliculus.

Young, J. Z. (1951). *Proc. Roy. Soc. (London) Ser. B.* **139**:18–37. Growth and plasticity in the nervous system.

Young, J. Z. (1964). *A Model of the Brain*. Clarendon Press, Oxford.

Zanobio, B. (1975). Golgian memorabilia in the museum for the history of the University of Pavia, pp. 650–655. In *Golgi centennial symposium* (M. Santini, ed.). Raven Press, New York.

Ziehen, T. (1890). *Anat. Anz.* **5**:692–709. Zur vergleichenden Anatomie der Hirnwindungen mit spezieller Berücksichtigung der Gehirne von Ursus maritimus und Trichechus rosmarus.

Ziff, P. (1984). *Epistemic Analysis: A Coherence Theory of Knowledge*. Reidel, Dordrecht, Holland.

Ziman, J. (1968). *Public Knowledge. An Essay Concerning the Social Dimension of Science*. Cambridge Univ. Press, Cambridge.

Zimmer, H. (1935). *Zentralblatt für Psychotherapie* **8**:147. Indische Anschauungen über Psychotherapie.

Zimmerman, H. J. (1985). *Fuzzy Set Theory and Its Applications*. Kluwer-Nijhoff, The Hague.

Index

Reference to a figure or a footnote is indicated by the page number italicized followed by either *f* or *n*.

A priori knowledge, 20, 292, 315, 318, 320
Addison, Joseph (1672–1719), 262
Adrian, Edward Douglas (1889–), 36, 137
Acetylcholine, 65, 218–219
Agnoiology, 48, 115, 120
Allegorization, 5, 259
All-or-none law, *205n*
Altruism, 320
Amateurism, 264
Analogy
  definition of, 44
  false, 179, 259
  reduction by, 105–106
  theory formation by, 5, 6, 21, *25f*, 94, 153, 233
  versus homology, 44, 105, *162f*
Analytical method, 53, *175n*; *see also* Reduction; Reductionism
Anomaly, in theory, vii, 45–46, 80
Antivivisection movement, 264
Architectonics, 186
Aristotle (384–322 BC), 137, 187, *296n*
Art related to neuroscience, viii, 89, 251, *251n*, 258, 262
Artifacts
  effects of, 38, 48–51, 55, 78, 177, 198
  See also Errors
Artificial intelligence, 130, 137
Association of ideas, 224
Astrocytes
  discovery, 85
  histogenesis, 85–87
Augustine, Saint (354–430), *10f*, 288
Austen, Jane (1775–1817), 257

Automata
  analogy with nervous systems, 5, 22, 25, *25f*, 123, 137
  Déscartes on, 27, 136–138
  LaMettrie on, 24, 112, 122, 139
  T. H. Huxley on, 120
Axon: *see* Nerve fiber
Axonal transport (flow), 67, 72

Babbage, Charles (1792–1871), 264
Bain, Alexander (1818–1903), 25, 122, 156
Bastian, Henry Charlton (1837–1915), 40, 270
Bayesian theory, 300–301
Bechterew, Wladimir (1857–1927), *156n*
Behaviorism, 125, 147
Bell, Charles (1774–1842), 91, 156
Bell-Magendie Law, 91, 208
Bentham, Jeremy (1748–1832), 289, 300
Berlin, Rudolph (1833–1897), 186
Bernard, Claude (1813–1878), 218, 244, 265, *265n*
Bernstein, Julius (1839–1917), 36
Berzelius, Jons Jakob (1779–1848), 68
Bielschowsky, Max (1869–1940), 195, 240
Bias
  of techniques, 50–52
  in theoretical models, 2, 14, 50–51
  See also Methodological neutrality; Misrepresentation of reality
Bichat, Marie François Xavier (1771–1802), 67, 82
Bidder, Friedrich Heinrich (1810–1894), 54, 93, 198, 254
Blake, William (1757–1827), 291
Blumenbach, Johann Friedrich (1752–1840), 180
Body-brain relations, 68
Borelli, Giovanni Alfonso (1608–1679), 56, 174, 175

375

Boyle, Robert (1627–1691), 175
Bradley, Francis Herbert (1846–1924), 48
Brain
  anatomy, 183–184
  complexity, 120–121
  evolution, 104, 108, 113, 120, 183–184
  macrophages: see Microglia
  models, 1–29, 5, 7–9, *9f*
  size, 183
Brain-to-body relations, 68
Brain-to-mind relations, 110, 122, 127, 128–133, 144; see also Mind
Broca, Paul (1824–1880), 39
Brown-Séquard, Charles Édouard (1817–1894), *156n*, 319
Burdach, Karl Friedrich (1776–1847), 11, 184
Butler, Joseph (1692–1752), 288–289, 318

Cabanis, Pierre Jean George (1757–1808), 112, 122
Cajal: see Ramón y Cajal
Capillary electrometer, 36, 284, 285
Carlyle, Thomas (1795–1881), 229, 232
Carnap, Rudolf (1891–1970), 175
Carpenter, William Benjamin (1813–1885), *70f*, *276n*
Cartesian
  certainty, 227
  soul, 56, 122, 138–139, 148, 181, 301
  theory of mind, 133–135, 138, 144, 148
Carus, Carl Gustav (1789–1869), 179
Category mistake, 123
Causal connection
  definition, 102, 143–144
  shown by reduction, 97, 102, 110, 143–149
Causal explanation, 143–149, 156
Causal relations, 9, 143–145
Causation
  Déscartes on, 148
  downward, 146
  Hume on, 130
  of neural states, 145
  of perceptions, 129
  psychological perception of, 147
  purposeful: see Teleology
  reduction and, 97, 109–110
  top-down versus bottom-up, 144
  "true," 148
  of voluntary behavior, 147
Cell membrane
  as hypothetical entity, 52, 164, *164f*, 207
  theories about, 205–207
Cell theory
  anomalies in, 165, *165n*
  definition, 161

Cell theory (*Cont.*)
  relation to neuron theory, 160–161, 177
Cell-type specificity, 188
Cerebellar cortex
  Cajal on, 245, 250
  circuit logic, 146
  structure, *19f*
Cerebral cortex
  architectonics, *62f*, 63, 186
  congenital deficits, 322
  convolutions, 39, *41f*, *42f*, *43f*, 44
  dendritic spines, 58–66, *60f*, *61f*
  development, 36–37, 322
  function of, 26, 38–44, 128
  homologies, 44, 183–184, *183n*
  lesions, 39–44
  localization, 26, 38–45, *39f*, *41f*, *42f*, *43f*, 267, 268
  modules, 13–14, 141
  plasticity, 26
  reduction of, 141
Chinese philosophy, 157, *260n*, 301, *306n*
Chromatolysis, 71
Chuang Tzu (ca 399–295 BC), 97
Codes of conduct, 315–317
Coercion, 304–305
Cognitive map, 7
Cognitive states, reduction of, 119
Coleridge, Samuel Taylor (1772–1834), 313
Collingwood, Robin George (1889–1943), 157
Communication, 2, 11, 21, 232, 314, 315
Competition between theories, 3–4, 31, 80
Competition in nervous systems, 222–223
Computer
  analogy with brain, 5, 13, 26, 110, 137
  as brain surrogate, 115
Conceptual models: see Models
Condition-action rules, 308
Conflict resolution, 315–317
Connections between neurons: see Synapse; Theory, nerve connections
Consciousness
  brain localization, 134–136, *135f*
  development of, 298–299, 312, 321
  fundamental character of, 5, 97, 312
  probabilistic character, 132–133
  relation to memory, 134
  unity of, 101, 134–135
Consilience, vii, 94, 110, 165, 166, 265, 283; see also Unification of knowledge
Contact theory: see Theory
Context-dependence, 19, 110, 147; see also Relations
Coordination, principle of, 274–276

INDEX

Correlation, 102, 105
Corpus callosum, 134, *135f*
Critical period, 125
Critique
  aims of, viii, 6, 32, 151, 233
  and communication, 232, 233
  and dissent, 32
  relation to reflection, 313–315
Crucial experiments, 2, 80
Culture
  effects on making theories, 24, 157
  effects on scientists' styles, 4, 234–237, 246, *251n*, 288–289
Curare, 218
Cushing, Harvey Williams (1869–1939), 283
Cuvier, George Baron (1769–1832), 67, 68, 253, 274

Dale, Henry (1875–1968), 35, 218, 220
Darwin, Charles Robert (1809–1882)
  brain evolution, 183
  emotion theory, 230, 318–319
  evolution of morality, 318–320
  on fertilization of orchids, *78n*
  homology, 274
  ignorance of cell theory, 161
  on hypothesis, 56
  relation to Lamarckian theory, 254, 319
  teleology, 257, 274
Darwinian theory, 16–17, 56, 115, 240
Data
  relations to theory, 1–12, *9f*, *10f*, 35–37, 47–49, 57
  selection criteria, 14, 18
  values of, 293
Decerebrate rigidity, *263f*, 277–279, *278n*
Degeneration of nerves, 44, 45, 69, 71, 79, 216, 240
Deiters, Otto Friedrich Karl (1834–1863), 25, *171f*, 191, 202–203
del Río-Hortega, Pío (1882?–1945): *see* Río-Hortega
Dendrites
  conceptual elimination of, 8, *10f*, *23f*, 60–61, *249f*
  forms and functions, 202–204
Dendritic spines, 58–66
Déscartes, René (1596–1650)
  animal machine, 24, 27, 144, 265
  causality, 148
  causes of error, 301, 318
  cogito ergo sum, 134
  molecular mechanisms, 175, 176
  nerve fibers, *156n*, *168f*, *278f*

Déscartes, René (*Cont.*)
  picture books, 157
  psychophysical dualism, 134, 144, 148
  reciprocal innervation, 277, *278f*
  reflex action, 265
  reduction, 175, *175n*
  soul, 56, 122, 138–139, 148, 181, 301
Dewey, John (1859–1952), 18
Dissent from authority, 32, 81
Distributed neural processing, 121, 130, 141
Döllinger, Ignaz (1770–1841), 92
Dualism
  arguments against, 136–137
  arguments for, 137
  psychophysical, 99, 133–137
  psychophysical versus epistemological, 133
du Bois-Reymond, Emil (1818–1896)
  anti-vitalism, 52, 180
  chemical neuromuscular transmission, 218
  electrophysiological techniques, 284
  limits of understanding, 115
  mechanistic reductionism, 180
  nerve conduction, 89
  specific nerve energies, 254
Duhem, Pierre Maurice Marie (1861–1916), 18, 51, 265
Duval, Mathias (1844–1907), *164f*, 207
Dynamic polarization of neuron, *61f*, 45, 72, *177f*, *203n*, 218, 254, 260

Eccles, John Carew (1903–   ), 138, 234, 267, 282, 219–221, *220f*
Ecker, Alexander (1816–1887), 92
Economic conditions affecting science, 36, *177n*, 246, 263–264
Edinger, Ludwig (1855–1918), 11–12, 92, *156n*
Ego, 28
Ehrenberg, Christian Gottfried (1795–1876), 177, 179
Ehrlich, Paul (1854–1915), 195
Electroencephalography, 50, 322
Electron microscopy, 35, 51, 78, 221
Electrophysiological techniques, 35–36, 52, 190, 219, 241, 284–285
Emergence, dialectics of, 142
Emergent properties, 117, 119, 142; *see also* Epiphenomenalism
Emerson, Ralph Waldo (1803–1822), 232
Emotion
  communication, 321–322
  definition, 302
  development, 317–318, 324
Empathy
  definition, 312

Empathy (*Cont.*)
  development, 320–321
Empriricism, 95–96, *96f*; *see also* Logical empiricism; Logical positivism
Ephapse, 214
Epiphenomenalism, 117, 120–121
Epistemology and agnoiology, 48, 113
Erlanger, Joseph (1874–1965), 36
Error
  causes of, 37–38, 49–50, 123
  in cerebral localization, 39–44
  effects of, 49, 77–78
  elmination of, 37
  logical, 123
  relation to progress, 6, 38
Espronceda, José de (1808–1842), *251n*
Ethical
  codes of conduct, 316
  naturalism, 295
  neutrality, 298, 299–300
  realism, 295–308, 311–312
  relativism, 305–306, 310, 311
  See also Moral
Ethical-to-neural reduction, 308
Ethics
  deontological, 306
  prescriptive, 316
  relations to science, 306–324
  utilitarian, 306
Evolution
  of complexity, 108
  fitness concept, 145
  mental, 113, 145
  molecular, 104, 111
  of nervous systems, 104, 108, 113, 120, 161, *162f*, 183–184, 319
Existentialism, *296n*, 302
Explanation: *see* Meaning; Representation; Unification of knowledge
Extrapolation between species, 99, 105–106

False
  data, 48–49
  theory, 1, 24, 48
Falsification of theory, 4, 46, 95–96, *96f*
Fechner, Gustav Theodore (1801–1887), 112, 122
Ferrier, David (1843–1928), 40–41, *41f*, *42f*, 269
Ferrier, James Frederick (1808–1864), 48
Flechsig, Paul Emil (1847–1929), 79
Flemming, Walter (1843–1905), 178
Flourens, Marie Jean Pierre (1794–1867), 43, 44, 274–275
Fontana, Felice (1730–1805), 177
Forel, August-Henri (1848–1931), 93, 199–200

Foster, Michael (1836–1907), 69, 209, 264, 275
Free will, 99, 122; *see also* Mind, causal efficacy
Freud, Sigmund (1856–1939), 28, *28f*, 205, *205n*
Fritsch, Gustav Theodor (1838–1927), 35, 40, *41f*
Functional equivalence, 26, 106, 145
Functionalism, 26–27, 108, 127, 146–147
Function-structure relation, 13, 26, 128, 253–254; *see also* Teleology
Functional isomorphism, 26, 106, 138
Functional localization, 38–44, *39f*, *41f*, *42f*, 267

Gall, Franz Josef (1758–1828), 35, 38–39, *39f*, 183
Galvani, Luigi (1737–1798), 49
Galvanometers, 36, 284
Gap junctions, 162, *163n*
Gaskell, Walter Holbrook (1847–1914), 277–278
Gassendi, Pierre (1592–1655), 137, 138, 174, 175
Gasser, Herbert Spencer (1888–1963), 36
Gehuchten: *see* van Gehuchten
Geoffroy Saint-Hilaire, Étienne (1772–1844), 27, 253, 274
Gerlach, Joseph von (1820–1896), 85, 208–209
Geulincx, Arnold (1624–1669), 139
Glial cells, 81–90, 121, 128; *see also* Astrocytes; Microglia; Oligodendrocytes
Goals of neuroscience, viii, 6, 20, 291, 295, 315–316; *see also* Progress
Goethe, Johann Wolfgang von (1749–1832), 67, 179, 225
Golgi, Camillo (1843–1926)
  axon collaterals, 165, *192n*
  and Cajal, 59, 63, 192, *192n*, 210, *210f*, 243, 250, *252n*
  dendritic forms and functions, *10f*, 203–204, 209
  dendritic spines, 58–59, 257
  and Koelliker, 193, *193n*, *194f*
  neuroglia, 85, *86f*
  reticular theory, *9f*, 80, 209–211, *211f*
  structure-function relations, 211
  unobservable entities, 212
  works, *192n*
Golgi apparatus, 52, 250
Golgi techniques, 58–59, 83, *84f*, 85, *86f*, 92, 192–193, 243, 245, 270
Goltz, Friedrich (1834–1902), 266, 268, 269, 277
Good versus bad, 289, 309; *see also* Right versus wrong
Gracián, Baltasar (1601–1658), 247–248
Graphic methods of recording, *284n*
Great War (1914–1918), effects on neuroscience, 240
Growth cone, 29, 199, *199f*, 201, 252

Grünbaum (Leyton), Albert Sidney Frankau (1869–1921), 235, 268
Gudden, Bernard Aloys von (1824–1886), 93, 199, 215

Haeckel, Ernst Heinrich (1834–1919), 257
Hall, Marshall (1790–1857), 269
Haller, Albrecht von (1708–1777), *156n*
Hannover, Adolph (1814–1894), 191
Harrison, Ross Granville (1870–1959), 197–198, 200–201, *200f*
Hartley, David (1705–1757), 224
Harvey, William (1578–1657), 56
Hebb, Donald Olding (1904–1991), 224
Heidegger, Martin (1889–1976), 302
Heisenberg, Werner (1901–), 114, 133, 140, 241
Held, Hans (1866–1942), 197, 216, *217f*
Helmholtz, Hermann Ludwig Ferdinand (1821–1894), 36, 49, 52, 93, 180, 275, 284
Henle, Friedrich Gustav Jakob (1809–1885), 92
Hermann, Ludimar (1838–1914), *25f*, 36, 89, 270
Hero worship, viii, 229–233, 248–249
Herz, Heinrich (1857–1894), 1, 6
Hierarchically ordered systems, 98–99, *99f*, 103, 104, 110–111
Hierarchy of theories, 20, 94–96, *99f*, 153, 159–160, 283; *see also* Intertheoretic reduction
His, Wilhelm (1831–1904)
 axon outgrowth theory, *172f*, 198, 215
 cerebral cortex development, 37
 gliogenesis, 82–85, *83f*
 mechanisms of morphogenesis, 200–201, 256–257
 neurepithelial germinal cells, 82, *83f*, 84
 neuron theory construction, 93, 200
 outgrowth of nerve fibers, 169, 198, 215
Histological artifacts, 50–51, 77–78, 198
Histological drawings, 9–13, *9f*, *10f*, *11f*, *12f*, 153
Histological techniques: *see* Techniques
Historiography of neuroscience, vii–viii, 81 *82n*, 155–158, 231–232
Hitzig, Eduard (1838–1907), 35, 40, *41f*
Hobbes, Thomas (1588–1679), 24, 312
Hodgkin-Huxley equations, 31
Holism, 81, 99, 140, 141, *164n*
Homology
 of brain parts, 13, 44, 105–106, 183–184
 definition of, 44
 reduction by, 105
 versus analogy, 44, 105
Horace (Quintus Horatius Flaccus, 65–8 BC), 237
Horsley, Victor Alexander Hadden (1857–1916), *236n*, 284–285
Hugo, Victor (1802–1873), *251n*

Human brain uniqueness, 183–184, *183n*
Hume, David (1711–1776), 124, 130, 266
Hutcheson, Francis (1694–1746), 300
Huxley, Thomas Henry (1825–1895), *78n*, 81, 120, 274
Hypotheses
 are guesses, inventions, vii, 2, *9f*, 22, 56, 94
 empirical versus rational, 95–96, *96f*
 hierarchy of, 94
 refutation, 38, 95–96
 role in ethics, 306–310
 role in science, 1–2, 4–5, 22, 94–96
 verification, 3–4, 38, 95
 See also Mental models; Models
Hypothetico-deductive method, 22; *see also* Logical empiricism

Identity theory, 110, 122, 127, 128–132, 221
Induction, role in theory building, 4, 20, 22
Inheritance of acquired characters, 253, 319–320
Inhibition
 drainage theory of, 22–23, *23f*, 155
 mechanism of, 22–23, 219–221, *220f*, 155, 280
 Sherrington on, 22–23, 276, 280, 283
Innate knowledge, 20, 292, 315, 318, 320
Intelligence and brain size, 183–184
Intentions of scientists, 299
Intertheoretic reduction, *99f*, 101, 113–114, 117–119, 158–159
Introspection, *19f*, 123–127, 137
Isomorphism, 106, 138

Jackson, John Hughlings (1835–1911), 41, 275, 289
James, Henry (1843–1916), 54
James, William (1842–1910), 18, 22, 24
Johnston, John Black (1868–1939), 163, 164
Just goals of science, viii, 57, 294–295, 315–316; *see also* Good versus bad; Right versus wrong
Justice, 315

Kant, Immanuel (1724–1804), 139, *175n*, 180, 257, 315
Kekule, Friedrich August (1829–1896), 28
Kingsley, Charles (1819–1875), *183n*
Knowledge
 critique of, 6, 151–152, 230, 233, 312–315
 fallibility, 20; *see also* Refutation
 limits of, 20, 115
 of mind, 122–127
 sensory origins of, 21, 124–125, 127
 subjective versus objective, 1, *53n*, 130
 theory of, 17–23, 48, 113

Knowledge (*Cont.*)
  unificaton of, 94, 111, 118, 165–166, 265, 283
Koch, Robert (1843–1910), 235, 268
Koelliker, Rudolf Albert (1817–1905)
  autobiography, 248
  and Cajal, 237, 248
  cell differentiation, 186
  cell polarity, 185, 201
  dendritic spines, 59
  and Golgi, 192, *193n*, *194f*
  Handbuch der Gewebelehre, 178, 185–186, *186n*, 196, 201, 214, 255
  molecular mechanisms, 176
  myelination, 75–76
  nerve cell types, *170f*, 185
  nerve network, *173f*, 212, 255
  neuroglia, 86, *86f*
  neuromuscular junction, 214
  neuron theory construction, 93, *170f*, 185, 212, 237
  and Remak, 202, *202n*
  structure-function relations, *252n*
Kuhn, Thomas Samuel (1922–   ), 2, 31, 33, 35, 37, 51
Kühne, Willy (1837–1900), 214–215

Lakatos, Imre (1922–1974), 4, 29–31, 46–48, 80–81, 157
Lamarck, Chevalier de (1744–1829), 253–254, 319
LaMettrie, Julien Offroy de (1709–1751), 24, 112, 122, 139
Langley, John Newport (1852–1925), 4, 23, 218, 223, 266, 268–269
Language
  brain localization of, 39
  development, 20–21
  limitations of, 5, 17, 20, 113, 118
  private, 127
Lashley, Karl Spencer (1890–1958), 26, 35
Laws of nature, 30, 31; *see also* Uniformity of nature
Laycock, Thomas (1812–1876), *169f*, 274, 275, 288
Learning
  mechanisms, 223–224
  moral, 317–324
  prenatal, 322–323
Leeuwenhoek, Antoni van (1632–1723), 56, 177
Leibniz, Gottfried Wilhelm (1646–1716)
  invention of calculus, *94n*
  law of identity, 111–112
  mind-brain relation, *102n*
  preestablished harmony, *102n*, 139
Lenhossék, Michael von (1863–1937), 84, *84f*, 199

Leuckart, Karl Georg Friedrich Rudolph (1822–1898), 256
Levels
  of explanation, 100, 101, 118
  of organization, 98–99, *99f*, 108–109
Lewes, George Henry (1817–1878), 156
Lichtenberg, Georg Christoph (1742–1799), 6
Lister, Joseph Jackson (1786–1869), 177
Localization of function in nervous system, 38–44, *41f*, *42f*, *43f*, 268
Locke, John (1632–1704), 224, 305, 309
Loewi, Otto (1873–1961), 218
Logical empiricism
  notion of theories, 22
  scientific explanation, 30
  unobservable entities, 4, 23, *53n*
Logical positivism
  requirements for reduction, 117–118
  and values, 303
  verification principle, 301–302
Lorente de Nó, Rafael (1903–1990), 13, *62f*
Lovejoy, Arthur Oncken (1873–1962), 133
Ludwig, Carl (1816–1895), 235
"Lumpers" versus "splitters," 100

Macaulay, Thomas Babington (1800–1859), *155n*
Mach, Ernst (1838–1916), 112, 122
Macrocosm–microcosm, 259–260, *260n*
Magendie, François (1783–1855), 68, 91, 156, *156n*, 180, 269
Malpighi, Marcello (1628–1694), 56, 176
Map
  cognitive, 7
  as model, 7, 24
  topographical, 7
Marchi, Vittorio (1851–1908), 79
Marchi technique, 50, 79, 195
Marey, Étienne Jules (1830–1904), 284, *284n*
Marinesco, Georges (1864–1938), 71
Marxist theory, 32–33, 142
Materialism
  dialectical, 142
  eliminative, 110
  mechanistic, 112
  mind-brain relation, 120, 122
  and reduction, 81, 98, 114
  versus epiphenomenalism, 119–120
Mathematics, 7, 175
Matteucci, Carlo (1811–1868), 36, 284
Maturity of theory, 54–58, *176n*
Maupertuis, Pierre Louis Moreau de (1698–1759), 56
Maxwell, James Clerk (1831–1879), 6
McDougall, William (1871–1938), 22, *23f*, 212

Meaning
  of concepts, 100
  derived from reflection, 174–175, 312–314
  of observations, 344–345
  obtained by reduction, 100, 110
  and verification, 301–302
  See also Representation; Unification of knowledge
Meaningfulness, 100
Memory, 134, 224–225, 322
Mendel, Gregor (1822–1884), 55–56
Mendelism, 35
Mental models
  cognitive, 7
  definition, 1, 6
  formation of, 312–313
  moral, 5, 292, 312–313
  representational, 6–7, 11, *12f*, 15, 18–19
  symbolic, 7
  See also models
Mental-neural identity, 122, 128–133
Mental-neural reduction, 99, *99f*, 101–105, 107, 122–133, 189; *see also* Reduction
Mental
  nihilism, 144
  realism, 144
  reductionism, 144
  See also Mind
Metaphor
  Cajal's use of, 29
  Freud's use of, 28, *28f*
  role in models of brain, 5, 27–28
Methodological essentialism, 154, 185
Methodological neutrality, 3, 50, 95, 154, 190
Methylene blue technique, *60f*, 64–65
Meynert, Theodore (1833–1892), 237
Microelectrode recording, 36, 190, 219
Microglia, 89–90
Microscopic techniques, 58–59, 64–65, 87, 88, *88n*, 177, 180, 190–195, 221
Microtome, 191
Mill, John Stuart (1806–1873), 287, 289, 300, 317
Milne-Edwards, Henri (1800–1885), 185
Mind
  causal efficacy, 144, 146, 147
  concept of, 100
  elimination of, 1, 144
  emergence of, 120, 142–143
  evidence of, 123–124, 134
  evolution of, 145, 286
  fallacy of, 123
  known by inference, 124
  known by introspection, 123–127, 137
  limitations of, 115, 120

Mind (*Cont.*)
  reduction to physical mechanisms, 100, 102–105, 109–112, 122–133
  reification of, 52, 123
Mind-brain complementarity, 132–133
Mind-brain relation, 123, 131, 154, 287; *see also* Mental-neural reduction
Misrepresentation of reality, 2, 8, *9f*
Models
  analogical, 5, 21–22, 25, *25f*, 94
  anomalies in, vii, 45–46, 80
  child's, 20–21
  differences between physical and biological, 16–17, 103–104
  ethical, 292, 312–313
  historical context, 151–152, 155–158, *156n*
  as maps, 7, 24
  multivariate, 9, *9n*
  natural, 21
  as principles of neuroscience, 23–27
  relation to language, 5, 17, 20–21, 113
  relation to theories, 257–258
  as representations, 18–19
  role of unobservable entities, 4, 22, *53n*, 207, 212
  sensory dependence, 21, 124–125, 127
Modules, 13–14, 256
Molecular biology, 95, 154, 185
Molecular neuroscience, 154, 185
Monism, 133; *see also* Identity theory
Moore, George Edward (1873–1958), 295, 309
Moral
  absolutism, 308–309
  behavior, brain models of, 5
  calculus, 300
  coercion, 304–305
  consciousness, 298–299
  development, 305
  inductive inference, 306–307
  intuition, 300, 306, 315, 318
  obligations and duties, 307–308, 315
  prescriptions, rules, 316
  reductionism, 308
  relativism, 306
  responsibility, 304–305, 316
  sense, 300, 318
  See also Ethical
More, Henry (1614–1687), 264
Motivation and intention
  as causes of behavior, 136, 147, 314
  as evidence of mind, 136
Motives
  definition, 299
  of scientists, 299

Motor functions
   integration, 272–276
   localization, 40–45, *41f*
Motor unit, 281–282
Müller, Johannes (1801–1858), 91, *156n*, 178, 181, 213, 254–255, 271
Multiple discovery, 90–91, 249; *see also* Priority
Munk, Hermann (1839–1912), 40–42, *42f*, *43f*
Muscle sense, 40, 270
Myelin, theories about, 51, 74–80, *76f*
Myography, 281

Nansen, Fridtjof (1861–1930), 78, 210
Nationalism, 91, 235, 250
Naturphilosophie, 27–28, 96, 179, 205, *260n*
Nerve fiber
   endings, 70–71, *70f*, 213–215, *217f*
   origins, 54
   outgrowth theory, 51, 54
Nerve growth factor (NGF), 67, 74
Neural networks, 99, 121, 128
Neurofibrils, *61f*, 240
Neuroglia
   functions, 81, *86f*, 87, 88
   origins, 82–87, *83f*, *84f*, 90
   staining, 85, 87, 88
   transformation, *84f*, 90
   See also Astrocytes; Microglia; Oligodendrocytes
Neuromimes, 138; *see also* Automata; Simulacra
Neuromuscular junction, 215
Neuron, status of the term, 16, *187n*
Neuron theory
   construction of, *69n*, 151–154, 167–174, *168f*, *169f*, *170f*, *171f*, *172f*, *173f*
   as high-level theory, 96, 152, 158–160
   incompleteness, 5, 154
   Johnston, J. B. on, 163, 164
   priority of discovery, 92–93
   Ramón y Cajal on, 92–93, 160
   relation to cell theory, 160–161
   relation to mind, 154
   Waldeyer on, *69f*, 163
Neuronal specificity, 154, 188, 201–202
Neuronal typology, 184–190
Neurotransmitter receptor molecules, as unobservable entities, 4
Neurotrophic theory, 66–74
Neurotropic theory, 66, 73–74
Nietzsche, Friedrich Wilhelm (1844–1900), *53n*, 237, 246
Nissl, Franz (1860–1919), 203, 212
Nissl technique, 203, 273, *273f*
Nucleus (cluster of nerve cells), origin of term, 13
Null-hypothesis, 46

Obersteiner, Heinrich (1847–1922), *12f*, *179f*, 193, 241
Objective knowledge: *see* Knowledge
Olfactory glomeruli, 106
Oligodendrocytes, 87
Ontogenetic method, 92
Ontology, 98, 112, 114
Open society, 34
Organicism, 81
Owen, Richard (1804–1892), 44, 91, 183

Pain, 289
Panpsychism, 259
Parallel distributed processes, 13, 121, 130, 141
Pavlov, Ivan (1849–1936), 241, 319
Peirce, Charles Sanders (1839–1914), 18, 307
Penfield, Wilder Graves (1891–1976), 89, 235
Philosophy of science, vii, 151–152, 310
Phrenology, 35, 38–39, *39f*, 183
Plasticity of brain, 26
Popper, Karl Raimund (1902–   ), 4, 30, 46–48, *53n*, 80, 157
Postmature theory, 57–58
Postsynaptic potentials, 219
Pragmatic justification, 8, 98, 118
Pragmatism, 18
Prediction from theory, 4, 14, 20, 115–116
Premature discovery, 56–57, 86, 90–91, 220
Premature theory, 54–56, 80, 90
Presynaptic endings, 216, *217f*
Printing techniques, 152, 177, 196
Priority of discovery, 90–94, *192n*
Process, 101, 103, 123
Procháska, Jiri (1749–1820), 66
Progress
   concept of, 5–6, 155
   ethical, 294–295, 297
   evolutionary versus revolutionary, 32–35
   Kuhnian theory, counter examples, 2, 31, 33, 35, 37, 51
   related to reductionism, 103
   role of techniques, 35–38
   scientific, viii, 2, 5–6, 80, 95, 174, 297
Psychoanalysis, 5, 28, *28f*, 96
Purkinje (Purkyne), Jan Evangelista (1787–1869)
   cell theory, 179
   Naturphilosophie, 27, 179
   neuron theory, *169f*, 180
Purkinje cell, *12f*, 58, 179

Quantitative analysis in neuroscience, 244
Quantum theory
   and causality, 114, 144

Quantum theory (*Cont.*)
  relation to mind, 99, 104, 119, 138, 144
Quantum wave function, 132

Radial glial cells, 84, *84f*, 86–87
Radioactive labelling and tracing techniques, 36–37
Ramón y Cajal, Santiago (1852–1934)
  analogical-correlative thinking, 29, 259–260, 274
  Anatomical Congress (1889), 193, 248
  artistic talent, 89, 251, *252n*, 258, 262
  assessment of his importance, 231, 236–237, 247
  autobiography, 247, *247n*, *249n*, 250, 251, *251n*
  brain evolution, 184
  and Castelar, *251n*
  cell membrane, 207
  cell theory, 242
  cellular competition, 223
  cerebellar cortex, *12f*, 245, 250
  cerebral cortex, 184
  claims to priority, 91–93, 160, 200, 242, 250
  contact theory, 216, 241–242
  definition of neuron, *187n*
  degeneration and regeneration of nervous system, 240
  dendritic spines, 58–66, *60f*, *61f*
  distortion of reality, *10f*, 14–15
  dynamic polarization of neuron, *61f*, 45, 72, *177f*, *203n*, 218, 254, 260
  on education, 265
  electrical conduction between neurons, 214
  experimental artifacts, 51, *83f*, 89
  and Forel, 93, 200
  glial staining technique, 87–88, *88n*, 240
  and Golgi, 59, 63, 192, *192n*, 210, *210f*, 250, 257
  Golgi techniques, 59, 92, 243, *252n*
  growth cone, 29, 199, *199f*, 201, 252
  and His, *83f*, 200, 257
  idealistic morphology, 256
  and Koelliker, 237, 248, *252n*
  limited interests, 246
  metaphorical thinking, 29, 257–258
  microscopic techniques, 65–66, 92, *243n*, 261–262
  myelin sheath in CNS, 78
  myelination, 76, 78
  narcissism, 234, *238f*, 247, 250, 251
  nationalism, 235, 246
  neglect of quantitation, 244
  neuron theory, 92–93, *187n*, 239, 242
  neurotrophic factors, 72, 73
  neurotropic theory, 73–74

Ramón y Cajal, Santiago (1852–1934) (*Cont.*)
  and Penfield, 89
  personality traits, 233, 235, 245–252
  portraits of, *238f*
  premature theories, 55
  principle of connections, 274
  problem solving, 234, 244
  publication practices, 244–245
  and Ranvier, 75, 253
  reduced silver technique, 240, 246
  reductionism, 149
  retina, models of, *10f*, 14–15, 64, 245, 248
  and Retzius, *203n*, 240
  and Río-Hortega, 89–90, 240
  romanticism, 251–253, 260–261
  and Schopenhauer, 259
  Schwann cells, 73, 75, 77, 78
  and Sherrington, 163, *250n*, *251n*, 252, 256, 259, 261, 262, 275–276
  scientific career, 239–245, *251–252n*
  selection of data, 14, 257–258
  simplification, 14
  Spain of, 246
  structure-function dichotomy, 72–73, 252
  style, 239, 262
  synapse misrepresentation, 8, *10f*, 14–15, 60, 226, 245, 255–256
  on technical virtuosity, 191
  tenacity, 64, 246
  "third element," 88, *88f*
  typological thinking, *243n*, 258, 274
  values, 234, 245–247, 249–251, *251n*
  and van Gehuchten, 254
  and Virchow, 242
  visual projections, 275–276
  and Waldeyer, *69n*, 160, 163
Ranson, Stephen (1880–1942), 195
Ranvier, Louis-Antoine (1835–1922), 75, 77, 253
Rationalism, 2, 18
Rawls, John (1921–), 315
Realism
  definition, 7, 113
  ethical, 295–308, 311–312
  relation to reductionism, 112–117
  and uniformity of nature, 113
  and unobservable entities, 17, 113
  versus simplification, 8, 107–109
  See also Knowledge; Models; Representation; Theory
Reciprocal innervation, *10f*, 22, *23f*, 219, *220f*, 277–279, *278f*, 281
Reduction
  by analogy, homology, or comparative extrapolation, 7–8, 105–107, 118

Reduction (*Cont.*)
  by causal connection, 94, 97, 102, 109
  by correlation, 94, 97, 102
  by identity, 94, 97, 102, 109
  causal explanation, 110
  cross-categorical, 118, 123, 308
  definition, 94, 174
  ethical-to-neural, 308
  experimental versus intertheoretic, 117–119
  interspecies, 99, *99f*, 105–107
  intertheoretic, *99f*, 101, 113–114, 117–119
  limits, 94, 98, 115, 117, 142, 148
  macro-to-micro, 101, 112, 113, 174
  materialism and, 94, 98
  mental-to-neural, 99–100, *99f*, 101–105, 107, 110, 122–133, 188–189
  methodology, 94–95, 98, 101
  opposition to, 140–143
  prediction from, 115–116
  relation to neuron theory, 174–184
  simplification and, 8, 107–109
  to genes, 111
  to molecules, 109, 113
  top-down versus bottom-up, 98, 144, 148
Reductionism, varieties of, *99f*, 114–117
Redundancy in nervous system, 26, 117
Reflection for deriving meanings, 174–175, 312–314
Reflex
  arc, 255, 280
  as hypothetical entity, 53–54, 166
  inhibition, 22–23, *23f*, 155, 219–221, *220f*, 276–279, 281, 283
  interactions, 281–282, *282f*
  mechanisms, 22, 155, 218–222, 281, 282, *282f*
Refutation of theory, 4–5, 45–48, 95, 153, 167
Regeneration of neurons, 26, 142
Regressive events in development, 92, 225
Reification, 52–53, *53n*
Reil, Johann Christian (1759–1813), 13
Relations
  components of process, 110
  irreducibility, 9, 110
  shown by reduction, 97, *99f*, 101–105
Religion and science, vii, 288
Remak, Robert (1815–1865)
  axons different from dendrites, 202
  discovery of myelin sheath, 74, 76
  neuron theory construction, 93
  relation of nerve fibers to cells, 68–69, *202n*
  theory of cell formation, 178, 182
Representation
  of "body image," 145
  of reality, 1, 7–8, 112–114, 130

Representation (*Cont.*)
  pictorial, 9, *9f*, *10f*, *11f*, 11, *12n*, 14–15, 153, 195–196, 261–262, *284f*
  versus misrepresentation, 8–9, *9f*, *10f*, *11f*, 226–227
  See also Bias; Models; Ramón y Cajal, synapse misrepresentation; Realism; Theory
Research program
  definition, 29, 195
  origins of, 18
  relations to theory, 3, 30
  role of subjective bias, 30
Reticular theories, 52, 80, *171f*, *172f*, 208–213, *211f*, 255, 270
Retina
  Cajal on histological structure, *10f*, 14, 64, 245, 248
  interspecies reduction, 105–106
Retzius, Gustaf (1842–1919), *12f*, 59, 71, 161, *162f*, *203n*, 207, 240, 254
Revolutionary techniques, 2, 35–37, 284–285
Revolutions in science, 2, 32–35, 225
Right versus wrong, 298; *see also* Good versus bad
Río-Hortega, Pío del (1882–1945), 77, 89–90
Rohon-Beard neurons, 201
Romanes, George John (1848–1894), 161, 286
Roux, Wilhelm (1850–1924), 185, *200n*, 222–223, 250, 257
Russell, Bertrand Arthur William (1872–1970), 110, 122, 139
Ryle, Gilbert (1900–1976), 123

Saint-Hilaire: *see* Geoffroy Saint-Hilaire
Sarton, George Alfred Léon (1884–1956), 32
Schäfer (Sharpey-Schafer), Edward Albert (1850–1935), 255–256, 264
Schiff, Moritz (1823–1896), *156n*
Schleiden, Matthias Jacob (1804–1881), 179
Schmidt-Lanterman incisures, 77
Schopenhauer, Arthur (1788–1860), 259
Schultze, Maximilian Johann Sigismund (1825–1874), 15, 75, 191–192, 206
Schwann, Theodor (1810–1882), 75, 161, 177–179, 182, *192f*, 197, 206
Schwann cells, 73, 75
Scientific communication, 193, 195–196
Scientific realism, 2, 17–18, 113
Scientific terms: *see* Terms
Scientific revolutions, 2, 32–35
Scientist, origin of term, *265n*
Sechenov, Ivan Mikhailovich (1829–1905), 255
Sensory-motor integration, 273–277, *276n*
Shaftesbury, Third Earl of (1671–1713), 288, 300, 318

Sharpey-Schafer, Edward Albert (1850–1935), *169f*, 264
Sherrington, Charles Scott (1857–1952)
  and Cajal, 163, *250n*, *251n*, 252, 256, 259, 261, 262, 275–276
  and Carpenter, *276f*
  causal-analytic mode of thought, 258
  cell theory, 218
  cerebral cortex, 268–269
  concept of theoretical entities, 4, 22, 23, 24, 28, *53n*, 113
  and Cushing, 283
  and Cuvier, 274
  decerebrate rigidity, *263f*, 277–279, *278n*
  and Déscartes, 264–265, 276, 277, *278f*, 286
  dualism, 265, 276, 286
  and Eccles, 234, 267, 282
  empiricism, 265
  England of, 263–264, 288
  evidence for synapse, 216–218, *217f*, 226
  experimental techniques, 284–285
  and Ferrier, 268
  and Flourens, 274–275
  and Foster, 264, 275
  and Fulton, 233, 267
  functional equivalence of structures, 26, 106
  and Gaskell, 277–278
  and Goltz, 266, 268, 269, 277
  and Granit, 235, 283, 286
  and Grünbaum (Leyton), 235, 268
  and Horsley, 236, *236n*
  and Hughlings Jackson, 236, 267
  inhibition, 22–23, 219, *220f*, 276, 280, 283
  Integrative Action of the Nervous System, 53, 235, 267, 273, 279
  and Koch, 235, 268
  and Langley, 266, 267, 268
  lecturing style, 285–286
  and Liddell, 235, 282
  limited interests, 236, 267–268
  literary style, 285
  and Ludwig, 235
  "Man On His Nature," 287–288
  on cerebral cortex, 267, 268, 275
  on mind-brain relationship, 286–287
  on visual coordination, 275–276
  personality traits, 234, 236, 262–263, 285–286
  portrait, *263f*
  problem solving, 233, 265
  reductionism, 149, 189–190
  reflex as abstract concept, 53–54
  reflex mechanims, 218–221, *220f*, 277–279
  scientific career, 234, *250n*, 266–267
  spinal reflexes, 266, 269–273, *271f*

Sherrington, Charles Scott (1857–1952) (*Cont.*)
  structure-function integration, 253–254
  synapse named, 209, 217–218
  synaptic mechanisms, 22–23, 209, 255–256, 279–283, 284
  synthetic theory of nervous function, 142
  telephone exchange analogy, 25–26
  terminology, 268
  values, 287–289
  and Virchow, 235
  and Waldeyer, 163, 189
  work practices, 267
Simplification
  uses of, 8, 14
  concept of, 107–109
  dangers of, 229
Simulacra, 137–138; *see also* Automata; Neuromimes
Sleep, 322
Socio-economic effects on science, 4, 33, 94, 177, *177n*, 232, 240, 264, 340
Soemmering, Samuel Thomas (1755–1830), 183
Soul
  Cartesian, 136, 138, 148, 181
  religious belief, 133
Specific nerve energies law, *156n*, 213, 254, 271
Specificity of nerves, 223–224; *see also* Specific nerve energies law
Spencer, Herbert (1820–1903), 224, 225, 265
Spinal cord
  anatomy, *10f*, 270, *271f*, *273f*
  models, *10f*, *210f*
  reflexes, 266, 269–273
Spines: *see* Dendritic spines
Spinoza, Baruch (1632–1677), 122
Split brain experiment, 134–136, *135f*
Spurzheim, Johan Caspar (1776–1832), *39f*, 183
Staining: *see* Techniques, histological
Statistical analysis in neuroscience, 244
Sterne, Laurence (1713–1768), 38
Stilling, Benedikt (1810–1879), 13, 184, 191, 237
Structuralism, 157
Structure-function relationship, 13, 67, 72–73, 128, 211, 252, 253
Style of scientific work, 236, 262, 285
Supervenience, 99, 142; *see also* Emergent properties; Epiphenomenalism
Symbolization, 8
Sympathy
  and altruism, 320–321
  and critique, 233
  definition, 312
  development of, 312, 320–321
  and ethics, 298

Sympathy (*Cont.*)
  and reciprocity, 311
Synapse
  chemical transmission, 65, 218–221
  competition between, 222–224
  demonstrating causality, 144
  electrically coupled, 35, *163f*, 219, 221
  electron microscopy, 221
  elimination of, 222, 225
  evidence for, 216–218, *217f*
  excitatory versus inhibitory, 60, 219, *220f*
  modification by use, 222–223, 224
  presynaptic endings, 215, *217f*
  receptor molecules, 218
  representation by Cajal, 8, *10f*, 14–15, 60, 226, 245, 255–256
  representation by Sherrington, 22, 209, 216–218, *217f*, 226
  specificity of connections, *156n*, 223
  transmitter molecules, 65, 218–219
  as unobservable entity, 4, 22

Techniques
  artifacts, 38, 48–50, 58, 77–78, 83, *83f*, 177, 198
  bias, 2, 14, 50
  electronmicroscopic, 35, 51, 78, 221
  electrophysiological, 35–36, 52, 219, 241, 284–285
  enslavement to, 190
  graphic, *9n*, 195–196, *284n*
  histological, 58–59, 64–65, 85, 87, *88n*, 92, 190–195, 198
  limits of accuracy, 48
  methodological neutrality, 3, 50, 95
  microelectrode, 36, 219
  microscopic, 58–59, 64–65, 87, 88, *88n*, 177, 198
  molecular biological, 95, 154, 185
  monoclonal antibody, 46, 185, 188
  nerve pathway tracing, 78, 199
  powerful, 2, 35–36, 78
  printing, 152, 177, 196
  radioactive tracer, 36–37
  revolutionary, 2, 35–37, 190, 284–285
  role in research programs, 2–3, 190
Teleology, 14, 136, 141, 179, 253, 257, 274
Tenacity in holding theories, 154, 158
Terms
  explanatory value, 16, 17, 166
  meaning of, 19–20
  precision of, 118
  relations to data, theory, 16, 17
  Sherrington's invention of, 268

Theoretical models, 1–29; *see also* Model; Theory
Theory
  analogical, 5, 6, 21, *25f*, 94, 153; *see also* Allegorization; Analogy
  anomalies in, vii, 45–46, 80
  artifacts in, 38, 48–51, 55, 78, 177, 198
  artificial intelligence, 130, 137
  Bayesian, 300–301
  Cartesian of mind, 133–135, 138, 144, 148
  competition between, 3–4, 30–31, 80, 95, 158
  construction of, 4–5, 17–23, 94–96, *96f*, 164–167
  Darwinian, 16, 56, 115, 240
  difference between dendrites and axons, 168, 182, 202–204
  difference between neuroscience and physics, 4, 17, 95, 116
  empirical versus rational, 95–96, *96f*
  ethics, 306–310
  evolution, 104–105, 108, 113, 120, 145, 161, *162f*, 183–184, 319
  false, 50–51, 155
  functional adaptation, 222
  generality of, 18, 152, 164–165, 165–167
  hierarchy of, 20, 54, 95–96, *96f*, 153, 159–160, 283
  identity of mind-brain, 110, 122, 127, 128–132
  inclusivity, 54, 152, 164–165
  maturation of, 54–58, 94–96
  Mendel's on inheritance, 35, 55–56
  mental-neural identity, 110, 122, 127, 128–132
  metaphors in, 5, 25–29
  multiplicity of, 51
  myelination, 51, 74–80, *76f*
  nerve cell membrane, 205–207
  nerve connections, 48, 50, 60–61, 207–227
  nerve fiber development, 169, *172f*, 197–201
  nerve fiber endings, 70–71, *70f*, 216, *217f*
  nerve fiber networks, 80, *172f*, *173f*, 207–212, 270
  neural networks, 99, 121, 128
  neurobiology versus physics, 4, 17, 95, 116
  neuron, 152, 165–174
  neurotrophic, 66–74
  neurotropic, 66, 73–74
  of ignorance, 48, 115
  of knowledge, 17–23, 48, 113
  postmature, 57–58
  prediction from, 4, 14, 20, 115–116
  premature, 54–56, 80, 90
  quantum, 99, 104, 114, 119, 132, 138, 144
  refutation of, 4–5, 45–48, 95–96, *96f*, 152–153, 167
  relation to data, 1–12, *9f*, *10f*, 35–37, 47–49, 57, 152

Theory (*Cont.*)
    relation to values, 4, 57, 182–183
    reticular: *see* nerve fiber networks
    scientific revolutions, 2, 32–35, 225
    specific nerve energies, 271
    synaptic competition, 223–225
    synaptic plasticity, 222–225
    synaptic transmission, 51–52, 216–222
    trophic effects, 66–74
    truth of, 48, 158
    underdetermination, 18, 51
    unity of nerve fiber and cell, 167, *169f*, *170f*, *171f*, *172f*, 180–181
    unobservable entities in, 4, 15, 22, 24, 28, *53n*, 113
    verification and refutation, 46–48, 80, 95–96, *96f*
Thudichum, Johann Ludwig Wilhelm (1829–1901), *75n*
Tiedemann, Friedrich (1781–1861), 91, 183
Tissue culture, *200n*
Tract tracing, 79, 199, 216, 246
Trophic factors, 66–74
Truth
    contingent, 48–49
    correspondence theory of, 157–158
    definition, 48
    probable, 37–38
    and style, 262
Turing, Alan Mathison (1912–1954), 137
Type concept, 67
Typologism
    definition, 154, 184
    mode of thinking, *243n*, 258, 274
    opposition to, 187–188
    and reductionism, 100, 114–115
    relation to development, 185–189
    and terminology, 187
    versus population theory, 142

Uncertainty principle, 131–133, 143
Unification of knowledge, 94, 98, 111, 118
Uniformity of nature, 17, 30–31, 98, 103, 114
Unity of sciences, 17, 103
Unobservable entities, 4, 15, 22, 24, 28, *53n*, 113, 207, 211, 280

Unzer, Johann August (1727–1799), 269
Utilitarianism, 289, 294, 300

Vaihinger, Hans (1852–1893), *53n*
Valentin, Gustav Gabriel (1810–1883), 27, 179, 214
Value-free science, 291, 300–304
Values
    calculation of, 300–301
    definition, 296, *296n*
    effects on research programs, viii, 156, 231, 250, 283–289
    relation to ethics of science, 3–4, 296, 303
    relation to theories, 3–4, 182–183, 242–243
van Gehuchten, Arthur (1861–1914), 196, 218, *271f*
Verification
    of theory, 46
    principle, 301
Virchow, Rudolf (1821–1902), 74, 81–82, 265
Vitalism, 81, 140–141, 143
von Gudden, Bernard Aloys (1824–1886): *see* Gudden
von Haller, Albrecht (1708–1777): *see* Haller
Vulpian, Edme-Félix-Alfred (1826–1887), 69

Wagner, Rudolph (1805–1864), 183, 202
Waldeyer (-Harz), Wilhelm (1836–1921), *10f*, *69f*, 160, 163
Walker, Alexander (1779–1852), *156n*
Waller, August Volney (1816–1870), 69, 71, 79
Weigert, Carl (1845–1904), 88
Weigert stain, 50, 88, 193–195
Weismann, August (1834–1914), 56, 178, 319
Whewell, William (1794–1866), 165, 265, *265n*, 283
Whig history, 155, 157, 174
Whitehead, Alfred North (1861–1947), 111, 123, *296n*
Whytt, Robert (1714–1766), 213
Wittgenstein, Ludwig Josef Johann (1889–1951), 94, 125, 127, 301
Wolf, Hugo (1860–1903), 251
Wundt, Wilhelm (1832–1920), 156, 226

Zeitgeist, 225
Zorrilla y Moral, José (1817–1893), *251n*